2nd International Conference on Particle Physics 2011 (ICPP 2011)

In Memoriam Engin Arik and Her Colleagues

Journal of Physics: Conference Series Volume 347

Istanbul, Turkey
20-25 June 2011

ISBN: 978-1-62276-693-2
ISSN: 1742-6588

Printed from e-media with permission by:

Curran Associates, Inc.
57 Morehouse Lane
Red Hook, NY 12571

Some format issues inherent in the e-media version may also appear in this print version.

Printed by Curran Associates, Inc. (2013)

Institute of Physics
Dirac House, Temple Back
Bristol BS1 6BE UK

Phone: 44 1 17 929 7481
Fax: 44 1 17 920 0979

techtracking@iop.org

Additional copies of this publication are available from:

Curran Associates, Inc.
57 Morehouse Lane
Red Hook, NY 12571 USA
Phone: 845-758-0400
Fax: 845-758-2634
Email: curran@proceedings.com
Web: www.proceedings.com

TABLE OF CONTENTS

Status of the Large Hadron Collider 1
M Lamont

Physics Results from CMS 9
P Spagnolo

Top Quark and Higgs Boson Physics at the Tevatron 19
S Jabeen

Overview of Experimental Results in PbPb Collisions at $\sqrt{SNN}$ = 2.76 TeV by the CMS Collaboration 31
P Kurt

Current Status of the T2K Experiment 41
Y Oyama

ATLAS Detector Performance Status, Improvements During Winter 2010 Shutdown and Results with Initial 2011 Data Taking Period 51
N C Benekos

The SuperB Project: Status and the Physics Reach 70
N Neri

The Status and Road Map of Turkish Accelerator Center (TAC) 76
Ö Yavas

OPERA Experimental Results 80
A Di Crescenzo

CONTRIBUTED TALKS

Early Higgs Search with the ATLAS Data 87
T Yamamura

Recent Results from CMS on SUSY Searches in Leptonic Final States 101
R Schöfbeck

Study of Short-lived Resonances with the ALICE Experiment at the LHC 109
A K Uysal

Search for New Physics in Rare Heavy Flavour Decays at LHCb 114
F Dettori

Recent B-Physics Results from CMS 122
F Simonetto

The Heavy Baryon Masses and Spin-Isospin Dependence 129
Z Ghalenovi, A A Rajabi, A Tavakolinezhad

Search for Exotic New Physics at BABAR 135
A Cervelli

Pion Fragmentation Functions from E+E- Production 145
M Soleymaninia, A Khorramian, M Moosavinejad

The Study of Heavy Contribution in Diffractive Parton Distribution Functions 150
S T Monfared, A Khorramian, S A Tehrani

Top Quark As a Gate to New Physics: Z' Production in Top Events at an E^+E^- Collider 154
A Espargilière

The Electromagnetic Calorimeter of the SuperB Experiment 164
D Pinci

Performance of the ATLAS Inner Detector at the LHC 170
T Cornelissen

Electroweak and QCD measurements at LHCb 175
V Coco

Electroweak Results with the ATLAS 2010 Data 181
L Dell'Asta

Collective Behaviour of Partons Could Be a Source of Energetic Hadrons 195
M K Suleymanov

The ATLAS TRT and its Performance at LHC 202
A Bingül

Recent Charm Physics Results from BABAR 209
R M White

Properties of Bound States Containing Fourth Family Quarks 226
V Bashiry, K Azizi, S Sultansoy

Search for the decay $\mu \rightarrow e\gamma$ in the MEG experiment 236
Y Nishimura

Performance of Micromegas Detectors in the CAST Experiment 245
S C Yildiz, S Aune, T Dafni, M Davenport, G Fanourakis, E Ferrer-Ribas, J Galán, J A García, A Gardikiotis, J G Garza, T Geralis, I Giomataris, H Gómez, F J Iguaz, I G Irastorza, T Papaevangelou, A Rodríguez, J Ruz, A Tomás, T Vafeiadis

The T-REX project: Micromegas for Rare Event Searches 254
T Dafni, S Aune, J Castel, S Cebrián, G Fanourakis, E Ferrer-Ribas, D Fortuño, J Galán, J A García, A Gardikiotis, J G Garza, T Geralis, I Giomataris, H Gómez, D C Herrera, F J Iguaz, I G Irastorza, G Luzón, J P Mols, A Ortiz, T Papaevangelou, A Rodríguez, J Ruz, L Segui, A Tomás, T Vafeiadis, S C Yildiz

The CUORE Experiment: a Search for Neutrinoless Double Beta Decay 265
M Carrettoni

QCD and Top Physics Studies in Proton-proton Collisions at 7 TeV Centre-of-Mass Energy with the Atlas Detector 273
M Barisonzi

Measurements of Forward Energy Flow and Forward Jet Production with CMS 283
D S Cerci

Decay Constants of Heavy Vector Mesons at Finite Temperature 293
E V Veliev, K Azizi, H Sundu, G Kaya, A Türkan

Analyses of $g_{D^{*}sDK}$ and $g_{B^{*}sBK}$ Vertices in QCD Sum Rules 298
J Y Süngü, H Sundu, K Azizi

PHIN Photo-injector As the CLIC Drive Beam Source 307
E Chevallay, M Csatari, A Dabrowski, S Doebert, D Egger, V Fedosseev, O Mete, M Olvegaard, M Petrarca

Start-to-end Simulation of TAC SASE FEL Facility 318
H D Yildiz

Author Index

Journal of Physics: Conference Series **347** (2012) 011001 doi:10.1088/1742-6596/347/1/011001

2nd International Conference on Particle Physics *in memoriam* Engin Arık and her Colleagues (ICPP-Istanbul II)

PREFACE

The 2nd International Conference on Particle Physics *in memoriam* Engin Arık and her Colleagues: Fatma Şenel Boydağ, İskender Hikmet, Mustafa Fidan, Berkol Doğan and Engin Abat was held at Doğuş University, İstanbul, Turkey on 20-25 June, 2011. The conference was organized jointly by the Doğuş and Boğaziçi Universities, with support from CERN and the Turkish Academy of Sciences.

This was the second International Conference on Particle Physics (ICPP) organized in memory of Engin Arık and her Colleagues who lost their lives in the tragic plane accident on November 30th, 2007 on their way to the workshop of the Turkish Accelerator Center (TAC) Project. The first of this conference series was held on 27-31 October, 2008 at Boğaziçi University, İstanbul, Turkey. The conference is intended to be repeated every two years in Istanbul as a Conference Series under the name "ICPP-Istanbul".

Professor Engin Arık had a pioneering role in experimental particle physics in Turkey, and was an inspiring teacher to many colleagues. She led the Turkish participation in experiments at CERN such as CHARMII, SMC, CHORUS, ATLAS and CAST. One of her latest involvements was in the national project to design the Turkish Accelerator Center with the collaboration of 10 Turkish universities including Doğuş and Boğaziçi. Our dear colleagues not only participated in the TAC project but also collaborated in the ATLAS (E Arık, E Abat and B Doğan) and CAST (E Arık, F Şenel Boydağ, İ Hikmet and B Doğan) experiments. We believe that the ICPP-Istanbul conference series has been, and will always be, a way to commemorate them in a most appropriate context.

The topics covered in ICPP-Istanbul-II were "LHC Physics and Tevatron Results", "Neutrinos and Dark Matter", "Particle Factories", and "Accelerator Physics and Future TeV Scale Colliders". The main emphasis was on the recent experimental results in high energy physics with discussions on expectations from existing or future experiments.

There were 20 plenary and 35 contributed talks at the conference, and a majority of these presentations are included in this proceedings. We are grateful to all speakers, the collaborations represented, and all members of the advisory and organizing committees for their invaluable contributions which enabled the conference to reach such a high scientific quality with many exciting results and discussions, making it a big success.

Serkant Ali Çetin, Chair of the Organizing Committee
Peter Jenni, Chair of the Scientific Advisory Committee

Journal of Physics: Conference Series **347** (2012) 011001 doi:10.1088/1742-6596/347/1/011001

Scientific Advisory Committee

Ovsat Abdinov	ANAS, Azerbaijan
Metin Arık	Boğaziçi U., Turkey
Albert De Roeck	CERN, Switzerland
Daniel Denegri	CEA, France
Samim Erhan	UCLA, USA
Dan Green	FNAL, USA
Erhan Gülmez	Boğaziçi U., Turkey
Rolf Heuer	CERN, Switzerland
Peter Jenni*	CERN, Switzerland
Max Klein	Liverpool U., UK
Livio Mapelli	CERN, Switzerland
Tatsuya Nakada	EPFL, Switzerland
Yaşar Önel	Iowa U., USA
Gülsen Önengüt	Çukurova U., Turkey
Ken Peach	Oxford U., UK
Christoph Rembser	CERN, Switzerland
Leonid Rivkin	PSI, Switzerland
Yannis Semertzidis	BNL, USA
Saleh Sultansoy	TOBB ETU, Turkey
Gökhan Ünel	UCI, USA
Konstantin Zioutas	Patras U., Greece

Organizing Committee

Kazem Azizi	Doğuş U.
Serkant Ali Çetin*	Doğuş U.
Zuhal Kaplan	Boğaziçi U.
Özgül Kurtuluş	Doğuş U.
Erkcan Özcan	Boğaziçi U.
Nefer Şenoğuz	Doğuş U.
İsmail Uman	Doğuş U.

* Committee Chairs

Organizing Institutions

Doğuş University

Boğaziçi University

Supporting Institutions

CERN - European Organization for Nuclear Research

TÜBA - The Turkish Academy of Sciences

Journal of Physics: Conference Series **347** (2012) 011002 doi:10.1088/1742-6596/347/1/011002

Short Biography of Engin Arık and Her Colleagues

	Engin Arık, Ph.D. Professor at Boğaziçi University Born: 1948 B.Sc.: İstanbul University, 1969 M.Sc.: University of Pittsburg, 1971 Ph.D.: University of Pittsburg, 1976 Latest projects: CHORUS, ATLAS, CAST, Turkish Accelerator Center Project
	Fatma Şenel Boydağ, Ph.D. Professor at Doğuş University Born: 1947 B.Sc.: İstanbul University, 1969 Ph.D.: İstanbul University, 1976 Latest projects: CAST, Turkish Accelerator Center Project
	İskender Hikmet, Ph.D. Associate Professor at Doğuş University Born: 1964 B.Sc.: Joseph Fourier University, 1986 M.Sc.: Joseph Fourier University, 1988 Ph.D.: Joseph Fourier University, 1991 Latest projects: CAST, Turkish Accelerator Center Project
	Mustafa Fidan, M.Sc. Research Assistant at Doğuş University, Ph.D. Student at Boğaziçi University Born: 1978 B.Sc.: Hacettepe University, 2001 M.Sc.: Boğaziçi University, 2005 Latest projects: Turkish Accelerator Center Project

	Berkol Doğan, M.Sc. Research Assistant and Ph.D. Student at Boğaziçi University Born: 1980 B.Sc.: Boğaziçi University, 2003 M.Sc.: Boğaziçi University, 2006 Latest projects: ATLAS, CAST, Turkish Accelerator Center Project
	Engin Abat, B.Sc. M.Sc. Student at Boğaziçi University Born: 1979 B.Sc.: Eskişehir Anadolu University, 2005 Latest projects: ATLAS, Turkish Accelerator Center Project

2nd International Conference on Particle Physics IOP Publishing
Journal of Physics: Conference Series **347** (2012) 012001 doi:10.1088/1742-6596/347/1/012001

Status of the Large Hadron Collider

M Lamont
CERN, Geneva, Switzerland
E-mail: Mike.Lamont@cern.ch

Abstract.
The key LHC operating parameters and their impact on potential performance are recalled and a brief summary of the 2010 beam commissioning program is presented. Progress in 2011 has been good and the present performance is discussed. The short and medium term plans and their potential are outlined.

1. Introduction

2010 was the first full year of LHC commissioning and it saw a number of important operational milestones. These included: first collisions at 3.5 TeV; commissioning of the squeeze; the move to physics with nominal bunch intensity; the move to bunch trains followed by a phased increase in intensity. A peak luminosity of 2×10^{32} cm^{-2}s^{-1} was achieved with an integrated luminosity of 6 pb^{-1} per day delivered in the final week of proton operations. The year culminated in a successful ion run.

2011 has seen a rapid re-commissioning of the machine after the Christmas stop with the main milestones listed below.

- 19th February: circulating beams re-established.
- 3rd March: first collisions at 3.5 TeV, β^* =1.5 m with pilot beams.
- 5th March: nominal bunches at 3.5 TeV, start of collimator set-up program.
- 13th March: first stable beams, 3 bunches per beam, initial luminosity 1.6×10^{30} cm^{-2}s^{-1}

By 21st March 2011, the LHC was back up to a peak luminosity of 1×10^{32} cm^{-2}s^{-1} and by the end of June had achieved 1.26×10^{33} cm^{-2}s^{-1} and already achieved the year's modest integrated luminosity target of 1 fb^{-1} in time for the summer conferences.

2. Overview of 2010

The clear priority of 2010 was to lay the foundations for 2011 and the delivery of 1 fb^{-1}. The peak luminosity target was 1×10^{32} cm^{-2}s^{-1}. Solid operational experience of injecting, ramping, squeezing and establishing stable beams was gained and a large number issues were addressed along the way. A period of steady running at or around 1 MJ for an extended period was used to fully verify machine protection and operational procedures before performing a safe, phased increase in intensity with validation and a running period after each step-up in intensity.

The luminosity delivery in 2010 can be divided into three main periods: the initial luminosity run with low bunch currents; nominal bunch operation with up to 48 bunches; and the performance ramp up period after the commissioning of bunch trains. The main milestones of 2010 commissioning are outlined in Table 1. Some notable achievements of the year's commissioning are shown in Table 2.

Table 1. Main phases of 2010 commissioning.

Date	Commissioning phase
28th February	Injection of both beams, rough RF capture
30th March	First colliding beams at 3.5 TeV
March	Initial commissioning leading to first collisions
April	Squeeze commissioning
May	Physics 13 on 13 bunches with 2×10^{10} protons per bunch
June	Commissioning of nominal bunch intensity
July	Physics 25 on 25 bunches with 9×10^{10} protons per bunch
August	3 weeks running with a stored beam energy of 1 to 2 MJ
September	Bunch train commissioning
Oct -Nov	Phased increase in total beam intensity

Table 2. 2010 commissioning: some notable records. Data courtesy of ATLAS.

Peak stable luminosity	2.07×10^{32} cm^{-2}s^{-1}
Maximum luminosity delivered in one fill	6.3 pb^{-1}
Maximum luminosity delivered in one day	6.0 pb^{-1}
Maximum luminosity delivered in 7 days	24.6 pb^{-1}
Maximum colliding bunches	348
Maximum average events/bunch crossing	3.8
Longest time in Stable Beams – one fill	30.3 hours
Longest time in Stable Beams – one day	22.8 hours (94.9%)
Longest time in Stable Beams – 7 days	69.9 hours (41.6%)
Faster turnaround (protons)	3.7 hours
Maximum stored beam energy at 3.5 TeV	28 MJ
Maximum stored beam energy in physics	24 MJ

3. Machine status at the end of 2010

2010 was a intense year of commissioning with many issues and problems addressed [1]. When the dust had settled the following key features related to beam based operation may be noted.

- The machine is magnetically and optically well understood with excellent agreement between the magnetic and optics models and measured beam parameters.
- Given a rigorous cycling strategy the LHC is magnetically reproducible. This has proved important because set-up remains valid from fill to fill, and indeed from month to month.
- The aperture has been measured carefully and is as expected.
- The operational sequence, after a lot of hard work, allows the beams to be taken through the ramp, squeeze and into collision essentially without loss.
- Better than nominal beam intensity and beam emittance was delivered by the injectors resulting in excellent luminosity performance.

- It was possible to collide nominal bunch currents with smaller that nominal emittances with no serious problems from head-on beam-beam. This important result is another key reason behind the impressive luminosity performance.
- There was excellent cleaning by the collimator system and good control of beam losses at all stages of operation. There were no accidental beam induced quenches above injection energy.

The key machine parameters used in towards the end of the 2010 proton run are shown in Table 3.

Table 3. Main machine parameters in use in 2010 compared with the nominal 7 TeV values.

Parameter	2010	Nominal
Energy [TeV]	3.5	7.0
$\beta*$ [m]	3.5, 3.5, 3.5, 3.5	0.55, 10, 0.55, 10
Normalized emittance [μm] (start of fill)	2.0 - 3.5	3.75
Bunch current	1.2×10^{11}	1.15×10^{11}
Maximum number of bunches	368 348 collisions/IP	2808
Maximum stored energy [MJ]	28	360
Peak luminosity [$cm^{-2}s^{-1}$]	2×10^{32}	1×10^{34}

The LHC has excellent single beam lifetime at 3.5 TeV before collisions of over 300 hours. At the start of a fill the luminosity lifetime is initially in the range 15 to 20 hours lengthening to 25 to 30 hours later. The luminosity lifetime is reasonably well given by the observed emittance growth and the intensity decay. There is minimal drifts in beam overlap during physics and the beams are generally very stable.

The 2010 run was driven mainly by commissioning, and not operations for physics. In this regard, any analysis of operational efficiency should be regarded with some latitude. However for a first year the signs are very encouraging. Some very extensive equipment systems performed above expectations (considering mean time between failures etc.), and the equipment groups are aware of the weak points and are working to improve them. Technical stops certainly caused problems initially but things got better during the course the year and overall there was impressive availability for a first full year of operations with the final number for machine availability being around 65% of the scheduled time [2].

4. 2010: the heavy ion run

The 2010 run finished with a switch from protons to lead ions. The operations team successfully leveraged the experience gained with protons to rapidly push through the ion commissioning program, relying on the fact that the magnetic machine for ions is near identical to that used for protons. The early ion parameters applicable to the 2010 run are shown in table 4 [5].

The first injection of beam took place on 4 November 2010. Stable beams were declared for physics on 7 November. In the following days, the number of bunches per beam was increased through 2, 5, 17, 69, 121 bunches, injecting single bunches or batches of 4 from the SPS in variants of the "Early" filling scheme (see Table 4). In the last few days of the run, injection of batches of 8 bunches gave a total of 137 bunches. The SPS supplied around 1.2×10^{8} ions per bunch which gave a peak luminosity of just over 3×10^{25} $cm^{-2}s^{-1}$. An integrated luminosity of 9.7 μb^{-1} was delivered to ALICE, ATLAS and CMS with some interesting results.

Journal of Physics: Conference Series **347** (2012) 012001 doi:10.1088/1742-6596/347/1/012001

Table 4. Parameter list for early (2010/2011) and nominal ion running.

Parameter	units	Early	Nominal
$\sqrt{s}$ per nucleon	TeV	2.76	5.5
Initial luminosity	$cm^{-2}s^{-1}$	1.25×10^{25}	1×10^{27}
Number of bunches		62	592
Bunch spacing	ns	1350	99.8
β^*	m	2	0.5
Pb ions per bunch		7×10^7	7×10^7
Transverse norm. emittance	μm	1.5	1.5
Luminosity half life (1,2,3 expts.)	hours	$3< \tau_{IBS} <70$	8, 4.5, 3

5. Issues

The splices in the interconnects remain a worry and still limit the beam energy to 3.5 TeV for the moment. Among a swathe of issues, two of the main concerns for 2011 are outlined below.

5.1. UFOs

Many sudden local beam losses have been recorded. The rise time of the losses is of the order 1 ms. There is no danger of a quench but preventive dumps are performed if the beam losses go above the threshold of the beam loss monitors implicated.

A potential explanation is dust particles falling into beam creating losses via elastic and inelastic scattering. The showers generated propagate downstream. This explanation has given rise to the moniker 'Unidentified Falling Object' or UFO. The UFOs are distributed around the ring and are seen in the arcs, inner triplets, and insertion regions. Most do not generate beam losses severe enough to dump the beam.

There are more UFOs at higher beam intensities and the UFOs observed at 450 GeV have a lower beam loss amplitude than those observed at high energy. The timescale of the losses are shorter with increasing total intensity but it seems that the signal amplitude does not increase with total beam intensity.

Additional quench tests will be used to benchmark the BLM thresholds, perhaps allowing more leeway in the setting of the thresholds. A revision of the BLM thresholds around the ring has been made in an attempt to make the effect of UFOs on operational efficiency acceptable and this has been partially successful.

Given that they tend to dump the beam at high energy, the straightforward cost of above threshold UFOs is machine efficiency and the 2 to 3 hour turnaround from high energy to high energy means that they can have a significant effect if they dump the beam on the order of once per day.

5.2. Electron cloud

Electron cloud effects in particle accelerators are well understood and have been very well studied over the years [3]. In the LHC electron cloud effects are predicted to occur both in the warm and cold regions. The most notable effects are: vacuum pressure rises (potentially causing background in experiments); single-bunch instabilities; multi-bunch instabilities; incoherent emittance growth; heat load in cold arcs (with quenches in the limit); and perturbation of beam diagnostics.

Experience of electron cloud in 2010 showed that it was very much dependent on bunch spacing: 150 ns saw local signs of electron cloud in the common beam pipe; 75 ns showed some first signs around the ring; and 50 ns induced enough vacuum activity to bring the vacuum valves in. Vacuum activity started off in regions with common beam pipe at 450 GeV as the number of bunches with 150 ns spacing was pushed up. Tests with solenoids wrapped around the beam pipe locally cured the problem, confirming the presence of electron cloud.

When the 50 ns bunch spacing was tried, the effects were very clearly seen. There was high vacuum activity in the warm regions (where the beams travel in separate beam pipes) and significant heat load in cold regions. Instabilities and beam size growth were observed. However, surface conditioning (scrubbing) was also observed with associated gas desorption rates and an implied drop in secondary emission yield (SEY). The time constant for the scrubbing processes was less than a day. The situation was a lot cleaner with 75 ns, however incoherent effects were seen and emittance blow-up was observed when 800 plus bunches in both beams were injected [4].

2010 observations are certainly due to a high SEY between 2 and 2.5, whereas 1.7 was usually the maximum value studied in the past. In 2011 it proved possible to reach 200 to 300 bunches with 75 ns without scrubbing. Thereafter a 5 day scrubbing run with large emittance ($\geq 3.5\ \mu$m), 50 ns bunch spacing, and high bunch intensity ($> 1.2 \times 10^{11}$) paved the way for 50 ns operations. In addition solenoids have been installed at many locations during the 2010/11 Christmas technical stop.

6. Operation in 2011

The beam energy remained at 3.5 TeV in 2011. The β^*s for the run are: 1.5 m in ATLAS and CMS; 3 m in LHCb; and 10 m in ALICE.

The official target for 2011 was 1 fb^{-1} delivered to each of ATLAS, CMS and LHCb at 3.5 TeV. Given the performance in 2010 the 1 fb^{-1} target for ATLAS and CMS was conservative and was indeed reached in June 2011. 1 fb^{-1} is less obvious for LHCb given their maximum acceptable luminosity of between 2 to 3×10^{32} cm^{-2}s^{-1} and associated pile-up limitations. Luminosity levelling via separation is in use to maximize the delivered luminosity. ALICE's requirements for the proton run are modest with a demanded luminosity for the proton run of between 5 and 50×10^{29} cm^{-2}s^{-1} and a "pile-up" of around 0.05.

The baseline operational scenario for 2011 unfolded as more-or-less as planned.

- Re-commissioning with beam after the Christmas technical stop took around 3 weeks. The exit condition from this phase was stable beams with low number of bunches.
- There was a ramp-up to around 200 bunches (75 ns) taking about 2 weeks. Multi-bunch injection commissioning also took place during this phase.
- An 5 day intermediate energy run (beam energy 1.38 TeV) took place towards the end of March. (Here the proton-proton collision energy is equivalent the nuclean-nuclean collision energy in the lead ion run.)
- There was a scrubbing run of 10 days which included 50 ns injection commissioning.
- A staged ramp-up in the number of bunches then took place with 50 ns bunch spacing up to a maximum of 1380 bunches.

6.1. Scrubbing run

The dedicated scrubbing run saw impressive progression in spite of several technical problems not related to the scrubbing. Eventually 5 days of scrubbing took place between 5th and 10th April. All solenoids were switched off (both experiments' and vacuum solenoids in the warm sections). There was a careful increase in intensity (in steps of 200 bunches) while monitoring cryogenics, vacuum, machine protection and particularly the RF system. The run was limited to 72 bunches/train by injection performance, but managed to reach 1020 bunches per beam at the end of the scrubbing run with a total of more than 10^{14} protons per ring. However, there was an impressive reduction of the heat load in the cold sectors of the machine in only a few hours. The results are consistent with a SEY reduction from 2.5 to to less than 1.8. After this encouraging performance the decision was taken to go with 50 ns bunch spacing.

6.2. Operating with higher intensities

Running with higher total beam intensity continues to provoke a number of issues including: UFOs as described above; the effects of radiation to electronics in the tunnel; and increased vacuum activity possibly related to residual electron cloud. The RF team has had to carefully monitor the effects of higher beam intensity and there have been issues with protection of RF Power Couplers and false wave-guide flash-overs (probably linked to beam losses induced by vacuum activity). Beam induced heating of injection kickers, beam screens, and collimators has been observed with a clear dependence on total intensity and bunch length.

6.3. Machine Protection

Machine protection is, as always, a concern. Performance of the machine protection system has been excellent, however some very occasional incidents do give pause for thought.

The SPS is now injecting 144 bunches in the LHC and beam quality from injector is very important with over 1 MJ per injection coming down the transfer line. An injection kicker flash-over in April 18th provoked heavy beam loss which was properly caught by the appropriate protection devices. Also in April there was a quench of a high temperature superconducting current lead which revealed a loss of redundancy in the associated protection system. This provoked machine wide check of similar systems.

6.4. Performance 2011

So far in 2011 there has been impressive progress in re-commissioning with beam and following the move to regular operation. Improvements and consolidation across the board took place during the Christmas technical stop. The machine operation is now reasonably well optimized, a faster ramp and squeeze is in place, and the essential mechanics for driving the machine through the nominal cycle well bedded in.

By June the LHC was running with 1092, 1236, and finally 1380 bunches per beam with a peak luminosity of up to 1.27×10^{33} cm^{-2}s^{-1}.

- The luminosity lifetime remains very acceptable at around 25 to 30 hours.
- Luminosity levelling in LHCb via transverse displace of the beams at the collision point is operational.
- The daily luminosity production rate held at around 30 pb^{-1} through June.
- Machine availability is dictated by the exposure to the intersecting failure space of a number of complex systems with huge number of components. This failure space is clearly inflated by high intensity. Nonetheless, despite the inherent complexity of the LHC, availability remains acceptable.
- Precise measurement of the luminous region (via Van der Meer scans) and beam intensity have given a useful handle on the absolute luminosity for cross section calibration and the obtained accuracy is currently at the 5% level.

The parameters in use at the end of June 2011 are shown in table 5. Performance related records, also at the end of June, are shown in table 6.

Given the a peak luminosity of 1.3×10^{33} cm^{-2}s^{-1}, around 90 days of delivery at or around this level, and reasonable operational efficiency reflected in the Hübner factor of around 0.2, the LHC should be able to deliver at total of 3 to 4 fb^{-1} in 2011. The main known unknowns with the potential to disrupt the delivery of this total are UFOs and radiation to electronics. There is the possibility to gently increase the bunch intensity and lower the emittances delivered by

Journal of Physics: Conference Series **347** (2012) 012001 doi:10.1088/1742-6596/347/1/012001

the injectors with the potential for correspondingly higher peak luminosities. This strategy will be pursued in the second half of the year.

The 2011 lead ion run will play on the number of bunches and the bunch population. There is the possibility to reach 540 bunches but this scheme is likely to be limited to the nominal bunch population. The optics used will take over ATLAS and CMS's β^* from the proton run and squeeze ALICE to same value. This will required 1 to 2 days to set-up. There is the potential to reach peak luminosities in the range of 1 to 1.4×10^{26} $cm^{-2}s^{-1}$ and to integrate 30 to 50 μb^{-1}.

Looking further ahead, physics data-taking is foreseen until the end of 2012. Following measurements of the copper stabilizers resistances during the Christmas stop, the maximum beam energy will be re-evaluated. The baseline bunch spacing for 2012 is 50 ns to take advantage of the smaller emittances and higher bunch intensity available from injectors. However, 25 ns will be evaluated first to verify its potential before a decision is made. The author declines to make any predictions on possible integrated luminosity in 2012.

Table 5. Performance related parameters for operation in 2011 (as of end June 2011).

Parameter	Value
Energy	3.5 TeV
β^* in ATLAS and CMS	1.5 m
Bunch spacing	50 ns
Number of bunches	1380
Maximum intensity in collisions	1.6×10^{14} protons per beam
Bunch intensity	1.2×10^{11}
Stored beam energy	≈93 MJ
Emittance [mm.mrad]	≈2.5

Table 6. Operational high points as of end June 2011. Data courtesy of ATLAS.

Integrated luminosity	1.06 fb^{-1} (0.36 fb^{-1} in LHCb)
Peak stable luminosity	1.26×10^{33} $cm^{-2}s^{-1}$
Maximum luminosity delivered in one day	63 pb^{-1}
Maximum luminosity delivered in 7 days	242 pb^{-1}

7. Conclusions

The LHC performance in 2010 and 2011 has been very encouraging: the magnetic model and optics look excellent; the beam instrumentation is in good shape; beam cleaning and collimation works reliably with predicted efficiency; the machine protection system's performance has been excellent; and the machine aperture looks good. The performance with beam (losses, lifetimes, luminosity, emittance growth) has been surprisingly forgiving.

2011 has seen an increase in number of bunches to 1380 using 50 ns bunch spacing and a peak luminosity of between around 1.3×10^{33} $cm^{-2}s^{-1}$. This should allow the LHC to deliver an integrated luminosity of around 3 to 4 fb^{-1} in 2011. Increasing intensity has brought with it a number of issues. Of these, UFOs and SEUs, perhaps, present the biggest threat to operational efficiency and they will both have to be carefully monitored and mitigation anticipated if their effects prove too serious. The destructive power of the stored energy of the beam at high energy is an ever present concern and diligent attention to the machine protection system is mandatory.

References

[1] *Proceedings of the second 2010 Evian workshop on LHC beam operation*, CERN-ATS-2011-017.
[2] W. Venturini, LHC operational efficiency in 2010, *Proceedings of the second 2010 Evian workshop on LHC beam operation*, CERN-ATS-2011-017.
[3] *Proceedings ECLOUD'02*, CERN, Geneva, 15-18 April 2003, CERN-2002-001.
[4] G. Arduini, 50 and 75 ns operation, *Proceedings of the second 2010 Evian workshop on LHC beam operation*, CERN-ATS-2011-017.
[5] J. Jowett, Ions in the LHC, *Proceedings of Chamonix 2009 workshop on LHC Performance*, CERN-ATS-2011-017.

Physics results from CMS

Paolo Spagnolo[1]
INFN Pisa, Italy
E-mail: Paolo.Spagnolo@cern.ch

Abstract.
These proceedings report the results from the CMS experiment with data collected in 12 months of 7 TeV LHC collisions at CERN. With the first data collected all the major Standard Model processes have been studied: W, Z and other electroweak physics results, top and B physics. The main results are achieved with the di-leptons and di-jets reconstruction analyses. First results on searches for Higgs boson are then discussed with the prospects for the current 2011-12 running period. Search for new physics are also presented as deviations from the Standard Model distributions. First limits on new vector bosons, extra-dimensions and microscopic black holes are set depending on the theoretical models.

1. Introduction

The Compact Muon Solenoid (CMS) is one of the two "general-purpose" detectors at the Large Hadron Collider (LHC) [1]. It is located at the experimental Point 5 of the LHC near Cessy (France). The main characteristic features of CMS, shown in fig. 1, are a large superconducting solenoid magnet, which creates a strong field of 3.8 Tesla, a state-of-the-art silicon tracker, a highly granular crystal electromagnetic calorimeter, fully hermetic hadronic calorimeters and a sophisticated and redundant muon system. The detector has been built thanks to the collective effort of the CMS Collaboration consisting of more than 3000 scientists and engineers from 182 institutes distributed in 39 countries all over the world.

Prior to collecting pp collisions the detector has been accurately calibrated using cosmic muons. A large data set of cosmics was recorded in successive campaigns of data taking from 2008 to 2009. As a result of these studies it was possible to achieve a good understanding of the initial alignment constants of the major detector components and a detailed map of the magnetic field. They led to an excellent control of the momentum resolution and absolute scale. The commissioning of the detector was then completed using the first LHC pilot runs at 0.9 and 2.36 TeV collision energies at the end of 2009. The LHC started 7 TeV operations in spring 2010, at very low luminosity, in the range of $\mathcal{L} = 10^{27}$ cm^{-2}s^{-1}, but reached quickly instantaneous luminosities exceeding $\mathcal{L} = 10^{32}$ cm^{-2}s^{-1}. In total, an integrated luminosity of 47 pb^{-1} has been delivered by the LHC in 2010. With the LHC running in pp mode CMS has collected 43.2 pb^{-1} of data corresponding to an overall data-taking efficiency of about 92%. The uncertainty in the luminosity determination is estimated to be 4%. The overall operational status of CMS during this data taking was excellent: for all the sub-systems more than 98% of the channels were operational. In 2011 about 1 fb^{-1} has been collected so far (June 2011). These proceedings

[1] On behalf of the CMS Collaboration

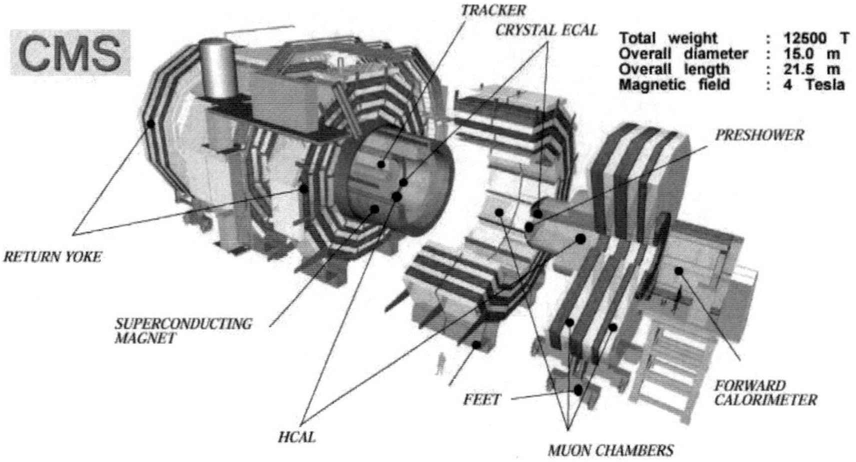

Figure 1. Schematic view of the CMS detector.

describe some of the most important CMS results published to date. The majority of these results are obtained with the 2010 data and will be soon updated with the 2011 luminosity.

2. QCD: Soft Physics and Jets

Taking advantage of the low luminosity phase at the beginning of the LHC running, many important measurements have been achieved by CMS. The first paper published by the collaboration using only 1 μb^{-1}, corresponding to 2 hours of data taking, was about the transverse-momentum and pseudorapidity distributions of charged hadrons [2]. The Bose-Eistein Correlations were observed for the first time at the LHC [3] and a very interesting excess in the near-side long range (small ϕ and large η) has been found in the charged particle distribution for high multiplicity events, with number of tracks above 110. This ridge, similar to the one observed in the ion-ion collisions, was measured for the first time in proton collisions [4]. Many analyses were also performed in the jet physics sector. Studying the inclusive jet cross-section [5] and the distributions of the di-jets events [6] it is possible to probe for new physics beyond the Standard Model. As rule of thumb, the sensitivity to a contact inetraction Λ is roughly 4 times the transverse energy of the most energetic jet. So far, up to jet transverse energy of about 1 TeV, all the measurements are in good agreement with the Standard Model.

3. B Physics

Cross sections for heavy quark production in hard scattering interactions provide an interesting testing ground for QCD calculations. Theoretically, large uncertainties remain due to the dependence on the renormalization and factorization scales. Measurements from the LHC at the center of mass energy of 7 TeV provide new opportunities to test our understanding of the heavy quark production mechanisms.

The first heavy flavor production measurements at CMS were made by reconstructing J/ψ and Υ mesons in their decays to two muon final states. Candidates are formed by fitting pairs of oppositely-signed muons to a common vertex and event yields are obtained by fitting invariant mass distributions. The fraction of J/ψ mesons produced from longlived B decays is also measured by fitting the lifetime distribution of the reconstructed J/ψ [7]. Results are shown in fig. 2. The three lowest Υ states are all visible due to the excellent mass resolution

Journal of Physics: Conference Series **347** (2012) 012002 doi:10.1088/1742-6596/347/1/012002

of the CMS detector, and their cross-sections have been measured as function of rapidity and transverse-momentum [8].

Two independent techniques are used to measure inclusive beauty production. The first technique makes use of semi-muonic decays of B hadrons [9] and the second relies on the identification of displaced secondary vertices within reconstructed jets to tag them as b jets [10].

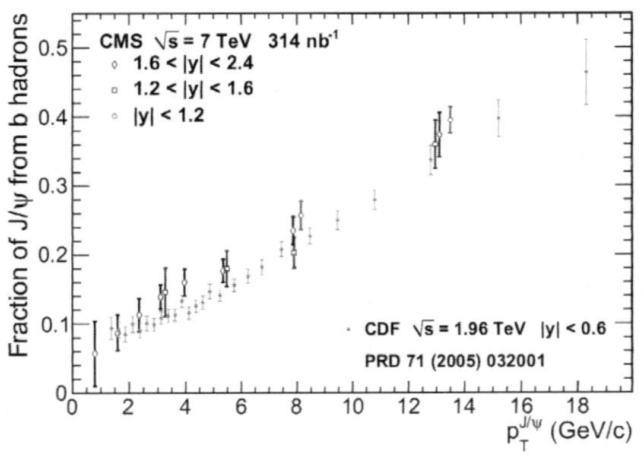

Figure 2. Fraction of J/ψ mesons produced from B decays.

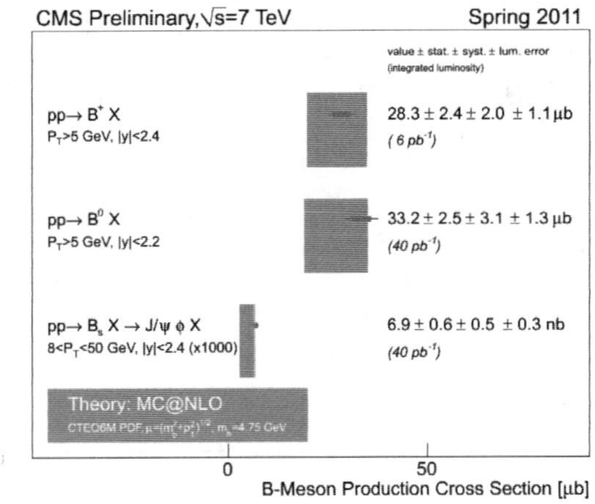

Figure 3. Cross-sections measurements for B meson production.

A fully exclusive reconstruction technique is used to measure the cross sections for B^+ [11], B^0 [12], and B_s^0 [13] mesons and Λ_b baryon decaying into a J/ψ and K^+, K_S^0, ϕ and Λ respectively, and to measure the relative production cross-sections. Fig. 3 shows the results for the cross-section measurements of the B mesons.

4. Electroweak Results: W and Z

The W and Z boson production is a benchmark process at the LHC and the selection of these boson candidates is particularly important for CMS. The Higgs boson at intermediate or high mass is expected to decay with high branching fractions to pairs of W and Z and, in general, the intermediate vector bosons are among the main sources of background to new physics processes. W candidate events are characterized by a prompt, energetic ($E_T > 25$ GeV), isolated lepton, and significant missing transverse energy. The main backgrounds are QCD multijet events and Drell-Yan events in which one lepton fails the selection. Simple selection cuts lead to the distributions of missing transverse energy that are used to extract the $W \to \ell\nu$ event yield. It is worth noticing that at the LHC, due to the quark content of the colliding protons, we expect to measure a production yield for W^+ larger than the corresponding yield for W^-. The $Z \to \ell^+\ell^-$ candidate events are required to have two opposite sign leptons satisfying the same selection criteria used for the $W \to \ell\nu$ sample. The inclusive cross section values are then measured in the data using data-driven methods for controlling the lepton efficiency, energy and momentum scale, resolution and all major sources of background. Fig. 4 shows the invariant mass distribution of the di-muon pairs in logarithmic scale to demonstrate the high purity of the selected sample of Z candidates, while Fig. 5 shows the ratio between results and theoretical predictions. It is worth noticing the high experimental precision achieved, 1%, and the excellent agreement between data and NNLO calculations performed adopting current parton distribution functions. The largest uncertainty for the cross section measurement comes from the uncertainty in the measurement of the luminosity that, however, cancels out in the cross-section ratios of

W to Z. To complete the picture, we very recently produced the measurement of the lepton charge asymmetry in W decays and the first measurement of the W polarization at a hadron collider. The lepton charge asymmetry in W events has been measured both with electrons and with muons in a large pseudo-rapidity range and for two different thresholds on the minimum transverse momentum of the W. The values of the charge asymmetry measured with electrons and muons are in good agreement with each other and the precision of the measurement is such that it is challenging the PDF predictions [15]. The measurement of the W polarization, as performed for $P_T^{\mathrm{W}} > 50\mathrm{GeV}/c$ [16] shows that, the LHC preferentially produces left handed W^+ and W^-as expected. The muon fit result yields the most precise measurement for the three polarization fractions, (f_L, f_R, f_0) left-handed, right-handed and longitudinal: $(\mathrm{f}_L - \mathrm{f}_R)^- = 0.240 \pm 0.036(\mathrm{stat.}) \pm 0.031(\mathrm{syst.})$ and $\mathrm{f}_0^- = 0.183 \pm 0.087(\mathrm{stat.}) \pm 0.123(\mathrm{syst.})$ for negatively charged W bosons, and $(\mathrm{f}_L - \mathrm{f}_R)^+ = 0.310 \pm 0.036(\mathrm{stat.}) \pm 0.017(\mathrm{syst.})$ and $\mathrm{f}_0^+ = 0.171 \pm 0.085(\mathrm{stat.}) \pm 0.099(\mathrm{syst.})$ for positively charged W bosons. These complex measurements are important benchmarks to prove that precision electroweak measurements are being already performed at the LHC. The list of all the public results in Electroweak Physics obtained by CMS can be found in [17].

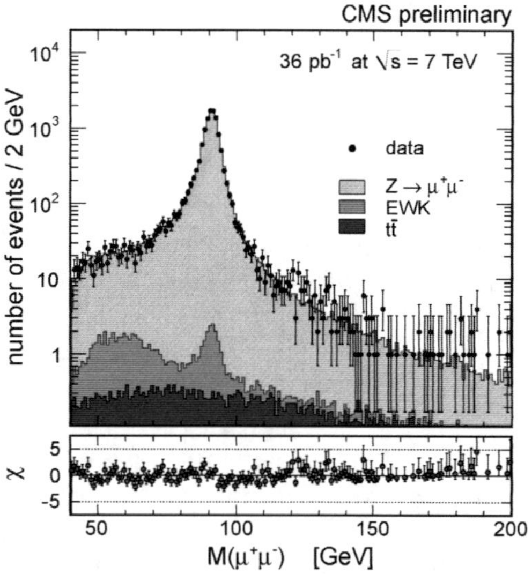

Figure 4. Invariant mass distribution for Z in dimuons (log scale).

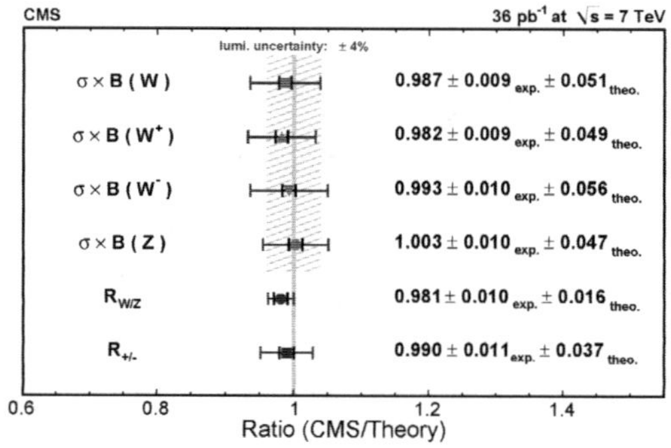

Figure 5. Ratio of measured to theoretical predicted values for W and Z cross-sections.

5. Top Physics

LHC can be considered as a top-factory for the huge rate of top quarks produced at TeV collisions. Searches in CMS for top quark candidates have been made looking at channels with high p_T leptons or di-leptons, jets with at least 1 jet b-tagged and missing E_T. Going through the full statistics collected so far corresponding to a luminosity of 36 pb^{-1}, we have been able to measure the top pair production cross section using different techniques and various decay channels leading to a combined measurement of $\sigma_{t\bar{t}} = 158 \pm 18 \pm 6$ (lum.) pb [18, 19]. This value is in good agreement with the most recent NLO and approximate NNLO predictions. These methods also lead to a precise measurement of the top mass with a combined result of $m_t = 173.4 \pm 1.9(stat) \pm 2.7(syst)\mathrm{GeV}/c^2$.

The excellent performance of the CMS detector and the effectiveness of the reconstruction algorithms have also been proven through the first measurement of the single top production cross section [20]. The measurement is particularly challenging as a consequence of the tiny cross section expected for the process and for the presence of important sources of background mainly due to W+jets and $t\bar{t}$ events (Fig. 6, 7). The measured cross-section for single top in the t-channel obtained using 36 pb^{-1} of LHC data is $\sigma_t = 83 \pm 29.8(stat + syst) \pm 3.3(lumi)$pb.

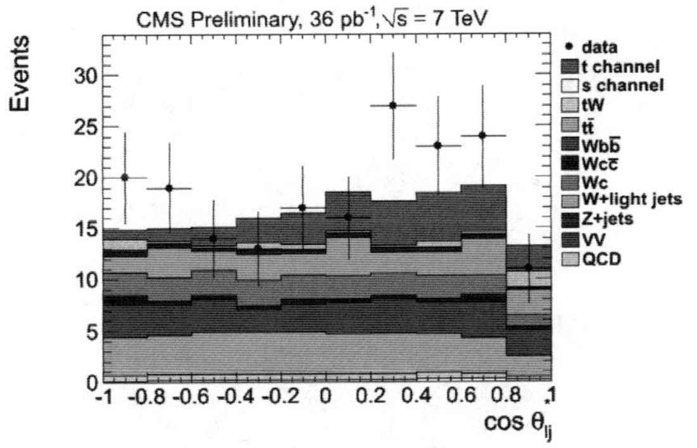

Figure 6. Extraction of the single top signal in the t-channel using the angle between lepton and jet.

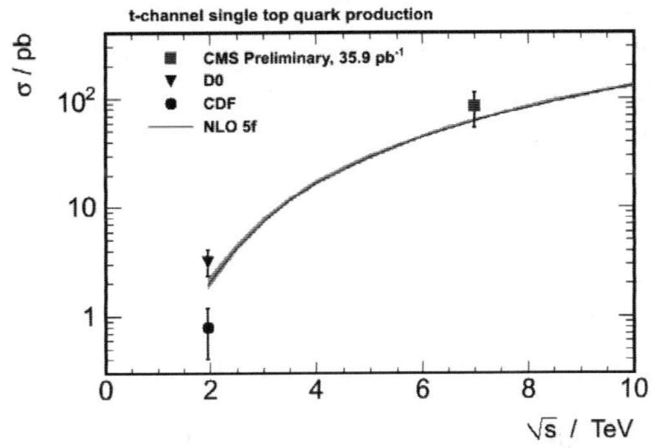

Figure 7. Comparison of the single top measured cross-section with the NLO theoretical predictions.

6. Higgs Physics

The search for the Higgs boson is one of the most important drivers of the LHC experiments. The amount of data collected in 2010 was not large enough to perform a complete and exhaustive search that can yield, in general, competitive results with respect to the Tevatron experiments. This has been possible, so far, only for a few analyses. One of these is the search for the SM Higgs in the W^+W^-channel with the Higgs production cross section enhanced by the presence of a fourth generation, a possible extension of the SM. For large lepton and quark masses, this extension has not been excluded by existing constraints. The presence of another family of fermions would produce an enhancement of the dominant gluon fusion cross-section. The irreducible background for $H \rightarrow W^+W^-$ production is the SM non-resonant production of W^+W^-. A good understanding of this process and of its properties is needed anyway since the W^+W^-channel is particularly sensitive in the intermediate mass range (114 - 200 GeV/c^2) and is therefore considered a candle channel for the search of the SM Higgs boson. This is why the search has been performed in conjunction with the first measurement of the W^+W^-production cross section at the LHC. W^+W^-candidates are selected in events with two high p_T leptons (electrons or muons) and large missing E_T. Leptons originating from $H \rightarrow W^+W^-$ decays tend to have a relatively small opening angle, while those from WW backgrounds are preferentially emitted back-to-back. The opening angle between the two leptons, $\Delta\phi_{ll}$, is therefore a variable providing the best discriminating power between the Higgs boson signal and the majority of the backgrounds in the low mass range. Fig. 8 shows the distribution of $\Delta\phi_{ll}$, after applying the W^+W^-selections, for a SM Higgs boson signal with m(H) = 160 GeV/c^2, and for the major sources of backgrounds. Since no excess above the SM expectations is seen in the lower $\Delta\phi_{ll}$ region, upper limits on the Higgs boson production cross section have been derived. In the presence of a sequential fourth family of fermions with very high masses, a Higgs boson with Standard Model couplings and a mass between 144 and 207 GeV/c^2 has been excluded at 95% condence level [21].

The second analysis yielding new results on Higgs is the search for minimally supersymmetric Standard Model (MSSM) Higgs decaying to τ pairs. The observed τ pair mass spectrum reveals no evidence for neutral Higgs boson production, and an upper bound on the Higgs boson production cross section is determined. These results, interpreted in the MSSM parameter space, exclude a previously unexplored region (Fig. 9) [22]. The complete set of published analyses by CMS on the Higgs search can be found in [23]. The prospects for the Higgs search at the LHC are very promising. Assuming that in 2011-12 the LHC will deliver to the experiments an integrated luminosity in a range of 5-10 fb^{-1}, as it appears to be possible extrapolating on last year's performance of the machine, we are confident that we will be able to exclude the SM Higgs boson in the mass range between 114 and 600 GeV/c^2, or to discover it with the combination of ATLAS and CMS results.

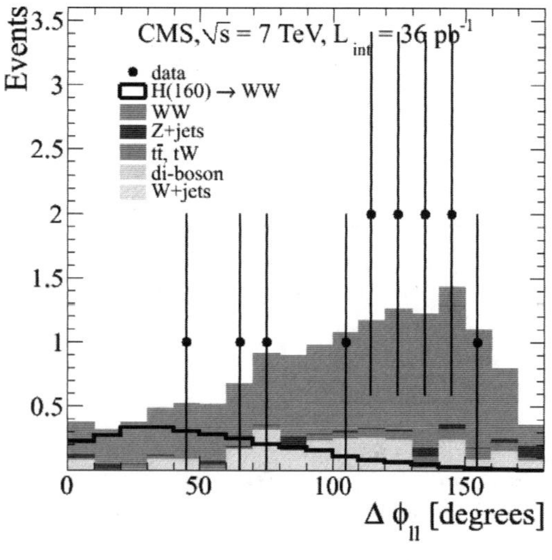

Figure 8. Opening angle between the two leptons in WW candidate events for the Higgs search.

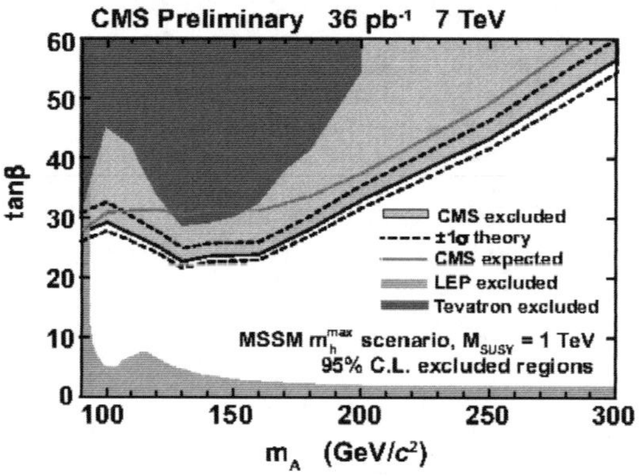

Figure 9. Exclusion limit in the search of MSSM neutral Higgs decaying to τ pairs.

7. Searches: SUSY and Exotica Physics

In this early phase of the LHC program many inclusive searches for beyond the Standard Model events have been performed. Many of these new particles have similar signature: high invariant mass of the final state, isolated high P_T leptons, large missing E_T, very energetic photons or jets. With the first data collected in 2010, even if is difficult to disentangle between different models, many analyses are already sensitive in regions not covered so far.

Supersymmetry (SUSY) searches at CMS have been performed using many different topological signatures: di-photons and large missing E_T, same sign and opposite sign di-leptons, single leptons and large missing E_T, multi-leptons and fully hadronic final states with large missing E_T. None of these searches have so far produced any hints of production of SUSY particles at the LHC. All the SUSY results can be found in [24]. Using well established statistical tools [25] we have extracted limits from the experimental data producing new results exceeding significantly the best measurements performed so far by the Tevatron experiments. Fig. 10 summarizes the exclusion limits produced by these analyses for a particular choice of SUSY parameters. The highest exclusion limits are obtained using the fully hadronic final states.

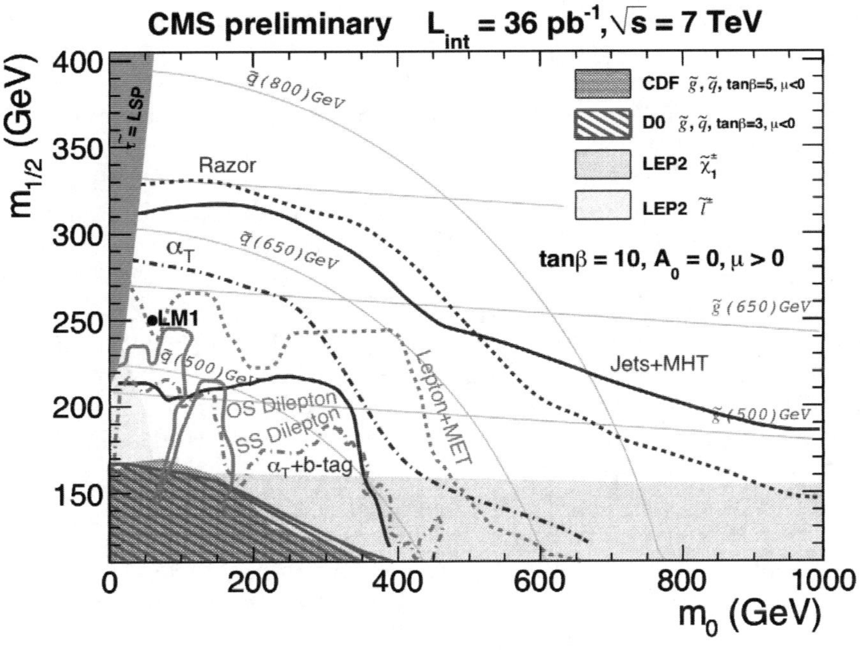

Figure 10. Exclusion limits for SUSY searches

All the exotica results are summarized in [26]. Among these analyses a particular focus is put here on new heavy bosons, extra-dimensions and microscopic black holes. New heavy gauge bosons, generally indicated as Z′ and W′, are predicted in various extensions of the Standard Model. The search for these particles is usually performed in the context of the benchmark models where these bosons are considered as heavy analogues of the SM W and Z bosons with the same fermionic couplings. Thus the decay modes and branching fractions are similar to those of the W and Z bosons. In this context the search is performed looking for anomalies in the tail of the distribution of the reconstructed transverse mass of the W and the invariant mass of the Z. W′ boson would imply an excess of events in the tail of the transverse mass distribution in the W leptonic decay. Since no excess is visible in our data, we can extract limits on the production of heavy W′ vector bosons at the LHC. Assuming Standard-Model-like couplings and decay branching fractions and combining together the decay modes in electrons and in muons, it is possible to exclude a W′ with mass lower than 1.58 TeV/c^2 [27], a value that exceeds the current limits set by the Tevatron experiments. The most stringent limits to date, around 1 TeV depending on the production model, have been obtained also for the search of Z′, where the analysis is conceptually similar [28]. The most important source of background for both analyses are multi-jet and $t\bar{t}$ events that must be carefully understood to set limits on new phenomena.

Compact large extra-dimensions are an intriguing proposed solution to the hierarchy problem of the Standard Model, which refers to the puzzling fact that the fundamental scale of gravity, $M_P = 10^{19}\mathrm{GeV}/c^2$, is so much higher than the electroweak scale $M_{EW} = 10^3\mathrm{GeV}/c^2$. With such a difference in scales, it is difficult to protect the Higgs mass from radiative corrections without a very high degree of fine-tuning. The original proposal to use extra dimensions to solve the hierarchy problem assumed a scenario where the SM is constrained to the common 3+1 space-time dimensions, while gravity is free to propagate through the entire multidimensional space. Because of this, the gravitational force is effectively diluted, having undergone a Gauss-

Law reduction in the flux. Phenomenologically, this scenario results in s-channel production of massive Kaluza-Klein (KK) graviton states, which decay into a di-photon final state that can be detected in modern, hermetic detectors like CMS. A search for large extra-dimensions via virtual graviton exchange in the di-photon channel has been performed by CMS looking for an excess of events in the high mass tail of the distribution of the di-photon invariant mass (Fig. 11). The new limits, obtained in the range of 1.6 - 2.3 TeV/c^2, depending on the number of extra-dimensions, can be interpreted as the lower limits on the effective Planck scale, M_D, in these models, and are to date the most restrictive limits on the existence of two or more large extra-dimensions [29].

Another possible manifestation of the fact that the effective Planck scale, M_D, could be brought to the TeV scale by the presence of compactified extra dimensions could be the production of microscopic black holes. Partons colliding in the LHC, once they approach each other to a distance comparable to the size of extra dimensions, could be sensitive to the full strength of gravity and may collapse into a microscopic black hole. The production cross section can be as high as 100 pb for M_D of 1 TeV/c^2. Once produced, the microscopic black holes evaporate almost instantaneously by emitting energetic particles. About three quarters of the emitted particles are expected to be quark and gluons; the rest is accounted for by leptons, photons, W/Z bosons, and possibly Higgs particles. We look therefore for events with high multiplicity of energetic objects. Since the main background comes from copious production of multi-jets that are not well described in QCD predictions, we must use data-driven methods. We have found that a variable, S_T , which is defined as the scalar sum of transverse momenta of all the energetic objects in the event (reconstructed hadronic jets, leptons, photons and missing transverse energy) can be used to describe the multi-jet QCD background (Fig. 12). The resulting background predictions for the inclusive multiplicities of 3, 4, and 5 or more objects in the final state agree with the observed spectra in the data. As a result, we have been able to exclude black holes with the masses up to between 3.5 and 4.5 TeV/c^2, for the values of M_D in the range of 1.5-3.5 TeV/c^2 and various other model parameters [30]. These limits are the first direct limits on black hole production at particle colliders and go well beyond potential reach of the Tevatron or cosmic-ray experiments.

8. Conclusions

In these proceedings the first physics results obtained using pp collisions at $\sqrt{s} = 7$ TeV are presented. After a few months of data taking a good understanding of the detector performance has been achieved and the Standard Model properties of the pp collisions studied. Soon afterwards started the systematic exploration of the new energy regime in the quest for signals of new physics. New limits have been produced in many searches: supersymmetry, new vector bosons, extra dimensions. The first studies of the Higgs boson in the LHC data have been presented together with the prospects for the current 2011-12 running period.

Acknowledgments

I would like to thank the LHC accelerator team for achieving an impressive performance of the machine in its first year of running. I am grateful to all colleagues of the CMS Collaboration for their huge collective effort in constructing and running the experiment and analyzing quickly and efficiently data collected recently.

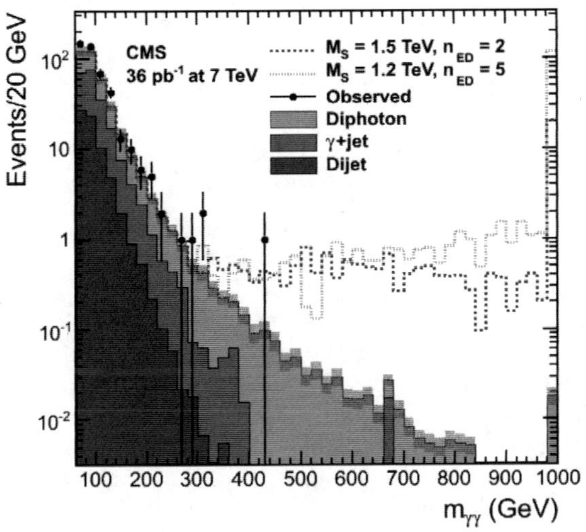

Figure 11. Invariant mass distribution for di-photon events with superimposed the expected results for two different extradimensional models.

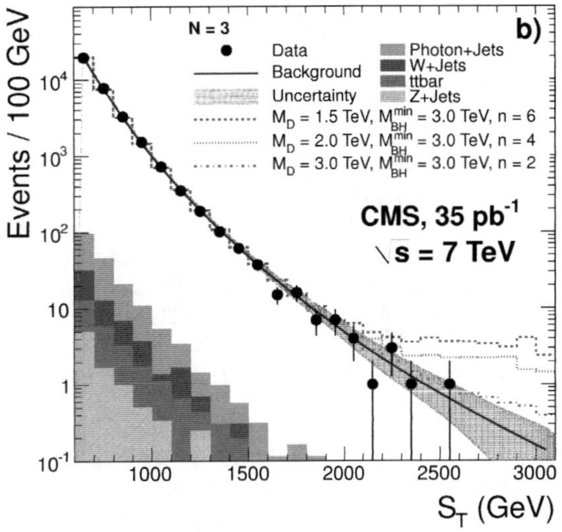

Figure 12. S_T distribution in the data compared with the simulation of the effect in case of microscopic black-holes production.

References

[1] "The CMS experiment at the CERN LHC" [CMS Collaboration], *JINST* 0803:S08004 (2008)

[2] "Transverse momentum and pseudorapidity distributions of charged hadrons in pp collisions at $\sqrt{s} = 0.9$ and 2.36 TeV" [CMS Collaboration], *JHEP* **1002** (2010) 041

[3] "Measurement of Bose Einstein Correlations in pp Collisions at $\sqrt{s} = 0.9$ and 7 TeV at the LHC" [CMS Collaboration], *JHEP* **1105** (2011) 029

[4] "Observation of Long-Range, Near-Side Angular Correlations in Proton-Proton Collisions at the LHC" [CMS Collaboration], *JHEP* **1009** (2010) 091

[5] "Measurement of the Inclusive Jet Cross Section in pp Collisions at 7 TeV" [CMS Collaboration], arXiv:1106.0208 ; CMS-QCD-10-011 ; CERN-PH-EP-2011-053, submitted to *Phys. Rev. Lett.*

[6] "Measurement of the differential dijet mass cross section in proton-proton collisions at $\sqrt{s} = 7$ TeV" [CMS Collaboration], *Phys. Lett.* B **700** (2011) 187

[7] "J/ψ prompt and non-prompt cross sections in pp collisions at $\sqrt{s} = 7$ TeV" [CMS Collaboration], *Eur. Phys. J.* C **71** (2011) 1575

[8] " Υ production cross section in pp collisions at $\sqrt{s} = 7$ TeV" [CMS Collaboration], *Phys. Rev.* D **83** (2011) 112004

[9] "Inclusive b-hadron production cross section with muons in pp collisions at $\sqrt{s} = 7$ TeV" [CMS Collaboration], *JHEP* **1103** (2011) 090

[10] "Measurement of $B\bar{B}$ Angular Correlations based on Secondary Vertex Reconstruction at $\sqrt{s} = 7$ TeV" [CMS Collaboration], *JHEP* **1103** (2011) 136

[11] "Measurement of the B^+ Production Cross Section in pp collisions at $\sqrt{s} = 7$ TeV" [CMS Collaboration], *Phys. Rev. Lett.* **106** (2011) 112001

[12] "Measurement of the B^0_d Production Cross Section in pp collisions at $\sqrt{s} = 7$ TeV" [CMS Collaboration], *Phys. Rev. Lett.* **106** (2011) 252001

[13] "Measurement of the B^0_s Production Cross Section in pp collisions at $\sqrt{s} = 7$ TeV" [CMS Collaboration], CERN-PH-EP-2011-063, arXiv:1106.4048, submitted to *Phys. Rev.* D

[14] "Measurement of inclusive W and Z cross sections in pp collisions at $\sqrt{s} = 7$ TeV" [CMS Collaboration], CMS PAS EWK-10-005 (2011)

[15] "Measurement of the Lepton Charge Asymmetry in Inclusive W Production in pp collisions at $\sqrt{s} = 7$ TeV" [CMS Collaboration], arXiv:1103.3470 ; CMS-EWK-10-006 ; CERN-PH-EP-2011-024.(2011), accepted by *JHEP*

[16] "Measurement of the Polarization of W Bosons with Large Transverse Momenta in W+Jets Events at the

LHC" [CMS Collaboration], arXiv:1104.3829 ; CMS-EWK-10-014 ; CERN-PH-EP-2011-043 (2011)
[17] https://twiki.cern.ch/twiki/bin/view/CMSPublic/PhysicsResultsEWK
[18] "Measurement of the $t\bar{t}$ production cross section and the top quark mass in the dilepton channel in pp collisions at $\sqrt{s}$ = 7 TeV" [CMS Collaboration], CERN-PH-EP-2011-055, arXiv:1105.5661, accepted by *JHEP*
[19] "Measurement of the $t\bar{t}$ production cross section in pp collisions at $\sqrt{s}$ = 7 TeV using the kinematic properties of events with leptons and jets" [CMS Collaboration], CERN-PH-EP-2011-060, arXiv:1106.0902, submitted to *Eur. Phys. J.* C
[20] "Measurement of the t-channel single top quark production cross section in pp collisions at $\sqrt{s}$ = 7 TeV" [CMS Collaboration], CERN-PH-EP-2011-066, arXiv:1106.3052, submitted to *Phys. Rev. Lett.*
[21] "Measurement of W^+W^- Production and Search for the Higgs Boson in pp Collisions at $\sqrt{s}$ = 7 TeV" [CMS Collaboration], *Phys. Lett.* B **699** (2011) 25-47
[22] "Search for Neutral MSSM Higgs Bosons Decaying to Tau Pairs in pp Collisions at $\sqrt{s}$=7 TeV" [CMS Collaboration], arXiv:1104.1619 ; CMS-HIG-10-002 ; CERN-PH-EP-2011-027, accepted by *Phys. Rev. Lett.*
[23] https://twiki.cern.ch/twiki/bin/view/CMSPublic/PhysicsResultsHIG
[24] https://twiki.cern.ch/twiki/bin/view/CMSPublic/PhysicsResultsSUS
[25] K. Nakamura et al. (Particle Data Group), *J. Phys.* G **37** (2010) 075021
[26] https://twiki.cern.ch/twiki/bin/view/CMSPublic/PhysicsResultsEXO
[27] "Search for a W′ boson decaying to a muon and a neutrino in pp collisions at $\sqrt{s}$=7 TeV" [CMS Collaboration], CMS-EXO-10-015, CERN-PH-EP-2011-012, arXiv:1103.0030v1, submitted to *Phys. Lett.* B
[28] "Search for Resonances in the Dilepton Mass Distribution in pp collisions at $\sqrt{s}$=7 TeV" [CMS Collaboration], CMS-EXO-10-013, CERN-EP-PH-2011-002, arXiv:1103.0981v3, accepted by *JHEP*
[29] "Search for Large Extra Dimensions in the Diphoton Final State at the Large Hadron Collider" [CMS Collaboration], CMS-EXO-10-026, CERN-PH-EP-2010-020, arXiv:1103.4279v1, accepted by *JHEP*
[30] "Search for Microscopic Black Hole Signatures at the Large Hadron Collider" [CMS Collaboration], *Phys. Lett.* B **697** (2011) 434-453

Journal of Physics: Conference Series **347** (2012) 012003 doi:10.1088/1742-6596/347/1/012003

Top quark and Higgs boson physics at the Tevatron

Shabnam Jabeen[1]
Physics Department, Brown University, RI, USA
E-mail: jabeen@fnal.gov

Abstract. The search for the Higgs boson and the study of the heaviest known fundamental particle, the top quark, have been at the center of the Tevatron research program. The Higgs boson is yet to be discovered and the top quark was discovered in 1995. Both of these particles have a very special place in the "periodic table" of fundamental particles. With Tevatron having delivered more than 11 fb^{-1} of data at 1.96 TeV and the Large Hadron Collider also rapidly collecting data at 7 TeV, we are entering a very exciting era where many interesting questions about these intriguing particles will be answered. Here I summarize the current status of Higgs boson and top quark studies at the Tevatron.

1. The top quark

The top quark program at the Tevatron is very broad. Since its discovery in 1995 at Fermilab, study of properties of this interesting particle and search for new physics effects in top quark decay and production has been one of the major goals of the Tevatron Run II. Here I present only few of the main and the recent results from CDF and DØ experiments. For a full list of recent results please check the webpages with the public results [1].

1.1. Top quark pair production cross section

The dominant production channel at the Tevatron in $p\bar{p}$ collisions is $q\bar{q}$ annihilation which happens in around 85% of collisions and gluon-gluon fusion which contributes only 15%. Within the Standard Model (SM) of particle physics, the top quark decays to a W boson and a b quark almost 100% of the time. The final state $W^+W^-b\bar{b}$ from top quark pair production is classified according to how the W boson decays, i.e. alljets, lepton+jets or dileptonic decay channels. The top quark pair production cross section has been determined by CDF and DØ experiments in different decay channels, with many different analysis techniques such as, simple event counting, b-jet identification (b-tagging), multivariate analysis techniques and combinations of these techniques. These measurements are consistent with theoretical predictions and a summary of these measurements done by CDF and DØ are shown in Figures 1 and 2 respectively.

1.2. Top quark mass measurement

Tevatron experiments, CDF and DØ, have measured top quark mass (m_t) to a very high precision in all decay modes with many different techniques. The most recent DØ results are in the lepton+jets channel using 3.6 fb^{-1} of data with in situ jet energy scale (JES) calibration [2]. Compared to previous measurements at DØ, a major improvement in this new measurement

[1] On behalf of the CDF and DØ Collaborations

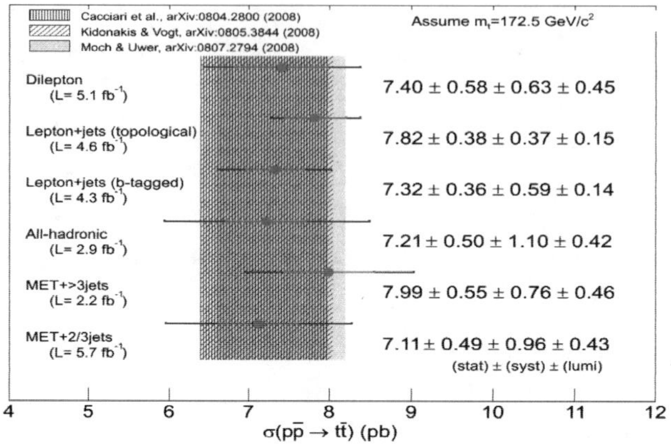

Figure 1. Summary of CDF measurements for $t\bar{t}$ production cross section.

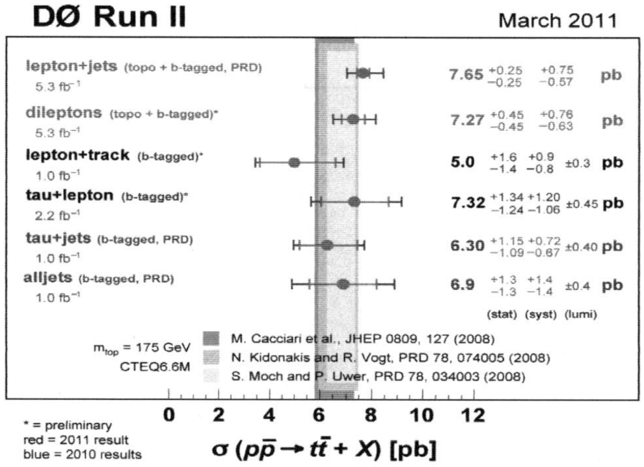

Figure 2. Summary of DØ measurements for $t\bar{t}$ production cross section.

is the significant reduction of the uncertainty associated with the modeling of differences in the calorimeter response to b-quark and light-quark jets originating from the introduction of a new flavor-dependent jet energy response correction. The measurement gives $m_t = 174.9 \pm 0.8(\text{stat}) \pm 0.8(\text{JES}) \pm 1.0(\text{syst})$ GeV. This is the best top-quark mass measurement at DØ. The second result comes from the dilepton channel using 5.4 fb^{-1} of data[3]. Here the dominant uncertainty comes from JES systematic uncertainties. The measurement gives $m_t = 174.0 \pm 1.8(\text{stat}) \pm 2.4(\text{syst})$ GeV. This is the world's best top-quark mass measurement in the dilepton channel. The two most recent CDF results both use the template method with in situ JES calibrations. The first one is obtained in the all hadronic channel using 5.8 fb^{-1} of data [4]. The measurement gives $m_t = 172.5 \pm 1.4(\text{stat}) \pm 1.0(\text{JES}) \pm 1.1(\text{syst})$ GeV. This is the second best measurement at CDF. The second new CDF result comes from the MET+jets channel using 5.7 fb^{-1} of data [5]. The measurement gives $m_t = 172.3 \pm 1.8(\text{stat}) \pm 1.5(\text{JES}) \pm 1.0(\text{syst})$ GeV. This measurement is complementary to the other top-quark mass measurements adding to the overall sensitivity. A summary of the direct mass measurements from CDF and DØ are shown

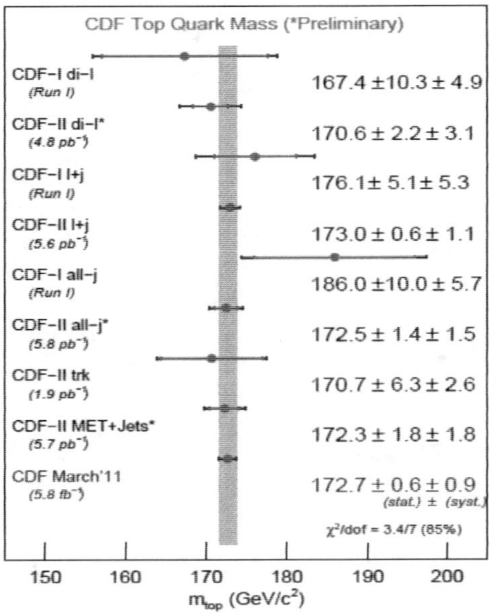

Figure 3. Summary of CDF top quark mass measurements.

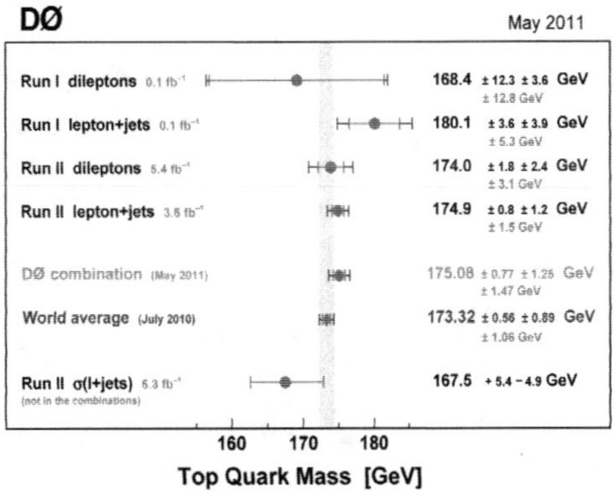

Figure 4. Summary of DØ top quark mass measurements.

in Figures 3 and 4 respectively.

1.3. Top quark mass from top quark pair production cross section

The mass of the top quark has been measured with a precision of <1%. But the top quark mass, beyond leading-order QCD, is a convention-dependent parameter. Therefore, in order to use this measured mass in higher-order QCD calculations or in fits of electroweak precision observables, it is important to know how to interpret this experimental result in terms of renormalization conventions. DØ has recently measured top quark mass in the pole mass scheme and the MSbar scheme using top quark pair production cross section measured with 5.3 fb^{-1} of integrated

Journal of Physics: Conference Series **347** (2012) 012003 doi:10.1088/1742-6596/347/1/012003

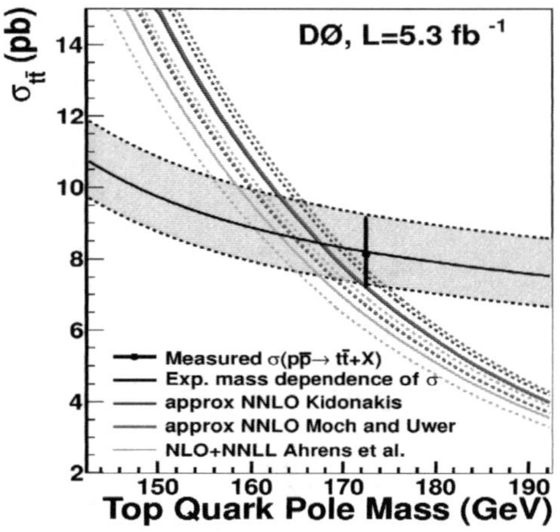

Figure 5. Measured $t\bar{t}$ cross section and theoretical NLO+NNLL and approximate NNLO calculations of cross section as a function of top quark mass, assuming that MC mass is pole mass.

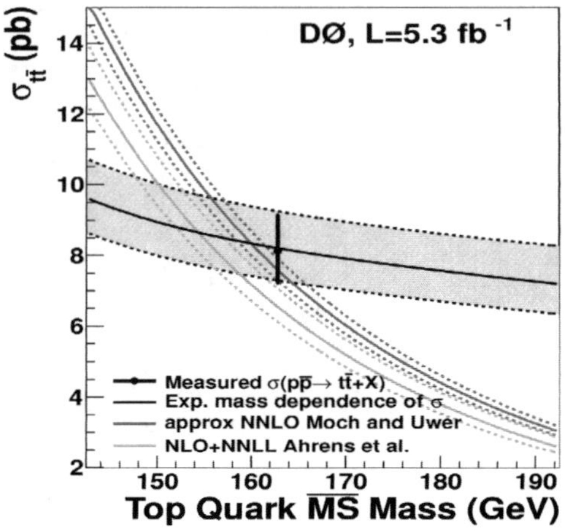

Figure 6. Measured $t\bar{t}$ cross section and theoretical NLO+NNLL and approximate NNLO calculations of cross section as a function of top quark mass, assuming that MC mass is MSbar mass.

luminosity [6]. The top quark cross section is compared with theoretical calculations assuming what we measure is the pole mass or the MSbar mass. The uncertainty on extracted top quark mass, dominated by systematic uncertainties in the measured cross section, is quite large to make some definite statement but the top mass extracted under the assumption that the Monte Carlo (MC) mass is equal to the pole mass, is found to agree with the average value of top mass from the Tevatron combination, while the mass extracted assuming that top quark mass in MC is equal to the top quark MSbar mass is found to be different from the average Tevatron value. Also, this is the first time that top quark mass has been extracted in MSbar scheme. Figure 5(6) shows measured $t\bar{t}$ cross section and theoretical NLO+NNLL and approximate NNLO calculations of cross section as a function of top quark mass, assuming that MC mass is pole mass (MC mass is MSbar mass). The colored dashed lines represent the uncertainties for the two theoretical calculations. The point shows the measured top pair cross section for MC top quark mass = 172.5 GeV, the black curve is the fit to the measured cross section, and the gray band corresponds to the total experimental uncertainty.

1.4. Top and anti-top quark mass difference

DØ has recently updated its top-antitop quark mass difference measurement with 3.6 fb^{-1} of data and found $\Delta M_{top} = 0.8 \pm 1.8 \pm 0.5$ (stat+syst) GeV [7]. This result is consistent with the expectation of CPT invariance $\Delta M_{top} = 0$ GeV. Recently, CDF has also measured this difference using a template method with 5.6 fb^{-1} of data, and found $\Delta M_{top} = 3.3 \pm 1.4(stat) \pm 1.0(\text{syst})$ GeV [8]. The result deviates from the expectation of CPT invariance by about 2 standard deviations.

1.5. Top quark width

CDF has performed a direct measurement of Γ_t, and set a limit $\Gamma_t < 7.64$ GeV at 95% CL [9]. The DØ collaboration extracted the total decay width of the top quark [10], $\Gamma_t = \Gamma_{t\to Wb}/B_{t\to Wb}$, from the partial decay width $\Gamma_{t\to Wb}$ measured using the t-channel cross section for single top

Journal of Physics: Conference Series **347** (2012) 012003 doi:10.1088/1742-6596/347/1/012003

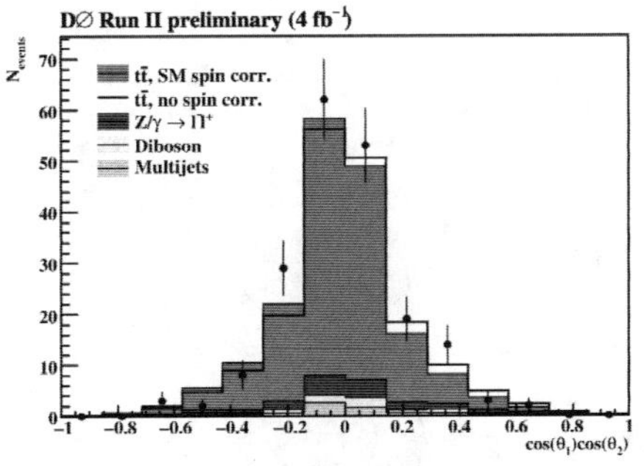

Figure 7. The distribution in $\cos\theta_1 \cos\theta_2$ for the entire dilepton sample with a dataset of 5.4 fb^{-1} compared with $t\bar{t}$ signal with and without NLO QCD spin correlations.

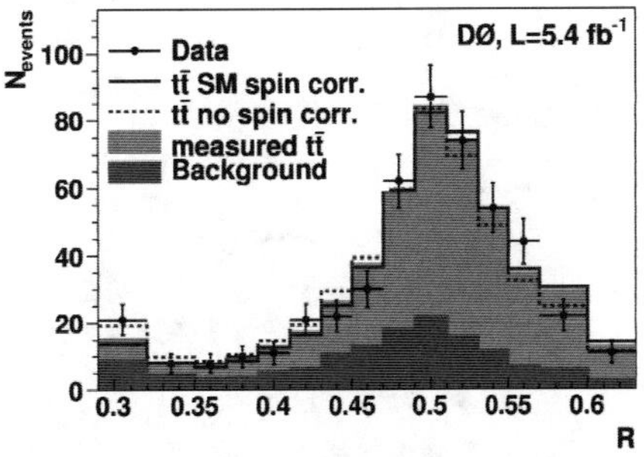

Figure 8. The ditribution of Matrix Element based discriminant R for data and backgrounds and signal estimation with and without assuming spin correlation.

quark production, and from the branching fraction $B_{t\to Wb}$ measured in top pair events. DØ finds $\Gamma_t = 1.99^{+0.69}_{-0.55}$ GeV. This constitutes the worlds most precise indirect determination of Γ_t and translates into a top quark lifetime of $\tau_t = (3.3^{+1.3}_{-0.9}) \times 10^{-25}$s.

1.6. Top quark spin correlation

In top pair production at hadron colliders, while the top quarks are unpolarised, the orientation of their spins is correlated. In contrast to other quarks, this correlation is not affected by fragmentation due to the short life time of the top quark, and is thus reflected in its decay products. Observation of spin correlation will place upper limit on top quark life time and scenarios beyond the SM predict changes in production and decay dynamics to change effect of spin correlation. Tevatron being a $p\bar{p}$ collider, the information on top-antitop correlation is complementary to LHC which is a pp collider. The NLO QCD prediction for spin correlation is C $= 0.777^{+0.027}_{-0.042}$. CDF has investigated this correlation using template method and measures the correlation to be C $= 0.7 \pm 0.7$. DØ has recently measured this correlation in two different analyses. One, choosing the beam momentum vector as the quantization axis in the dilepton channel, using $cos\theta_1 cos\theta_2$ distribution (Figure 7), results in spin correlation $C = 0.1 \pm 0.5$. $\theta_1(\theta_2)$ is the angle between the line of flight of the lepton (antilepton) with the polarization direction. Two, using the matrix element technique and building a discriminant R (Figure 8) using event probabilities with and without spin correlations. In this case the spin correlation is measured to be $C = 0.6 \pm 0.3$. This analysis, for the first time, excludes uncorrelated spin case at 97.7% CL.

1.7. Top quark forward-backward asymmetry

In the SM, the pair production of top quarks in $p\bar{p}$ collisions at LO is symmetric under charge conjugation. Next-to-LO calculations predict a small forward-backward asymmetry A_{fb} of the order of few %, which is due to contributions from the interference of diagrams for initial and final state radiation, and from the interference of box and tree-level diagrams. DØ measured A_{fb} in the $t\bar{t}$ rest frame in lepton+jets final states on a dataset corresponding to 4.3 fb^{-1} using $t\bar{t}$ event candidates fully reconstructed with a kinematic fitter, and found $A_{fb} = 8\% \pm 4\%$ [14]. DØ result, shown in Figure 9, is about 2 standard deviations away from the MC@NLO prediction of $1 \pm 2\%$. A similar measurement is performed by CDF in lepton+jets and dilepton final states [15] and the distributions corrected back to parton level. After all corrections, CDF finds

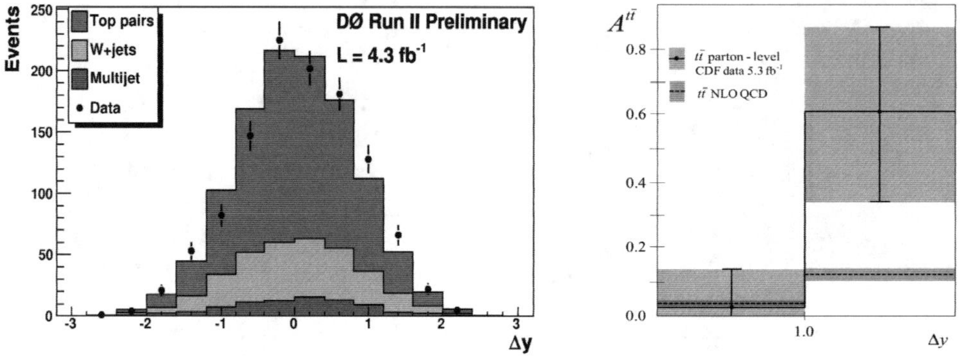

Figure 9. Left(right) The distribution in Δy for DØ (CDF) data compared to MC predictions.

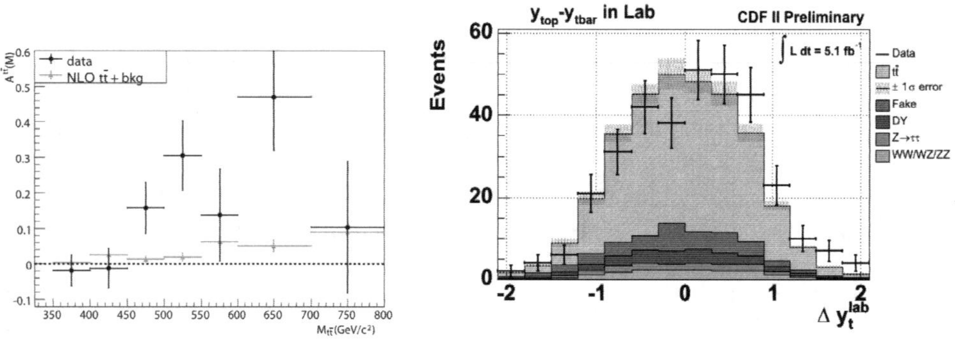

Figure 10. Left: Dependance of A_{fb} on $t\bar{t}$ mass; Right: CDF Δy distribution in dilepton events.

$A_{fb} = 16\% \pm 7\%$, which is about 1.5 standard deviations away from the MC@NLO prediction of $6\% \pm 1\%$. CDF also investigated the dependence of A_{fb} on the invariant mass of the $t\bar{t}$ system, $M_{t\bar{t}}$, which is compared to the prediction of MC@NLO plus background in Figure 10. Motivated by the resolution in $M_{t\bar{t}}$, CDF measures A_{fb} in two bins of $M_{t\bar{t}} < 450$ GeV and $M_{t\bar{t}} > 450$ GeV. The $A_{fb} = 48\% \pm 11\%$ at parton level and after all corrections in the $M_{t\bar{t}} > 450$ GeV bin is more than 3 standard deviations away from the NLO SM prediction of $A_{fb} = 9\% \pm 1\%$. Another measurement, carried out by CDF in the dilepton channel using 5.1 fb^{-1} of data, yields $A_{fb} = 42\% \pm 16\%$ parton level after all corrections, which is about 2.5 standard deviations away from the SM NLO prediction. The above results indicate tension between the measurement and the NLO MC predictions.

1.8. W helicity in top pair decays and search for Wtb anomalous couplings

In the SM, the top quark decays into a W boson and a b quark with a probability of $> 99.8\%$, where the on-shell W boson can be in a left-handed, longitudinal, and right-handed helicity state. NLO calculations within the SM of the corresponding helicity fractions predicts $f_- = 0.301$, $f_0 = 0.698$, and $f_+ = 4.1 \times 10^{-4}$, respectively. A significant deviation from the SM expectation would indicate a contribution from new physics. CDF performed a model

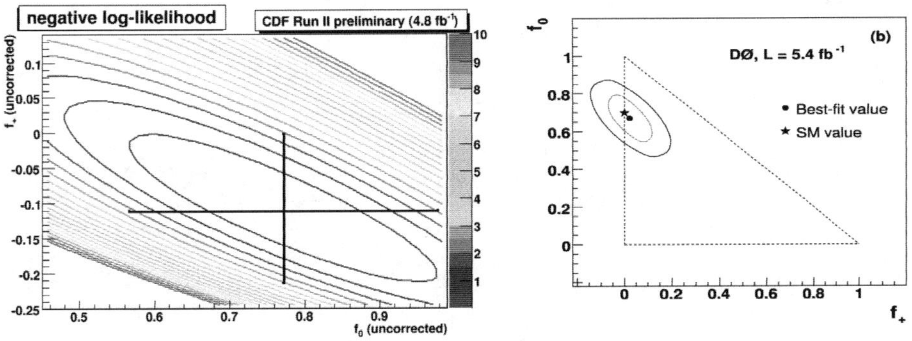

Figure 11. Left: the result of CDF W boson helicity measurement in the f_0, f_+ plane. The ellipses correspond to the negative log-likelihood profile of two-dimensional simultaneous fit of both f_0, f_+. Right: the result of the DØ boson helicity measurement in the f_0, f_+ plane. The ellipses indicate the 68% and 95% CL contours.

independent measurement [12] and found $f_0 = 0.78 \pm 0.21$ and $f_+ = 0.12 \pm 0.12$ as shown in Figure 11. DØ has also measured the f_0 and f_+ helicity fractions in a model independent measurement [11], observing $f_0 = 0.669 \pm 0.102$ and $f_+ = 0.023 \pm 0.053$ as shown in Figure 11, in agreement with the SM expectation. DØ has also combined the helicity fraction measurement with the measurement of anomalous Wtb couplings in single top quark production and decay. Combining these complementary measurements helps DØ set even better limits on the anomalous Wtb couplings [13].

1.9. Search for fourth generation quarks
Search for pair production of fourth generation t' has been carried out by both CDF and D0. DØ searched for t' and its antiparticle, followed by their decays to a W boson and a jet, based on an integrated luminosity of 5.3 fb^{-1} and set an upper limit on the $t'\bar{t}'$ production cross section that exclude at the 95% C.L. a t' quark that decays exclusively to W+jet with a mass below 285 GeV [16]. CDF performed a similar search using 4.6 fb^{-1} and exclude a fourth-generation t' quark with mass below 335 GeV at 95% C.L. [17]. Both limits are shown in Figure 12.

1.10. Sigle top quark production
Within the SM, top quarks are produced through two different types of interactions at hadron colliders: $t\bar{t}$ pairs, produced via the strong interaction, the mode of production in which top quark was discovered first; the electroweak production of single top quarks, observed by the DØ and CDF collaborations in 2009 [18]. The single top quark production rate is about half that for $t\bar{t}$ pairs and the signal-to-background ratio is worse, so the observation of this process is more difficult. At the Tevatron, single top quarks are produced through W boson exchange, and accompanied by a b quark in the s-channel (denoted as "tb"), or by both a b and a light quark in the t-channel (denoted as "tqb"). A third process, usually called associated production, in which the top quark is produced together with a W boson, has negligible cross section at the Tevatron. Until now DØ and CDF have measured the production cross section for the sum of tb and tqb while assuming the relative rate between the two processes as predicted by the SM. This relative rate can however be different in many new physics scenarios, for example, in models with 4th generation quarks, or heavy bosons or with Flavor Changing

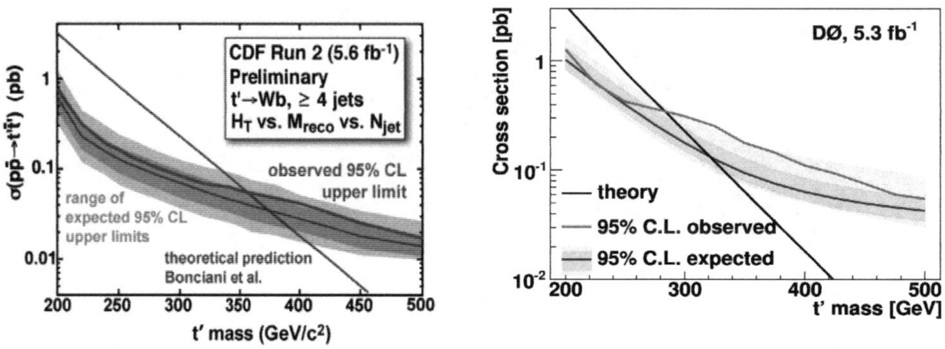

Figure 12. Observed and expected upper limits and predicted values for the $t'\bar{t}'$ production cross section as a function of the mass of the t' quark for CDF (left) and DØ (right). The shaded regions around the expected limit represent the 1 σ and 2 σ bands.

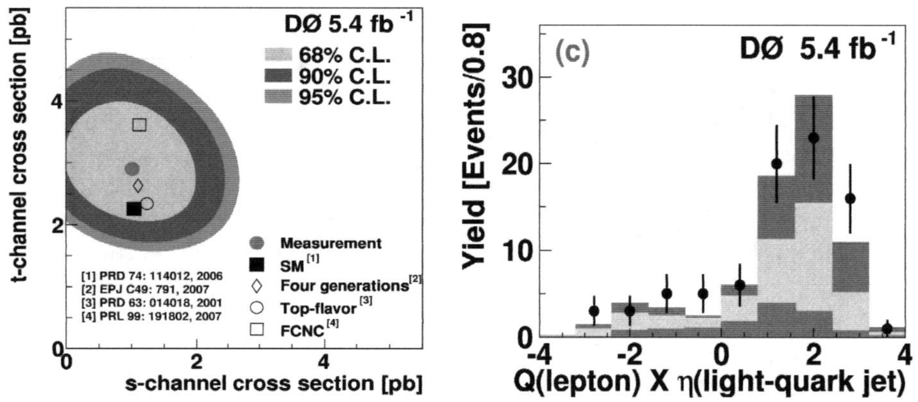

Figure 13. Left: Posterior probability density for tqb vs tb single top quark production in contours of equal probability density. The measured cross section and various theoretical predictions are also shown. Right: lepton charge multiplied by the pseudorapidity of the leading non-b-tagged quark jet. The t-channel signal is shown in purple, and the main backgrounds, W+jets and $t\bar{t}$, are shown in green and red, respectively.

Neutral Currents (FCNC) or anomalous top quark couplings. Recently DØ has performed a model-independent measurement of t-channel production of single top quarks using 5.4 fb^{-1} of integrated luminosity in events containing an isolated electron or muon, missing transverse energy and one or two jets originating from the fragmentation of b quarks. DØ measured a cross section $\sigma(p\bar{p} \rightarrow tqb + X) = 2.90 \pm 0.59$ (stat + syst) pb for a top quark mass of 172.5 GeV [19]. The probability of the background to fluctuate and produce a signal as large as the one observed is 1.6×10^{-8}, corresponding to a significance of 5.5 standard deviations. Figure 13 shows the 2D posterior probability density for tqb vs tb single top quark production (left) and comparisons between the data, the background model, and the t-channel signal for a sample from final combined discriminant bins selected with signal-to-background ratio $S/B > 0.32$ (right).

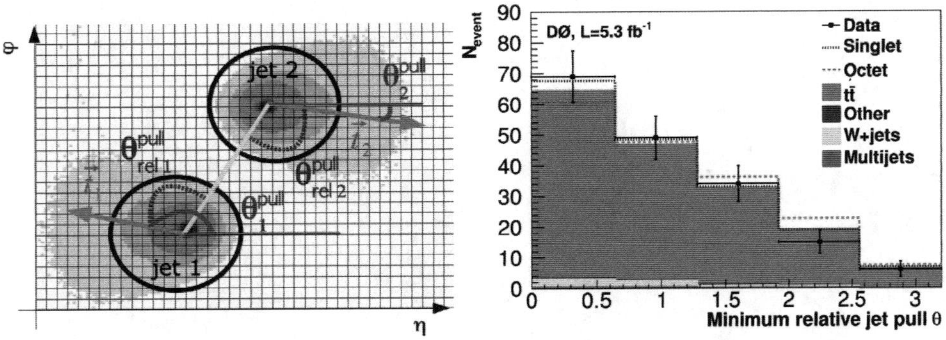

Figure 14. Left: Diagram showing two jets in the $\eta - \phi$ plane, and the reconstruction of the jet pull vectors (t), jet pull angles (θ^{pull}), and relative jet pull angles (θ^{pull}_{rel}). The red contours correspond to the jet transverse energy distribution in the calorimeter. Right: the discriminating color-flow variable, the minimum θ^{pull}_{rel} for the W-pair jets, for events passing the M_W requirement, with $\Delta R < 2$, and $\eta < 1.0$ for both jets. The $t\bar{t}$ MC shape is obtained using the measured value of $f_{Singlet}$.

1.11. Color flow or jet pull

Color charge is conserved in QCD. At leading order in the strong coupling constant α_s, color can be traced from initial partons to final-state partons in high-energy hadron collisions. Two final-state partons on the same color-flow line are color-connected and attracted by the strong force. As these colored states hadronize, the potential energy of the strong force between them is released in the form of hadrons. Thus, knowledge of the color-connections between jets can serve as a powerful tool for separating processes that otherwise appear similar. For example, in the decay of a Higgs boson to a pair of b quarks, the two b quarks are color connected to each other, since the Higgs is uncolored or color singlet, whereas in $g \to b\bar{b}$ background events, they are color-connected to beam remnants because the gluon is a color-octet and carries a color and an anti-color. DØ recently used $t\bar{t}$ events to measure the fraction of uncolored W bosons using a calorimeter-based vectorial variable, jet pull, which is sensitive to the color-flow structure of the final state. This variable is explained and the distribution is shown in Figure 14. We find that the fraction of uncolored W bosons is $f_{singlet}$ = 0.56 ±0.42(stat+syst) [20]. The fraction of events with light jets coming from the W boson (color) singlet is $f_{singlet} = 1$ in the SM.

2. Search for Higgs boson production

Searching for the BEHHGK mechanism and boson, named after the authors (Breit, Englert, Higgs, Hagen, Guralnik, Kibble), was proposed in 1964[21]. The search for the Higgs boson is considered to be the search for the "Holy Grail" in high energy physics and is one of the highest priorities at Run II of the Tevatron. There are direct and indirect limits on the SM Higgs boson mass from LEP, SLD and Tevatron constraining the Higgs boson mass to be between 114 GeV and 185 GeV at the 95% CL. The primary Higgs channels of $H \to W^+W^-, WH \to lnb\bar{b}, ZH \to \nu\nu b\bar{b}$, and $ZH \to l^+l^-b\bar{b}$ from CDF and DØ correspond to about 500 Higgs bosons produced at each mass, $114 < m_H < 185$ GeV. The search is divided in two categories: low mass ($m_H < 135$ GeV) and high mass ($m_H > 135$ GeV).

A low mass Higgs boson preferentially decays $H \to b\bar{b}$. It is most easily identified in events produced via associated Higgs production, WH and ZH, when the W and Z decay leptonically, into final states of $WH \to l\nu b\bar{b}$, $ZH \to \nu\nu b\bar{b}$, and $ZH \to l^+l^-b\bar{b}$ where $l = e$ or μ. In events

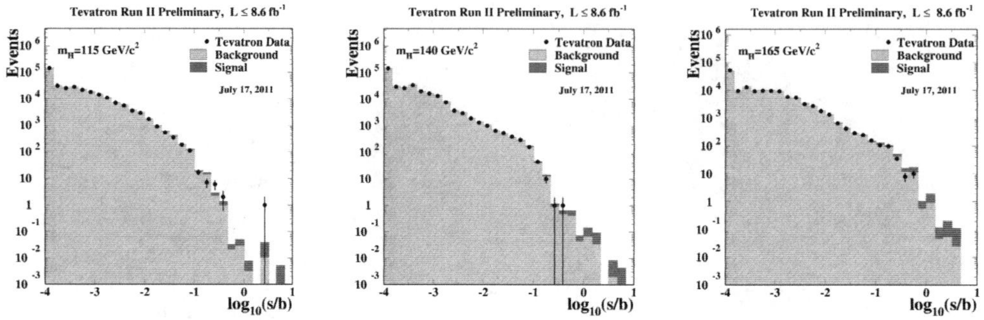

Figure 15. Distributions of log10(S/B), for the data from all contributing channels from CDF and DØ, for Higgs boson masses of 115 (left), 140 (middle), and 165 (right) GeV. The data are shown with points, and the expected signal is shown stacked on top of the backgrounds. Underflow and overflow entries are collected into the leftmost and rightmost bins.

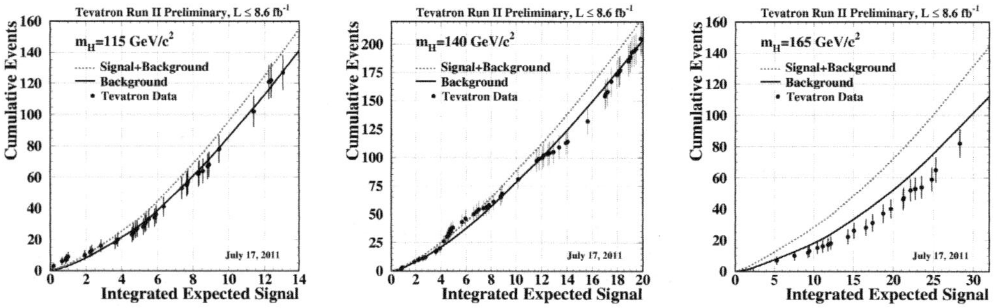

Figure 16. Integrated distributions of S/B, starting at the high S/B side, for Higgs boson masses of 115 (left), 140 (middle), and 165 (right) GeV. The total signal+background and background-only integrals are shown separately, along with the data sums. Data are only shown for bins that have data events in them.

with a reconstructed W or Z boson and two or more additional jets, the di-jet invariant mass is used to search for a resonance originating from $H \to b\bar{b}$. Expected sensitivity is improved by categorizing events according to the number of identified b quark jets. For details of the CDF and DØ Higgs results, please visit the public wepages of the experiments [22].

In searches for a high mass Higgs boson the most sensitive mode at the Tevatron is $gg \to H \to WW \to l\nu l\nu$ due to the high cross section and well identified final state. The high mass analysis benefits from separating events into categories according to the number of jets and the number of leptons because of the difference in the kinematics of the signal production mechanisms and background processes.

Since 2006, CDF and DØ experiments have been producing combined Higgs searches and constant efforts are made to improve signal acceptance by re-visiting the kinematic cuts and the use of sophisticated analysis techniques. The most recent combination combines up to 8.6 fb^{-1} [23]. Figure 15 shows combined discriminant outputs for three Higgs mass points from

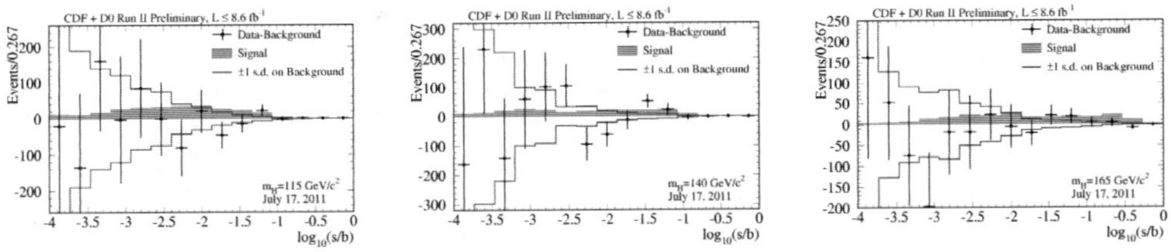

Figure 17. Background-subtracted data distributions for all channels, summed in bins of S/B, for Higgs boson masses of 115 (left), 140 (middle), and 165 GeV. The background has been fit, within its systematic uncertainties, to the data. The points with error bars indicate the background-subtracted data; the sizes of the error bars are the square roots of the predicted background in each bin. The unshaded (blue-outline) histogram shows the systematic uncertainty on the best-fit background model, and the shaded histogram shows the expected signal for SM Higgs boson.

the multiple search channels (dozens of them) summing up those bins with the same S/B into the same bins, and then sorting the bins from lowest to highest S/B. These distributions can be integrated from the high-S/B side downwards, showing the sums of signal, background, and data for the most pure portions of the selection of all channels added together. These integrals can be seen in Figure 16. An excess in these bins relative to the background prediction drives the Higgs boson cross section limit upwards, while a deficit drives it downwards. The lower-S/B bins show that the modeling of the rates and kinematic distributions of the backgrounds is very good. The integrated plots show a slight excess of events in the highest-S/B bins for the analyses seeking a Higgs boson mass of 115 GeV and 140 GeV, and a slight deficit of events in the highest-S/B bins for the analyses seeking a Higgs boson of mass 165 GeV. Figure 17 shows the distributions of the data after subtracting the expected background, and the comparison with the expected signal yield for a SM Higgs boson, after collecting all bins in all channels sorted by S/B. Finally Figure 18 shows the ratios of the 95% C.L. expected and observed limit to the SM cross section. We exclude $156 < m_H < 177$ GeV and $100 < m_H < 108$ GeV. The expected exclusion region, given the current sensitivity, is $148 < m_H < 180$ GeV and $100 < m_H < 109$ GeV and we did not study masses below $m_H < 100$ GeV. The Tevatron is scheduled to deliver up to 12 fb^{-1} , providing about 10 fb^{-1} of analyzable data for Higgs analysis per experiment. With this dataset, a full combination of Higgs searches will have at least 2.4 σ level sensitivity for $100 < m_H < 185$ GeV.

There are many beyond SM (BSM) scenarios or alternative Higgs boson models. Many of these models can be probed at the Tevatron energies and can be constrained in the absence of any excess observed. For both CDF and DØ searches of BSM Higgs, please see CDF and DØ public results webpages [22].

References

[1] CDF top quark results: http://www-cdf.fnal.gov/physics/new/top/top.html;
DØ top quark results: http://www-d0.fnal.gov/Run2Physics/top/top_public_web_pages/top_public.html.
[2] V. M. Abazov et. al. (DØ Coll.), *Phys. Rev.* D **84**, 032004 (2011).
[3] V. M. Abazov et. al. (DØ Coll.), *Phys. Rev. Lett.* **107**, 082004 (2011).
[4] T. Aaltonen et. al. (CDF Coll.), CDF public note 10456 (2011).
[5] T. Aaltonen et. al. (CDF Coll.), CDF public note 10433 (2011).
[6] V. M. Abazov et. al. (DØ Coll.), *Phys. Lett.* B **703**, 422 (2011).
[7] V. M. Abazov et. al. (DØ Coll.) *Phys. Rev.* D **84**, 052005 (2011).

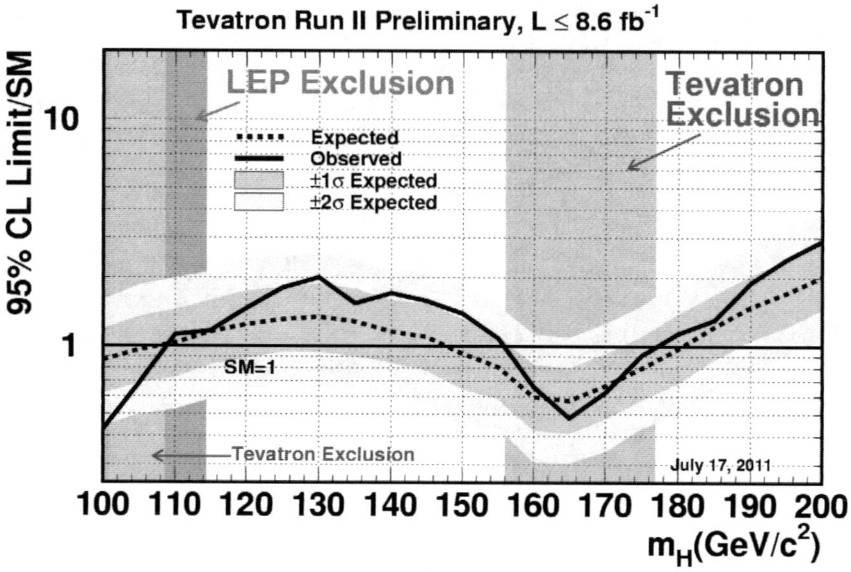

Figure 18. Observed and expected (median, for the background-only hypothesis) 95% C.L. upper limits on the ratios to the SM cross section, as functions of the Higgs boson mass for the combined CDF and DØ analyses. The limits are expressed as a multiple of the SM prediction for test masses (every 5 GeV) for which both experiments have performed dedicated searches in different channels. The points are joined by straight lines for better readability. The bands indicate the 68% and 95% probability regions where the limits can fluctuate, in the absence of signal. The limits displayed in this figure are obtained with a Bayesian calculation.

[8] T. Aaltonen et. al. (CDF Coll.), *Phys. Rev. Lett.* **106**, 152001 (2011).
[9] T. Aaltonen et. al. (CDF Coll.), *Phys. Rev. Lett.* **105**, 232003 (2010).
[10] V. M. Abazov et. al. (DØ Coll.), *Phys. Rev. Lett.* **106**, 022001 (2011).
[11] V. M. Abazov et. al. (DØ Coll.), *Phys. Rev. D* **83**, 032009 (2011).
[12] T. Aaltonen et. al. (CDF Coll.), CDF Conf. Note 10333 (2010).
[13] V. M. Abazov et. al. (DØ Coll.), *Phys. Rev. Lett.* **102**, 092002 (2009).
[14] V. M. Abazov et. al. (DØ Coll.), DØ Conf. Note 6062 (2010).
[15] T. Aaltonen et. al. (CDF Coll.), Fermilab-Pub-10-525-E, submitted to *Phys. Rev.* D, *Preprint* hep-ex/1101.0034 (2010).
T. Aaltonen et. al. (CDF Coll.), CDF Conf. Note 10436 (2011).
[16] V. M. Abazov et. al. (DØ Coll.), accepted by *Phys. Rev. Lett.*, *Preprint* hep-ex/1104.4522 (2011).
[17] T. Aaltonen et. al. (CDF Coll.), CDF Conf. Note 10110 (2010)
[18] T. Aaltonen et al. (CDF Coll.), *Phys. Rev.* D **82**, 112005 (2010);
T. Aaltonen et. al. (CDF Coll.), *Phys. Rev. Lett.* **103**, 092001 (2009);
CDF and DØ combination: *Preprint* hep-ex/0908.2171 (2009).
[19] V. M. Abazov et. al. (DØ Coll.), *Phys. Lett.* B **705**, 313 (2011).
[20] V. M. Abazov et. al. (DØ Coll.), *Phys. Rev.* D **83**, 092002 (2011)
[21] F. Englert and R. Brout, *Phys. Rev. Lett.* **13** 321 (1964);
P. W. Higgs, *Phys. Rev. Lett.* **13** 508 (1964);
G. S. Guralnik, C. R. Hagen and T. W. B. Kibble, *Phys. Rev. Lett.* **13**, 585 (1964).
[22] CDF public results: http://www-cdf.fnal.gov/physics/new/hdg/Results.html ;
DØ public results: http://www-d0.fnal.gov/Run2Physics/WWW/results/higgs.htm
[23] Tevatron Higgs Combination, *Preprint* hep-ex/1107.5518 (2011).

Journal of Physics: Conference Series **347** (2012) 012004 doi:10.1088/1742-6596/347/1/012004

Overview of experimental results in PbPb collisions at $\sqrt{s_{NN}} = 2.76$ TeV by the CMS Collaboration

Pelin Kurt[1]
Vanderbilt University, TN, USA
E-mail: pkurt@cern.ch

Abstract.
We present first results of the CMS experiment from PbPb collisions at $\sqrt{s_{NN}} = 2.76$ TeV, probing quark and gluon matter at unprecedented values of energy density. The capabilities of the CMS detector allow us to investigate various hard probes, as well as bulk particle production and collective phenomena, using the calorimetry, muon and tracking systems covering a large range in pseudorapidity, complemented by a flexible two-level trigger system for both proton-proton and ion-ion collisions. The data collected during the November 2010 PbPb run at $\sqrt{s_{NN}} = 2.76$ TeV, was analyzed and multiple measurements of the properties of the hot and dense matter were obtained. Global event properties, detailed study of jet production and jet properties, isolated photons, quarkonia and weak bosons were measured and compared to pp data and Monte Carlo simulations.

1. Introduction

The Large Hadron Collider (LHC) first produced collisions of heavy ions in November of 2010 at a nucleon-nucleon center of mass energy of $\sqrt{s_{NN}} = 2.76$ TeV. Nuclear interactions at these energies are expected to produce matter at energy densities exceeding any previously explored in experiments conducted at particle accelerators. The CMS detector was used to study the properties of the matter created in this new regime. CMS is a general purpose particle detector very well suited to study high energy nuclear collisions [1]. High precision tracking and calorimetry with fine granularity are augmented with a sophisticated multi-level triggering system. The initial analysis of CMS heavy ion data revealed information including global particle production which is related to the initial state formed in the collision, various observables sensitive to the hydrodynamical expansion of the system, and detailed probes of the produced medium by studying the propagation of a broad range of particles. More details on the analysis of specific results can be found in the measurements on global properties of charged particle production [2], transverse energy [3], high multiplicity [4] and HBT [5] in pp collisions, elliptic flow [6], dihadron correlations [7, 8], charged particle R_{AA} [9], jet quenching and fragmentation [10, 11], photons [12], quarkonia suppression [13, 14] and weak bosons [15].

The CMS detector incorporates a broad suite of high precision subsystems which were originally optimized for very high energy pp collisions but which also provide unprecedented capabilities for studying nuclear collisions. The CMS uses a right-handed coordinate system with z-axis parallel and x, y-axis perpendicular to the LHC beam direction, azimuthal angle ϕ

[1] On behalf of the CMS Collaboration

and polar angle θ. Particles with energies E and momenta $\vec{p}$ are characterized by transverse momentum $p_T = |\vec{p}| \sin\theta$, rapidity $y = \frac{1}{2} \ln \frac{E+p_z}{E-p_z}$ and pseudorapidity $\eta = -\ln \tan \frac{\theta}{2}$. A central feature of the CMS apparatus is the superconducting solenoid of 6 m internal diameter, providing a uniform magnetic field of 3.8 T. The inner tracking system is composed of a pixel detector with three barrel layers at radii between 4.4 and 10.2 cm and a silicon strip tracker with 10 barrel detection layers extending outwards to a radius of 1.1 m. Each system is completed by two endcaps, extending the acceptance up to a pseudorapidity $|\eta| = 2.5$. The nominal momentum resolution is typically 0.7 (5.0)% at 1 (1000) GeV/c in the central region [16].

The calorimeters inside the magnetic coil consist of a lead-tungstate crystal electromagnetic calorimeter (ECAL) and a brass-scintillator hadronic calorimeter (HCAL). These calorimeters provide a hermetic coverage over a large range of pseudorapidity ($|\eta| < 3$ for ECAL and $|\eta| < 5$ for HCAL). The return field is large enough to saturate 1.5 m of iron, allowing 4 layers of muon stations located outside the solenoid. Each muon station consists of several layers of aluminum drift tubes (DT) in the barrel region surrounding the solenoid and cathode strip chambers (CSC) in the endcap region, complemented by resistive plate chambers (RPC). All of these detectors have sufficient granularity and resolution to function well even in the highest multiplicities encountered in PbPb collisions. The CMS trigger and data acquisition systems were specially configured for the 2010 PbPb run to prepare for possible surprises related to the large multiplicities expected in these collisions. In particular, the silicon micro strip tracker and the ECAL and HCAL calorimeters were read out in non-zero suppressed mode, recording information from all channels. The zero suppression was done during the offline processing using algorithms optimized for PbPb environment. The maximum inelastic PbPb collision rate achieved during the run was about 220 Hz while the maximum rate that could be written to tape was about 140 Hz. CMS employed its flexible triggering system to select and write to tape all events containing high p_T jets, photons and muons while recording a scaled-down random sample of minimum bias events. For analysis of PbPb events, it is important to determine the overlap or impact parameter of the two colliding nuclei, usually called "centrality". Centrality in CMS was determined using the total sum of transverse energy in reconstructed towers from both positive and negative Hadronic Forward (HF) calorimeters. Centrality for specific event classes is expressed as a percentage of the inelastic nucleus-nucleus interaction cross section.

The multiplicity of charged particles produced in the central rapidity region is a key observable to characterize the properties of the matter created in heavy-ion collisions. The pseudorapidity dependence of a measurement of charged particle multiplicity density $dN_{ch}/d\eta$ is shown in Figure 1(left) in various centrality bins. The η dependence of the results is weak, varying by less than 10% over the η average. The slight dip at η=0 is a trivial kinematic effect owing to the use of pseudorapidity rather than rapidity. The data for this measurement was taken without a magnetic field in order to include particles with transverse momenta down to about 30 MeV/c. The number of charged hadrons was obtained by two methods based on the inner silicon pixel system. One technique involved counting the number of reconstructed single particle hits in the pixel detector, while the other formed hit pairs ("tracklets") from the different detector layers. Another key observable to study the medium properties which is sensitive to the energy density achieved in heavy-ion collisions is the transverse energy E_T. Measurement of the transverse energy provides very basic information on heavy-ion collisions about the dynamics of the collisions and the total entropy created. Figure 1(right) shows the η dependence of E_T production. For the most central collisions, $\sim$2.1 TeV of transverse energy is created at $\eta = 0$. This is about three times higher than at Relativistic Heavy Ion Collider (RHIC).

The hydrodynamic expansion of matter produced in non-central heavy-ion collisions as well as the fluctuation of initial state lead to an azimuthal anisotropy in particle production, and the azimuthal anisotropy of charged particles is an important feature of the hot and dense medium produced in heavy-ion collisions. There are various methods to estimate the magnitude

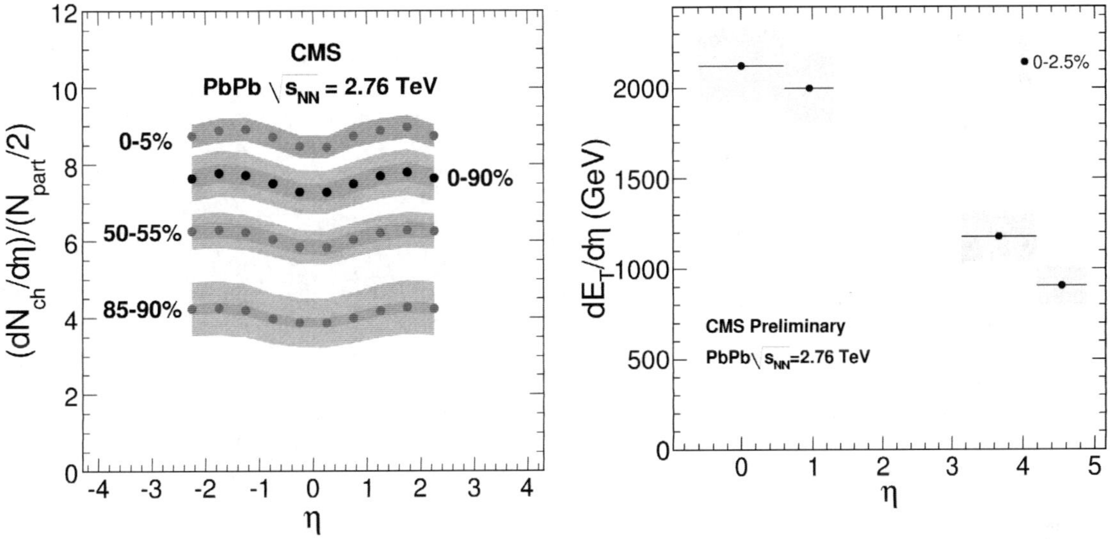

Figure 1. (left) Charged particle multiplicity density $dN_{ch}/d\eta$ for different centralities as a function of $|\eta|$. (right) Transverse energy density $dE_T/d\eta$ as a function of $|\eta|$ for most central events (0-2.5%)

of anisotropic flow, including event plane, 2- and 4-particle cumulants, Lee-Yang Zeros (LYZ) as well as dihadron correlations in the full CMS detector acceptance. These methods have different sensitivities to the initial conditions of the system and the interplay between soft-particle production and jet quenching effects. Figure 2 shows a comparison of $v_2(p_T)$ between CMS and PHENIX [17] using event plane method in mid-rapidity range. Despite a dramatic energy increase from RHIC to LHC, v_2 is only slightly larger than that at RHIC. The observed increase is within the systematic uncertainties. To better understand the pseudorapidity dependence of v_2 as a function of centrality, the results for the event plane method are shown for each centrality bins on Figure 3, in which negative and positive rapidities were averaged. The more peripheral collisions plotted in the right panel seem to indicate a stronger pseudorapidity dependence.

The charged particle p_T spectrum is an important observable for studying the properties of the hot and dense medium produced in the collisions of heavy nuclei. The modification of charged particle p_T spectrum, compared to nucleon-nucleon collisions at the same energy can shed light on the detailed mechanism by which hard partons lose energy traversing the medium. It has been already observed at RHIC that high p_T particle production is suppressed relative to expectations from an independent superposition of nucleon-nucleon collisions. This observation is typically expressed in terms of the nuclear modification factors R_{AA}. The R_{AA} was measured in CMS for all charged particles with p_T up to 100 GeV/c. Figure 4 shows the charged particle invariant differential yields in six centrality bins and compared to the corresponding quantity taken from the interpolated pp reference spectrum [18]. By comparing the points to the dashed lines from the scaled pp reference spectrum, it is very clear that the charged particle spectrum is strongly suppressed in central PbPb events. Figure 5 shows the nuclear modification factor for charged particles as a function of transverse momentum for the same centrality bins. In the most peripheral events (70-90%), a moderate suppression of about two (R_{AA}=0.5) is observed at low p_T while R_{AA} rises gently with increasing transverse momentum.

High transverse energy prompt photons in nucleus-nucleus collisions are produced directly

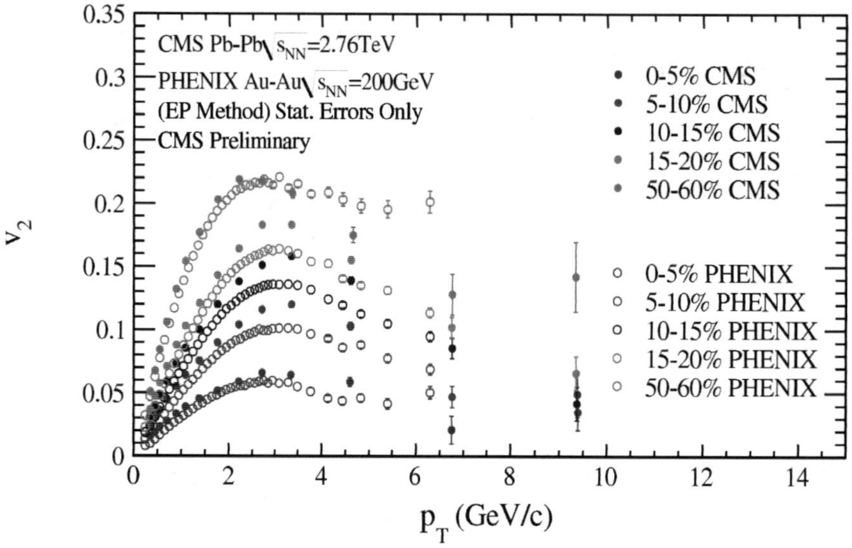

Figure 2. Comparison of v_2 using the event plane method as a function of p_T from CMS (closed symbols) at $\sqrt{s_{NN}} = 2.76$ TeV and PHENIX (open symbols) at $\sqrt{s_{NN}} = 200$ GeV for mid-rapidity events ($|\eta|$ <0.8, and $|\eta|$ <0.35, respectively).

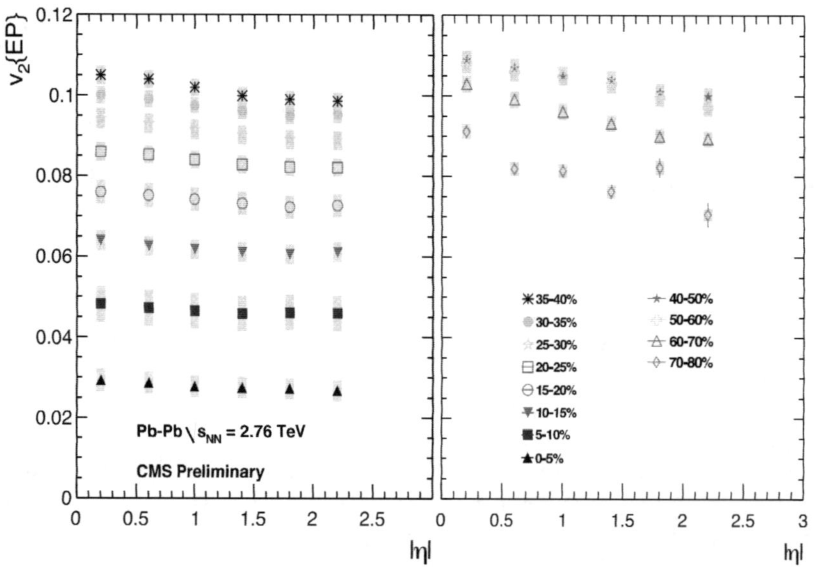

Figure 3. Dependence of v_2 on $|\eta|$ for the event plane method in 12 centrality classes. The error bars show the statistical uncertainties, and boxes give the systematic uncertainties.

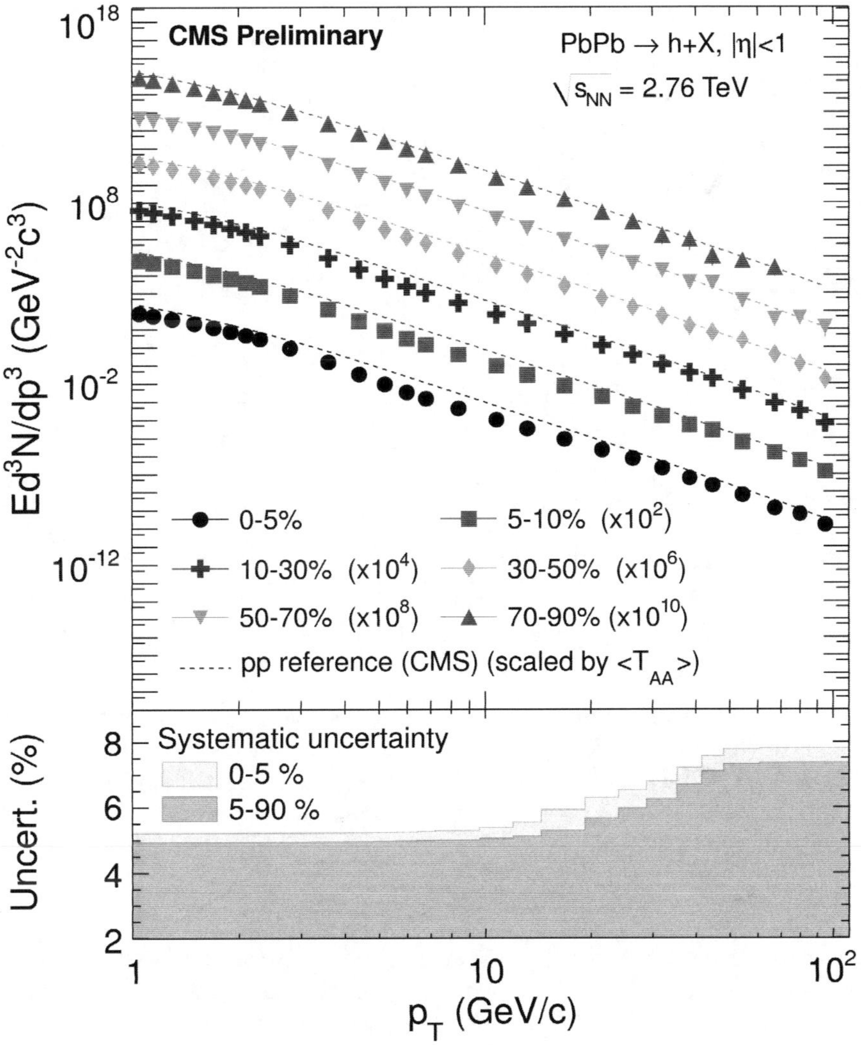

Figure 4. (Upper) Invariant differential yield in bins of collision event centrality (symbols), compared to a pp reference spectrum, scaled by the corresponding number of binary nucleon-nucleon collisions (dashed lines). The spectra for different centrality bins have been scaled by the arbitrary factors shown in the figure for easier viewing. (Lower) The systematic uncertainties on the PbPb differential yields as a function of p_T for the 0-5% and 5-90% centrality intervals.

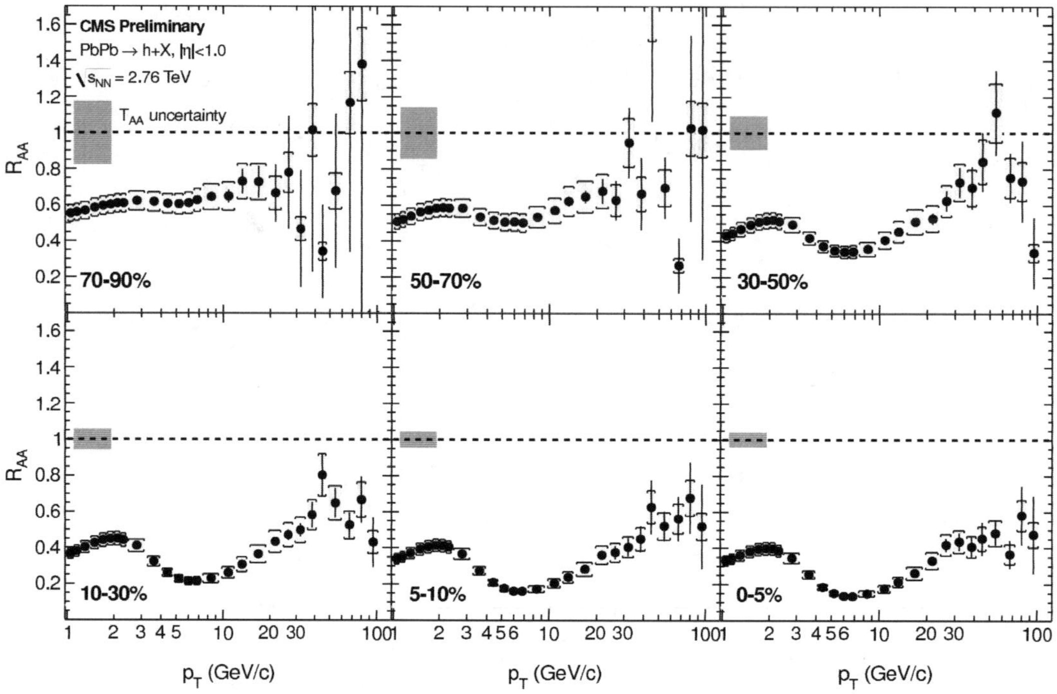

Figure 5. Nuclear modication factor R_{AA} (filled circles) as a function of p_T for six centrality intervals. The error bars represent the statistical uncertainties, and the yellow boxes the p_T-dependent systematic uncertainties on the R_{AA} measurements. An additional systematic uncertainty from the normalization of T_{AA} [9], common to all points, is shown as the shaded band around unity in each plot.

from the hard scattering of two partons. The photons provide a direct test of perturbative Quantum Chromodynamics (pQCD) and the nuclear parton densities when they pass trough the hot and dense medium without any interaction. However, the measurement of prompt photons is complicated by the much larger background coming from the electromagnetic decays of neutral mesons produced in the fragmentation of other hard scattered partons. One can suppress a large fraction of these decay photon backgrounds by imposing isolation cuts on the reconstructed photon candidates. The centrality and p_T dependence of the direct photon production in PbPb collisions are compared to the NLO pQCD predictions [19]. Figure 6(left) shows the normalized yields $(dN/dE_T)/T_{AA}$. The normalized yields are compared to the pp spectrum predicted by JETPHOX calculation [19]. The normalized PbPb data and the predicted pp results are in agreement within the quoted statistical and systematic uncertainties.

With the increase of center-of-mass energy at the LHC, electroweak probes are accessible for the first time in heavy-ion collisions. Electroweak bosons (Z, W) are of interest because they go through the Quark Gluon Plasma (QGP) unaffected. Among the leptonic decays of the electroweak bosons, the study of the Z in the $\mu^+\mu^-$ channel is the cleanest one experimentally. The dimuon decay channel is of special interest because the muons are unaffected by the strongly interacting QGP. Figure 6(right) shows the nuclear modification factor R_{AA} as a function of transverse mass for Z bosons, isolated photons and charged particles for the most central events. As expected, no modification is observed in Z and isolated photon production while large suppression is observed in the charged particle spectra which is due to the final state

Journal of Physics: Conference Series **347** (2012) 012004 doi:10.1088/1742-6596/347/1/012004

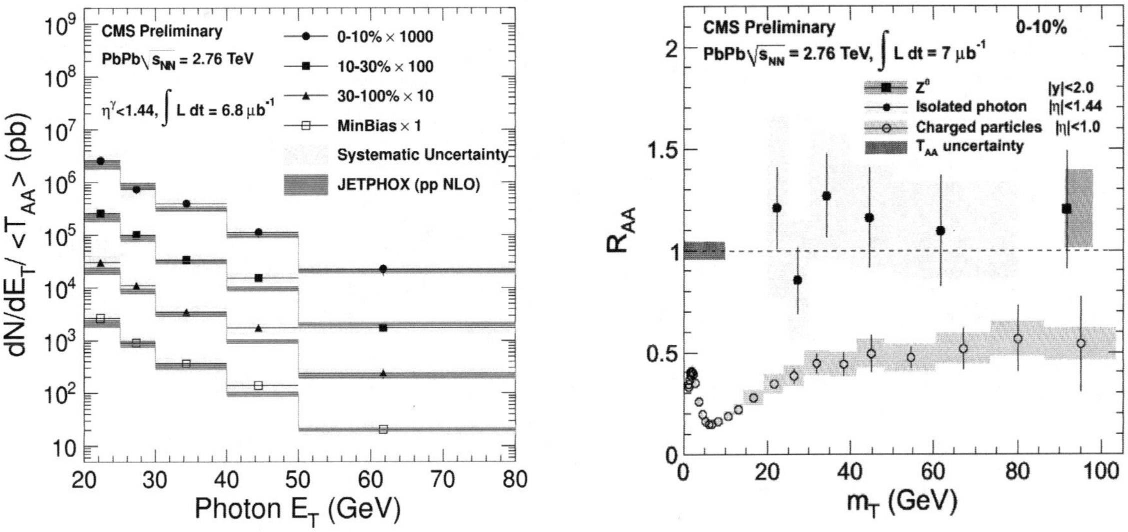

Figure 6. (left) Normalized isolated photon yields $(dN/dE_T)/T_{AA}$ as a function of photon E_T in 0-10%, 10-30%, 30-100% and MinBias PbPb collisions at $\sqrt{s_{NN}} = 2.76$ TeV. The results are compared with the pp prediction using JETPHOX, with its scale uncertainty. (right) The ratio R_{AA} for Z, isolated photon and charged particles for the most central events (0-10%).

medium modification.

Studying the modification of jets as they are formed from high p_T partons passing through the hot and dense medium has been proposed as a particularly useful tool for probing QGP properties. The energy lost by a parton in this medium is often referred to as "jet quenching" and provides fundamental information on the thermodynamical and transport properties of the traversed medium. Of particular interest are the dominant "dijets", consisting of the most energetic ("leading") and second most energetic ("sub-leading") jets. To study medium effects, dijets are selected with a leading (sub-leading) jet with p_T of at least 120 (50) GeV/c, both within $|\eta| < 2$ and with an opening angle $\Delta\phi > 2\pi/3$. The most striking observation is the large, centrality-dependent, imbalance in the energy of the two jets, as measured in the CMS calorimeters (See Figure 7). While their energies are very different, the two jets are observed to be very close to back-to-back in the azimuthal plane, implying little or no angular scattering of the partons during their traversal of the medium. To find the missing energy, the calorimetric measurement was complemented by a detailed study of low p_T charged particles in the tracker and by using missing p_T techniques. The apparent missing energy was found among the low p_T particles, predominantly with $0.5 < p_T < 2$ GeV/c, emitted outside of the sub-leading jet cone. To further study jet properties in the PbPb environment, the hard component of the fragmentation function was compared to the fragmentation of jets produced in pp collisions at the same energy. The comparison of fragmentation functions for pp and PbPb events for different centralities is shown in Figure 8. The distribution of charged particle momenta within the jet, normalized to the measured jet energy, is strikingly the same, within uncertainties, to that seen in the equivalent jets energy produced in pp events.

Quarkonia are important for studying the QGP since they are produced early in the collision and their survival is affected by the surrounding medium. The bound states of charm and bottom quarks are expected to be suppressed in heavy ions, as compared to pp. The magnitude

Journal of Physics: Conference Series **347** (2012) 012004 doi:10.1088/1742-6596/347/1/012004

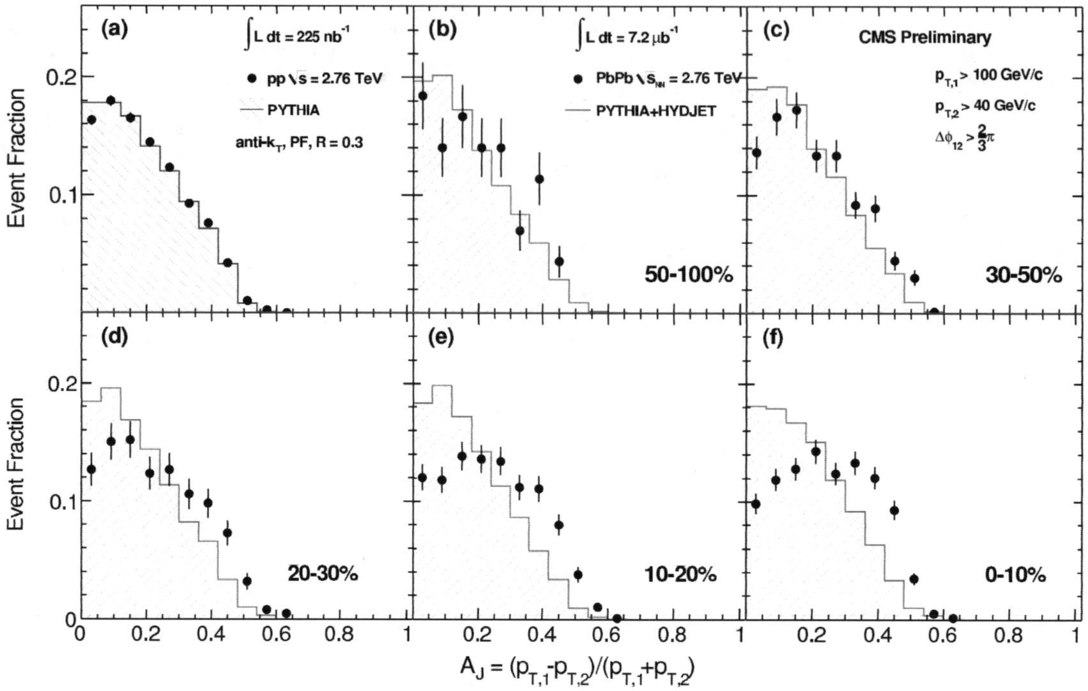

Figure 7. Calorimetric jet imbalance in dijet events as a function of collision centrality for pp and PbPb events.

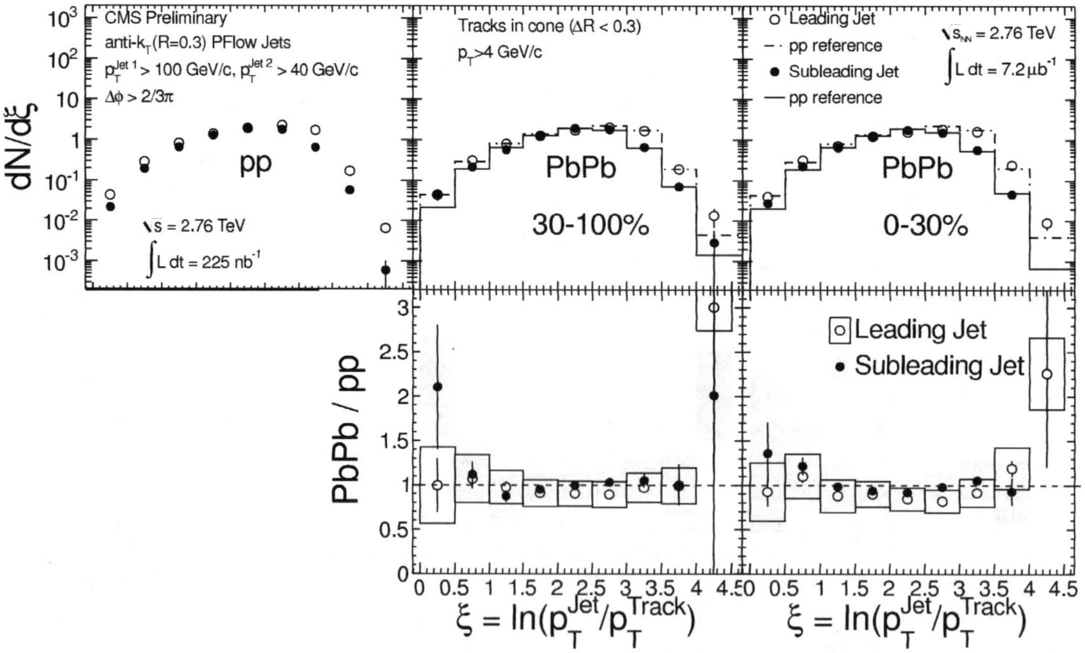

Figure 8. Fragmentation functions for jets produced in pp and in both peripheral (30-100%) and central (0-30%) PbPb collisions.

Journal of Physics: Conference Series **347** (2012) 012004 doi:10.1088/1742-6596/347/1/012004

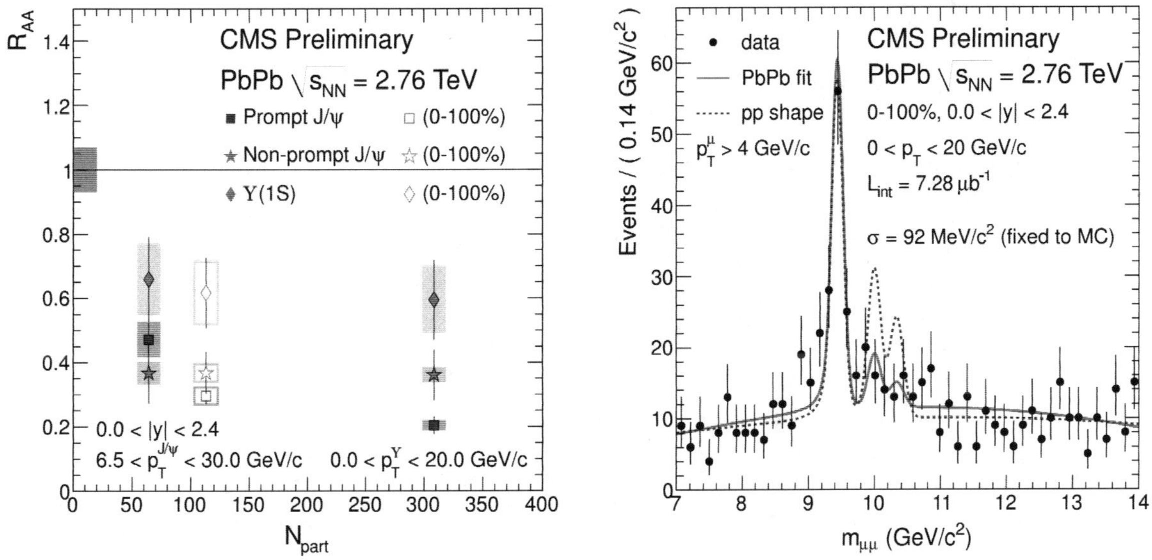

Figure 9. (left) Suppression of quarkonia states, R_{AA} as a function of centrality, for prompt J/Ψ, non-prompt J/Ψ and Υ(1S). (right) Mass spectrum of Υ family in PbPb collisions compared to the obtained in pp events at the same energy.

of the suppression for different quarkonium states is expected to depend on their binding energy. By selecting events with opposite-sign dimuons, CMS obtained production rates of J/Ψ mesons and of the Υ family. Non-prompt J/Ψs (those produced from B-meson decays) are identified by their displaced decay vertex. The suppressions of prompt and non-prompt J/Ψ particles were measured separately. The non-prompt J/Ψ suppression is one measure of the quenching of b-quarks. The R_{AA} as a function of the number of participants N_{part} indicates that high transverse momentum J/Ψs are strongly suppressed as shown in Figure 9(left). The excellent dimuon mass resolution allowed good separation of the three bound states of the Υ family. Figure 9(right) shows that the excited states, Υ(2S) and Υ(3S), are suppressed as compared to the Υ(1S). This is compatible with differential melting of quarkonium states in the high temperatures produced by PbPb collisions.

The CMS collaboration has measured the properties of heavy-ion collisions at the highest energies available with the CMS detector performing very well in all major sectors. Measurements of charge multiplicity, azimuthal asymmetry, dihadron correlations, photons, jets, quarkonia and weak bosons were conducted over a wide azimuthal and rapidity range and with high resolution. It has been shown that a strongly interacting medium can be well described by hydrodynamics. The detailed pattern of suppression was measured using reconstructed quarkonium states with varying binding energies. Large suppression of strongly interacting probes has been observed while the photons and weak bosons appear not to be suppressed. These results provide quantitative input to models of the transport properties of the medium created in the heavy-ion collisions.

2. References

[1] CMS Collaboration, "The CMS experiment at the CERN LHC", *JINST.* **3** (2008) S08004. doi:10.1088/1748-0221/3/08/S08004 .

[2] CMS Collaboration, "Dependence on pseudorapidity and centrality of charged hadron production in PbPb collisions at a nucleon-nucleon centre-of-mass energy of 2.76 TeV", arXiv:1107.4800v1 (2011).

[3] CMS Collaboration, "Pseudorapidity and centrality dependence of energy flow in $\sqrt{s_{NN}} = 2.76$ TeV PbPb collisions", *CMS PAS HIN-11-003* (2011).

[4] CMS Collaboration, "Long-range and short-range dihadron angular correlations in central PbPb collisions at $\sqrt{s_{NN}} = 2.76$ TeV", *CMS PAS HIN-11-001* (2011).

[5] CMS Collaboration, "Measurement of Bose-Einstein Correlations in pp Collisions at $\sqrt{s}=$ 0.9 and 7 TeV", arXiv:1101.3518 (2011).

[6] CMS Collaboration, "Azimuthal correlations of charged hadrons in PbPb collisions $\sqrt{s_{NN}} = 2.76$ TeV ", *CMS PAS HIN-10-002* (2011).

[7] CMS Collaboration, "Centrality and multiplicity dependence of dihadron correlations in PbPb and pp collisions", *CMS PAS HIN-11-006* (2011).

[8] CMS Collaboration, "Long-range and short-range dihadron angular correlations in central PbPb collisions at a nucleon-nucleon center of mass energy of 2.76 TeV", arXiv:1105.2438.

[9] CMS Collaboration, "Centrality dependence of the nuclear modification factor for charged particle transverse momentum spectra in PbPb collisions at $\sqrt{s_{NN}} = 2.76$ TeV" , *CMS PAS HIN-10-005* (2011).

[10] CMS Collaboration, "Observation and studies of jet quenching in PbPb collisions at $\sqrt{s_{NN}} = 2.76$ TeV", arXiv:1102.1957. Accepted by Physical Review C.

[11] CMS Collaboration, "Measurement of the hard component of jet fragmentation functions in PbPb collisions at 2.76 TeV", *CMS PAS HIN-11-004* (2011).

[12] CMS Collaboration, "Isolated photon production in $\sqrt{s_{NN}} = 2.76$ TeV PbPb collisions as a function of transverse energy and reaction centrality", *CMS PAS HIN-11-002* (2011).

[13] CMS Collaboration, "Quarkonium production in PbPb collisions at $\sqrt{s_{NN}} = 2.76$ TeV", *CMS PAS HIN-10-006* (2011).

[14] S. Chatrchyan et al. [CMS Collaboration], "Indications of Suppression of Excited Υ States in Pb-Pb Collisions at $\sqrt{s_{NN}} = 2.76$ TeV", arXiv:1105.4894, *Phys. Rev. Lett.* **107**, *052302* (2011).

[15] S. Chatrchyan et al. [CMS Collaboration], "Study of Z boson production in PbPb collisions at nucleon-nucleon centre of mass energy = 2.76 TeV", *Phys. Rev. Lett.* **106**, *212301* (2011), arXiv:1102.5435[nucl-ex]. doi:10.1103/PhysRevLett.106.212301.

[16] CMS Collaboration, "Tracking and Primary Vertex Results in First 7 TeV Collisions", *CMS PAS TRK-10-005* (2010).

[17] PHENIX Collaboration, "Formation of dense partonic matter in relativistic nucleus-nucleus collisions at RHIC: Experimental evaluation by the PHENIX collaboration ", *Nucl.Phys.* **A 757** *(2005) p.184*, doi:10.1016/j.nuclphysa.2005.03.086.

[18] CMS Collaboration, "Charged particle transverse momentum spectra in pp collisions at $\sqrt{s}=$ 0.9 and 7 TeV", arXiv:1104.3547.

[19] S. Catani, M. Fontannaz, J.P. Guillet and E. Pilon, "Cross section of isolated prompt photons in hadron-hadron collisions", *JHEP* **05** (2002) 28, arXiv:0204023.

Current status of the T2K experiment

Yuichi Oyama[1]
KEK, Oho 1-1, Tsukuba Ibaraki 305-0801, Japan
E-mail: yuichi.oyama@kek.jp

Abstract. The T2K long-baseline neutrino-oscillation experiment was started in January 2010 for the purpose of physics data-taking. Until the massive earthquake on March 11, 2011 in Japan, 1.43×10^{20} pot data were accumulated. In this data, 6 possible ν_e appearance candidates are found for expected background of 1.5 ± 0.3(syst.) if $\sin^2 2\theta_{13} = 0$ is assumed. The probability that 6 or more events can be observed for 1.5 ± 0.3 expected events is 7×10^{-3}, equivalent to 2.5σ significance. This is the first indication of a ν_e appearance or, in other words, the first indication of a non-zero $\sin^2 2\theta_{13}$.

1. Introduction

In the presently known three-flavor framework of neutrinos, the flavor eigenstate is a mixture of mass eigenstates:

$$\begin{pmatrix} \nu_e \\ \nu_\mu \\ \nu_\tau \end{pmatrix} = \mathbf{U_{MNS}} \begin{pmatrix} \nu_1 \\ \nu_2 \\ \nu_3 \end{pmatrix}.$$

The 3×3 unitary mixing matrix, $\mathbf{U_{MNS}}$, is known as the Maki-Nakagawa-Sataka (MNS) mass matrix[1]. It has 6 independent parameters: 2 square mass differences (Δm^2_{12} and Δm^2_{23}), 3 mixing angles (θ_{12}, θ_{23}, and θ_{13}), and 1 CP-violating phase ($\delta_{\rm CP}$). Among these parameters, Δm^2_{12} and θ_{12} were determined through solar[2, 3] and reactor[4] neutrino oscillation experiments:

$$\Delta m^2_{12} \sim 8 \times 10^{-5}{\rm eV}^2 \qquad \text{and} \qquad \theta_{12} \sim 34^\circ.$$

On the other hand, Δm^2_{23} and θ_{23} were measured through atmospheric[5] and long-baseline[6, 7] neutrino-oscillation experiments:

$$\Delta m^2_{23} \sim 2.4 \times 10^{-3}{\rm eV}^2 \qquad \text{and} \qquad \theta_{23} \sim 45^\circ.$$

The remaining parameters, θ_{13} and $\delta_{\rm CP}$, are unknown.

The oscillation parameter θ_{13} can be determined by studying the following transition probability:

$$P(\nu_\mu \leftrightarrow \nu_e) \sim \sin^2 2\theta_{13} \sin^2\theta_{23} \sin^2\left(\frac{\Delta m^2_{23}L}{4E_\nu}\right) \sim \frac{1}{2}\sin^2 2\theta_{13} \sin^2\left(\frac{\Delta m^2_{23}L}{4E_\nu}\right).$$

[1] On behalf of the T2K collaboration

Two different types of experiments can be performed to determine non-zero θ_{13}. One is the ν_e disappearance experiment using $\sim$3MeV $\bar{\nu}_e$ from reactors with baseline distance of a few kilometers. At present, the upper limit on $\sin^2 2\theta_{13}$ is known to be 0.15, as reported from the CHOOZ[8] reactor experiment. The other type of experiment is the ν_e appearance experiment using an almost pure $\sim$GeV ν_μ beam from accelerators with more than hundreds kilometer long baseline. From the MINOS[9] experiment, it was reported that $\sin^2 2\theta_{13} < 0.12$.

T2K (Tokai to Kamioka)[10] is a new long-baseline neutrino-oscillation experiment in Japan. A high-intensity neutrino beam from the J-PARC Main Ring is shot toward the Super-Kamiokande (SK) detector, 295 km away. It is claimed [10] that the T2K experiment is sensitive for $\sin^2 2\theta_{13} > 0.006$ in 90% C.L.

2. T2K neutrino beam line and detectors

A schematic view of the T2K neutrino beamline and detector components are shown in Figure 1. Details on the beamline and the detector components have been comprehensively discussed in [11]. In this article, some of the important features are presented below.

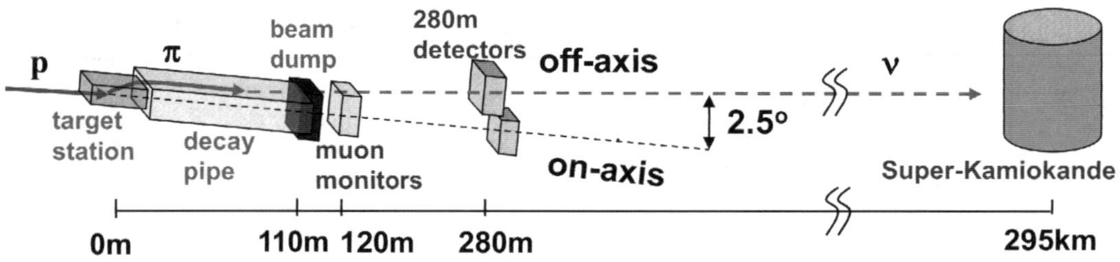

Figure 1. A schematic view of the T2K neutrino beamline and detectors. Beamline components are indicated by the words in red, and the detectors are indicated by words in blue.

2.1. Neutrino beam line

The proton beam for the T2K experiment is extracted from the J-PARC 30-GeV Main Ring proton synchrotron. This beam is delivered to the target station where it strikes the carbon target, resulting in the production of pions and kaons; the pions and kaons are bent toward the direction of SK by a magnetic field produced by magnetic horns. Neutrinos are produced as decay products of pions and kaons. All particles except neutrinos and high-energy (> 5 GeV) muons are absorbed by a beam dump located 110 m downstream of the target.

The most important feature of the T2K neutrino beamline is the off-axis beam, which was originally proposed in [12]. The details of the concept of the off-axis angle in T2K have been reported in [13]. When the beamline construction was started, the off-axis angle was tunable[13] because the optimum off-axis angle was unknown. The value of Δm^2_{23} from other experiments[5, 14] was considered, and the off-axis angle was accordingly adjusted to be 2.5° in 2007. The corresponding peak energy of the neutrino flux is approximately 600 MeV, and the oscillation study is the most sensitive around $\Delta m^2_{23} = 2.5 \times 10^{-3}\mathrm{eV}^2$.

2.2. Muon monitors

Two types of muon monitors are installed downstream of the beam dump: the Ionization Chamber and the Semiconductor Array. They can measure the intensity distribution of muons that escape from the beam dump. Since these muons are mainly produced along with neutrinos

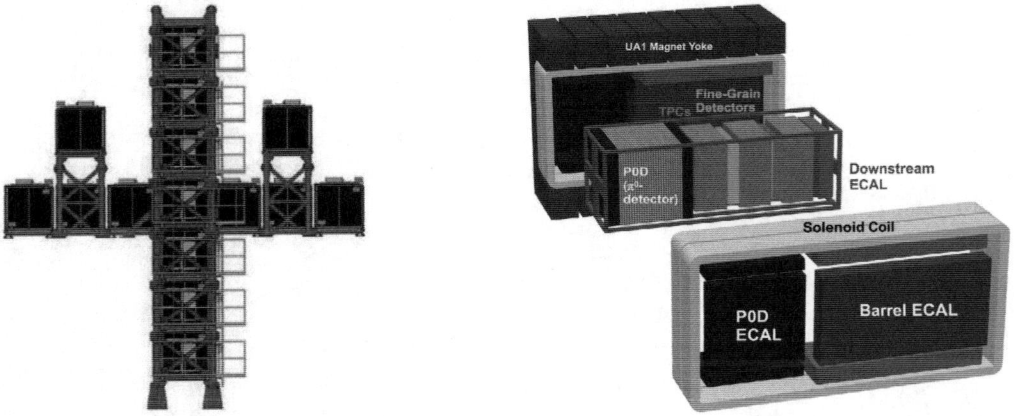

Figure 2. Schematic view of the INGRID on-axis detector (left) and the ND280 off-axis detector (right).

from the π^+ decays, the beam center of muons is in accordance with the beam center of neutrinos. The position of the beam center can be monitored on a bunch-by-bunch basis, within a 3-cm resolution, from the peak of the muon intensity distribution. This position resolution corresponds to a 0.25-mrad accuracy of beam direction.

2.3. Near detectors

The near detectors were constructed in the underground experimental hall at a depth of 33.5 m, with a diameter of 17.5 m, and 280 m downstream from the target. Two detectors were installed: the on-axis detector in the direction of the neutrino beam center and the off-axis detector in the direction of SK.

The on-axis detector, referred to as the INGRID detector, consists of 16 1m×1m×1m cubic modules, as shown in Figure 2 (left). Each module is a "sandwich" of 11 scintillator layers and 10 iron layers. They are surrounded by 4 veto planes. The modules are arranged in such a way that seven are horizontally aligned, seven are vertically aligned, and two are placed in an off-diagonal fashion. The neutrino beam center can be measured from the horizontal/vertical distribution of the neutrino event rate.

The off-axis detector is named ND280. The purposes of the ND280 detector are

- measurement of the energy spectrum of muon neutrinos in the SK direction,
- measurement of fraction of electron neutrinos, and
- study of neutrino interactions, especially, neutral current π^0, which is background for the ν_e appearance signal.

A schematic view of the ND280 detector is shown in Figure 2 (right). All the detector components except SMRD are placed inside a 0.2 T magnetic field produced by the recycled UA1 magnet from CERN. The Pi zero Detector (P0D) is a subdetector component placed in the upstream inside the magnet. It is a "sandwich" of a water target, scintillator planes, and lead plates. It is customized for measurement of neutral π^0 production. γ-rays from $\pi^0 \to 2\gamma$ are converted to electromagnetic showers using lead plates and are detected by scintillators.

Three Time Projection Chambers (TPCs) and 2 Fine Grained Detectors (FGDs) are placed downstream of the P0D. The TPCs can measure the momentum of muons from the curvature of the muon track in the magnetic field. The momentum resolution is better than 10% at 1 GeV.

FGDs consist of scintillator bars. They provide the target material for neutrino interactions and are optimized for detecting the proton recoils. By combining the TPCs and FGDs, the energy spectrum of ν_μ can be precisely measured from charged current quasi-elastic (CCQE) neutrino interactions.

Other detector components, Side Muon Range Detector (SMRD) and Electromagnetic Calorimeter (ECAL) are installed surrounding the P0D and TPC/FGD. They are reported in detail in [11].

2.4. Super-Kamiokande

The far detector, SK is a 50-kton water Cherenkov detector[15]. It is located underground at a depth of 1000 m at the Kamioka mine, Japan. The distance of SK from J-PARC is 295 km. In the inner detector (ID), 22.5 kton fiducial volume is viewed using 11129 20-inch diameter PMTs. The outer detector (OD), surrounding the ID, is also a water Cherenkov detector. It is used to veto particles that enter or exit the ID. SK started its operations in April 1996. After an accident involving PMTs in 2001, it was fully recovered by June 2006.

The most important characteristic of SK as the far detector of the T2K experiment is its particle identification capability between muons and electrons. This particle identification directly relates to identification between parent ν_μ and ν_e. The principles of particle identification have been reported in [16]. It was verified that the μ/e misidentification probability is less than 1%[16].

3. Status of data collection

The first proton beam was delivered to the target station on April 23, 2009. Generation of neutrinos was confirmed by the signal of the muon monitors. The first neutrino event in INGRID was detected on November 22, 2009. The physics data collection started in January 2010, and the first neutrino event in SK was observed on February 24, 2010.

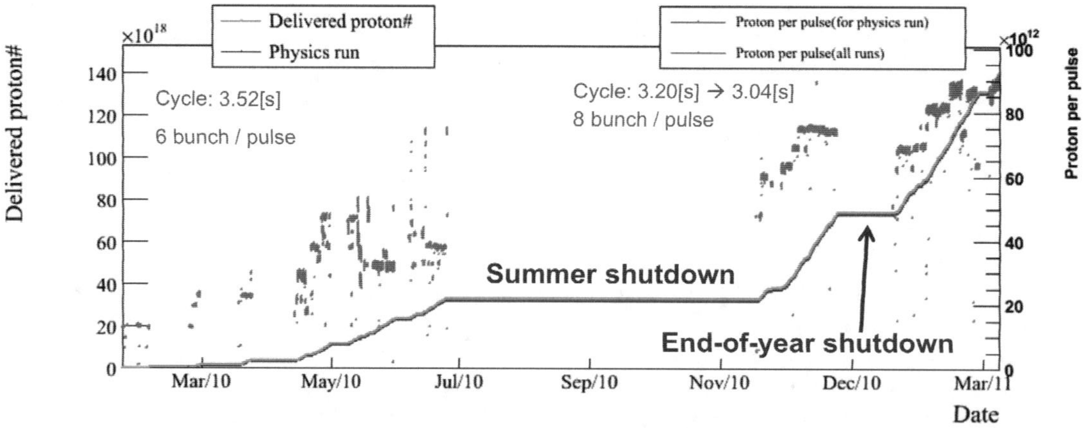

Figure 3. History of primary proton beam intensity in the T2K experiment. Red dots indicate the number of protons per pulse, and the scale is given in the right vertical axis. The green solid line indicate the integrated number of delivered protons from the beginning of the experiment. The scale is given in the left vertical axis.

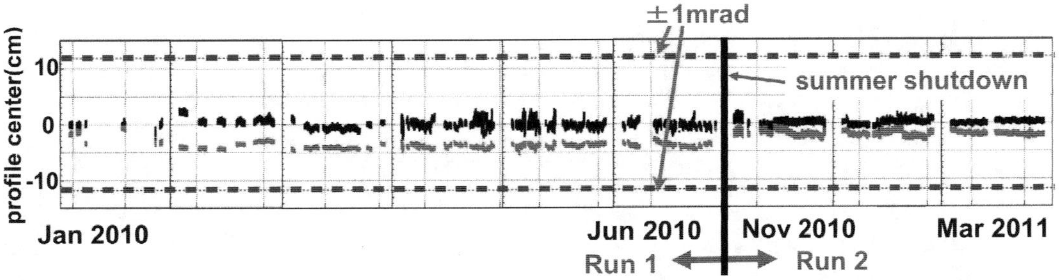

Figure 4. Muon beam direction measured using muon monitors downstream of the beam dump. The black dots represent the horizontal profile center, and the red dots represent the vertical profile center. The beam direction is well controlled and is stable within 1 mrad.

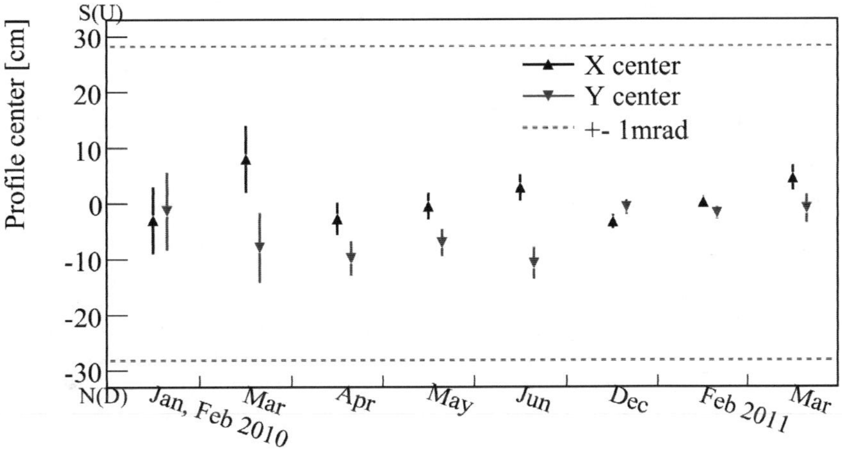

Figure 5. Neutrino beam direction measured using the INGRID detector at 280 m downstream from the target[17]. The beam direction is well controlled and is stable within 1 mrad.

The history of proton beam delivery is shown in Figure 3. The beam duration of the T2K beam is ~5 μs. In early 2010, the beam cycle was 3.52 s, and each pulse contained 6 bunches. This was changed to a beam cycle of 3.04 s, and 8 bunches per pulse in order to increase the beam power. The number of protons per pulse (ppp) was improved from $\sim 2 \times 10^{13}$ ppp to $\sim 9 \times 10^{13}$ ppp. The last beam power corresponds to about 150 kW. This value is about 1/5 of the design value, 750 kW. We have accumulated $1.43{\times}10^{20}$ protons on target (pot) until March 11, 2011. This is about 2% of our goal, i.e., $\sim 78{\times}10^{20}$ pot, which can be attained by the beam operation for 5 years with 750 kW.

The stability of the beam direction is continuously monitored using the muon monitors (Figure 4) and the INGRID detector (Figure 5). The beam direction is stable and well controlled within an accuracy of ±1 mrad. Note that if the beam direction is changed by 1 mrad, the neutrino flux at SK is changed by 2% at $E_\nu = 0.5 \sim 0.7$GeV. Therefore, the accuracy of the beam direction is satisfactory.

Table 1. Number of selected events after each condition. T2K data from 1.43×10^{20} pot, and expectations from a Monte Carlo simulation with systematic errors. The assumptions in the simulation are $\Delta m^2_{23} = 2.4 \times 10^{-3}\text{eV}^2$, $\sin^2 2\theta_{23} = 1$, and $\sin^2 2\theta_{23} = 0$.

	Data	Expected
1)Fully Contained (FC)	121	109
2)Fully Contained Fiducial Volume (FCFV)	88	74.1
3)Single ring e-like ($E_{\text{vis}} > 100$MeV)	7	5.8 ± 2.2
4)Additional ν_e cut	6	1.5 ± 0.3

4. Electron neutrino appearance analysis

The possible ν_e appearance signal was searched for using the 1.43×10^{20} pot data[17]. The results of the event selection and expected background are summarized in Table 1. Details of the event selection are given below.

4.1. Event selection

Beam-related fully contained (FC) events are selected to satisfy the following 3 conditions.

1-1) The total energy deposit in the ID to be larger than 30 MeV

1-2) No OD activity

1-3) The SK event time to be within the timing range from -2 to 10 μs around the beam trigger time

The third condition involves an examination of the correlations between the GPS time recorded in Tokai and that recorded in SK. After these three conditions, 121 FC events are selected. The accidental contamination from non-beam related events is estimated to be 0.023.

A spatial reconstruction algorithm is applied to the 121 FC events, and the following condition is applied.

2) The vertex position in the 22.5 kton of the fiducial volume

After this condition, 88 events are selected as fully contained fiducial volume (FCFV) events.

A ring counting algorithm and the particle identification algorithm are applied to the 88 FCFV events. The events are categorized as single-ring e-like events, single-ring μ-like events, and multi-ring events. The following events are selected.

3) Single ring e-like events with visible energy larger than 100 MeV

As presented in Table 1, 7 single-ring e-like events remain. Note that the above analysis procedures are almost the same as the well-established atmospheric neutrino analysis procedures[5], except 1-3).

To enhance the purity of the ν_e appearance signal, 3 additional conditions are required. They are

4-1) No delayed-electron signal

4-2) Non-π^0-like event

4-3) Reconstructed neutrino energy to be less than 1250 MeV

From condition 4-1), events with a delayed-electron signal are rejected because such a signal indicates the generation of muons. One event is rejected considering this condition. Condition 4-2) is for suppression of misidentified π^0. The reconstruction of the second ring is forced, and the 2-ring invariant mass, M_{inv}, is calculated. The M_{inv} distribution is shown in Figure 6 (left).

The condition $M_{inv} < 105$ MeV is imposed. No events are rejected considering condition 4-2). From condition 4-3), the neutrino energy E_ν^{rec} is calculated by assuming quasi-elastic kinematics. To suppress the contribution from the intrinsic ν_e component arising primarily from kaon decays, the event is rejected if $E_\nu^{rec} > 1250$ MeV. As shown in Figure 6 (right), no events are rejected by condition 4-3). Finally, 6 events remain after considering all selection criteria.

4.2. Expected background and its systematic error

The expected background for ν_e appearance is carefully calculated through simulations of neutrino productions and their interactions.

From the interactions between the primary proton beam and the carbon target, secondary pions and kaons are generated. Pions and kaons propagate in the secondary beamline. Neutrinos are produced as decay products of pions and kaons. All of these neutrino production processes are simulated using standard simulation programs, FLUKA[18, 19], GEANT3[20], and GCALOR[21]. To reduce the systematic uncertainties in the pion and kaon production simulation, some of the T2K members join CERN NA61 experiment: "Study of hadron productions in hadron-nucleus and nucleus-nucleus collisions in CERN SPS". Pion production data[22] for a 30-GeV proton beam incident on a carbon target was obtained with 5~10% error and is employed in the simulation. The estimated uncertainties of the intrinsic ν_e flux below 1 GeV is around 14%

The neutrino interactions are simulated using the NEUT[23] neutrino interaction generator. The GENIE[24] generator is also used for the evaluation of uncertainty due to the simulation program. The present uncertainty of neutrino interaction comes from limited experimental data. For example, the uncertainty of CCQE interactions relative to the CCQE total cross section is energy dependent and is about 7% at 500 MeV.

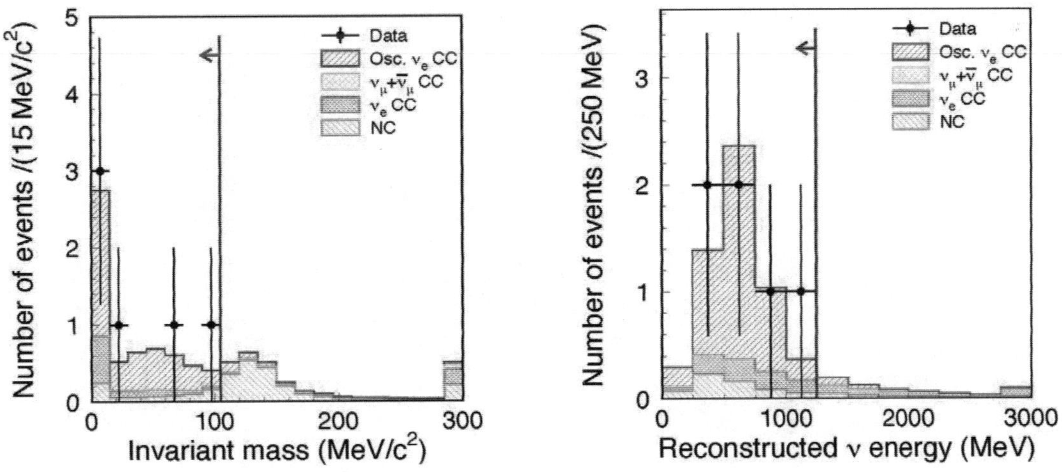

Figure 6. (left) Distribution of invariant mass M_{inv} when each event is forced to be reconstructed into two rings[17]. The data are shown using points with error bars (statistical only). The MC predictions are in shaded histograms, corresponding to the oscillated ν_e CC signal and various background sources for $\sin^2 2\theta_{13} = 0.1$. The vertical line shows the applied cut at 105 MeV/c^2. (right) Reconstructed neutrino energy spectrum of the events that satisfy all conditions of ν_e appearance signal selection[17]. The vertical line shows the applied cut at 1250 MeV.

To reduce the uncertainty on the absolute event rate, the neutrino event rate at ND280 is used. The charged-current inclusive muon events generated in FGD are selected and compared with the simulation data. The agreement of the total event number and its momentum distribution is excellent, as shown in Figure 7. The ratio, $R_{\mathrm{data/MC}}$, is calculated to be

$$R_{\mathrm{data/MC}} = 1.036 \pm 0.028(\mathrm{stat.})^{+0.044}_{-0.037}(\mathrm{det.syst.}) \pm 0.038(\mathrm{phys.syst.})$$

This factor is used as a correction for the normalization of the SK event rate. After employing $R_{\mathrm{data/MC}}$ for the near/far cancellation, the systematic uncertainty on an expected background for the ν_e appearance signal due to the uncertainty of the neutrino interaction is 14%.

The event selection algorithm presented in the previous subsection was also applied to the events generated by the simulation. The results are summarized in Table 1. The final background events for the ν_e appearance signal are 1.5±0.3(syst.). Of these, the contribution from the beam intrinsic ν_e is 0.8 and that from neutral current interaction is 0.6.

4.3. Results of electron neutrino appearance and discussion

T2K observed 6 possible ν_e appearance signal for the expected background of 1.5±0.3, as reported in the previous subsections. The chance probability that 6 or more events are observed for the expectation of 1.5 ± 0.3 is calculated to be 7×10^{-3}, equivalent to a 2.5σ significance. Although this statistical significance is still far from discovery or even from evidence, it should be reported that it is an **indication** of ν_e appearance.

The constraints on $\sin^2 2\theta_{13}$ for assumed δ_{CP} are shown in Figure 8. If normal hierarchy ($m_3^2 >> m_2^2 > m_1^2$) and $\delta_{\mathrm{CP}} = 0$ are assumed, the best fit is $\sin^2 2\theta_{13} = 0.11$, and the 90% C.L. interval is $0.03 < \sin^2 2\theta_{13} < 0.28$. On the other hand, in the case of inverted hierarchy ($m_2^2 > m_1^2 >> m_3^2$) and $\delta_{\mathrm{CP}} = 0$, the best fit value is $\sin^2 2\theta_{13} = 0.14$ and the 90% C.L. interval

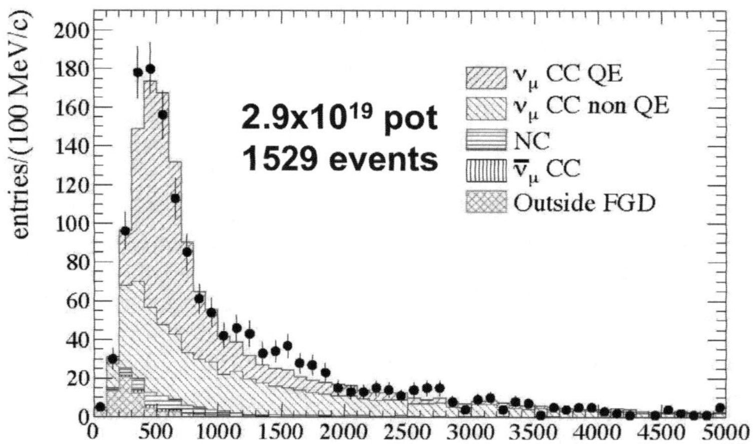

Figure 7. Measured muon momentum of ν_μ charged current candidates reconstructed in the FGD target[17]. The data are shown using points with error bars (statistical only), and the MC predictions are represented by histograms.

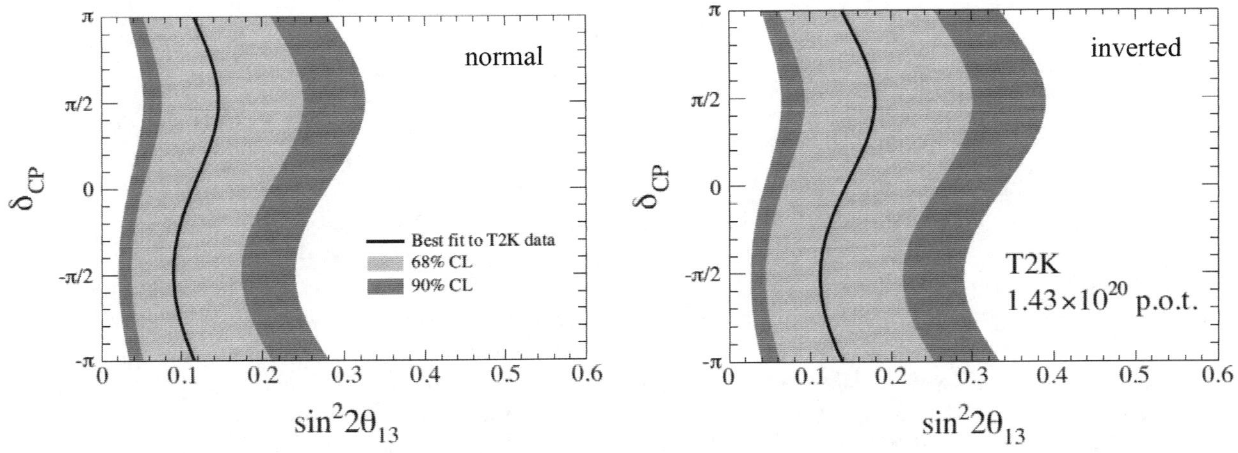

Figure 8. The 68% and 90% C.L. regions for $\sin^2 2\theta_{13}$ for each value of δ_{CP}, consistent with the observed number of events in the three-flavor oscillation case for normal (left) and inverted (right) mass hierarchy[17]. The best fit values are indicated by solid lines.

is $0.04 < \sin^2 2\theta_{13} < 0.34$. These results do not contradict with the 90% C.L. upper limits by the CHOOZ[8] experiment and the MINOS[9] experiment.

Just after the ICPP-II Istanbul conference, the MINOS collaboration announced new preliminary results[25] on the ν_e appearance analysis with 8.2×10^{20} pot data. They observed 62 events whereas the expected background is $49.5 \pm 2.8(\text{syst}) \pm 7.0(\text{stat})$. The excess is 1.7σ. Results from T2K and MINOS are summarized in Table 2. The signal-to-background ratio in T2K is quite larger than that in MINOS. This is the effect of the off-axis beam in T2K. The peak of the neutrino energy is adjusted to the oscillation maximum, and high-energy neutrinos that produce a neutral current π^0 background are suppressed.

At present, 3 reactor experiments[26, 27, 28] are under construction or just started. The indication of non-zero $\sin^2 2\theta_{13}$ will be examined by the reactor experiments and/or T2K

Table 2. Comparisons of the ν_e appearance analysis between T2K and MINOS

	T2K	MINOS
Proton on target ($\times10^{20}$)	1.43	8.2
Expected number of background events	1.5 ± 0.3	$49.5\pm2.8(\text{syst.})\pm7.0(\text{stat})$
Observed number of possible ν_e appearance events	6	62
Excess	2.5σ	1.7σ
Constraints on $\sin^2 2\theta_{13}$ (Normal hierarchy, $\delta_{CP} = 0$)	$\sin^2 2\theta_{13} = 0.11$ (Best fit)	$\sin^2 2\theta_{13} = 0.04$ (Best fit)
	$0.03 < \sin^2 2\theta_{13} < 0.28$ (90% C.L. interval)	$\sin^2 2\theta_{13} < 0.12$ (90% C.L. interval)
Constraints on $\sin^2 2\theta_{13}$ (Inverted hierarchy, $\delta_{CP} = 0$)	$\sin^2 2\theta_{13} = 0.14$ (Best fit)	$\sin^2 2\theta_{13} = 0.08$ (Best fit)
	$0.04 < \sin^2 2\theta_{13} < 0.34$ (90% C.L. interval)	$\sin^2 2\theta_{13} < 0.19$ (90% C.L. interval)

Journal of Physics: Conference Series **347** (2012) 012005 doi:10.1088/1742-6596/347/1/012005

additional data within a couple of years.

5. The earthquake

On March 11, 2011, Japan encountered a 9.0 Magnitude earthquake. It was the 4th largest earthquake in the world since 1900. The earthquake and followed tsunamis resulted in the death of more than 15000 people. In addition, the Fukushima nuclear power stations, which was damaged by the tsunamis, were found to have crucial radiation leaks.

J-PARC is located about 260 km from the seismic center, and about 100 km from the Fukushima nuclear power stations. No one at J-PARC was injured by the earthquake and tsunamis, but some of the accelerator and beamline components were damaged. After the radiation leak from Fukushima, the radiation dose in the J-PARC area was considerably higher than the normal value. However, the radiation dose has already returned to the normal value, and recoveries of the accelerator/beamline components are in progress. We will resume the J-PARC operation in December 2011 and restart the T2K data-taking as soon as possible.

References

[1] Z.Maki, M.Nakagawa and S.Sakata, Prog.Theor.Phys. 28, 870 (1962).
[2] J.Hosaka *et al.*, Phys.Rev.D 73, 112001 (2006).
[3] B.Aharmim *et al.*, Phys.Rev.C 72, 055502 (2005).
[4] S.Abe *et al.*, Phys.Rev.Lett. 100, 221803 (2008).
[5] Y.Ashie *et al.*, Phys.Rev.D 71, 112005 (2005).
[6] M.H.Ahn *et al.*, Phys.Rev.D 74, 072003 (2006).
[7] P.Adamson *et al.*, Phys.Rev.Lett. 106, 181801 (2011).
[8] M.Apollonio *et al.*, Phys.Lett.B 466, 415 (1999).
[9] P.Adamson *et al.*, Phys.Rev.D 82, 051102 (2010).
[10] Y.Itow *et al.*, hep-ex/0106019 (2001).
[11] K.Abe *et al.*, Nucl.Instrum.Meth.A in press, arXiv:1106.1238 (2011).
[12] D.Beavis *et al.*, BNL AGS E-889 (1995).
[13] Y.Oyama, hep-ex/0512041 (2005).
[14] D.G.Michael *et al.*, Phys.Rev.Lett. 97, 191801 (2006).
[15] S.Fukuda *et al.*, Nucl.Instrum.Meth.A 501, 418 (2003).
[16] S.Kasuga *et al.*, Phys.Lett.B 374, 238 (1996).
[17] K.Abe *et al.*, Phys.Rev.Lett. Phys. 107, 041801 (2011).
[18] A.Ferrari, P.R.Sala, A.Fasso, and J.Ranft, CERN-2005-010 and SLAC-R-773 and INFN-TC-05-11 (2005).
[19] G.Battistoni *et al.*, AIP Conf. Proc. 896,31 (2007).
[20] R.Brun, F.Carminati and S.Giani, CERN-W5013 (1994).
[21] C.Zeitnitz and T.A.Gabriel, In Proc. of International Conference on Calorimetry in High Energy Physics, Tallahasse, FL, USA, February 1993.
[22] N.Abgrall *et al.*, Submitted to Phys.Rev.C, arXiv:1102.0983 (2011).
[23] Y.Hayato, Nucl.Phys.(Proc. Suppl.) B112, 171 (2002).
[24] C.Andreopoulos *et al.*, Nucl.Instrum.Meth. A 614, 87 (2010).
[25] `http://www-numi.fnal.gov/pr_plots/nue2011.pdf`
[26] `http://doublechooz.in2p3.fr/`
[27] `http://neutrino.snu.ac.kr/RENO/`
[28] `http://dayawane.ihep.ac.cn/twiki/bin/view/Public/`

Journal of Physics: Conference Series **347** (2012) 012006 doi:10.1088/1742-6596/347/1/012006

ATLAS Detector Performance status, improvements during winter 2010 shutdown and results with initial 2011 data taking period

Nektarios Chr. Benekos[1]

University of Illinois at Urbana Champaign, Department of Physics Loomis Laboratory, 1110 West Green Street, Urbana, Illinois US

E-mail: Nektarios.Benekos@cern.ch

Abstract. After the successful 2010 LHC run, where the ATLAS detector recorder $45pb^{-1}$ of proton-proton collision data with 93.6% data taking efficiency and during the recent LHC shutdown period, ATLAS performed vital maintenance and improvements on the various sub-detectors. Maintenance on the Muon Spectrometer included repairs on the readout system as well as updates and leak checks in the gas system. Six TGC chambers have been also replaced. For the Calorimeters, repairs were carried out on the front-end electronics and power supplies to recover detector coverage that have been lost since the last maintenance period. Repairs were also performed on the Inner Detector, but at a smaller scale. Finally the ALFA luminosity detector was installed along the beam line and is being commissioned. This talk summarizes the above repairs and their expected improvement for physics performance and reliability of the ATLAS for the 2011 LHC run.

1. Introduction

The LHC and ATLAS are motivated by seeking answers to some of the most fundamental questions of particle physics, among them the mystery of the origin of mass (Higgs, Brout and Englert mechanism), the search for an underlying symmetry between fermions and bosons (Supersymmetry, SUSY), and the quest for further unification of forces including gravity (Extra-dimensions). There is a great potential that the LHC physics results could have deep implications in the understanding of the evolution of the Universe, and help solving questions like the origin of the Dark Matter, if SUSY would be experimentally established. The ATLAS detector was designed to be able to address these physics topics, and to be ready to explore also unknown New Physics at the new frontier in energy and luminosity offered by the LHC.

2. The ATLAS Detector description

ATLAS (A Toroidal LHC ApparatuS) [1] is a general−purpose detector primarily designed to study pp collisions at LHC energies. It is characterized by cylindrical symmetry, high granularity in the readout cells and coverage close to 4π, which makes it fit to study aspects of heavy−ion ($Pb - Pb$) collisions as well. ATLAS uses a coordinate system where the nominal interaction point is at the center of the detector, the beam direction defines the z-axis and the x-y plane is

[1] On behalf of the ATLAS Collaboration

Journal of Physics: Conference Series **347** (2012) 012006 doi:10.1088/1742-6596/347/1/012006

transverse to the beam. The positive x-axis is defined as pointing to the center of the ring and the positive y-axis is upwards. Side-A of the detector is in the positive z direction and side-C is in the negative z direction. The azimuthal angle ϕ is measured around the beam axis. The pseudorapidity η is defined as η = - $ln(tan\theta/2)$ where θ is the polar angle. Various subsystems are arranged in layers around the beam axis, as shown in Figure 1 .

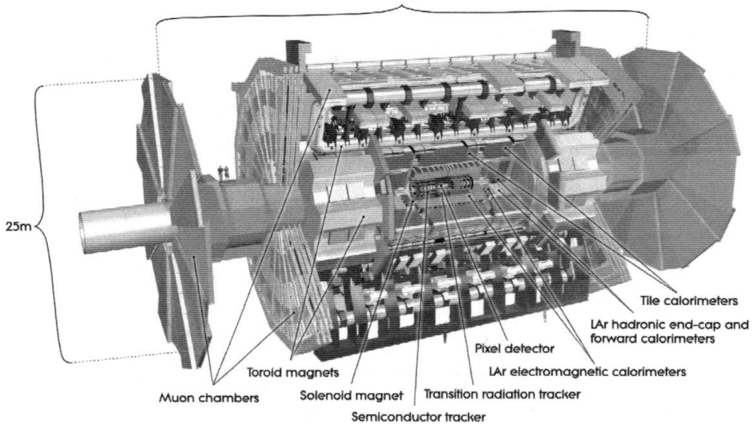

Figure 1. The ATLAS detector.

The ATLAS detector is forward-backward symmetric with respect to the interaction point. Pattern recognition, momentum and vertex measurements, and electron identification are achieved with a combination of discrete, high-resolution semiconductor pixel and strip detectors in the inner part of the tracking volume, and straw-tube tracking detectors with the capability to generate and detect transition radiation in its outer part. High granularity liquid-argon (LAr) electromagnetic sampling calorimeters, with excellent performance in terms of energy and position resolution, cover the pseudorapidity range $|\eta| < 3.2$. The Hadronic Calorimetry in the range $|\eta| < 1.7$ is provided by a scintillator-Tile Calorimeter (TileCal), which is separated into a large barrel and two smaller extended barrel cylinders, one on either side of the central barrel. In the end-caps ($|\eta| > 1.5$), LAr technology is also used for the Hadronic Calorimeters, matching the outer $|\eta|$ limits of end-cap Electromagnetic Calorimeters. The LAr forward calorimeters provide both electromagnetic and hadronic energy measurements, and extend the pseudorapidity coverage to $|\eta| = 4.9$. The Calorimeter is surrounded by the Muon Spectrometer (MS). The air-core toroid system, with a long barrel and two inserted end-cap magnets, generates strong bending power in a large volume within a light and open structure. Multiple-scattering effects are thereby minimized, and excellent muon momentum resolution is achieved with three layers of high precision tracking chambers. The muon instrumentation includes trigger chambers with timing resolution of the order of 1.5-4 ns. The MS defines the overall dimensions of the ATLAS detector. Two forward detectors have been added to the original ATLAS design: LUCID and the Zero Degree Calorimeters (ZDC), located respectively at 17 and 140 m from the ATLAS Interaction Point (IP), one arm per side. LUCID is dedicated to the measurement of the luminosity delivered to ATLAS by LHC, the ZDC measure neutral particles emitted at zero degrees with respect to the beam axis.

Two magnetic systems are present in ATLAS to bend the trajectory of charged particles and allow the measurement of their momentum: the ID is inside in a thin superconducting solenoidal magnet providing a 2 T axial field, while three large superconducting toroids (one barrel and two end-caps) arranged with an eight-fold azimuthal symmetry around the calorimeters, provide B−fields with a peak value of 4.1 T in the regions of the MS. A detailed description of the various subsystems can be found in [1]. The overall dimensions of ATLAS are quite impressive: it is 44 m long and has a diameter of 25 m. The ATLAS Collaboration is huge as well, spanning over all inhabitated continents and gathering together 174 Institutes, for a total of about three thousand physicists.

3. LHC Running in 2010

The Large Hadron Collider (LHC), [2],[3],[4] the largest particle accelerator in the world, is located beneath the border between France and Switzerland, in the underground tunnel which used to host the former Large Electron-Positron (LEP) Collider. LHC has two different modes of operation: one is as proton collider with a design centre-of-mass energy of 14 TeV, the other is as $Pb-$ion collider, accelerating fully ionized lead atoms. For lead ions, the design centre-of-mass energy is 2.76 TeV/nucleon. The rate, R, of events provided by accelerators is proportional to the machine instantaneous luminosity, L, and the event total cross section, $\sigma : R = \sigma \times L$. LHC is expected to reach a luminosity $L = 2 \times 10^{33} cm^{-2} s^{-1}$ during the first years of operations with protons, while the design luminosity $L = 10^{34} cm^{-2} s^{-1}$ should be reached later. The design luminosity for $Pb - Pb$ beams is $L = 10^{27} cm^{-2} s^{-1}$, substantially lower than the former because the large ion charge makes it more difficult to control beam−beam interactions. A large number of particles are produced when complex nuclei collide, and the inelastic cross section is a strong function of the atomic number A: $\propto A^{\frac{2}{3}}$.

3.1. Luminosity Determination

The LHC luminosity is determined in real time approximately once per second using a variety of detectors and algorithms that are described later in this note. These measurements are displayed in the ATLAS control room and results for a single "preferred" method are sent to the LHC control room over the network, providing real-time feedback for machine tuning. The basic time unit for storing luminosity information for later use is the Luminosity Block (LB). The length of a LB is approximately two minutes with start and end times set by the ATLAS data acquisition system. All ATLAS data quality information, as well as luminosity, are stored in a relational database for each LB. Offline users therefore can select data samples with specific data quality requirements (for example, pixel detector ON) and then calculate the luminosity for that selection. Because the LB number is updated whenever a trigger prescale is changed and because trigger scalars are also recorded for each LB, offline users can correct the delivered luminosity for data acquisition deadtime and prescale rates. Furthermore, since the list of processed LB is included as in-file metadata for all ATLAS data formats, users can calculate the integrated luminosity properly for subsets of the full data sample. The luminosity tables in the offline database allow for storage of multiple luminosity methods and are versioned so that updated calibration constants can be applied. Corrections for data acquisition deadtime and other sources of data loss are performed when the integrated luminosity is calculated. A tool is provided as part of the ATLAS software release that takes as its input the list of LB that have been analysed (after data quality selection), the trigger used for the analysis (so that corrections for deadtime and trigger prescale can be made) and the luminosity method requested and returns the integrated luminosity of the data sample.

3.2. LHC Delivered Luminosity

From March to November 2010 LHC collided protons at a centre-of-mass energy of 7 TeV, more than 3 times larger than the one achieved by the Tevatron ($p\bar{p}$). The instantaneous luminosity continuously increased over this period, passing from $L = 10^{27} cm^{-2} s^{-1}$ at the beginning of March to the peak value $L = 2.1 \times 10^{32} cm^{-2} s^{-1}$ in October. Such an improvement was accomplished both by increasing the number of circulating bunches and thus reducing the time between bunch crossings and increasing the number of protons stacked in each bunch as well as squeezing them into a smaller and smaller area. The average number of additional *pp* interactions taking place at each bunch crossing, also known as pile-up, was negligible at the beginning and reached a maximum value of about four at the end of the data-taking period, against an expected value of 25 at full luminosity and $\sqrt{s} = 14$ TeV. The last month of operations was dedicated to $Pb - Pb$ collisions. The centre-of-mass energy reached the value of 2.76 TeV per nucleon pair, more than one order of magnitude larger than the one provided by RHIC, and the instantaneous luminosity reached the value $L = 3 \times 10^{25} cm^{-2}s^{-1}$. The total integrated luminosity delivered by LHC in 2010 was about $48pb^{-1}$ with proton beams and $9.7\mu b^{-1}$ with $Pb-$ion beams. The delivered luminosity accounts for the luminosity delivered from the start of stable beams until the LHC requests ATLAS to turn the sensitive detector off to allow a beam dump or beam studies.Very recently the total integrated luminosity recorded by ATLAS also exceeds $1fb^{-1}$ (including $959pb^{-1}$ in 2011 + $45pb^{-1}$ in 2010. (Figure 2). This is a very important milestone for the LHC and ATLAS, which demonstrates the outstanding performance of the accelerator, of the experiment and of the involved teams. The LHC luminosity has been measured in ATLAS using different detectors and methods [5]. The absolute luminosity calibration was obtained using the van der Meer beam separation scans method [6], where the two beams are scanned against each other in the horizontal and vertical planes to measure their overlap function, with a systematic uncertainty of 3.4% [7]. All the sub-detectors of ATLAS operated with a very high efficiency > 97%. In Figure 3 the denominator is the luminosity delivered between the declaration of stable beams and the LHC request to turn the sensitive detectors off to allow a beam dump or beam studies. The numerator is the luminosity recorded by ATLAS. Each bin represents a week. The empty bins are due to weeks in which no stable beams were delivered by the LHC. The inefficiency accounts for the turn-on of the high voltage of the Pixel, SCT and some of the muon detectors and any inefficiencies due to deadtime or due to individual problems with a given subdetector that prevent the ATLAS data taking to proceed.

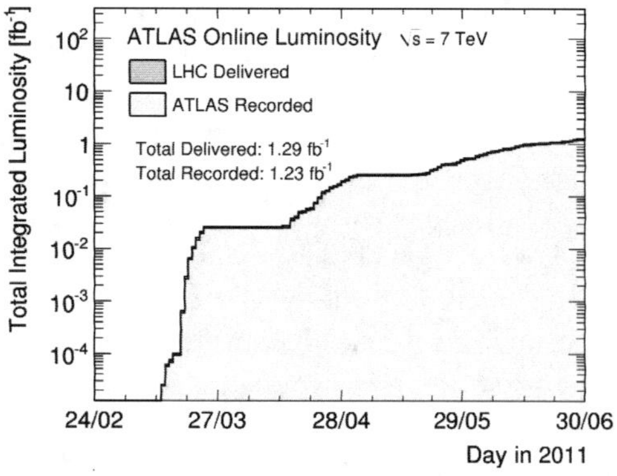

Figure 2. Cumulative luminosity versus day delivered to (green), and recorded by ATLAS (yellow) during stable beams and for *pp* collisions at 7 TeV centre-of-mass energy in 2011.

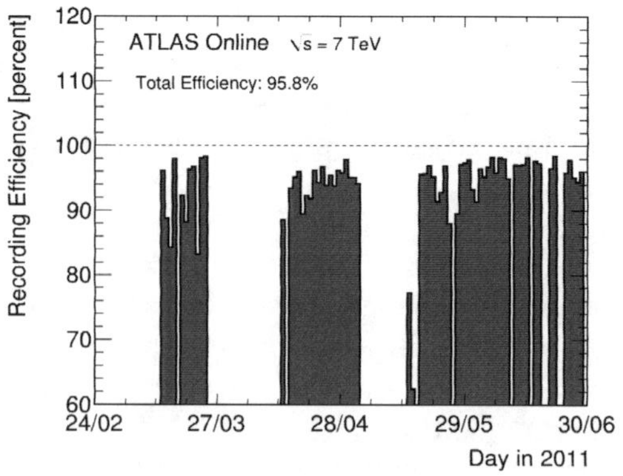

Figure 3. ATLAS data taking efficiency in 2011.

3.3. Data Quality

ATLAS data taking is segmented by run and luminosity block (LB). A run corresponds nominally to an LHC fill, and a LB is a length of time over which detector conditions are assumed to be constant. A decision comes in the form of a 'traffic light' coloured flag, whose evolution across LBs is stored in the conditions database. Red and green mean bad and good for analysis respectively, whilst yellow is a temporary value used to defer a decision to a later time. Other states exist which are only meaningful as diagnostic information. Before the flags become inputs to analyses, experts converge on the 'good' and 'bad' states. Example decisions include those coming from the DQMF (Data Quality Monitoring Framework) [8], [9], the DCS (Detector Control System) [10], and the control room shifter. They are summarized into a single decision per physical subdetector region (such as 'pixel detector barrel' or 'monitored drift tube endcap') by the 'Data Quality Calculator'. An expert reviews the automated decision and may decide to override it. Regions are summarized across subdetectors into 'combined performance' flags used to select ranges of good runs and luminosity blocks. Along with information about the LHC state, these flags can be used to build a 'Good Run List', an XML document describing ranges of data suitable for particular types of physics analyses. Events collected from the experiment are filtered into different streams for specific uses. There exists three types of stream: physics, calibration and express. The physics streams will ultimately end up in user analyses, the calibration stream is used to compute calibration constants and the express stream undergoes rapid processing (usually 1-2 hours after the data is recorded). The histogram analysis framework produces, per active stream, 20,000 histograms (100 MB) per run, along with 700 histograms (6 MB) per ten luminosity blocks. Originally, these were all dumped as 'png' images so that they could be served over the web to shifters and experts as static content, but the disk space and CPU usage per run were not sustainable. Given that the vast majority of histograms are not looked at for each run, it is better to generate the image files as they are requested and to cache them for a short time. This has been implemented and has lead to a saving in disk space, whilst not impacting the load of the server.

3.4. Trigger and Data Acquisition Performance

Reading out a detector may involve some dead time, during which further events are lost. In case of very high acquisition rates, data storage can also become a serious issue. In 2010 the average bunch-crossing rate of LHC in *pp* operations was close to 1 MHz (it will approach $\approx 40MHz$ at design luminosity), and an online way to identify interesting events was clearly needed. When studying *pp* collisions, we are mostly interested in events where hard interactions between proton constituents occur. These produce particles at high momentum transverse to the beam axis. The other, much more frequent, events on the contrary produce most of the activity close to the beam line. The ATLAS online event selection is realized in three stages. At each level, events are kept in pipelines waiting for the decision of the following step. Loose granularity information from the calorimeters and the muon spectrometer are the main input to the first trigger level, whose average rate was about 20 kHz in 2010. At the second level, portions of each event are studied at full granularity. In 2010, the rate of events accepted by the second trigger level was decreased to about 3.5 kHz. Finally, the third trigger level is based on full analysis of complete events. The typical rate of *pp* collision events written to disk in 2010 was about 300 Hz, higher than the design rate of 200 Hz. Due to lower luminosity, the trigger configuration for $Pb-Pb$ running was aimed at avoiding biasing the selected sample. The main input to the first trigger level came from ZDC. The other trigger levels were used to reject background like beam−gas interactions, but no selection based on physics objects was applied.

Journal of Physics: Conference Series **347** (2012) 012006 doi:10.1088/1742-6596/347/1/012006

4. Inner Detector Performance

The ID consists of three subsystems: a pixel detector, a silicon strip tracker (SCT) and a transition radiation straw-tube tracker (TRT) that also distinguishes between electrons and hadrons by measuring the amplitude of signals due to both ionization and transition radiation. The Pixel detector consists of three layers of pixels both in the barrel and in the end-cap regions, for a total of 80 million channels, with the innermost layer located at 5 cm distance from the beam line. The design values of the spatial resolution are $10\mu m$ in the bending plane (R-ϕ) and $115\mu m$ in the beam direction (z). The SCT consists of 4 (9) double layers of silicon strips in the barrel (end-cap) with a spatial resolution of $17\mu m$ (R-ϕ) and $580\mu m$ (z in the barrel, or R in the end-caps). In order to match the required performance in tracking and momentum resolution, the position (alignment) of the Pixel and SCT modules must be known with an accuracy of a few microns. With ATLAS 2010 collision data, the track based detector alignment is performed using an iterative procedures that minimizes the hit residuals (distance between the hit and the reconstructed track). Figure 4 shows the local x residual distribution (the projection of the residual onto the module local x direction) for all hits-on-tracks in Pixel barrel modules. Here the residual is calculated by re-fitting the track with the hit-on-track under study removed. For the aligned geometry the measured width of $25\mu m$ is already close to the value of $18\mu m$ from simulations.

The TRT detector is composed of 4 mm diameter straw tube layers, interleaved with transition radiation material allowing for $e-\pi$ identification in the range 0.5 - 150 GeV. In the barrel region the tubes are assembled in 73 layers, parallel to the beam. Each of the end-cap TRT is composed of 160 layers of radial tubes (perpendicular to the beam).

The TRT provides substantial discrimination between electrons and pions over the wide energy range between 1 and 200 GeV by utilizing transition radiation in foils and fibres. The readout discriminates at two thresholds, the lower set to register minimum-ionising particles and the higher intended for transition radiation (TR) photon interactions. The fraction of high-threshold TR hits as a function of the relativistic γ factor is shown in Figure 5 for particles in the forward region. This region is displayed because there are more conversion candidates and they have higher momenta than in the barrel.

The high part of the distribution is constructed using electrons from photon conversions while the low component is made using charged particle tracks with a hit in the B-layer and treating them as pions. All tracks are required to have at least 20 hits in the TRT. To ensure high purity (about 98%), the conversion candidates are also required to have a vertex more than 40 mm away from the beam axis. The pion sample excludes any photon conversion candidate tracks.

The nominal transverse momentum resolution:

$$\frac{\sigma_{p_T}}{p_T} \simeq 3.8 \times 10^{-4} p_T(GeV) \oplus 0.015 \tag{1}$$

and the tracking performance of the Inner Detector was assessed via the observation of well measured particle decays, like K_s^0, ϕ, D mesons and Ω, Ξ, Λ baryons. Clear signals of all these resonances were obtained in data. Figure 6 shows the reconstruction of K_s^0 decaying to a pair of pions and Figure 7 shows the J/ψ decaying to a pair of muons. The signal lineshape fits are both Gaussian with a third-order polynomial to model the background. The masses and the widths of the two signals are free parameters of the fit. The uncertainty on the yield is statistical only, whereas the uncertainty on the fitted mass corresponds to a $\sim$1 per mille uncertainty on the p_T scale calibration.

Those studies allowed the momentum scale to be determined at the per mill level for the low transverse momentum (p_T) region, and for higher momentum at the % level (for momentum up to 100 GeV), as verified from the resolution in the measured invariant mass of the mentioned particles. The resolution was found as expected to be dominated by multiple scattering in the

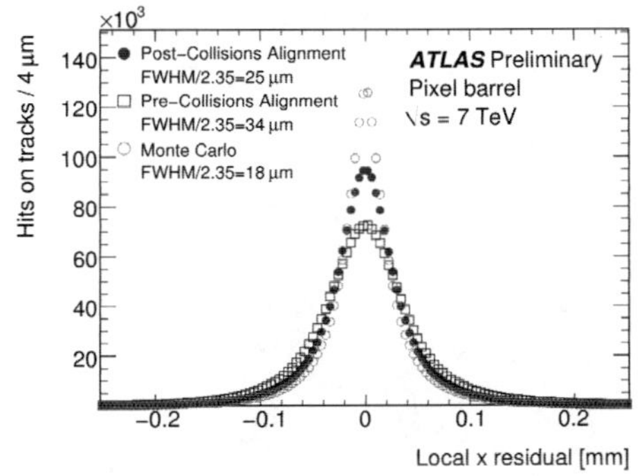

Figure 4. Pixel local x residual distributions integrated over all hits-on-tracks in barrel modules.

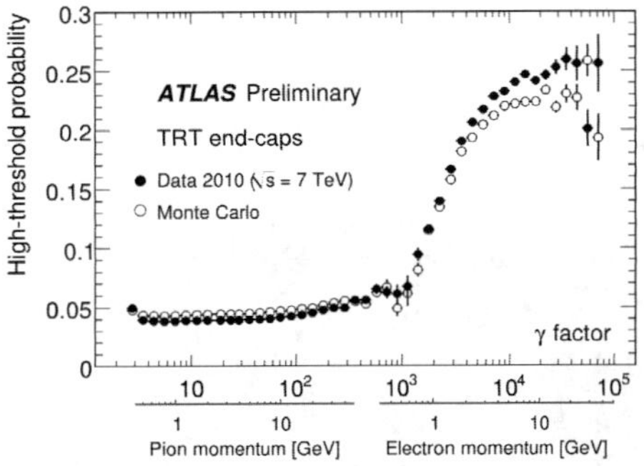

Figure 5. probability of a TRT high-threshold (HT) hit as a function of the Lorentz factor, $\gamma = \frac{E}{m}$, for the TRT end-cap regions, as measured in 7 TeV collision events.

low p_T region. Photon conversions are particularly suitable to track the distribution of dead material and have been used to tune the Inner Detector simulation program

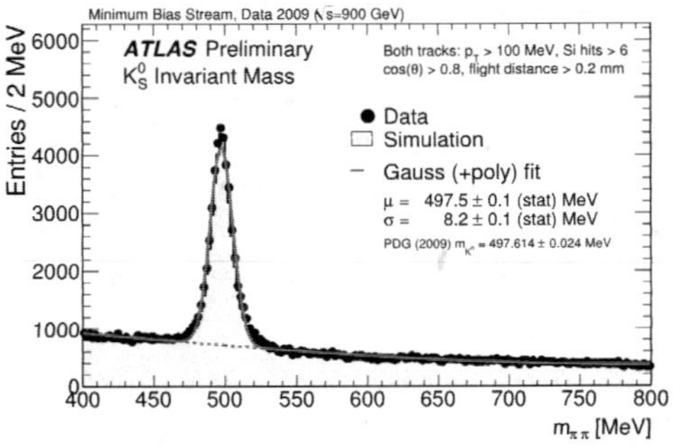

Figure 6. Reconstruction of K_s^0 invariant mass in the barrel region (both tracks are in $|\eta| < 1.2$). The black circles are the data, while the histograms show the Monte Carlo simulation (normalized to the data). The red line is the line-shape function fitted to data [12].

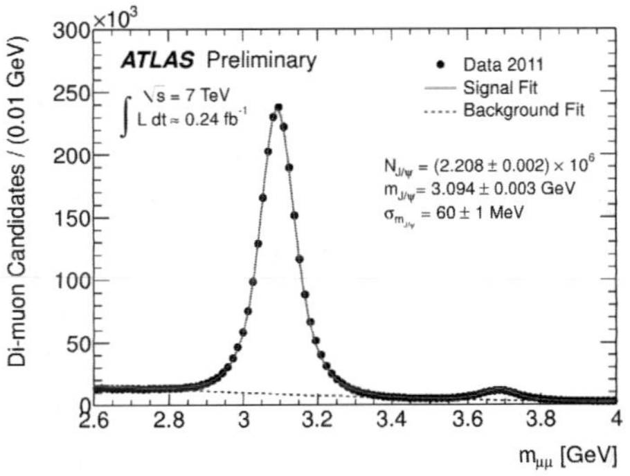

Figure 7. Di-muon invariant mass spectrum at the J/ψ and $\psi(2s)$ mass range: All oppositely charged di-muon pairs passing vertexing with invariant masses between 2.6 and 4.0 GeV. The plot corresponds to an integrated luminosity of around 0.24 fb^{-1}.

The tracking efficiency as a function of transverse momentum (p_T), averaged over all pseudorapidity, rises from $\simeq$ 10% at 100 MeV to $\simeq$ 86% for p_T above a few GeV. The resolution on the position of the primary vertex enables the identification of multiple interactions in the same bunch crossing on an event-by-event basis, as made clear in Figure 8. Secondary vertices such as those produced by B-hadrons, that typically travel a few mm before decaying, can be resolved as well.

Events produced in Pb−ion collisions are characterised by particle multiplicities much larger than the ones found in pp events. As a consequence, the performance of the ID is not as good as in pp collisions, and different tracking algorithms are used. Nevertheless, the agreement between data and simulated events with respect to the average number of hits found in the various subsystems of the Inner Detector is rather good both for *pp* and Pb−ion collision events, as shown in Figure 9 meaning that we have a good understanding of the detector behaviour in both conditions.

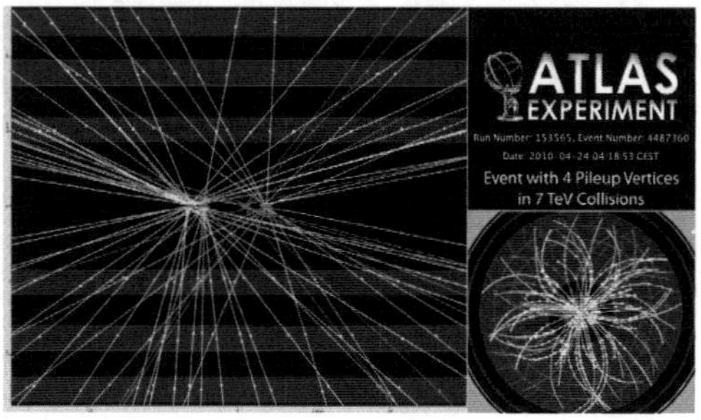

Figure 8. A *pp* collision event at 7 TeV recorded by ATLAS where four different interactions are piling up. The zoomed image shows the reconstructed primary vertices.

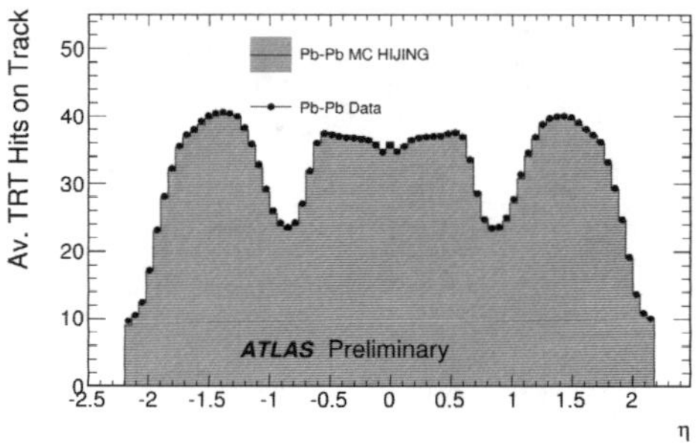

Figure 9. Number of TRT hits on track (integrated and averages as function of η) for Pb−ion collision and comparison with Monte Carlo expectations.

5. Calorimeter Performance

The ATLAS Calorimeters are located just outside the Inner detector. They must absorb and measure the energy of photons, electrons, isolated hadrons and hadron jets. They also allow the measurement of the missing transverse energy carried by high momentum neutrinos or, possibly, new weakly interacting particles. The Electromagnetic Calorimeter uses lead as absorber and liquid argon as the active medium, which is robust against radiation doses. An accordion geometry was chosen to minimize dead areas. By neglecting the noise term, the ultimate energy resolution can be written as:

$$\frac{\sigma_E}{E} \simeq \frac{10\%}{\sqrt{E}} \oplus 0.7\% \tag{2}$$

The Hadron Calorimeter is made of iron and scintillating tiles in the central region, where radiation doses are moderate. More radiation−hard technologies are used in the forward region, where liquid−argon is again used as the sensitive material. The ultimate energy resolution can be described as:

$$\frac{\sigma_E}{E} \simeq \frac{50\%}{\sqrt{E}} \oplus 3\% \tag{3}$$

Electron and photon identification is based on the presence of energy clusters in the electromagnetic calorimeter, and little or no energy in the hadron calorimeter. A track in the Inner Detector pointing to the electromagnetic cluster is expected for electrons, whereas no tracks are present for photons. As an example of the photon identification capability of ATLAS, Figure 10 shows the measured invariant mass of diphoton candidates selected in *pp* collision events in the region around the π^0 mass [13] with the first data of 2009 at $\sqrt{s}$ = 900 GeV. From the comparison of its mass to the Particle Data Group (PDG) value the energy scale was checked within 2%.

Good electron identification performance allowed us to reconstruct all known e^+e^- resonances from a few GeV (J/ψ) up to the Z boson. The em scale has been determined on 7 TeV centre-of-mass-energy data using the Z boson decays to a pair of electrons. Events with two opposite charge electrons reconstructed with transverse energy $E_T > 20$ GeV and constraint their mass

Journal of Physics: Conference Series **347** (2012) 012006 doi:10.1088/1742-6596/347/1/012006

to follow the well known Z line shape derived from simulation. This method was applied in different em regions. Figure 11 shows the distribution of this calibrated Z mass, after applying the corrections to the electron energy. Electrons from Z boson decays have p_T in the tens of GeV range and can be easily identified even in the high occupancy environment of Pb−ion collisions. As an example, Figure 12 shows a Z boson candidate produced in a Pb−ion collision at $\sqrt{s} = 2.76$ TeV. Massive gauge boson production had never been observed at other heavy−ion colliders.

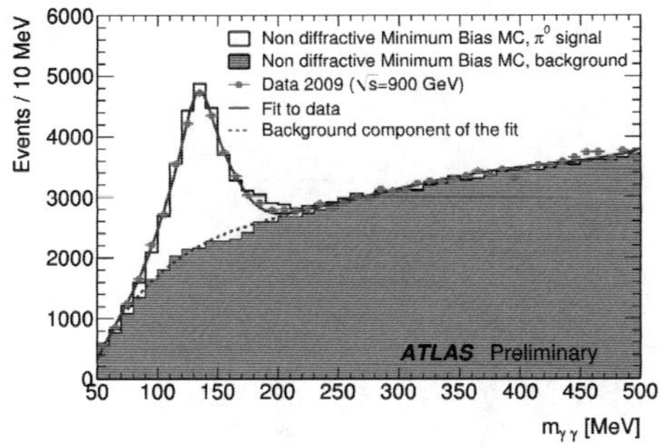

Figure 10. Diphoton invariant mass spectrum in the region around the π^0 mass. A fit is superimposed to the points representing data. The histogram refers to Monte Carlo prediction, where the number of entries is normalized to the one found in data.

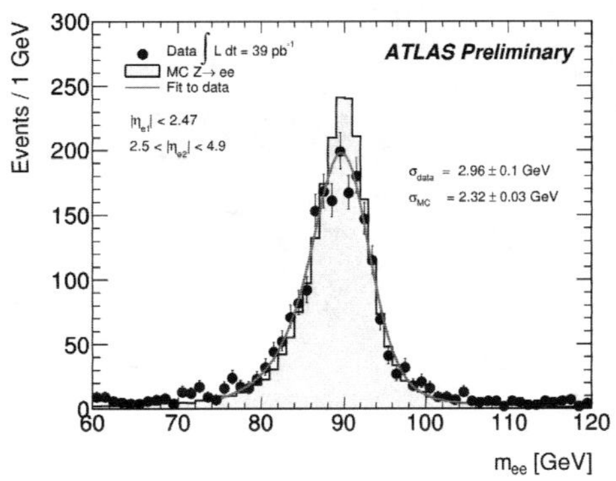

Figure 11. Invariant mass distribution of e^+e^- candidates in the region of the Z mass. Black points represent data, the histogram Monte Carlo prediction.

Diphoton and dielectron invariant mass spectra covering the range from tens of MeV to hundreds of GeV are exceptionally sensitive calibration data. They were used to understand the alignment and energy resolution of the Electromagnetic Calorimeter.

Understanding and measuring the performance of jets is crucial for many physics analyses at the LHC. Hadron jets are reconstructed with the anti−kt algorithm [14], with distance parameters of 0.6 and 0.4, starting from clusters of cells of energy above a set threshold in the hadron calorimeter. The uncertainty of the jet energy calibration is the dominant experimental uncertainty for numerous physics results with jets. The energy resolution depends on both the jet transverse momentum and the jet pseudorapitidy. It has been evaluated on data for jets with transverse momenta above 20 GeV up to the kinematic limit of 3.5 TeV and η up to 4.5. To give an idea of the p_T range of jets produced at LHC, Figure 13 shows its distribution as measured with the data and as expected from Monte Carlo simulation. A remarkable agreement over six orders of magnitude is apparent.

Figure 14 shows the final fractional jet energy scale systematic uncertainty as a function of jet p_T for the barrel region. For example it is found to be less that 2.5% in the region of jet transverse momenta between 60 and 800 GeV. The overall precision on the energy scale is currently $\simeq$ 7% [15] . It is expected to reach the 3−4% level once ongoing studies with beam data on jet−balance and E/p measurements for isolated hadrons will have been finalized. The impact of pile−up will also be quantitatively evaluated.

Missing transverse energy, E_T^{miss}, in an event is a measure of the energy imbalance in the transverse plane and it indicates the presence of undetected particles (like neutrinos). Such imbalance may also rise from detector effects or relative miscalibration of the scales of the many detectors, as well as for beam backgrounds and cosmics. Being measured from all particles seen

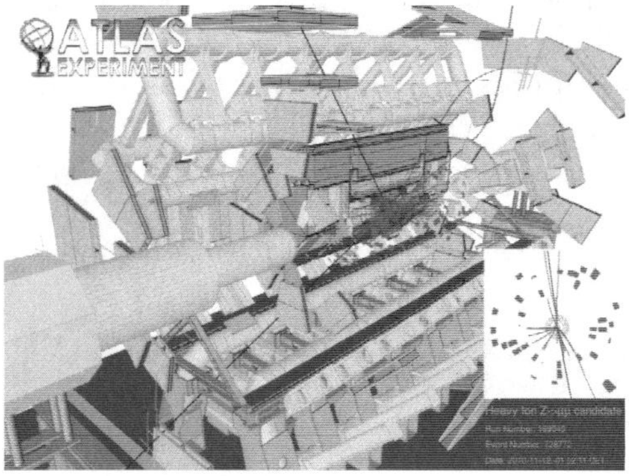

Figure 12. A heavy−ion collision with a candidate Z to $\mu^+\mu^-$ decay. The two muons shown in purple (first display) or red (second display) are the candidates to originate from the Z decay.

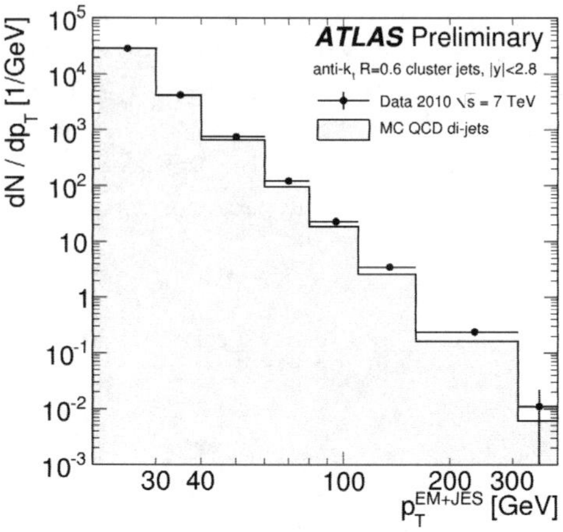

Figure 13. Distribution of the transverse momentum of jets as reconstructed in the data (black points) and as expected from Monte Carlo simulation (histogram).

in the detector, it is expected to scale with the amount of energy in the event. The calorimeters allow for the measurement of the missing transverse energy E_T^{miss}. The key performance figure is the transverse energy resolution as a function of the total transverse energy. It has been measured in the ATLAS detector with the 7 TeV collision data [16] and found to be in reasonable agreement with the simulation (Figure 15.)

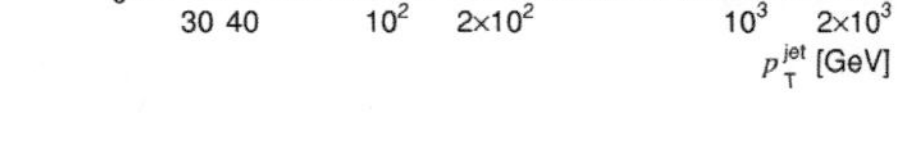

Figure 14. Fractional jet energy scale systematic uncertainty as a function of p_T for jets in the pseudorapidity region $0.3 < |\eta| < 0.8$ in the calorimeter barrel. The total uncertainty is shown as the solid light blue area. The individual sources are also shown, with uncertainties from the fitting procedure if applicable.

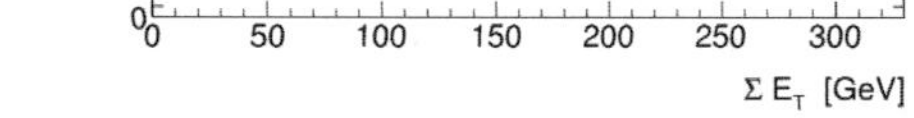

Figure 15. Resolution of the transverse missing energy (E_x^{miss} ,E_y^{miss}) as a function of the total transverse energy, $\sum E_T$, for pp collision data at $\sqrt{s}=7$ TeV. Full dots represent the data while the line is a fit to the resolution obtained in Monte Carlo simulations. ATLAS-CONF-2010-057

Journal of Physics: Conference Series **347** (2012) 012006 doi:10.1088/1742-6596/347/1/012006

6. MuonSpectrometer Performance

Muons are identified through their capability of penetrating through the Calorimeters and reach the Muon Spectrometer. The deflection of muons by the magnetic field generated by a system of air core toroid coils in the muon spectrometer is measured by three layers of precision gaseous detector up to $|\eta| < 2.7$. The position in the non bending plane is provided by the muon trigger detectors. A minimum momentum of $\simeq 3$ GeV is required for muons to emerge from the upstream material. The detector has the capability of measuring the momentum of muons without the help of the Inner Detector, with nominal resolution $\sigma p_T/p_T < 3\%$ up to 200 GeV and $\sigma p_T/p_T \simeq 10\%$ at 1 TeV. Matching to the Inner Detector can obviously improve the momentum resolution and reduce the background from kaon decays as well as non−containment of hadrons within the calorimeters. Relative efficiencies have been studied on single muons exploiting the redundancy of detector technologies and reconstruction algorithms and the minimum ioninizing particles identification in the calorimeters. Fake rates studies are based on the mis-tag rate for pions and kaons identified in hadronic resonances (K_s^0 and D^*) and on the momentum unbalance between inner detector and muon spectrometer measured momenta, due to late decays of pions and kaons.

The four technologies adopted in the MS have been fully commissioned and a fraction of channels very close to 100% is operational for MDT and TGC, and more than 97% and 98% for RPC and CSC, respectively.

At high momentum, the transverse momentum resolution of stand-alone muons is dominated by the calibration of the muon precision chambers and the internal alignment of the muon spectrometer. The precision muon drift chambers (MDTs) measure the drift time which is then converted into a drift distance using the conversion function, which is determined from data in a calibration procedure. Each drift channel has a constant time offset ('t0') for which the measured time has to be corrected to obtain the drift time. The t0s are determined from data. Chamber alignment and sagitta resolution are studied using cosmic-ray tracks collected without magnetic field. The segment sagitta is defined as the distance from the Middle-station segment to the straight line connecting the segments in the Inner and Outer stations. The segment sagitta distribution for each sector is fitted to a double Gaussian (see Figure 16)). The mean of the narrow Gaussian is used for track-based alignment of the spectrometer, while the sigma corresponds to the sagitta resolution. The sagitta resolution is parametrized into two separate components: multiple scattering and intrinsic resolution, respectively dominating at high and low momenta. Using the solenoidal magnetic field of the Inner Detector to determine the momentum of the muon tracks, the intrinsic component of the sagitta resolution is isolated and found to be between 80 and 100 μm.

To measure the momentum resolution of the MS without requiring a comparison with the ID, the top and bottom sections of a cosmic-ray track traversing the whole detector are compared. The momentum resolution is the width of the fitted distribution of relative p_T differences ($\Delta p_T/p_T$) between the top and bottom halves of the track. Fitting the momentum resolution against the momentum of the tracks (see Figure 17) allows the extraction of its three components: energy loss correction (P_0), multiple scattering (P_1), and intrinsic resolution (P_2). Extrapolating the fitted function to 1 TeV momenta gives a resolution of $11 \pm 2\%$ for tracks crossing small MDT chambers and $25 \pm 2\%$ for tracks crossing large ones. The difference between small and large chambers is due to the difference in integrated magnetic field along the muon paths. The design goal for 1 TeV muon tracks is a p_T resolution of approximately 10%.

$$\frac{\sigma_{p_T}}{p_T} = \frac{P_0}{p_T} \oplus P_1 \oplus P_2 \times p_T \qquad (4)$$

In the ATLAS detector, four kinds of muons are distinguished depending on the way they are reconstructed:

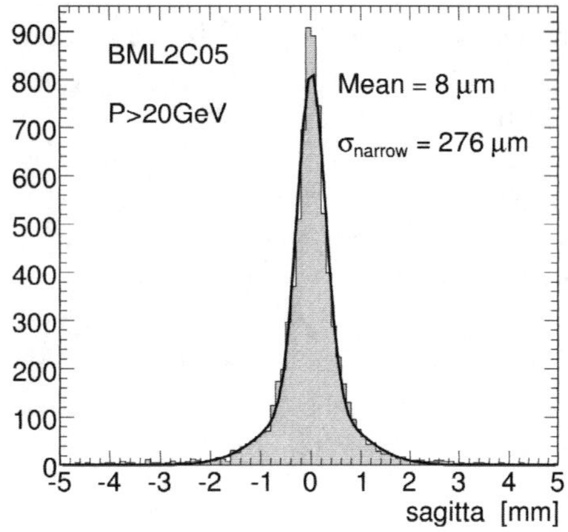

Figure 16. Sagitta resolution in the Small Barrel Sectors of the MS as a function of muon momentum. Muon momentum is taken from ID measurements. Data is fitted by the function containing multiple scattering term and intrinsic resolution term.

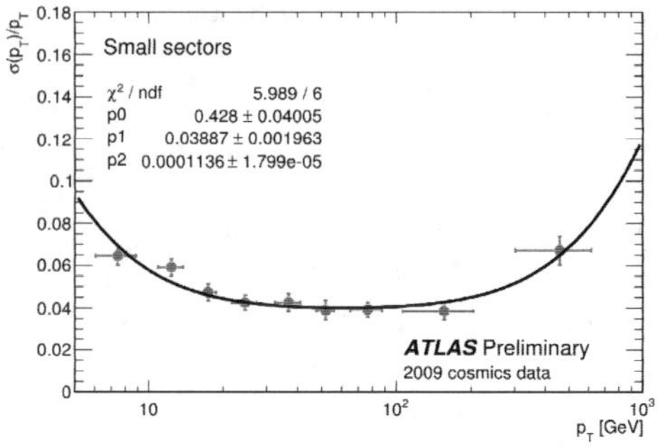

Figure 17. Transverse momentum resolution evaluated with the top bottom method as a function of p_T for small sectors as measured with the Muonboy algorithm. The fitted curve is a phenomenological description of the stand-alone momentum resolution; it is the quadratic sum of an energy loss fluctuation term p0/pT, a multiple scattering term p1, and a spectrometer resolution term p2*pT. (See equation 4)

Stand-alone muon: The muon trajectory is only reconstructed in the muon spectrometer, and is extrapolated to the beam line while correcting for the energy loss in the calorimeter.

Combined muon: The measurement of the stand-alone muon is combined with the measurement in the inner detector.

Segment tagged muon: A track in the inner detector is identified as a muon if its extrapolation to the muon spectrometer can be matched to (a) track segment(s) in the muon chambers.

Calorimeter tagged muon: A track in the inner detector is identified as a muon if the associated energy deposition in the calorimeters is compatible with a minimum ionizing particle (mip).

As a demonstration of the good performance of the spectrometer, Figure 18 shows the invariant mass of two opposite charge muons of $p_T > 15$ GeV that are coming from a common vertex. The J/ψ and Z resonances are used to study in detail the performance of the MS in the low and high p_T regime respectively. The muon reconstruction efficiency as measured on data with "tag and probe" techniques [17],[18] was found to be consistent with the simulation predictions; for example the average efficiency for muons of $p_T > 5$ GeV is 98%. The J/ψ decays gave a unique opportunity to probe the efficiency measurement down to very low values of p_T . The K_s^0 decays to a pair of pions were used to study the muon fake rate (i.e the muon background from decays in flight or reconstruction ghosts) which was found to be of the order of 1 per mill [19].

7. Forward detectors in 2010

7.1. LUCID

LUCID is a Cherenkov light detector primarily dedicated to online luminosity monitoring. LUCID is a relative luminosity detector. Its main purpose is to detect inelastic proton-proton scattering in the forward direction, in order to both measure the integrated luminosity and to provide online monitoring of the instantaneous luminosity and beam conditions. LUCID consists

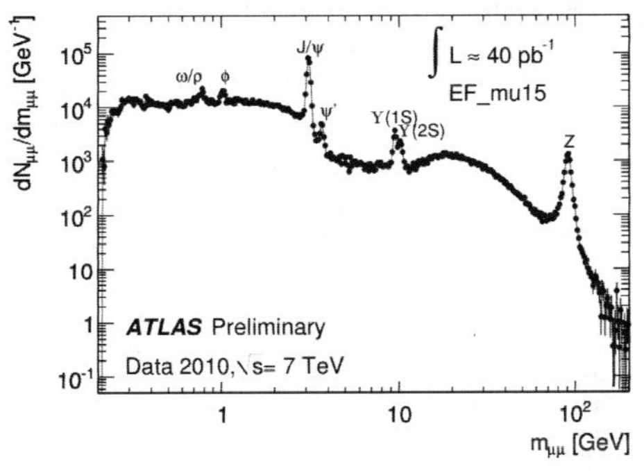

Figure 18. Dimuon invariant mass spectrum of two opposite charge muons of $p_T > 15$ GeV that are coming from a common vertex.

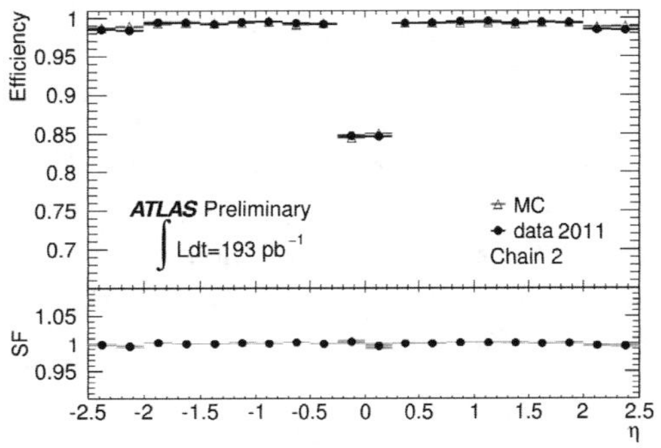

Figure 19. Combined muon reconstruction efficiency with respect to the inner tracking efficiency as a function of the pseudorapidity of the muon for muons with $p_T > 20$ GeV. The panel at the bottom shows the ratio between the measured and predicted efficiencies.

of twenty aluminium tubes which surround the beam pipe and point toward the interaction point. The 1.5 m long mechanically polished tubes with a diameter of 15 mm are placed in a lightweight aluminium gas vessel which ensures that the tubes are filled with C_4F_{10} gas at a constant pressure of about 1.2 bar, providing a Cerenkov threshold of 2.8 GeV for pions and 10 MeV for electrons. The two LUCID detectors are installed in the ATLAS end-cap regions, at a distance of approximately ± 17 m from the interaction point, at a radial distance of approximately 10 cm from the beam line, and cover $5.61 < |\eta| < 5.93$. The electronics of LUCID implement various algorithms to monitor the luminosity delivered by LHC (before any trigger decision) in individual bunches. All algorithms need an independent calibration to provide absolute values of delivered luminosity. The LUCID detector was optimized for high luminosity, but performed very well in early data taking with a luminosity of less than $L = 10^{27} cm^{-2} s^{-1}$.

7.2. ZDC

The primary purpose of the ATLAS Zero-Degree Calorimeters (ZDC) is to detect forward neutrons and photons with $|\eta| > 8.3$ in both proton-proton and heavy−ion collisions. The ZDCs play a key role in determining the centrality of heavy− ion collisions and contribute to minimum bias and diffractive physics in proton-proton collisions. With a time resolution of about 100 ps, a coincidence of the two ZDCs can be used to greatly reduce backgrounds from beam-gas and beamhalo effects. The ZDC consists of two arms located at 140 m from the IP in slots in the LHC TAN (Target Absorber Neutral) [20], occupying space that would otherwise contain inert copper shielding bars. ZDCs is consisted of four modules on each arm: one electromagnetic (EM) module (about 29 radiation lengths thick), and 3 hadronic modules (each about 1.14 interaction lengths thick). The modules are composed of tungsten with an embedded matrix of quartz rods which are read out by photomultiplier tubes. The modules are 9 cm by 9 cm laterally to the beam, and 15 cm thick.

7.3. ALFA

The ALFA Roman pots and detectors are located at 240 m from the interaction point with a purpose of measuring elastic scattering at small angles in the Coulomb-Nuclear Interference region. The system consists of four Roman pot stations, two on each side of the interaction point, each housing two vertically movable detectors. The Roman pots house scintillating fiber detectors arranged in 20 u/v detection planes, providing a spatial resolution of about 30μm. The luminosity calibration will be obtained during special runs using a dedicated high β^* optics which features a parallel-to-point focusing to allow for a direct reconstruction of the scattered proton kinematics.

Figure 20 shows the location of each of these detector systems relative to the ATLAS central detector (only one arm is shown, but all the detector systems are symmetrically located on each side of the interaction region).

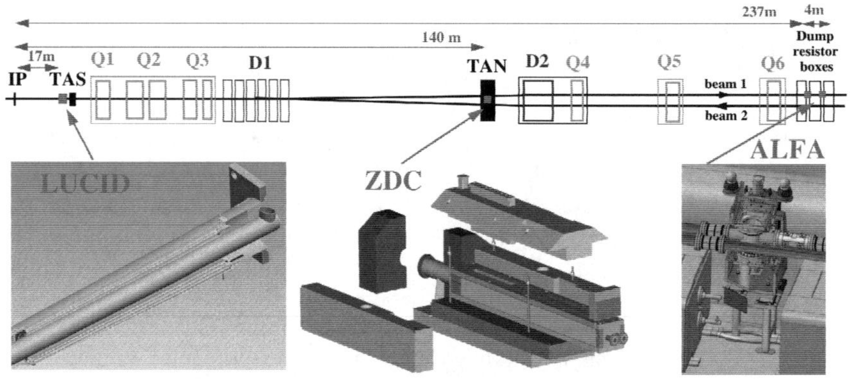

Figure 20. The layout of the ATLAS Forward detectors.

8. Highlights from physics results

With all the ingredients described so far, we can start to talk about physics. The selection of results presented in the following sections are to a large extent focussed on data collected in the 2010 and early 2011 running periods of LHC, when the pile–up probability was negligible. The delay between data collection and availability of results is due to the need to understand the detector and the effects of pile–up, as well as tune Monte Carlo simulations using real data. However, updates based on full statistics are expected soon.

8.1. Results from Proton-Proton collisions

The total cross section for pp collisions at 7 TeV amounts to $\simeq 75mb$. Most of the interactions result in events characterized by the production of particles with transverse momentum not exceeding a few GeV. Hard jets, with transverse momentum of tens of GeV or more, are produced in a small fraction of the collisions ($10^{-5}\%$). Vector boson production is equally rare, and even rarer is the production of top quarks. Given the luminosity delivered by LHC in 2010, the physics studied so far is often referred to as "rediscovery of the Standard Model", although searches for new physics are starting to improve limits on production of exotic particles. The full rediscovery of the Standard Model in just one experiment was at any rate quite an exciting exercise.

8.1.1. Production of top quarks $t\bar{t}$ The top quark is by far the heaviest quark, with a measured mass of 172 GeV. At LHC top quarks are mostly produced in pairs. Their mass is so high that there is no time for them to hadronize before decaying. The branching fraction to a W boson and a b quark is close to 100%, so that the final $t\bar{t}$ event topology is fully determined by the decay of the W bosons. Events where at least one of the W bosons decays *leptonically* can be

Journal of Physics: Conference Series **347** (2012) 012006 doi:10.1088/1742-6596/347/1/012006

separated from the high multi−jet QCD background with the best signal−to−noise ratio. The capability to identify jets originating from b−quarks (b tagging) is the key to rejecting W+jets events where no top quarks are produced. The *semileptonic* channel, where one of the W boson decays leptonically and the other decays into a jet pair, can be used to measure the top mass via the invariant mass of the three jets coming from the $bq\bar{q}$ system.

The measurement of the top-quark pair-production cross-section $\sigma_{t\bar{t}}$ is one of the milestones for the early LHC physics program. A precise measurement of $\sigma_{t\bar{t}}$ allows for precision tests of perturbative QCD, where uncertainties on $\sigma_{t\bar{t}}$ are now at the level of 10% [21]. In addition $t\bar{t}$ production is an important background to the search for the Higgs boson and various searches for physics beyond the Standard Model. ATLAS has measured the production cross section of $t\bar{t}$ [22] combining the results of two topologies: single lepton (electron e or muon μ) with large missing transverse energy and at least four jets, and dilepton (ee, $\mu\mu$ or $e\mu$) with large missing transverse energy and at least two jets. The result is $\sigma_{t\bar{t}} = 189 \pm 9(stat.) \pm 15(syst.) \pm 6(luminosity)p$, which is in excellent agreement with the Standard Model prediction. Figure 21 shows various cross-section measurements from Tevatron and LHC results overlaid with the theoretical predictions as a function of centre-of-mass energy. Figure 22 shows the summary measured $\sigma_{t\bar{t}}$ using several analyses in each decay channel, including errors bars for both statistical uncertainties only and all systematics.

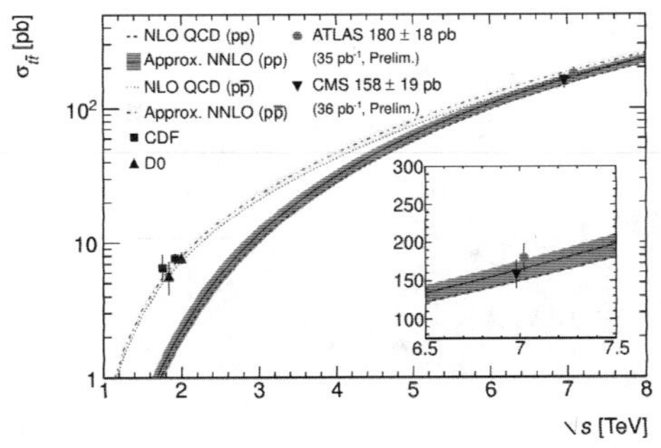

Figure 21. $t\bar{t}$ production cross−section at hadron colliders as measured by CDF and D0 at Tevatron ($p\bar{p}$ collisions), CMS and ATLAS at LHC (pp collisions). The theoretical predictions for $p\bar{p}$ and pp collisions include the scale and PDF uncertainties and assume a top−quark mass of 172.5 GeV. The present result is indicated by the red circle.

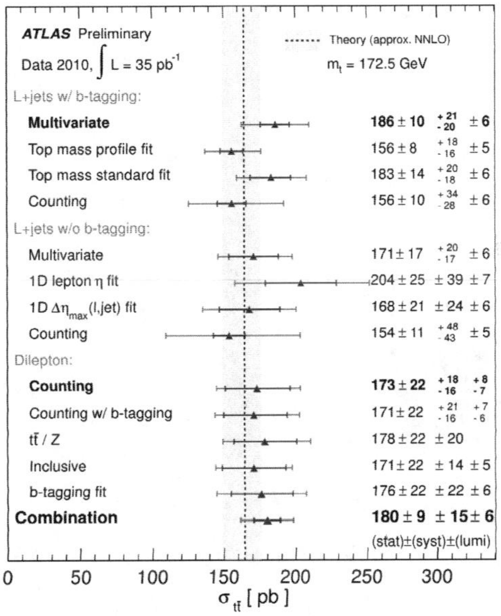

Figure 22. Plots of measured $\sigma_{t\bar{t}}$ using several analyses in each decay channel, including errors bars for both statistical uncertainties only (blue) and all systematics (red). The combined result is based on the L+jets b-tag multivariate and the dilepton counting analyses. The approximate NNLO prediction is shown as a vertical dotted line with its error in yellow.

8.1.2. First Searches for New Physics First searches for new physics included heavy particles decaying into jets, for which the signature would be a bump in the dijet invariant mass spectrum, as well as new interactions giving rise to special features in the dijet angular distributions. In fact, QCD calculations predict that high$-p_T$ dijet production is dominated by $t-channel$ gluon exchange, leading to angular distributions that are peaked at $|cos\theta^*|$ close to 1, where θ^* is the polar scattering angle in the two$-$parton center$-$of$-$mass frame. By contrast, models of new processes characteristically predict angular distributions that would be more isotropic than those of QCD. Many models of new physics predict the existence of additional heavy gauge bosons, where the charged ones are commonly denoted $W^{'}$. Such particles are most easily searched for in their decay to a charged lepton (either electron or muon) and a neutrino.

Search for $Z^{'}$: ATLAS detector has been used to search for high mass resonances decaying into e^+e^- or $\mu^+\mu^-$ pairs in the invariant mass spectrum above 110 GeV with $\sim 167pb^{-1}$ in the e^+e^- and $\sim 236pb^{-1}$ in the $\mu^+\mu^-$ final state of 7 TeV pp collision data. Among the possibilities for such resonances, this note focuses on new heavy neutral gauge bosons $Z^{'}$ [23],[24],[25]. The benchmark model for $Z^{'}$ bosons is the Sequential Standard Model (SSM) [23], in which the $Z^{'}$ ($Z^{'}_{SSM}$) has the same couplings to fermions as the Z boson. Figure 23 presents the invariant mass ($m_{e^+e^-}$) distribution after final selection. The dielectron invariant mass distribution is well described by the prediction from SM processes. Similarly, results for the $\mu^+\mu^-$ channel are presented in Figure 24. A good agreement between the data and the SM prediction are again observed. No evidence for such a resonance is found. Limits on the cross section times branching ratio $\sigma \times B(Z^{'} \to l^+l^-)$ are calculated, as well as mass limits on the $Z^{'}_{SSM}$ (1.407 TeV) and E6-motivated $Z^{'}$ bosons (in the range 1.116$-$1.259 TeV).

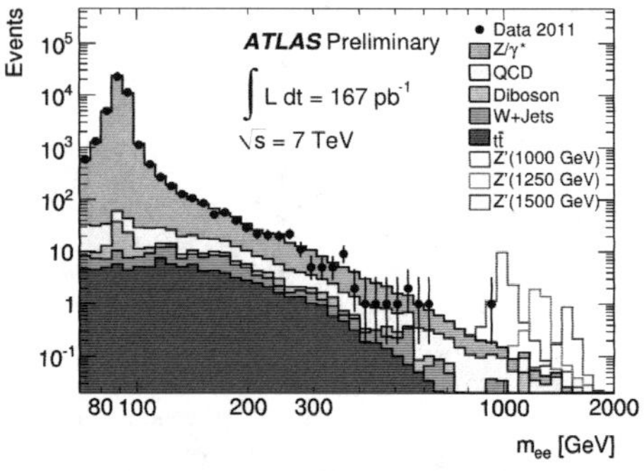

Figure 23. $t\bar{t}$ Dielectron invariant mass ($m_{e^+e^-}$) distribution after final selection, compared to the stacked sum of all expected backgrounds, with three example $Z^{'}_{SSM}$ signals overlaid.

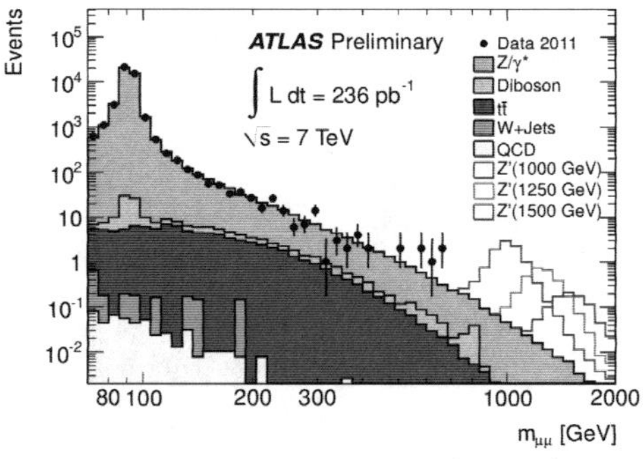

Figure 24. Dimuon invariant mass ($m_{\mu^+\mu^-}$) distribution after final selection, compared to the stacked sum of all expected backgrounds, with three example $Z^{'}_{SSM}$ signals overlaid.

Search for $W^{'}/W^*$: A first limit on the production of $W^{'}$ was set by assuming the same couplings as those of the Standard Model W boson (Sequential Standard Model) and considering the $W^{'} \to l^{\pm}\nu$ ($l = e$ or μ) decay channels. Combined limits assuming flavor independence has been set by ATLAS. The kinematic variable used to identify the $W^{'}/W^*$ is the transverse mass:

$$m_T = \sqrt{2p_T E_T^{miss}(1 - cos\phi_{l\nu})} \tag{5}$$

which displays a Jacobian peak that, for $W^{'} \to l^{\pm}\nu$, falls sharply above the resonance mass. Here p_T is the lepton transverse momentum, E_T^{miss} is the magnitude of the missing transverse momentum (missing E_T), and $\phi_{l\nu}$ is the angle between the p_T and missing E_T vectors. Figures 25 and 26 show the m_T spectrum in $e\nu$ and $\mu\nu$ channels, respectively, after final selection for the data, for the expected background, and for three examples of W' signals at different masses. Both direct production of leptons and indirect from $\tau-$leptons are included. The agreement between the data and expected background is good. No excess beyond SM expectations is observed. A W' with SSM couplings is excluded for masses below 1490 GeV at 95% CL. The exclusion for W^*, the charged partner of the chiral bosons described in [26], with couplings set in accordance with reference [27] is 1470 GeV.

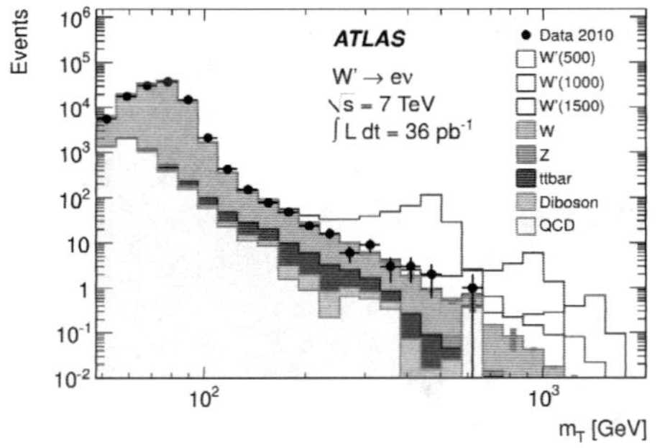

Figure 25. $t\bar{t}$ Spectrum of m_T for the electron channel after final event selection. The points represent ATLAS data and the filled histograms show the stacked backgrounds.

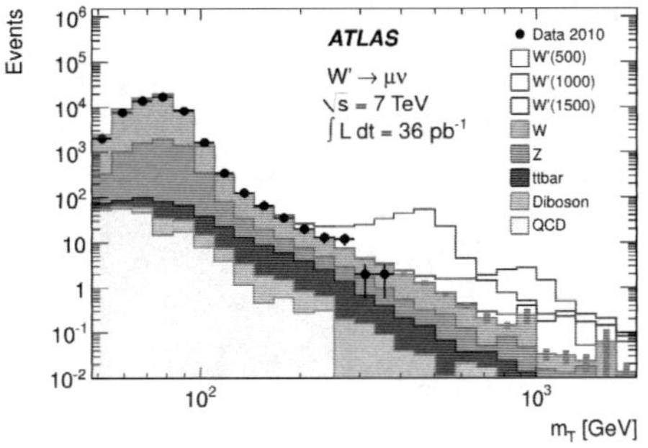

Figure 26. Spectrum of m_T for the muon channel after final event selection. The points represent ATLAS data and the filled histograms show the stacked backgrounds.

8.2. Results from Pb − Pb collisions

Collisions of lead ions at ultra−relativistic energies are expected to produce a dense state with temperatures exceeding two trillion Kelvin. If deconfinement is indeed reached, the relevant degrees of freedom are not hadrons, but quarks and gluons. In this medium, high−energy quarks and gluons are expected to transfer energy to the medium by multiple interactions with the ambient plasma. The study of events produced at ultra−relativistic heavy−ion collisions is typically carried out by labelling the events according to their geometry, or centrality: the larger the centrality, the larger the number of interacting nuclei and the event activity. At ATLAS, the level of event activity is characterized by using the total transverse energy ($\sum E_T$) deposited in the Forward Calorimeters as shown in Figure 27. Bins are defined in centrality according to fractions of the total lead−lead cross−section selected by the trigger and are expressed in terms of percentiles ($0-10\%$, $10-20\%$, $20-40\%$ and $40-100\%$) with 0% representing the upper end of the $\sum E_T$ distribution.

An interesting observation, the so called "jet quenching", which is the production of highly unbalnaced jets with one jet produced at the periphery of the collision, gives hints of how the parton shower develops in this dense medium. The jet quenching is expected in central heavy−ion collisions and actually at Relativistic Heavy Ion Collider experiments (RHIC) quenching in two particle correlation is observed. With the data collected by the ATLAS detector observations have been made of a centrality dependent dijet asymmetry in these collisions [17]. Events with two jets were selected requiring the leading jet to have transverse energy $E_T > 100$ GeV and the second jet $E_T > 25$ GeV. The transverse energies of dijets in opposite hemispheres were observed to become systematically more unbalanced with increasing event centrality leading to a large number of events which contain highly asymmetric dijets as it is shown in Figure 28.

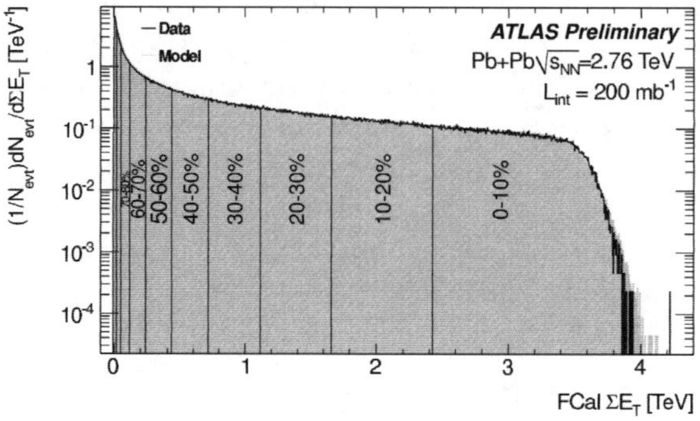

Figure 27. Distribution of uncorrected $\sum E_T$ in the Forward Calorimeter (FCal). Bins in event activity or "centrality" are indicated by the alternating bands and labeled according to increasing fraction of lead−lead total cross section starting from the largest measured $\sum E_T$.

This was the first observation of an enhancement of events with such large dijet asymmetries, not observed in proton-proton collisions, which may point to an interpretation in terms of strong jet energy loss in a hot, dense medium.

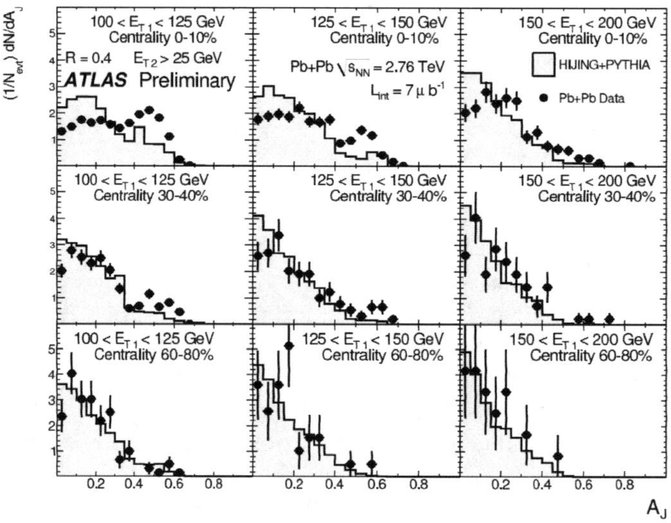

Figure 28. Dijet asymmetry distributions for $R = 0.4$ jets in three collision centrality bins: $0-10\%$ (top), $30-40\%$ (middle), $60-80\%$ (bottom) for $\sqrt{s_{NN}} = 2.76$ TeV $Pb+Pb$ collisions with a leading jet with $100 < E_{T1} < 125$ GeV (left), $125 < E_{T1} < 150$ GeV (middle), $150 < E_{T1} < 200$ GeV (right). For all of the plots, comparison to HIJING events with embedded PYTHIA dijets for the same conditions on reconstructed jets as the data is shown (yellow).

9. Summary and Conclusions

2010 was an extremely exciting year for LHC. The accelerator reached all the announced goals, and the detectors performed equally well. ATLAS data collection was very smooth, and data have been under deep scrutiny to understand the details of the detector behaviour as well as tune detector−related effects in simulations. The ATLAS collaboration has delivered a plethora of physics results with the 2010 data. Overall, 2010 was the year of Standard Model rediscovery, as is frequently said. No evidence for physics beyond the Standard Model has been found yet, although previous limits on the production of new particles start to be extended, even with limited statistics. A lot of new results are just about to come. Some model tuning is apparently needed for diffractive physics. Ultra−relativistic lead−ion collisions carried most of the surprises, with the first impressive evidence for jet−quenching. And we are just at the beginning of a very promising period to come. All these achievements were possible thanks to the fantastic LHC machine team and to the many years of dedicated work in commissioning of the ATLAS detector.

References

[1] The ATLAS Collaboration, *The ATLAS Experiment at the Large Hdron Collider*, JINST 3 (2008). S08003.
[2] O.Bruning et. al., LHC design report. Vol. I, *The LHC main ring*, CERN-2004-003-V-1.
[3] O.Bruning et. al., LHC Design Report. 2, *The LHC infrastructure and general services*, CERN-2004-003-V-2.
[4] O.Bruning et. al., LHC Design Report. 2, *The LHC injector chain*, CERN-2004-003-V-3.
[5] The ATLAS Collaboration, *Luminosity Determination Using the ATLAS Detector*, ATLAS-CONF-2010-060.
[6] S. van der Meer, *Calibration of the effective beam height in the ISR*, CERN-ISR-PO-68-31, 1968.
[7] The ATLAS Collaboration, *Updated Luminosity Determination in pp Collisions at* $\sqrt{s} = 7$ *TeV using the ATLAS detector*, ATLAS-CONF-2011-011.
[8] The ATLAS Collaboration, *ATLAS Offline Data Quality Monitoring*, Journal of Physics: Conference Series, 147 219(4):042018, 2010.
[9] N.Chr. Benekos et al., *Muon Offline Monitoring at Tier0 using cosmics and FDR data*, ATL-MUON-INT-2011-002.
[10] The ATLAS Collaboration, *Data quality from the Detector Control System at the ATLAS experiment*, Journal of 149 Physics: Conference Series, 219(2):022037, 2010.
[11] The ATLAS Muon Collaboration, *The ATLAS Muon Spectrometer Technical Design Report*, CERN/LHCC/97-22, 31 May 1997.
[12] The ATLAS Collaboration, *Kinematic Distributions of* K_s^0 *and* Λ^0 *decays in pp collisions at* $\sqrt{s} = 7$ *TeV*, ATLAS-CONF-2011-033.
[13] The ATLAS Collaboration, *Performance of the ATLAS electromagnetic calorimeter for* $\pi^0 \rightarrow \gamma\gamma$ *and* $\eta \rightarrow \gamma\gamma$ *events*, ATLAS-CONF-2010-006.
[14] M. Cacciari, G. P. Salam and G. Soyez, *The anti-kt jet clustering algorithm*, JHEP 04 (2008) 063.
[15] The ATLAS Collaboration, *Jet energy scale and its systematic uncertainty in ATLAS for jets produced in proton − proton collisions at* $\sqrt{s} = 7$ *TeV*, ATL-CONF-2010-058.
[16] The ATLAS Collaboration, *Performance of the Missing transverse Energy Reconstruciton and Calibration in proton − proton collisions at a Centre-of-Mass Energy of* $\sqrt{s} = 7$ *TeV with the ATLAS Detector*, ATLAS-CONF-2010-057.
[17] The ATLAS Collaboration, *A measurement of the ATLAS muon reconstruction and trigger efficiency using* J/ψ *decays*, ATLAS-CONF-2011-021.
[18] The ATLAS Collaboration, *Determination of the muon reconstruction efficiency in ATLAS at the Z resonance in proton − proton collisions at* $\sqrt{s} = 7$ *TeV*, ATLAS-CONF-2011-008.
[19] The ATLAS Collaboration, *Muon reconstruction Performance*, ATLAS-CONF-2010-064
[20] L. Evans, (ed.), P. Bryant, (ed.), *The CERN Large Hadron Collider: Accelerator and Experiments*, JINST 3 (2008) S08001.
[21] The ATLAS Collaboration, *Measurement of the top quark-pair cross section with ATLAS in pp collisions at* $\sqrt{s} = 7$ *TeV in the semileptonic channel using b-tagging*, ATLAS-CONF-2011-035.
[22] The ATLAS Collaboration, *A combined measurement of the top quark pair production cross-section using dilepton and single-lepton final states*, ATLAS-CONF-2011-040.
[23] P. Langacker, *The Physics of Heavy* Z' *Gauge Bosons*, Rev. Mod. Phys. 81 (2009) 1199-1228.
[24] J. Erler, P. Langacker, S. Munir, and E. R. Pena, *Improved constraints on* Z' *bosons from electroweak precision data*, JHEP 0908 (2009) 017.
[25] M. Chizhov, V. Bednyakov, and J. Budagov, *Proposal for chiral-boson search at LHC via their unique new signature*, Physics of Atomic Nuclei 71 (2008) 2096-2100.
[26] M. V. Chizhov, V. A. Bednyakov, J. A. Budagov, *Physics of Atomic Nuclei*, 71 (2008) 2096-2100.
[27] M. V. Chizhov, V. A. Bednyakov, J. A. Budagov, *Nuovo Cimento*,C33 (2010) 343.
[28] J.M. Campbell, R.K. Ellis, D. Rainwater, *Next-to-leading order QCD predictions for* $W + 2$ *jet and* $Z + 2$ *jet production at the CERN LHC*, Phys. Rev. D 68 (2003) 094021 [hep-ph/0308195].

The SuperB Project: Status and the Physics Reach

Nicola Neri[1]
INFN - Sezione di Milano, Via Celoria 16, 20133 Milano, Italy
E-mail: nicola.neri@mi.infn.it

Abstract. The SuperB experiment is a next generation Super Flavour Factory expected to accumulate 75 ab^{-1} of data at the $\Upsilon(4S)$ in five years of nominal running, and will be built at the recently established Cabibbo Laboratory on the outskirts of Rome. In addition to running data at the $\Upsilon(4S)$, SuperB will be able to accumulate data from the $\psi(3770)$ up to the $\Upsilon(6S)$. A polarized electron beam enables unique physics opportunities at SuperB. The large samples of B, D and τ decays that will be recorded at SuperB can be used to provide both stringent constraints on new physics scenarios, and over-constraints on the Standard Model. We present the status of the project as well as the physics potential of SuperB.

1. Introduction

The BaBar and BELLE experiments and the B factories (PEP-II and KEKB) at SLAC and KEK were designed and built with the primary goal of performing the first precision tests of the Cabibbo-Kobayashi-Maskawa (CKM) mechanism for flavor mixing and CP violation with 3 generations of quarks. From the unitarity of the CKM quark mixing matrix we obtain the relation $V_{ud}V_{ub}^* + V_{cd}V_{cb}^* + V_{td}V_{tb}^* = 0$, which is usually represented as a triangle in the complex plane (the "Unitarity Triangle"). BaBar and BELLE have made fundamental contributions to the measurements of the angles and sides of the Unitarity Triangle and have established the success of the CKM paradigm for describing CP violation within the Standard Model (SM). Kobayashi and Maskawa were awarded of the 2008 Nobel Prize in physics.

However, in flavor physics there are few tensions and puzzles that require larger data samples to be solved. As a lesson from history, I would like to mention an interesting anecdote that I learned at this conference and taken from L. B. Okun [1]: "A special search at Dubna was carried out by E. Okonov and his group. They have not found a single $K_L^0 \to \pi^+\pi^-$ event among 600 decays of K_L^0 into charged particles [2]. At that stage the search was terminated by administration of the Lab. The group was unlucky". It is hard to believe nowadays that we could have missed the discovery of CP violation in the kaon sector by Fitch and Cronin [3] if we had stopped the search at the 1% level of accuracy.

In this Letter I will discuss the perspectives on SuperB, a next generation asymmetric e^+e^- flavor factory with very high peak luminosity ($\mathcal{L} = 10^{36}$ cm^{-2} s^{-1}) to be built near Rome, whose main purpose is to search for evidence of physics beyond the SM and investigate its nature. The experiment is designed to collect an integrated luminosity exceeding 75 ab^{-1} at the $\Upsilon(4S)$ mass peak and about 500 fb^{-1} at the $\Psi(3770)$ mass peak in few years of running. The polarization of the electron beam will offer the unique possibility of performing precision

[1] on behalf of the SuperB collaboration.

Published under licence by IOP Publishing Ltd

electroweak measurements and will provide also extra handles for background rejection in lepton flavor violating decay searches, such as $\tau^- \to \mu^-\gamma$ [2] and $\tau^- \to e^-\gamma$.

I will also discuss the basic principles of the innovative accelerator scheme and a general overview of the detector.

2. The Physics Program and the Discovery Potential

SuperB is a next generation high luminosity e^+e^- collider that will accumulate a data sample of 75 ab^{-1} with five years of nominal data taking. This experiment has been approved for funding in 2011 and could start running as early as 2016, by which time the LHC will have accumulated a significant sample of data, and would be reporting the results of searches for or direct measurements of new physics (NP). Those results are limited in that they measure only flavour diagonal processes. In order to fully understand the nature of NP, one also has to measure the off-diagonal terms, in analogy to the CKM mixing matrix. In the scenario that LHC does not find NP particles, SuperB has the possibility to explore NP scale beyond the reach of the LHC (up to 10 TeV or more depending on NP models) looking for indirect signals.

In this sense the SuperB physics reach is complementary to the LHC one. In particular there is also a large complementarity with the flavor physics potential of the LHCb experiment. For example, rare decay modes with one or more neutrinos in the final state such as $B^- \to l^-\nu$ and $B^- \to K^{(*)-}\nu\bar{\nu}$, inclusive analyses of processes such as $b \to s\gamma$ and $b \to sl^+l^-$, measurements of the CKM matrix elements $|V_{ub}|$ and $|V_{cb}|$ and searches for lepton flavor violation such as $\tau^- \to \mu^-\gamma$ are unique to SuperB, where the environment of the e^+e^- collider is clean and relatively simple compared to the events at the hadronic machine [7].

The SuperB machine design includes the polarization of the electron beam (about 80% polarization) allowing the unique possibility of performing precision $\sin^2\theta_W$ measurements at $Q^2 = (10.58 \text{ GeV})^2$ with $\mu^+\mu^-$, $\tau^+\tau^-$ and $c\bar{c}$ events, as well as measurements of the neutral current vector coupling of the b. Such measurements are sensitive to a Z' and can probe neutral current universality at high precision. Sensitivity on $\sin^2\theta_W$ will be at the level 1%, if the systematic error on polarization is kept under control, which is competitive with the SLC measurement but at a much lower energy. In addition, the production of polarized τ pairs will allow a better separation between signal and background events for τ decays, thus improving the experimental sensitivity in lepton flavor violating (LFV) processes. In Table 1 are reported the expected experimental sensitivities (90% upper limits) for τ LFV processes at SuperB with 75 ab^{-1} of data along with the present upper limits.

Charm physics also plays an important role in the SuperB physics program. Constraints on flavor-changing neutral currents from NP in the up quark sector are much weaker than in the down quark sector. Thus, high sensitivity studies of rare charm decays offer the possibility of isolating NP effects in $D^0 - \bar{D}^0$ mixing, in CP violation and in rare decay branching fractions. The sensitivity on CP violation asymmetries in decay modes such as $D^0 \to K^+K^-$, $\pi^+\pi^-$ and $D^0 \to K^0_S\pi^+\pi^-$ is expected to reach the level of $\mathcal{O}(10^{-4})$ with 75 ab^{-1}, which will probe the SM predictions. The machine design includes the possibility to run at different center-of-mass energies, e.g. at the $\Psi(3770)$ for exclusive $D\bar{D}$ production, at the $\Upsilon(5S)$ for $B_s\bar{B}_s$ production and up to the $\Upsilon(6S)$, thus further enriching the physics program.

The SuperB physics case is extensively discussed in refs. [4, 5, 6]. In Table 2 are reported, as an example, the golden modes for different NP scenarios. These decays are very difficult or impossible to reconstruct at the LHC. Even at the clean environment of Super$B$ the selection is experimentally challenging and to suppress backgrounds to an acceptable level the recoil technique is often necessary, in which the other B in the $B\bar{B}$ event is reconstructed in either a semileptonic or hadronic decay. Table 3 reports the comparison of the experimental sensitivity

[2] charge conjugation is implied throughout the paper if not otherwise stated.

Table 1. Expected 90% CL upper limits on representative LFV τ lepton decays with 75 ab^{-1} and current upper limits.

Process	Sensitivity at Super$B$	Current limit
$\mathcal{B}(\tau \to \mu\gamma)$	2×10^{-9}	4.5×10^{-8}
$\mathcal{B}(\tau \to e\gamma)$	2×10^{-9}	1.1×10^{-7}
$\mathcal{B}(\tau \to \mu\mu\mu)$	2×10^{-10}	3.2×10^{-8}
$\mathcal{B}(\tau \to eee)$	2×10^{-10}	3.6×10^{-8}
$\mathcal{B}(\tau \to \mu\eta)$	4×10^{-10}	6.5×10^{-8}
$\mathcal{B}(\tau \to e\eta)$	6×10^{-10}	9.2×10^{-8}
$\mathcal{B}(\tau \to eK^0_S)$	2×10^{-10}	3.3×10^{-8}
$\mathcal{B}(\tau \to \mu K^0_S)$	2×10^{-10}	4.0×10^{-8}

Table 2. Golden modes in different New Physics scenarios. An "X" indicates the golden channel of a given scenario. An "O" marks modes which are not the "golden" one of a given scenario but can still display a measurable deviation from the Standard Model. The label CKM denotes golden modes which require the high-precision determination of the CKM parameters achievable at Super$B$.

	H^+ high tanβ	Minimal FV	Non-Minimal FV	NP Z-penguins	Right-Handed currents
$\mathcal{B}(B \to X_s\gamma)$		X	O		O
$A_{CP}(B \to X_s\gamma)$			X		O
$\mathcal{B}(B \to \tau\nu)$	X-CKM				
$\mathcal{B}(B \to X_s l^+ l^-)$			O	O	O
$\mathcal{B}(B \to K\nu\overline{\nu})$			O	X	
$S(K_S\pi^0\gamma)$					X
β			X-CKM		O

today and with 75 ab^{-1}, showing that in most cases Super$B$ is able to measure the observables with a few percent accuracy.

Mode	Sensitivity	
	Current	Expected (75 ab^{-1})
$\mathcal{B}(B \to X_s\gamma)$	7%	3%
$A_{CP}(B \to X_s\gamma)$	0.037	0.004–0.005
$\mathcal{B}(B^+ \to \tau^+\nu)$	30%	3–4%
$\mathcal{B}(B^+ \to \mu^+\nu)$	not measured	5–6%
$\mathcal{B}(B \to X_s l^+ l^-)$	23%	4–6%
$A_{\rm FB}(B \to X_s l^+ l^-)_{s0}$	not measured	4–6%
$\mathcal{B}(B \to K\nu\overline{\nu})$	not measured	16–20%
$S(K^0_S\pi^0\gamma)$	0.24	0.02–0.03

Table 3. Comparison of current experimental sensitivities with those expected at Super$B$ (75 ab^{-1}). Only a small selection of observables is shown. Quoted sensitivities are relative uncertainties if given as a percentage, and absolute uncertainties otherwise. For more details, see Refs. [4, 5, 6, 7].

Journal of Physics: Conference Series **347** (2012) 012007 doi:10.1088/1742-6596/347/1/012007

3. The Accelerator: the Innovative Scheme

SuperB is a second generation flavor factory aiming for a luminosity of 10^{36} cm^{-2}s^{-1}. This represents a step forward of about two order of magnitudes in luminosity with respect to the first generation B factories. The key advances in the design of this accelerator come from recent successes at the DAFNE collider at INFN in Frascati, Italy, at PEP-II at SLAC in California, USA, and at KEKB at KEK in Tsukuba Japan, and from new concepts in beam manipulation at the interaction region (IP) called "crab waist". This new collider comprises of two beam rings, one at 4.2 GeV and one at 6.7 GeV, a common interaction region, a new injection system at full beam energies, and one of the two beams longitudinally polarized at the IP. Most of the new accelerator techniques needed for this collider have been achieved at other recently completed accelerators including the new PETRA-3 light source at DESY in Hamburg (Germany) and the upgraded DAFNE collider at the INFN laboratory at Frascati (Italy), or during design studies of CLIC or the International Linear Collider (ILC).

The very high luminosity is obtained increasing the density of the bunches at the IP by demagnifying their vertical size to about 30 nm. To reach this goal the amplitude of the betatron oscillation must be kept at minimum. This requires optimal ring lattice design, precise magnet alignment and machine tuning, small emittance damping ring (about 5 pm $\times$ Rad) and a "crab waist" collision scheme in order to overcome the beam-beam luminosity limit. The first ingredient of the "crab waist" collision scheme is the large Piwinsky angle ϕ defined as:

$$\phi = \frac{\sigma_z}{\sigma_x} \tan\frac{\theta}{2}, \quad (1)$$

σ_x being the horizontal rms bunch size, σ_z the rms bunch length. It is worth noting that the luminosity of an e^+e^- collider is given by the expression:

$$\mathcal{L} = \frac{N^+N^-}{4\pi\sigma_y\sqrt{\left(\sigma_z \tan\frac{\theta}{2}\right)^2 + \sigma_x^2}} f_c = \frac{N^+N^-}{4\pi\sigma_y\sigma_x\sqrt{\phi^2+1}} f_c. \quad (2)$$

In SuperB, $N^+ = 5.1 \times 10^{10}$ ($N^- = 6.6 \times 10^{10}$) is the number of positrons (electrons) per bunch, $\sigma_y = 36$ nm the vertical rms size and $f_c = 233$ MHz is the collision frequency. In the limit of small collision angle $\theta = 66 \cdot 10^{-3} \ll 1$ and large Piwinsky angle $\phi \gg 1$ can be approximated by:

$$\mathcal{L} = \frac{N^+N^-}{2\pi\sigma_y\sigma_z\theta} f_c \simeq 1 \cdot 10^{36}\mathrm{cm}^{-2}\mathrm{s}^{-1}. \quad (3)$$

The large Piwinsky angle is obtained by decreasing the horizontal beam size and increasing the crossing angle. Several advantages follow: higher luminosity, reduction of the vertical tune shift and suppression of vertical synchro-betatron resonances. The "crab waist" scheme is realized by a couple of sextupole magnets on the two sides of the IP. The waist of the beam moves to the axis of the other beam. As a consequence, each beam collides with the other in the minimum waist region, with a net luminosity gain. Besides the geometrical gain just mentioned, because of the "crab waist" transformation the non linear component of the beam-beam forces decreases, hence reducing the emittance growth due to the collision.

The present machine design is based on a circumference of about 1.2 Km with beam currents of about 2 A and a relatively low power consumption of 16 MW. The estimated construction time for this collider is a total of about four years and will be built in Rome close to the "Laboratori Nazionali di Frascati" of INFN. The name of the new laboratory will be "Cabibbo Laboratory" in honor of Nicola Cabibbo.

Journal of Physics: Conference Series **347** (2012) 012007 doi:10.1088/1742-6596/347/1/012007

4. The Detector: a General Overview

The SuperB detector is based on the BaBar "prototype" and reuses several parts of it: fused silica bars and bar boxes of the Cherenkov detector (DIRC), the CsI(Tl) crystals for the barrel part of the electromagnetic calorimeter (EMC), and the superconducting coil and flux return yoke of the axial magnetic field. However, the detector requires moderate design improvement and R&D to cope with new machine interaction region, the higher luminosity and the high DAQ rates. A brief description of the detector novelties is reported below. A more detailed description of the SuperB detector can be found elsewhere [4, 9].

The drift chamber will be using He as ionizing medium but the gas mixture and the cell size will be optimized for operating at higher rates. The DIRC will replace the photon camera with a twelve-fold modular camera using multi-channel plate photon detectors in a focusing configuration using fused silica optics to reduce the impact of beam related backgrounds and improve performance. The forward EMC will feature cerium-doped LYSO (lutetium yttrium orthosilicate) crystals, which have a much shorter scintillation time constant, a lower Moliere radius and better radiation hardness than the current CsI(Tl) crystals, again for reduced sensitivity to beam backgrounds and better position resolution. The hermeticity of the SuperB detector, and, thus, its performance for certain physics channels will be improved by including a backwards "veto-quality" EMC detector comprising a lead-scintillator stack. The flux-return will be augmented with additional absorber to increase the number of interaction lengths for muons to roughly 7λ.

The smaller center-of-mass boost ($\beta\gamma = 0.24$) with respect to BaBar ($\beta\gamma = 0.55$) imposes improvements in decay vertex reconstruction in order to maintain adequate proper time resolution for time-dependent CP violation measurements. The vertex resolution will be improved by reducing the radius of the beam pipe, placing the innermost layer of the silicon vertex tracker (SVT) at a radius of roughly 1.5 cm, with a material budget less that 1% of X_0 and intrinsic resolution of about 10 - 15 μm.

According to the baseline design this innermost layer of the SVT will be constructed of silicon striplets. Other solutions based on Monolithic Active Pixel Sensors (MAPS) or other pixelated sensors are currently considered as options for detector upgrades, representing more robust solutions against possible high occupancy from beam-related backgrounds.

5. Conclusions

The case for Flavor physics in the LHC era is still compelling. It offers a unique opportunity for searching for NP and understanding its origin. The most sensitive measurements to NP are reported in Tables 1, 3 and are currently statistics dominated.

SuperB, a new generation super flavor factory, will accumulate a data sample of 75 ab^{-1} with five years of nominal data taking. The new facility has been recently approved for funding and is going to be built in Rome. The SuperB detector will based on the BaBar one and it will reuse part of it for the DIRC, the EMC and the superconducting coil. However the detector will require some improvements and R&D to cope with the new machine design and the higher luminosity.

The physics program is focused on the study of the flavor structure of NP by measuring the flavor couplings and also by exploring the NP scale beyond the LHC reach (up to 10 TeV or more), by looking for indirect signals. The golden modes for different NP scenarios, reported in Table 2, are difficult or even impossible to reconstruct at LHC. In this sense, the SuperB experiment can be considered as an alternative path for discovering NP with respect to the LHC.

References

[1] L. B. Okun, "Space-time and vacuum as seen from Moscow," Int. J. Mod. Phys. A **17S1**, 105 (2002) [arXiv:hep-ph/0112031].
[2] M. Anikina *et al.*, "Experimental study of some consequences of CP invariance in the decays of K_2^0-mesons," Sov. Phys. JTEP **15** (1962) 93; ZhETF **42** (1962) 130.
[3] J. H. Christenson, J. W. Cronin, V. L. Fitch and R. Turlay, "Evidence for the 2π Decay of the K_2^0 Meson," Phys. Rev. Lett. **13**, 138 (1964).
[4] M. Bona *et al.*, "SuperB: A High-Luminosity Asymmetric e^+e^- Super Flavor Factory. Conceptual Design Report," arXiv:0709.0451 [hep-ex].
[5] D. G. Hitlin *et al.*, "Proceedings of SuperB Workshop 6: New Physics at the Super Flavor Factory," arXiv:0810.1312 [hep-ph].
[6] B. O'Leary *et al.* [SuperB Collaboration], "SuperB Progress Reports – Physics," arXiv:1008.1541 [hep-ex].
[7] T. E. Browder, T. Gershon, D. Pirjol, A. Soni and J. Zupan, "New Physics at a Super Flavor Factory," arXiv:0802.3201 [hep-ph].
[8] M. E. Biagini *et al.* [SuperB Collaboration], "SuperB Progress Reports – The Collider," arXiv:1009.6178 [physics.acc-ph].
[9] D. Aston *et al.* [SuperB Collaboration], "SuperB Progress Reports – The Detector," arXiv:1007.4241 [physics.ins-det].

Journal of Physics: Conference Series **347** (2012) 012008 doi:10.1088/1742-6596/347/1/012008

The status and road map of Turkish Accelerator Center (TAC)

Ö Yavaş[1]
Ankara University, 06100 Ankara, Turkey

E-mail: yavas@ankara.edu.tr

Abstract. Turkish Accelerator Center (TAC) project is supported by the State Planning Organization (SPO) of Turkey and coordinatedby Ankara University. After having completed theFeasibility Report (FR) in 2000 and the Conceptual Design Report (CDR) in 2005, third phase of the project started in 2006 as an inter-universities project including ten Turkish Universities with the support of SPO. Third phase of the project has two main scientific goals: to prepare the Technical Design Report (TDR) of TAC and to establish an Infrared Free Electron Laser (IR FEL) facility, named as Turkish Accelerator and Radiation Laboratoy at Ankara (TARLA) as a first step. The facility is planned to be completed in 2015 and will be based on 15-40 MeV superconducting linac. In this paper, main aims, national and regional importance, main parts main parameters, status and road map of Turkish Accelerator Center will be presented.

1. TAC Collaboration and TAC Project

Accelerator physics research in Turkey formally started with the establishment of Ankara University Accelerator Physics Research Group in 1994 and the first paper is published on linac-ring type ep and γp colliders. Afterwards, a project to do the feasibility study on a national center was proposed to SPO in 1997 by Ankara University [1]. After completion of the feasibility report of TAC in 2001, the second phase of TAC project namely a General Design Report (GDR) was completed between 2002-2005. In the third phase of TAC Project, covering the 2006-2013 period, ten Turkish Universities (Ankara, Boğaziçi, Doğuş, Dumlupınar, Erciyes, Gazi, İstanbul, Niğde, S. Demirel and Uludağ Universities) collaborate under the coordination of Ankara University.

The TAC project has international collaborations with CERN, DESY, The Cockcroft Institute, BESSY and HZDR. A National Congress on Particle Accelerators and Their Applications (UPHUK) has been organized since 2001 triennially by the Turkish Physical Society together with the **Turkish Atomic Energy Authority (TAEK)** and Turkish Universities. In addition, National Summer School on Particle Accelerators and Detectors (UPHDYO) has been organized since 2005 annually.

TheInstitute of Accelerator Technologies at Ankara University [2] which will house TARLA as well, was established on February 26, 2010 at Gölbaşı campus of Ankara University (Figure 1). Three major fields of studies, namely, Particle Accelerators and Technologies, Accelerator Based Light Sources and Detector and Data Acquisition Technologies, are considered for the graduate studies at IAT.

[1] On behalf of TAC Collaboration (http://thm.ankara.edu.tr)

Figure 1. Buildings of IAT and TARLA Facility

The International Scientific Advisory Committee (ISAC) of TAC consists of 13 members from Turkey, USA, Italy, Canada, England, Switzerland, Germany and Japan.On the other hand, the International Machine Advisory Committee (IMAC) established for the TARLA facility includes 5 international members from Germany, Japan and USA. Furthermore, National Advisory Council of the TAC project consists of TAC directors, Rectors of collaborating Turkish Universities, Turkish Atomic Energy Authority (TAEK), The Scientific and Technological Research Council of Turkey (TÜBİTAK), Turkish Academy of Sciences (TÜBA), Turkish Union of Chambers of Commodity Exchanges (TOBB), Turkish Higher Education Council (YÖK), Ministry of Science, Industry and Technology, Ministry of Energy and Natural Resources, Ministry of Foreign Affairs and Secreteriat General for European Union Affairs.

2. The First Facility and proposed facilities of TAC

The TAC project has 4 main proposals [1], linac-ring type e^-e^+ collider (Super Charm Factory), synchrotron radiation facility, self amplified spontanous emission (SASE) free electron laser (FEL) facility and a proton accelerator facility. Before these proposals, an IR-FEL oscillator (TARLA) & Bremsstrahlung facility [3] was proposed for Turkish scientists and researchers to become acquainted with accelerator technologies and accelerator based light sources. The TARLA facility aims to produce free electron laser in oscillator mode between 2-250 microns range using 15-40 MeV electron beam. In order to have a wide research area, it is requested to have continuous wave electron beam with high average current, as well as pulsed beam with low current. Therefore to obtain FEL in 2-250 microns range using electron beam in the available energy range, one needs undulators with 2.5 cm and 9 cm period length inside independent optical resonator systems. Main parameters of electron beam and FEL of the TARLA are summarized in Tables 1 and 2 respectively. TARLA is also planned to be operated for Bremsstrahlung experiments to study nuclear spectroscopy.

Table 1. Electron beam parameters of the TARLA facility

Parameter	Unit	Value
Beam energy	MeV	15-38
Max. average beam current	mA	1.6
Bunch repetition rate	MHz	26-13
Bunch length	ps	1-10
Norm. rms transverse emittance	mm.mrad	< 15
Norm. rms longitudinal emittance	keV.ps	< 100
Micropulse repetition	Hz	1 to cw
Macropulse length	μs	10 to cw

Table 2. FEL parameters of the TARLA facility

Parameter	Unit	U25	U90
Wavelength	μm	2-30	15-250
Micropulse repetition	MHz	13	13
Max. peak power	MW	6	3
Average power	W	1-100	1-100
Max. pulse energy	μJ	3	3
Peak brightness	ph/s.mm^2.mrad^2.0.1%bw	~ 10^{30}	~ 10^{29}
Pulse length	ps	1-10	1-10

The main four proposed facilities of TAC are summarized as a list below.

- **Synchrotron Radiation Facility:** Energy of synchrotron is planned to be about 3 GeV with a beam current of 500 mA, and circumference of the ring will be about 546 m.

- **Free Electron Laser Facility:** The SASE FEL proposal was originally based on the collider, but after several feasibilty studies and ISAC recommendations, it is modified to be a stand-alone facility based on 1 GeV electron accelerator. The facility aims to scan soft X-ray range (down to a few nanometers) of the spectrum.

- **Particle Factory:** TAC project fundamentally consists of a particle factory. Thisparticle factory was first proposed as a Charm-Tau factory based on alinac-ring type e^-e^+ collider, but in recent years, it is modified to Charm factory based on an ERL-ring type collider as a super factory (Figure 2) to raise the luminosity up to ~ 10^{35} cm^{-2}s^{-1}.

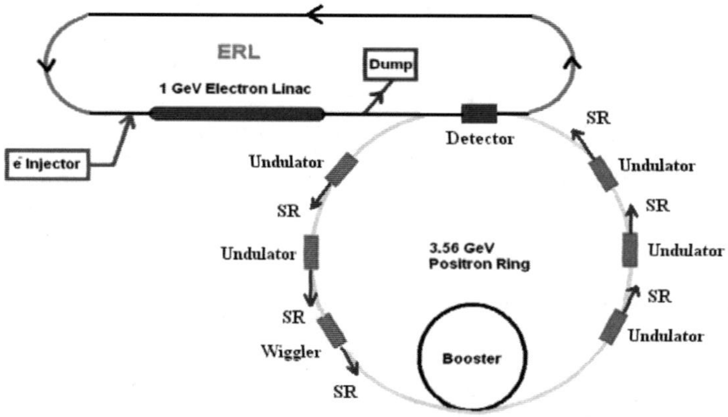

Figure 2. Schematic view of TAC ERL-Ring type Super Charm Factory

- **Proton Accelerator Facility:** A 1-3 GeV proton accelerator with a beam power of 1 MW is planned. In addition, a 3 MeV test stand and a 55 MeV DTL will be included as low energy part of the chain.

Journal of Physics: Conference Series **347** (2012) 012008 doi:10.1088/1742-6596/347/1/012008

3. Conclusion

The first beam in TARLA facility is expected at the end of 2013. Different schedules are forseen for different stages of the proposed TAC facilities up to the middle of 2020s. Meanwhile, IAT will provide permanent positions for qualified engineers & technicians and will have postdoc positions for scientists as well.

4. References

[1] http://thm.ankara.edu.tr
[2] http://hte.ankara.edu.tr
[3] http://www.tarla-fel.org

OPERA experimental results[1]

Antonia Di Crescenzo
Dipartimento di Fisica dell'Università Federico II di Napoli, 80126 Napoli, Italy
INFN Sezione di Napoli, 80126 Napoli, Italy
E-mail: antonia.dicrescenzo@infn.it

Abstract. The aim of the OPERA experiment is to search for the appearance of the tau neutrino in the quasi pure muon neutrino beam produced at CERN (CNGS). The detector, installed in the Gran Sasso underground laboratory 730 km away from CERN, consists of a lead/emulsion target complemented with electronic detectors. Runs with CNGS neutrinos were successfully carried out in 2008, 2009, and 2010. After a brief description of the beam and the experimental setup, we report on event analysis of a sample of events corresponding to 1.89×10^{19} p.o.t. in the CERN CNGS ν_μ beam that yielded the observation of a first candidate ν_τ CC interaction. The topology and kinematics of this candidate event are described in detail. The background sources are explained and the significance of the candidate is assessed.

1. Introduction

In the last decades solar and atmospheric neutrino experiments observed deficits in the measured fluxes which are all well reproduced in a neutrino oscillations model, implying non vanishing, not degenerate neutrino masses and neutrino mixing. Within such hypothesis weak interactions eigenstates differ from the mass eigenstates. The mixing can be parametrized in an unitary matrix whose parameters (3 angles and 1 or 3 phases depending on the Dirac or Majorana nature of neutrinos) associated to the square masses differences Δm^2 drive the amplitude of the disappearance ($P(\nu_\alpha \rightarrow \nu_\beta)$) or survival ($P(\nu_\alpha \rightarrow \nu_\alpha)$) probabilities.

The major experimental results for solar [1, 7], atmospheric [4, 9], reactor [10, 11] or accelerator [12, 13] neutrinos were obtained by observing the disappearance of neutrinos w.r.t. to a close position measurement or a predicted flux.

OPERA [14] has been designed to perform a unique appearance observation of the oscillation products to confirm (or infirm) the neutrino oscillation hypothesis in the atmospheric sector through the $\nu_\mu \rightarrow \nu_\tau$ channel and also to set limits on the θ_{13} angle through the $\nu_\mu \rightarrow \nu_e$ channel.

2. The OPERA experiment

2.1. The CNGS beam

The CNGS [15] programme of neutrino beam from CERN to Gran Sasso has been approved in 1999. The beam has been optimized to maximize the number of τ events in the detector (convolution of the neutrino flux, the appearance probability and the detection efficiency). The neutrino average energy is 17 GeV. Measured in the number of interactions in the detector, the $\overline{\nu_\mu}$ contamination is $\sim 2\%$, the $\nu_e(\overline{\nu_e})$ is $<1\%$ and the number of prompt ν_τ is negligible.

[1] On behalf of the OPERA Collaboration

Journal of Physics: Conference Series **347** (2012) 012009 doi:10.1088/1742-6596/347/1/012009

The beam configuration starts from the SPS protons directed to the target chamber where mainly pions and kaons are produced, which decay then into a 1 km long decay tunnel followed by hadron stops and muon detectors. From CERN to Gran Sasso the escaping neutrinos travel for 2.44 ms and their mean direction w.r.t. the horizontal in Gran Sasso forms a 3° angle due to the earth curvature. There are two fast extractions separated by 50 ms in each CNGS cycle (6 s long). These cycles are repeated 3 or 4 times during each SPS supercycle.

The goal is to accumulate a statistics of neutrino interactions corresponding to 22.5×10^{19} p.o.t. in 5 years. The 2008, 2009 and 2010 runs achieved a total intensity of 1.78×10^{19}, 3.52×10^{19} and 4.04×10^{19} p.o.t. respectively. Within these three years, neutrinos produced 9637 beam events. The processing of these events, particularly the scanning of emulsion films, is continuously going on. The 2011 run started on May 2011 and is still in progress.

2.2. The detector tecnique

The challenge of the experiment is to measure the appearance of ν_τ from ν_μ oscillations through CC τ interactions. The events induced by the shortlived τ have a characteristic topology (with a kink due to the presence of undetected neutrinos in the τ decay) but extends over $\sim$ mm^3 typical volumes. The detector should therefore match a large mass for statistics, a high spatial resolution and high rejection power to limit background contamination. These requirements are satisfied using the proven ECC (Emulsion Cloud Chamber) technique which already worked successfully in the DONUT experiment [16].

The passive target consists of lead plates. Particles are tracked in nuclear emulsions films with a sub-micrometric intrinsic resolution. 57 emulsions films are assembled and interspaced with 56 lead plates 1 mm wide in a detector basic cell called *brick*. An additional doublet of emulsion film (Changeable Sheets, CS) is attached on the downstream face of each brick to guide the tracks predictions inside the brick itself. A brick is a 12.7×10.2 cm^2 object with a thickness along the beam direction of 7.5 cm (about 10 radiation lengths). Bricks are assembled in 31 walls (52×64 bricks) separated by electronic detectors planes to trigger the event and identify the brick with the interaction vertex.

The total OPERA target contains 150000 bricks with a total mass of 1.25 ktons. The ECC bricks were assembled underground at an average rate of 700 per day by a dedicated fully automated Brick Assembly Machine (BAM); the OPERA target was filled using two automated manipulator systems (BMS). The passage of a m.i.p. in an emulsion film results, after development, in a set of aligned grains, 35 grains/100 μm.

2.3. The OPERA electronic detectors

OPERA is a hybrid detector made of two identical Super Modules (SM1 and SM2) consisting of a target section followed by a muon spectrometer (see Figure 1). The target section is made of the already mentionned 31 brick walls each one being followed by a highly segmented scintillator tracker plane. A large VETO plane is placed in front of the detector to further discriminate beam events from horizontal cosmics and beam neutrino interactions in the rock. The construction of the experiment started in Spring 2003.

The target tracker covers a total area of 7000 m^2 and is built of 32000 scintillator strips, each 7 m long and of 25 mm$\times$15 mm cross section. The muon spectrometer consists of a large 8×8 m^2 dipolar magnet delivering a magnetic field of 1.55 T and instrumented with RPCs and drift tubes. Each magnet arm consists of twelve 5 cm thick iron slabs, alternating with RPC planes. This sandwich structure allows the tracking in the magnetic field to identify the muons and to determine their momentum and sign. In addition the precision tracker measures the muon track coordinates in the horizontal plane. It is made of 8 m long drift tubes with an outer diameter of 38 mm. The charge misidentification is expected to be 0.1 % - 0.3 % in the relevant momentum range which is efficient enough to minimize the background originating from

Journal of Physics: Conference Series **347** (2012) 012009 doi:10.1088/1742-6596/347/1/012009

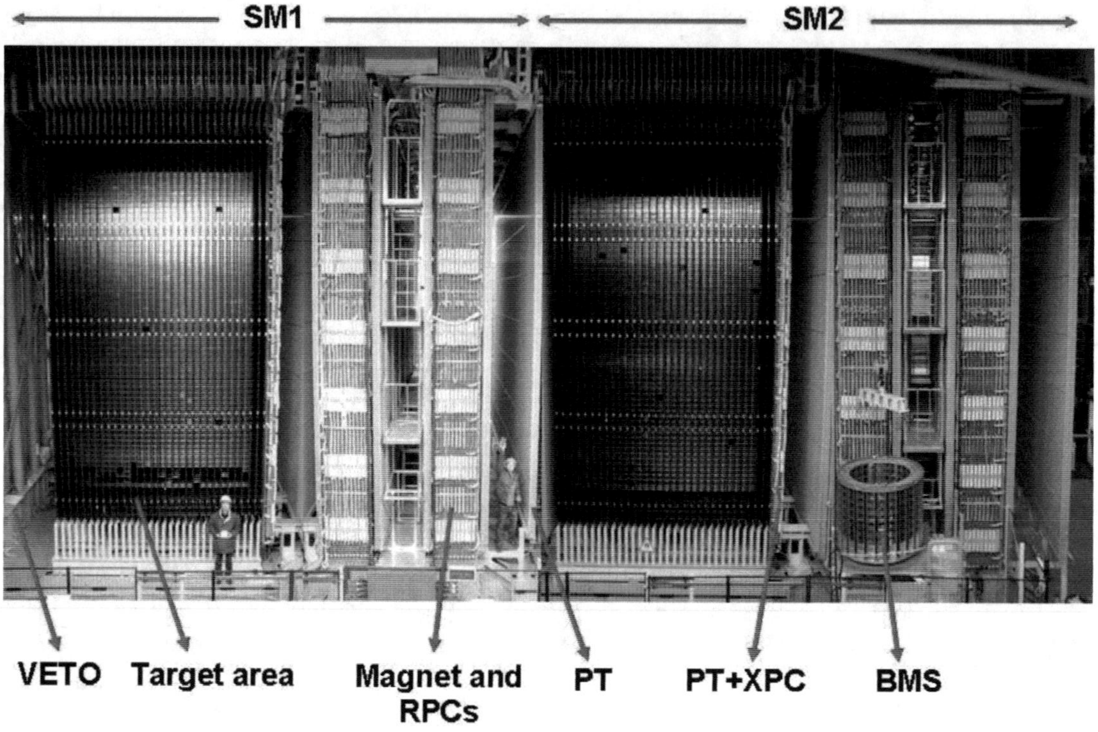

Figure 1. View of the OPERA detector; the neutrino beam enters from the left. Arrows show the position of detector components, the VETO planes, the target and TT, the drift tubes (PT) laid out along the XPC, the magnets and the RPC installed between the magnet iron slabs. The Brick Manipulator System (BMS) is partly shown.

the charmed particles produced in ν_μ interactions. With the muon spectrometer a momentum resolution of $\Delta p/p < 0.25$ for all muon momenta p up to a maximum of $p = 25$ GeV/c can be achieved.

3. Neutrino interaction location

Neutrino event analysis starts with the pattern recognition in the electronic detectors. Charged particle tracks produced in a neutrino interaction generate signals in the TT and in the muon spectrometer. A brick finding algorithm is applied in order to select the brick which has the maximum probability to contain the neutrino interaction. The brick with the highest probability is extracted from the detector for analysis. The efficiency of this procedure reaches 83% in a subsample where up to 4 bricks per event were processed.

After extraction of the brick predicted by the electronic detectors, its validation comes from the analysis of the CS films. The measurement of emulsion films is performed through high-speed automated microscopes [17, 18] with a sub-micrometric position resolution and angular resolution of the order of one milliradian. If no expected charged track related to the event is found in the CS, the brick is returned back to the detector with another CS doublet attached. If any track originating from the interaction is detected in the CS, the brick is exposed to cosmic rays (for alignment purposes) and then depacked. The emulsion films are developed and sent to the scanning laboratories of the Collaboration for event location studies and decay search analysis.

All the track information of the CS is then used for a precise prediction of the tracks in the most downstream films of the brick (with an accuracy of about 100μm). When found in this films, tracks are followed upstream from film to film. The scan-back procedure is stopped when no track candidate is found in three consecutive films and the lead plate just upstream the last detected track segment is defined as the vertex plate. In order to study the located vertices and reconstruct the events, a general scanning volume is defined with a transverse area of $1\times 1\text{cm}^2$ for 5 films upstream and 10 films downstream of the stopping point. All track segments in this volume are collected and analysed. After rejection of the passing through tracks related to cosmic rays and of the tracks due to low energy particles, the tracks produced by the neutrino interaction can be selected and reconstructed.

The present overall location efficiency averaged over NC and CC events, from the electronic detector predictions down to the vertex confirmation, is about 60%.

4. Decay search

Once the neutrino interaction is located, a decay search procedure is applied to detect possible decay or interaction topologies on tracks attached to the primary vertex. The main signature of a secondary vertex (decay or nuclear interaction) is the observation of a track with a significant impact parameter (IP) relative to the neutrino interaction vertex. The IP of primary tracks is smaller than 10 μm after excluding tracks produced by low momentum particles. When secondary vertices are found in the event, a kinematical analysis is performed, using particle angles and momenta measured in the emulsion films. For charged particles up to about 6 GeV/c, momenta can be determined using the angular deviations produced by Multiple Coulomb Scattering (MCS) of tracks in the lead plates [19] with a resolution better than 22%. For higher momentum particles, the measurement is based on the position deviations. The resolution is better than 33% on $1/p$ up to 12 GeV/c for particles passing through an entire brick.

A γ-ray search is performed in the whole scanned volume by checking all tracks having an IP with respect to the primary or secondary vertices lower than 800 μm. The angular acceptance is $\pm$500 mrad. The γ-ray energy is estimated by a Neural Network algorithm that uses the number of segments, the shape of the electromagnetic shower and also the MCS of the leading tracks.

5. Data analyisis

In the following, the analysis results [20] of about 35% of the 2008 and 2009 data sample, corresponding to the 1.89×10^{19} p.o.t are presented. The decay search procedure was applied to a sample of 1088 events of which 901 were classified as CC interactions. In the sample of CC interactions, 20 charm decay candidates were observed, in good agreement with the expectations from the Monte Carlo simulation, 16$\pm$ 2.9. Out of them 3 have a 1-prong topology where 0.8$\pm$0.2 was expected. The background for the total charm sample is about 2 events. Several ν_e-induced events have also been observed.

Moreover, a first CC ν_τ candidate has been detected. The expected number of ν_τ events detected in the analysed sample is about $0.54\pm 0.13(syst.)$ at $\Delta m^2 = 2.5\times 10^{-3}eV^2$ and full mixing.

6. The first tau neutrino candidate

In this section, the first tau neutrino candidate [20] will be described. The location and decay search procedure yielded a neutrino interaction vertex with 7 tracks. One track exhibits a visible kink with an angular change of 41 $\pm$ 2 mrad after a path length of 1335 $\pm$ 35 μm. The kink daughter momentum is estimated to be 12^{+6}_{-3} GeV/c by MCS measurement and its transverse momentum to the parent direction is 470^{+230}_{-120} MeV/c. The event is displayed in Figures 2 and 3.

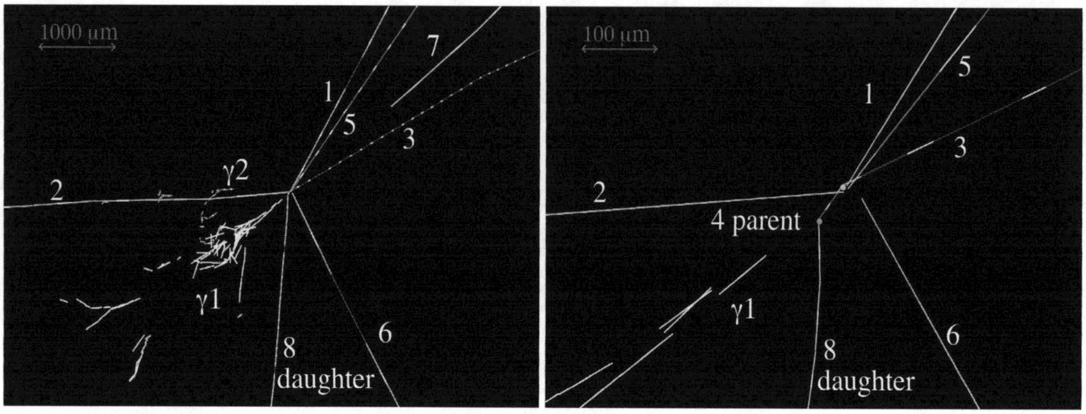

Figure 2. Display of the ν_τ candidate event. Left: view transverse to the neutrino direction. Right: same view zoomed on the vertices.The short track named *4 parent* is the τ^- candidate.

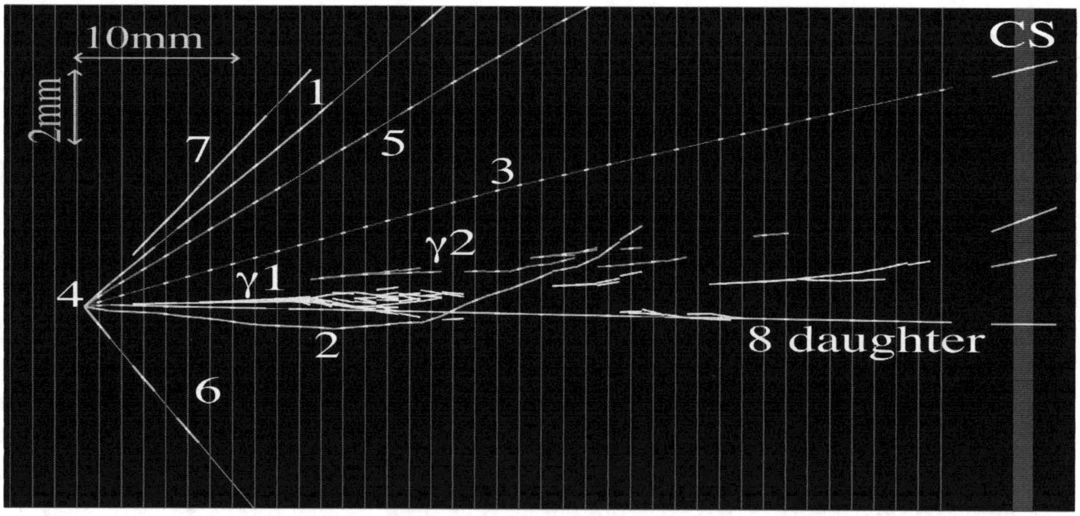

Figure 3. Longitudinal view of the ν_τ candidate.

All the tracks from the neutrino interaction vertex were followed until they stop or interact. The probability that one of them is left by a muon is estimated to be less than 10^{-3}. The residual probability for being a ν_μ CC event, with a possibly undetected large angle μ track, is about 1%; a nominal value of 5% is assumed. None of the tracks is compatible with being an electron.

Two electromagnetic showers caused by γ-rays, associated with the event, have been located and studied. The energy of γ_1 is $5.6\pm1.0(stat.) \pm 1.7(syst.)$ GeV and it is clearly pointing to the decay vertex. The γ_2 has an energy of $1.2\pm0.4(stat.) \pm 0.4(syst.)$ GeV and it is compatible with pointing to either vertex, with a significantly larger probability to the decay vertex.

All the selection cuts used in the analysis were those described in detail in the experiment proposal [21] and its addendum [22]. All the kinematical variables of the event and the cut applied are given in Table 1.

Table 1. Kinematical variables of ν_τ candidate event.

Variable	Measured	Selection criteria
Kink angle ($mrad$)	42 ± 2	> 20
Decay length (μm)	1335 ± 35	Within 2 plates
P daughter (GeV/c)	12^{+6}_{-3}	> 2
P_t daughter (MeV/c)	470^{+230}_{-120}	> 300 (γ attached)
Missing P_t (MeV/c)	12^{+320}_{-170}	< 1000
Angle ϕ (deg)	1335 ± 35	> 90

The invariant mass of the two observed γ-rays is $120\pm20(stat.) \pm 35(syst.)$ supporting the hypothesis that they are emitted in a π^0 decay. The invariant mass of the charged decay daughter assumed to be a π^- and of the two γ-rays amount to $640^{+125}_{-80}(stat.)^{+100}_{-90}(syst.)$ MeV/c, which is compatible with the $\rho(770)$ mass. So the decay mode of the candidate is consistent with the hypothesis $\tau \to \rho^- \nu_\tau$ (where the branching ratio is about 25%).

7. Background estimation

The two main sources of background to $\tau \to h^-(n\pi^0)\nu_\tau$ the channel where a similar final state may be produced are:

- the decays of charmed particles produced in ν_μ CC interactions where the primary muon is not identified as well as the $c\bar{c}$ pair production in ν_μ NC interactions where one charm particle is not identified and the other decays to a 1-prong hadron channel;
- the 1-prong inelastic interactions of primary hadrons produced in ν_μ CC interactions where the primary muon is not identified or in ν_μ NC interactions and in which no nuclear fragment can be associated with the secondary interaction.

The Monte Carlo expectation of the first background source is $0.007\pm0.004(syst.)$ event, the fraction produced in ν_e CC interactions is less than 10^{-3}events. The second type of background amounts to $0.011\pm0.006(syst.)$ event. The total background in the decay channel to a single charged hadron is $0.018\pm0.007(syst.)$ events. The probability that this background events fluctuate to one event is 1.8% (2.36 σ). As the search for τ^- decays is extended to all four channels, the total background then becomes $0.045\pm0.023(syst)$.

The probability that this expected background to all searched decay channels of the τ^- fluctuates to one event is 4.5% (2.01σ). At $\Delta m^2 = 2.5 \times 10^{-3} eV^2$ and full mixing, the expected number of observed τ^- events with the present analyzed statistics is $0.54 \pm 0.13(syst.)$ of which $0.16\pm0.04(syst.)$ in the one-prong hadron topology, compatible with the observation of one event.

8. Conclusions

During 2008, 2009 and 2010 runs, a total intensity of 1.78×10^{19}, 3.52×10^{19} and 4.04×10^{19} p.o.t. respectively, was achieved. Within these three years, 9637 beam events have been collected within the OPERA target. The neutrino interaction location and decay search are going on.

A first candidate ν_τ CC interaction in the OPERA detector at LNGS was detected after analysis of a sample of events corresponding to 1.89×10^{19} p.o.t. in the CERN CNGS ν_μ beam. The expected number of ν_τ events in the analysed sample is $0.54\pm0.13(syst.)$.

The candidate event passes all selection criteria, it is assumed to be a τ^- lepton decaying into $h^-(n\pi^0)\nu_\tau$. The observation of one possible tau candidate in the decay channel $h^-(n\pi^0)\nu_\tau$ has a significance of 2.36σ of not being a background fluctuation.

References

[1] R.Davis, Prog. Part. Nucl. Phys. **32**, 13(1994).

[2] T. Kirsten *et al.*, [GALLEX Collaboration], *Neutrino 98*, Nucl. Phys. **B** (Proc. Suppl.).

[3] V. Gavrin *et al.*, [SAGE Collaboration], *Neutrino 98*, Nucl. Phys. **B** (Proc. Suppl.).

[4] Y. Fukuda *et al.*, [Super-Kamiokande Collaboration], Phys. Rev. Lett., **81** 1562 (1998);
J. Hosaka *et al.*, Phys. Rev. D **74** 032002 (2006).

[5] K. S. Hirata *et al.*, [KAMIOKANDE-II Collaboration], Phys. Lett. B **205**, 416 (1998).

[6] R. Becker-Szendy *et al.*, Nucl. Phys. **B38**, 331 (1995).

[7] Q.R. Ahmad *et al.*, [SNO Collaboration], Phys. Rev. Lett. 87 (2001) 071301;
Q.R. Ahmad *et al.*, [SNO Collaboration], Phys. Rev. Lett. 89 (2002) 0113011&011302.

[8] W. W. M. Allison *et al.*, [SOUDAN2 Collaboration], Phys. Lett. B **449**, 137 (1999);
W. W. M. Allison *et al.*, Phys. Rev. D **72**, 052005 (2005).

[9] S. P. Ahlen *et al.*, [MACRO Collaboration], Phys. Lett. B **357**, 481 (1995);
M. Ambrosio *et al.*, Eur. Phys. J. C **36**, 323 (2004).

[10] M. Apollonio *et al.*, [CHOOZ Collaboration], Eur. Phys. J. C **27**, 331 (2003).

[11] T. Araki *et al.*, [KamLAND Collaboration], Phys. Rev. Lett. **94** (2003) 081801.

[12] M. H. Ahn *et al.*, [K2K Collaboration], Phys. Rev. D **74**, 072003 (2006).

[13] D. G. Michael *et al.*, [MINOS Collaboration], arXiv:hep-ex/0607088.

[14] M. Guler *et al.*, [OPERA Collaboration], CERN-SPSC-2000-028;
Y. Declais *et al.*, [OPERA Collaboration], CERN-SPSC-2002-029 SPSC-059.

[15] CNGS project: **http://proj-cngs.web.cern.ch/proj-cngs/**.

[16] K. Kodama *et al.*, [DONUT Collaboration], Phys. Lett. B **504**, 218 (2001).

[17] N. Armenise *et al.*, Nucl. Instrum. Methods A, **551** (2005) 261;
M. De Serio *et al.*, Nucl. Instrum. Methods A, **554** (2005) 247;
L. Arrabito *et al.*, Nucl. Instrum. Methods A, **568** (2006) 578.

[18] K. Morishima, T. Nakano, JINST, **5** (2010) P04011.

[19] M. De Serio *et al.*, Nucl. Instrum. Methods A, **512** (2003) 539.

[20] N. Agafonova *et al.* Phys. Lett. B **691**, 138 (2010);

[21] M. Guler *et al.*, [OPERA Collaboration], CERN-SPSC-2000-028; LNGS P25/2000.

[22] M. Guler *et al.*, [OPERA Collaboration], CERN-SPSC-2001-025; LNGS-EXP 30/2001 add. 1/01.

Early Higgs search with the ATLAS data

Taiki Yamamura[1]

International Center for Elementary Particle Physics (ICEPP), the University of Tokyo, 7-3-1 Hongo Bunkyo-ku, Tokyo 113-0033, Japan

E-mail: `Taiki.Yamamura@cern.ch`

Abstract. A search for the Higgs boson at the Large Hadron Collider (LHC) running at a center-of-mass energy of 7 TeV is reported, based on an early data collected by the ATLAS detector. For the Standard Model Higgs boson search, several decay channels are used : $H \to \gamma\gamma$, $H \to WW^{(*)} \to l\nu l\nu$, $H \to WW \to l\nu qq$, $H \to ZZ^{(*)} \to llll$, $H \to ZZ \to ll\nu\nu$, $H \to ZZ \to llqq$ (l is e, μ). After obtaining the search results for the individual channels, the combined exclusion limits on the Higgs boson production are determined in a mass range from 110 GeV to 600 GeV. A search for the Higgs boson generated by the extended models of the Standard Model is also performed. The exclusion limits for the MSSM $A/H/h \to \tau\tau$ and the NMSSM $a_1 \to \mu\mu$ are presented.

1. Introduction

Discovering the mechanism responsible for the electroweak symmetry breaking and the origin of mass for elementary particles is one of the key aims of the Large Hadron Collider (LHC) at CERN [1]. In the Standard Model (SM) this mechanism requires the existence of a scalar particle, the Higgs boson [2, 3, 4, 5, 6]. Prior to the LHC, the best direct information is a lower limit of 114.4 GeV, set using the combined results of the four LEP experiments [7], and an excluded band of 158 GeV to 175 GeV from the combined Tevatron experiments [8, 9]. There are also extended models of the Standard Model providing the mechanism of giving masses to all the particles. In the Minimal SuperSymmetric Standard Model (MSSM) [10, 11], two Higgs doublets of opposite hypercharge are required, resulting in five observable Higgs bosons, where three of the bosons are neutral (h, H and A) while two are charged ($H^{\pm}$). In the Next-to MSSM (NMSSM) [12], where the Higgs sector is expanded to generate a total of three CP-even scalars (h_1, h_2, h_3), two CP-odd scalars (a_1, a_2) and two charged scalars ($H^{\pm}$), the existence of the light CP-odd Higgs boson a_1 is predicted.

In this note, a search for the Higgs boson at the Large Hadron Collider (LHC) is reported, based on early data at the ATLAS experiment. The search for the Standard Model Higgs boson is described in Section 3. In this section, the results for individual Higgs boson decay channels $H \to \gamma\gamma$, $H \to WW^{(*)}$, $H \to ZZ^{(*)}$ are outlined, and the combined result in a mass range from 110 GeV to 600 GeV is also presented. A search for the Higgs boson in the extended model is also performed : MSSM $H \to \tau\tau$ and NMSSM $a_1 \to \mu\mu$, and their results are presented in Section 4. The conclusions are given in Section 5, where the prospect for the SM Higgs boson search is also discussed.

[1] On behalf of the ATLAS Collaboration

2. Limit extraction procedure

In the statistical interpretation, the profile-likelihood ratio is used as test-statistic. Exclusion limits are obtained using the power-constrained profile likelihood method which is known as the Power Constrained Limit, PCL [13, 14, 15]. This method is preferred to the more familiar CL_s technique [16] because the constraint is more transparently defined and it has reduced overcoverage resulting in a more precise meaning of the quoted confidence level. In some analysis channels, the limits from the CL_s method are also presented for the comparison with the results from other experiments using only CL_s.

3. Search for the Standard Model Higgs boson

At the LHC, the Standard Model Higgs boson is produced mainly from the following four production processes : gluon fusion (ggF, $gg \to H$), vector boson fusion (VBF, $qq' \to qq'H$), associated production with a vector boson ($q\bar{q} \to WH/ZH$) and associated production with a top-quark pair ($q\bar{q}/gg \to t\bar{t}H$). Their cross-sections have been gathered and summarized in Ref. [17]. The Higgs boson decay branching ratios used take into account the higher order QCD and EW corrections in each of Higgs boson decay channels [17, 18]. The cross section in pp collisions at $\sqrt{s} = 7$ TeV and the branching ratio for each final state are shown in figure 1 as a function of the Higgs boson mass.

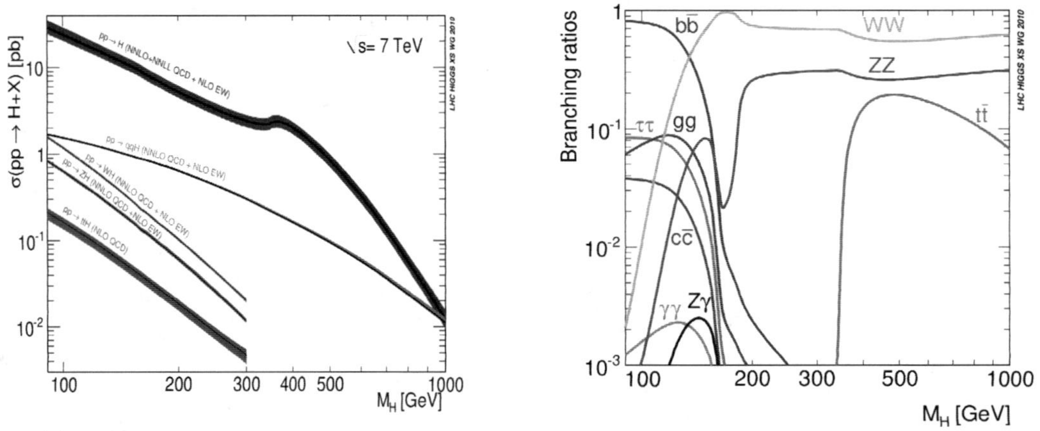

(a) The production cross section

(b) The branching ratio

Figure 1. The cross section and decay branching ratios for the Standard Model Higgs boson in pp collisions at a 7 TeV center-of-mass energy as a function of mass [17] : In (a), the cross section is shown for each of the main production processes ; gluon fusion (ggF, $gg \to H$), vector boson fusion (VBF, $qq' \to qq'H$), associated production with a vector boson ($q\bar{q} \to WH/ZH$) and associated production with a top-quark pair ($q\bar{q}/gg \to t\bar{t}H$). In (b), the branching fraction for each final state is shown.

In this section, the Standard Model Higgs boson is searched by using the six final states : $H \to \gamma\gamma$, $H \to WW^{(*)} \to l\nu l\nu$, $H \to WW \to l\nu qq$, $H \to ZZ^{(*)} \to llll$, $H \to ZZ \to ll\nu\nu$ and $H \to ZZ \to llqq$. The search regions for the Higgs boson mass are summarized in table 1. In the $H \to \gamma\gamma$, $H \to ZZ \to ll\nu\nu$ and $H \to ZZ \to llqq$, the exclusion limit is extracted from a fit of signal plus background contributions to the observed Higgs boson candidate mass distributions. In all other channels, limits are extracted from a comparison of the numbers of observed events to the numbers of estimated background events.

Table 1. The search region for the Higgs boson mass in each of the analysis channels.

Mode	Mass range (GeV)
$H \to \gamma\gamma$	110-140
$H \to WW^{(*)} \to l\nu l\nu$	120-200
$H \to WW \to l\nu qq$	220-600
$H \to ZZ^{(*)} \to llll$	120-600
$H \to ZZ \to ll\nu\nu$	200-600
$H \to ZZ \to llqq$	200-600

3.1. Search for $H \to \gamma\gamma$

The search for the Higgs boson using the decay channel $H \to \gamma\gamma$ is performed using a dataset corresponding to an integrated luminosity of 209 pb^{-1}. Events are selected by requiring the presence of at least two identified and isolated photons [19]. The leading photon is required to have $p_{\rm T} > 40$ GeV while the subleading photon is required to have $p_{\rm T} > 25$ GeV. The main background processes in this analysis are the irreducible prompt diphoton production ($\gamma\gamma$), the reducible photon-jet process (γ-jet), the dijet (jet-jet) process with one or more fake photons from jets fragmenting mainly into π^0 or η meson. The abundance of these background contributions is estimated from sideband control samples of the data. The background from Drell-Yan events, where both electrons are misidentified as photons, is estimated using Z decays to electrons where one electron is misreconstructed as a photon.

The di-photon invariant mass distribution after the full event selection is shown in figure 2. The background in the di-photon mass range between 100 and 150 GeV is modelled by fitting an exponential function to the data. The signal peak is modelled by the sum of a Crystal Ball function (for the bulk) and a Gaussian (for the tails) by using the Monte Carlo. The relative uncertainty on the resolution of the signal peak is $\pm 13\%$, where the dominant contributions come from those on the constant term of the cluster energy resolution ($\pm 11\%$), the photon energy calibration ($\pm 6\%$) and so on. The systematic uncertainty on the signal yield is also considered ($\pm 15\%$ in total), which mainly arises from the uncertainty on photon identification and photon isolation efficiencies. The signal search is performed in the Higgs boson mass range of $110 \le m_H \le 140$ GeV, and no significant evidence for an excess is found. The extracted limits by using PCL method is shown in figure 3.

3.2. Search for $H \to WW^{(*)}$

The search using the decay $H \to WW^{(*)}$, while either one of the W bosons may be off the mass shell ($W^{(*)}$), offers strong sensitivity in the mass range of $m_H \ge 120$ GeV. Two different decay modes of the W bosons are considered : the search for $H \to WW^{(*)} \to l\nu l\nu (l = e, \mu)$ is performed in the range between 120 and 200 GeV, and that for $H \to WW \to l\nu qq$ channel is performed in the range from 220 to 600 GeV.

3.2.1. $H \to WW^{(*)} \to l\nu l\nu$

The search for $H \to WW^{(*)} \to l\nu l\nu$ is performed by using a dataset corresponding to an integrated luminosity of 35 pb^{-1}. Events are selected requiring exactly two isolated leptons with opposite charge, where the leading lepton is needed to have $p_{\rm T} > 20$ GeV and the sub-leading lepton is required to have $p_{\rm T} > 15$ GeV. The large missing transverse energy in the event is also imposed to suppress the backgrounds ($E_T^{miss} > 30$ GeV).

Events are classified into $3 \times 3 = 9$ channels by depending on : (i) the lepton flavours (ee, $e\mu$

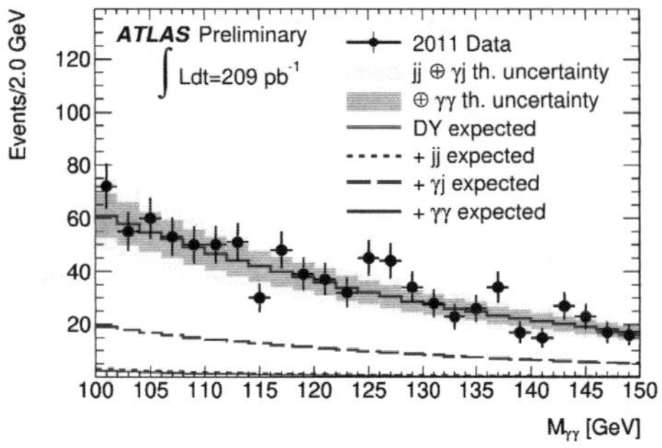

Figure 2. Di-photon invariant mass distribution after applying the full event selection in $H \to \gamma\gamma$ analysis : Distribution for data is presented with black plots. The cumulative predictions of the Drell Yan (red solid), di-jet (blue dotted), photon-jet (blue dashed) and diphoton (blue solid) are also shown. The light yellow band indicates the overall uncertainty on the expected total yield, while the dark yellow one is the uncertainty due to the reducible background contribution.

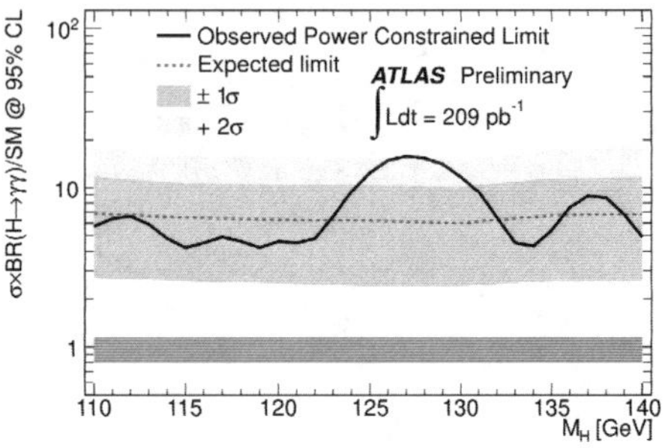

Figure 3. The 95% C.L. exclusion limits on the production cross section relative to the Standard Model cross section for the $H \to \gamma\gamma$, for an integrated luminosity of 209 pb^{-1} at $\sqrt{s} = 7$ TeV.

or $\mu\mu$) ; (ii) the number of jets of $p_T > 25$ GeV and $|\eta| < 4.5$ in the event (0/1/2-jet channels). In each channel, kinematic requirements are applied using the azimuthal opening angle between the two leptons ($\Delta\phi_{ll}$), the invariant mass of the two leptons (m_{ll}) and the transverse mass (m_{T}), where the transverse mass is defined as [20] :

$$m_{\mathrm{T}} = \sqrt{(E_{\mathrm{T}}^{ll} + E_{\mathrm{T}}^{\mathrm{miss}})^2 - (\mathbf{P}_{\mathrm{T}}^{ll} + \mathbf{P}_{\mathrm{T}}^{\mathrm{miss}})^2}.$$

$E_{\mathrm{T}}^{ll} = \sqrt{(\mathbf{P}_{\mathrm{T}}^{ll})^2 + m_{ll}^2}$, $|\mathbf{P}_{\mathrm{T}}^{\mathrm{miss}}| = E_{\mathrm{T}}^{\mathrm{miss}}$ and $\mathbf{P}_{\mathrm{T}}^{ll}$ is the transverse momentum of the dilepton system. Cut value for each selection is optimized depending on the Higgs boson mass hypothesis in order to increase the sensitivity as much as possible.

The major backgrounds come from WW, top quark ($t\bar{t}$ and single top) and W+jets events. Their contributions into the signal region are estimated by using dedicated control regions in data, while the expected contributions from other smaller backgrounds are estimated by using the Monte Carlo. The observed and expected m_T distributions for the combined $e\mu$, ee and $\mu\mu$ channels are shown in figure 4, which are obtained after all the selections except for the transverse mass cut. For masses of 120< m_H <200 GeV, no evidence for an excess is observed in any of the analysis channels.

The obtained exclusion limits are shown in figure 5. By using the PCL method, a SM-like Higgs boson with a production rate of 1.2 times the SM value is excluded at 95% C.L. for $m_H = 160$ GeV. In the figure, the Tevatron results are overlaid. As their results were calculated using the CL$_s$ method, the ATLAS results with CL$_s$ are also shown.

3.2.2. $H \to WW \to l\nu qq$

For the $H \to WW \to l\nu qq$, a dataset corresponding to an integrated luminosity of 35 pb^{-1} is used. Events with exactly one high-p_{T} lepton ($p_{\mathrm{T}} > 30$ GeV), and large missing transverse

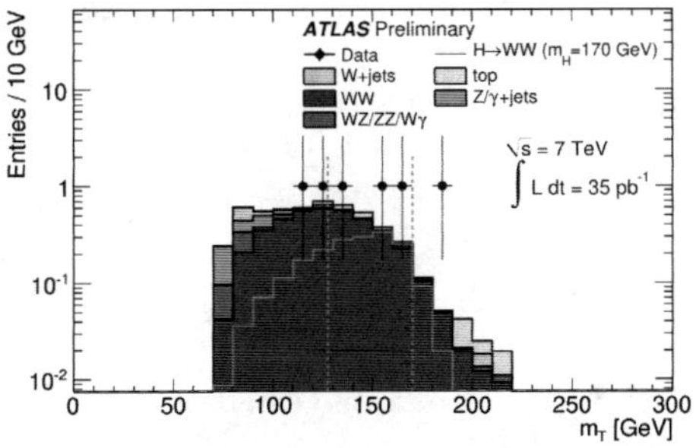

(a) The 0jet channel

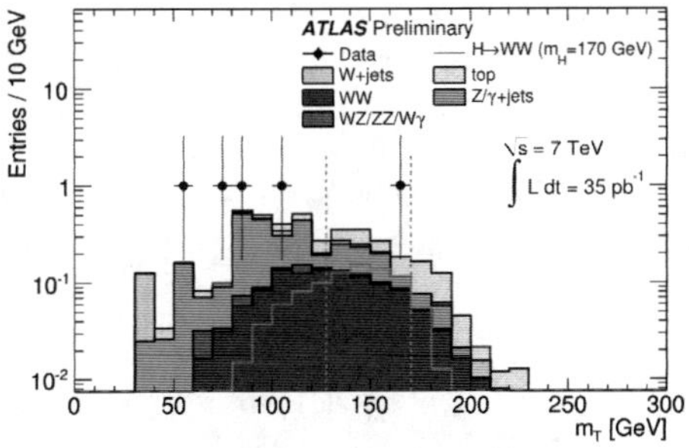

(b) The 1jet channel

Figure 4. Distributions of the transverse mass m_T for the combined $e\mu$, ee and $\mu\mu$ channels for the $H \to WW^{(*)} \to l\nu l\nu$ search : (a) the 0-jet channel ; (b) the 1-jet channel : The distributions are obtained after all the selections except for the transverse mass cut. The data plots are compared with the background expectation and an added hypothetical signal, where the Higgs boson mass is assumed as $m_H = 170$ GeV. The cut values for the transverse mass are indicated by the vertical dotted lines.

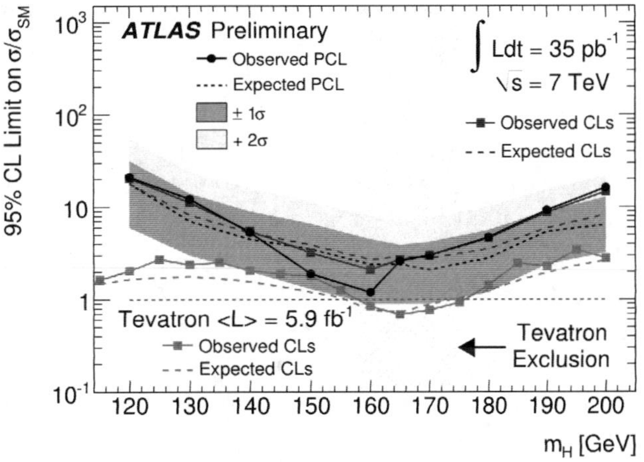

Figure 5. The 95% C.L. exclusion limits on the production cross section relative to the Standard Model cross section for the $H \to WW^{(*)} \to l\nu l\nu$, for an integrated luminosity of 35 pb^{-1} at $\sqrt{s} = 7$ TeV : The ATLAS limit using the PCL method is compared with the Tevatron limits. Since the Tevatron results were calculated using the CL$_s$ method, the ATLAS results with CL$_s$ are also shown.

energy ($E_{\mathrm{T}}^{\mathrm{miss}} > 30$ GeV) are selected. The number of jets in the event is required to be 2 or 3, and the selected events are separated into two channels depending on the jet multiplicity. The two jets with the invariant mass closest to the mass of the W boson are required to satisfy 71< m_{jj} <91 GeV. The final discriminant is the invariant mass of the Higgs boson candidate, $m_{l\nu jj}$, which is reconstructed with a W boson mass constraint on the lepton-neutrino system, giving rise to a quadratic equation. If there are two solutions, the one with the lower longitudinal momentum is chosen. And if a complex solution is obtained, the real part is used. The major backgrounds come from W/Z+jets and multijet events. Their background levels are determined using a fit to the $E_{\mathrm{T}}^{\mathrm{miss}}$ distribution, while other small contributions from the top and diboson events are estimated by using the Monte Carlo.

The $m_{l\nu jj}$ distributions are shown in figure 6. The background spectrum is modelled with a single exponential function. By using the fit result, the limit extraction is made in a mass range of 200$< m_H <$600 GeV. The result is shown in figure 7, where the Higgs boson with a production rate of 11.2 times the Standard Model prediction is excluded at 95% C.L. for $m_H = 400$ GeV.

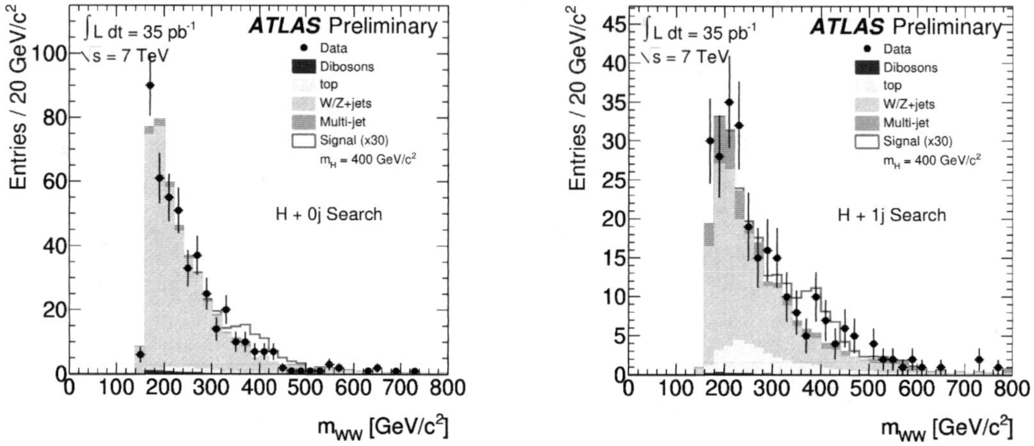

(a) 0 extra jet channel (b) 1 extra jet channel

Figure 6. Distributions of the invariant mass $m_{l\nu jj}$ for the $H \to WW \to l\nu qq$ search after applying all selection criteria. The plots are summed over lepton flavour. In (a), no extra jets are allowed and in (b), one additional jet is required. The data plots are compared with the background expectation and an added hypothetical signal, where the Higgs boson mass is assumed as $m_H = 400$ GeV. The expected signal yield is scaled up by a factor of 30 for illustration purposes.

3.3. Search for $H \to ZZ^{(*)}$

For searching $H \to ZZ^{(*)}$, three final states are analyzed : $H \to ZZ^{(*)} \to llll$, $H \to ZZ \to ll\nu\nu$ and $H \to ZZ \to llqq$.

3.3.1. $H \to ZZ^{(*)} \to llll$

A dataset with an integrated luminosity of 40 pb^{-1} is used for the $H \to ZZ^{(*)} \to llll$ search. Events are required to have two same-flavour and opposite-charge pairs of leptons. All leptons must be well separated and isolated. The cut values for p_{T} are 7 GeV and 15 GeV for muons and electrons, respectively. And at least two out of the four leptons are required to have $p_{\mathrm{T}} > 20$ GeV. The invariant mass needs to be close to the Z boson mass for at least one lepton pair, while the requirement on the invariant mass of the other lepton pair varies depending on the Higgs boson candidate mass (m_{llll}).

The main backgrounds come from the irreducible component $ZZ^{(*)} \to llll$ and from the reducible ones, $Z + jets$ and $t\bar{t}$ events. But after applying all the selection criteria, no candidate events remain in data at any Higgs boson mass, that is consistent with the expectation of very small background level.

Journal of Physics: Conference Series **347** (2012) 012010 doi:10.1088/1742-6596/347/1/012010

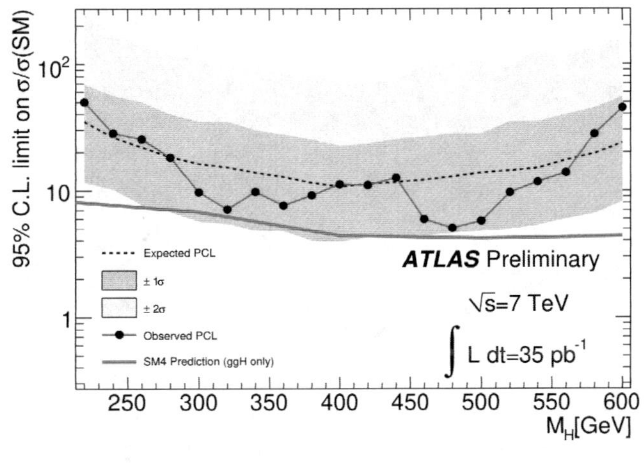

Figure 7. The 95% C.L. exclusion limits on the production cross section relative to the Standard Model cross section for the $H \to WW \to l\nu qq$, for an integrated luminosity of 35 pb^{-1} at $\sqrt{s} = 7$ TeV. The exclusion limit on a four-generation Standard Model (SM4) which assumes that the masses of the fourth-generation quarks are infinite [29] is also shown.

3.3.2. $H \to ZZ \to llqq$

Events with exactly two same-flavour leptons are selected, where the invariant mass of the lepton pair must be close to the Z mass ($76 < m_{ll} < 106$ GeV). The missing transverse energy is required to be less than 50 GeV to reduce background from top events. The leading and subleading jets also need to have an invariant mass near the Z mass ($70 < m_{jj} < 105$ GeV). In the case of searching for high mass Higgs boson, further kinematics cuts are applied by using the azimuthal opening angles between the two jets ($\Delta\phi_{jj}$), the two leptons ($\Delta\phi_{ll}$) and so on, where the cut values are optimized depending on the Higgs mass.

The dominant background comes from Z+jets, and other backgrounds are $t\bar{t}$, QCD and ZZ/WZ production. The contribution from each of Z+jets, $t\bar{t}$ and QCD events is well understood by using the side-band region in data, and that from ZZ/WZ is estimated using the Monte Carlo. The distribution of the Higgs boson candidate mass (m_{lljj}) for the events passing the full selection is shown in figure 8.

3.3.3. $H \to ZZ \to ll\nu\nu$

Events with exactly two leptons of the same flavour and with large $E_{\mathrm{T}}^{\mathrm{miss}}$ are selected. The lepton pair must have an invariant mass near the Z mass, and in order to reject top background, events with jets that have been identified as coming from b-quarks are rejected. Kinematic requirements on $\Delta\phi_{ll}$, $\Delta\phi_{ll}$ and so on are also applied, where the cut values are optimized depending on the Higgs mass.

The major background comes from Z+jets, $t\bar{t}$ and ZZ/WZ and QCD production. As in the $H \to ZZ \to llqq$ analysis, the contribution from each of Z+jets, $t\bar{t}$ and QCD events is well understood by using the side-band region of data, and that from ZZ/WZ is estimated using the Monte Carlo. The dilepton transverse mass distribution after applying all selection criteria for $m_H = 400$ GeV is shown in figure 9.

3.3.4. Exclusion limit

The exclusion limits from the $H \to ZZ^{(*)} \to llll$ channel is shown in figure 10, where a SM-like

Journal of Physics: Conference Series **347** (2012) 012010 doi:10.1088/1742-6596/347/1/012010

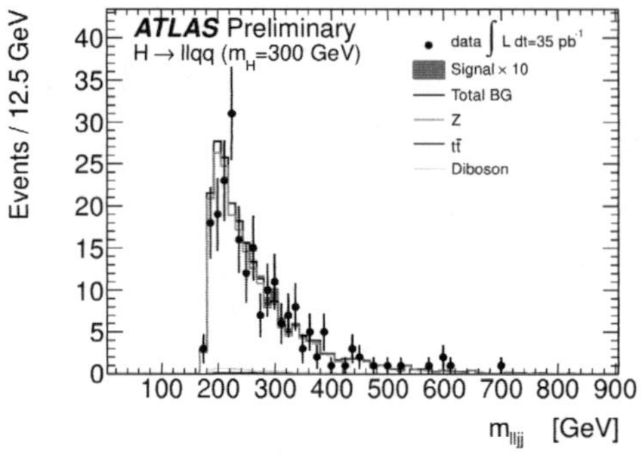

Figure 8. Distribution of the invariant mass m_{lljj} for the $H \to ZZ \to llqq$ search after applying all selection criteria. Data plots are compared with the background expectation (black histogram) and an added hypothetical signal (red filled histogram) where the Higgs boson mass is assumed as 300 GeV.The contribution from each background component is shown with pink (Z), blue ($t\bar{t}$) and green (Diboson) histograms. The plots are summed over lepton flavour.

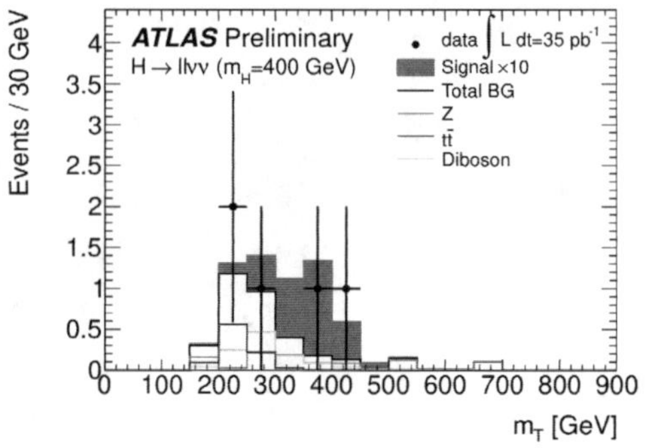

Figure 9. The dilepton transverse mass distribution for the $H \to ZZ \to ll\nu\nu$ search after all selection cuts. Data plots are compared with the background expectation (black histogram) and an added hypothetical signal (red filled histogram) where the Higgs boson mass is assumed to be 400 GeV. The contribution from each background component is shown with pink (Z), blue ($t\bar{t}$) and green (Diboson) histograms. The plots are summed over lepton flavour.

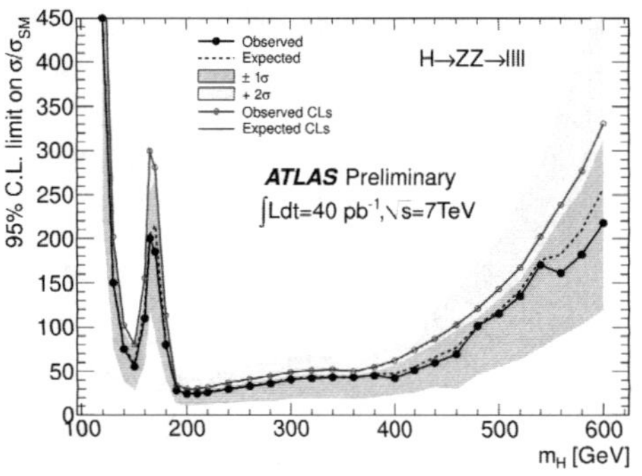

Figure 10. The 95% C.L. exclusion limits on the production cross section relative to the Standard Model cross section for the $H \to ZZ \to llll$, for an integrated luminosity of 40 pb^{-1} at $\sqrt{s} = 7$ TeV.

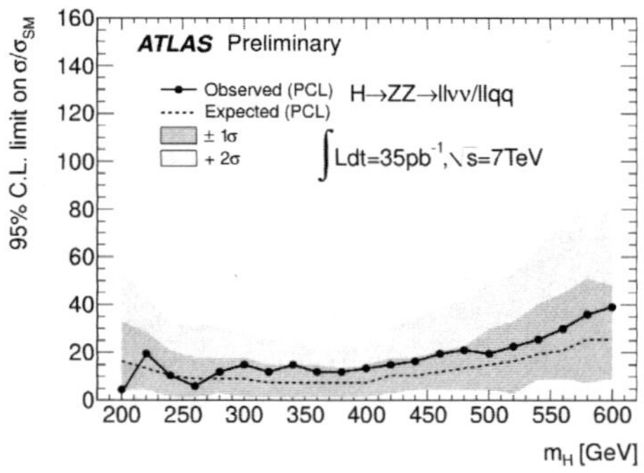

Figure 11. The 95% C.L. exclusion limits on the production cross section relative to the Standard Model cross section obtained from the combined $H \to ZZ \to llqq$ and $ll\nu\nu$ channels for an integrated luminosity of 35 pb^{-1} at $\sqrt{s} = 7$ TeV.

Higgs boson with a production rate of 24.0 times the SM prediction is excluded at 95% C.L. for $m_H = 200$ GeV. The combined result for $H \to ZZ \to llqq$ and $ll\nu\nu$ is also shown in figure 11. A SM-like Higgs boson with a production rate of 3.5 to 39 times the SM value is excluded at 95% C.L in the mass region of 200< m_H <600 GeV.

Journal of Physics: Conference Series **347** (2012) 012010 doi:10.1088/1742-6596/347/1/012010

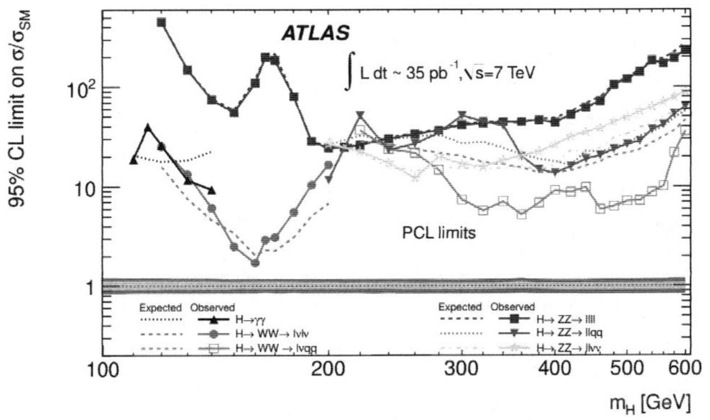

Figure 12. The 95% C.L. exclusion limits on the production cross section relative to the Standard Model prediction, as a function of the Higgs boson mass for the individual analysis channels. The gray horizontal bands show the uncertainty on the Standard Model cross section prediction, with the inner region highlighting the contribution of QCD scale uncertainties.

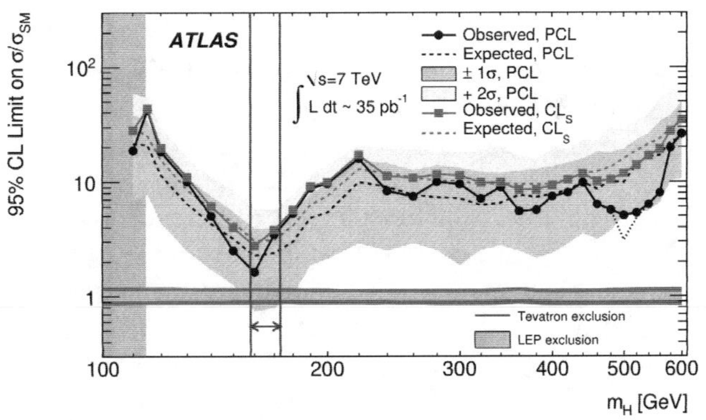

Figure 13. Combined result for the 95% C.L. exclusion limits on the production cross section relative to the Standard Model prediction. In addition to the PCL limits, CL_s limits are overlaid for comparison purposes. The excluded regions from the LEP [7] and the Tevatron experiments [9] are also shown.

3.4. Combined result for the exclusion limits on the Standard Model Higgs boson

The combined exclusion limits on the Standard Model Higgs boson, using the results for $H \to \gamma\gamma$, $H \to WW^{(*)}$ and $H \to ZZ^{(*)}$ channels, are extracted. As for $H \to \gamma\gamma$ channel, the result for 2010 data (an integrated luminosity of 38 pb^{-1}) is used in the combination, while the result for 2011 data (an integrated luminosity of $209\mathrm{pb}^{-1}$) is shown in Section 3.1. The exclusion limits for the individual channels are shown in figure 12, where the step with which the limit is investigated is 5-10 GeV. The combined limits, obtained using PCL, are shown in figure 13, where the limits with the CL_s method are also shown for comparison purposes. Further details of the combination method are well described in Ref. [21].

4. Search for the Higgs boson in the extended models of the Standard Model

The search for the Higgs boson generated by the extended models of the Standard Model is also performed, which is described in this section.

4.1. Search for neutral MSSM Higgs bosons decaying to $\tau\tau$

The search for neutral MSSM Higgs bosons in the decay mode $A/H/h \rightarrow \tau\tau$ is performed. As explained in Section 1, there are five observable Higgs bosons in the MSSM. At the tree level their properties such as masses, widths and branching ratios are described in terms of only two parameters : the mass of the CP-odd Higgs boson (m_A) and the ratio of the vacuum expectation values of the two Higgs doublets (tanβ). The Higgs boson is produced mainly via gluon fusion ($gg \rightarrow gg\phi$) or in association with b quarks ($bb \rightarrow bb\phi$), where ϕ denotes a neutral Higgs boson, and the latter process becomes important especially for large tanβ.

The decay into a $\tau\tau$ pair is a promising channel since the coupling of the Higgs bosons to third-generation fermions is strongly enhanced for a wide region in the MSSM parameter space. The search is performed by using two different final states : $A/H/h \rightarrow \tau\tau \rightarrow l\tau_{had}3\nu$ and $e\mu 4\nu$.

4.1.1. $A/H/h \rightarrow \tau\tau \rightarrow l\tau_{had}3\nu$

Events with one isolated lepton (e or μ) and one τ decaying hadronically (τ_{had}) are selected. Exactly one electron or muon with $p_{\mathrm{T}}^{e} > 20$ GeV or $p_{\mathrm{T}}^{\mu} > 15$ GeV and one oppositely-charged τ candidate with $p_{\mathrm{T}}^{\tau,\mathrm{vis}} > 20$ GeV are required in the event. In order to enhance the sensitivity, the large missing transverse energy ($E_{\mathrm{T}}^{\mathrm{miss}} > 20$ GeV), and small transverse mass of the e/μ-$E_{\mathrm{T}}^{\mathrm{miss}}$ system ($m_{\mathrm{T}} < 30$ GeV) are required.

In order to perform the signal search, the normalization and shape of the $m_{\tau\tau}^{\mathrm{visible}}$ distribution need to be determined. The main backgrounds in this analysis come from $Z \rightarrow \tau\tau$ (irreducible), QCD and W+jets. The $m_{\tau\tau}^{\mathrm{visible}}$ distribution for the total background is estimated by using events with same-sign charges of the electron/muon and the τ_{had} candidate, assuming that the distribution shape is the same between opposite-sign (OS) and same-sign (SS) events ; an assumption that has been verified with simulated events. When extrapolating from the SS to the OS regions, corrections are needed to compensate for the difference of the number of events between the two regions. Correction factor to be taken is well understood with data for W+jets and QCD events, while that for $Z \rightarrow \tau\tau$ is predicted from the Monte Carlo. Figure 14 shows the $m_{\tau\tau}^{\mathrm{visible}}$ distribution for the estimated background compared with data.

4.1.2. $A/H/h \rightarrow \tau\tau \rightarrow e\mu 4\nu$

Events with one isolated electron and one isolated muon are selected. Exactly one electron with $p_T^{\mathrm{e}} > 20$ GeV and one muon with $p_{\mathrm{T}}^{\mu} > 10$ GeV with opposite charge are needed in the event. To reduce background events, the scalar sum of the transverse momentum of the electron, that of the muon and $E_{\mathrm{T}}^{\mathrm{miss}}$ is required to be smaller than 120GeV, and the cut on the azimuthal opening angle between the electron and muon is also applied ($\Delta\phi_{e\mu} > 2.0$ rad.). The effective mass ($m_{\tau\tau}^{\mathrm{effective}}$) is used as a final discriminant to search for a potential Higgs boson signal, where $m_{\tau\tau}^{\mathrm{effective}}$ is defined as the invariant mass of the electron, muon and $E_{\mathrm{T}}^{\mathrm{miss}}$ system :

$$m_{\tau\tau}^{\mathrm{effective}} = \sqrt{(p_e + p_\mu + p_{\mathrm{miss}})^2}$$

where p_e and p_μ are the four-vectors of the electron and muon, while the missing energy four-vector is defined by $p_{\mathrm{miss}} = (E_{\mathrm{T}}^{\mathrm{miss}}, E_{\mathrm{x}}^{\mathrm{miss}}, E_{\mathrm{y}}^{\mathrm{miss}}, 0)$. Major backgrounds come from $Z \rightarrow \tau\tau$ and QCD events. QCD contribution is estimated from the control data sample, while the contribution of the irreducible component is obtained from the simulation. The effective mass distribution after applying all selection criteria is shown in figure 14, where data and the estimated background are compared.

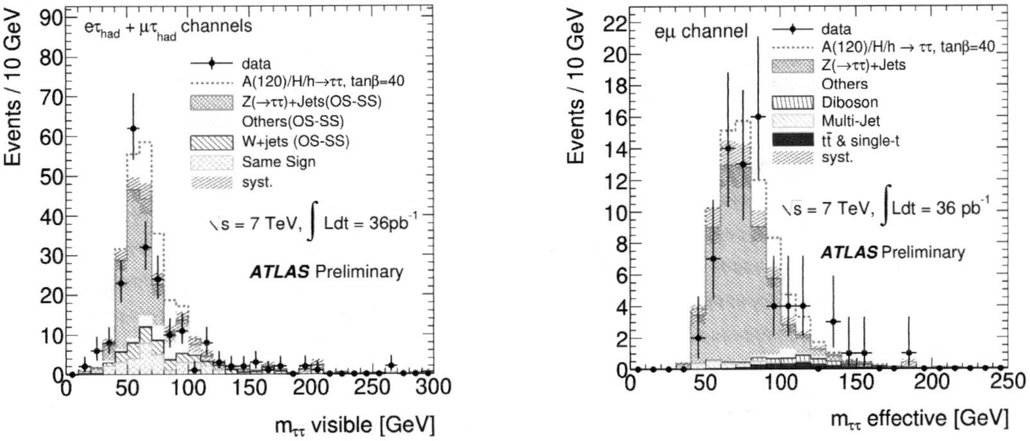

Figure 14. (left) : Distribution of visible mass $m_{\tau\tau}^{\text{visible}}$ after applying all selection criteria for the $A/H/h \to \tau\tau \to l\tau_{had}3\nu$ search. (right) : Distribution of effective mass $m_{\tau\tau}^{\text{effective}}$ after all selection criteria for the $A/H/h \to \tau\tau \to e\mu 4\nu$ search. The data plots are compared with the background expectation and an added hypothetical signal.

4.1.3. Exclusion limits

Figure 15 shows the obtained exclusion limits. The cross section limit is evaluated for signal acceptances of two different production processes ($gg \to \phi$ and $bb \to bb\phi$). The limits on the Higgs boson production cross-section times branching ratio into a pair of τ leptons is in the range between approximately 300 pb for a Higgs boson mass of 90 GeV and approximately 10 pb for a mass of 300 GeV. The limit in the tanβ-m_A plane is obtained by using the m_h^{max} scenario and Higgsino mass parameter $\mu > 0$. Limits from previous experiments at LEP [22] and the Tevatron [23] are shown for comparison.

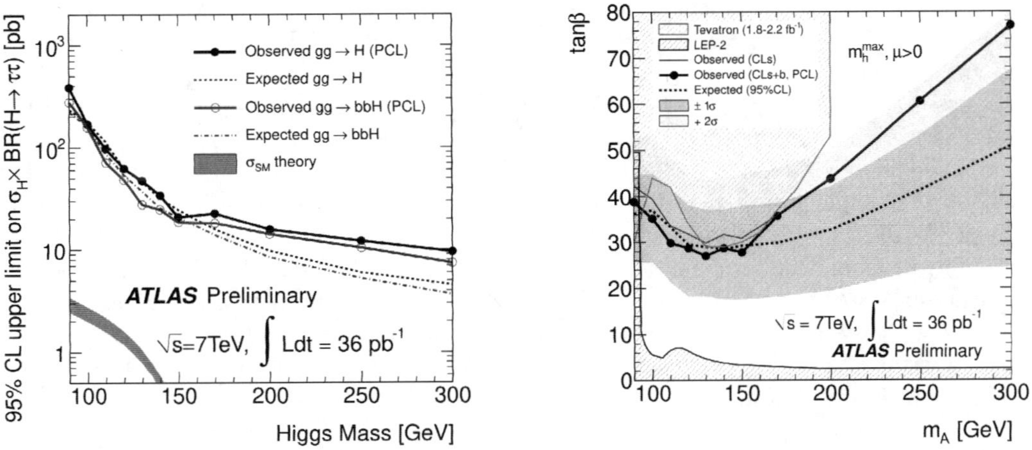

Figure 15. (left) : The 95% C.L. exclusion limits on the production cross section for the MSSM $A/H/h \to \tau\tau$ search for an integrated luminosity of 36 pb^{-1}. For comparison, the Standard Model cross section is also shown. (right) : The 95% C.L. exclusion limits in the m_A-tanβ plane. The limits from LEP and Tevatron are also overlaid. For a direct comparison with the Tevatron limit, the observed limit based on CL$_s$ is shown in addition to the one based on CL$_{s+b}$.

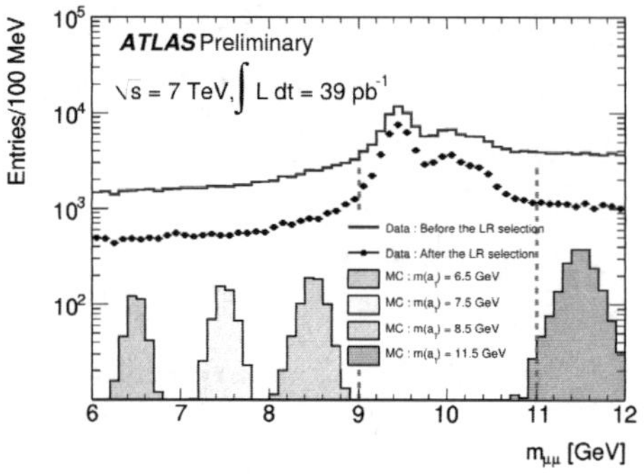

Figure 16. The $m_{\mu\mu}$ spectrum after the dimuon selection (blue line) and after the Likelihood Ratio selection (points) for the NMSSM $a_1 \to \mu\mu$ search. The predicted spectra for signal events (with arbitrary normalization) are also shown for various a_1 masses.

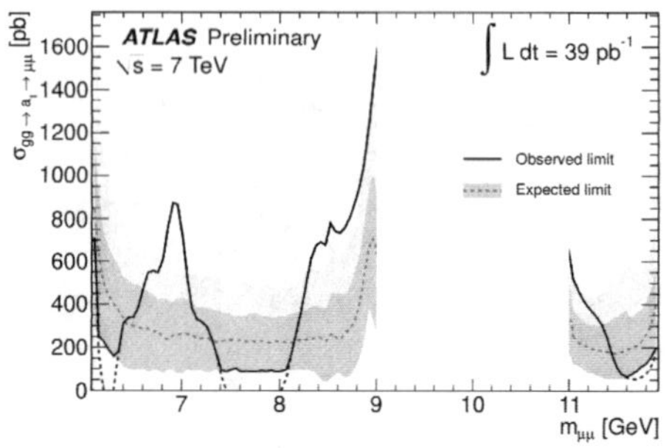

Figure 17. The 95% C.L. exclusion limits on the cross section for the NMSSM $a_1 \to \mu\mu$ process, as obtained in a search with an integrated luminosity of 39 pb^{-1}.

4.2. Search for a light CP-odd Higgs boson decaying to $\mu\mu$

The next-to-minimal SSM (NMSSM) solves the μ-term problem by introducing a complex singlet scalar field S. This generates the μ-term as the vacuum expectation value of S through spontaneous symmetry breaking, allowing it naturally to reach the electroweak scale without the fine-tuning required in the MSSM [12]. As a result of introducing a new scalar field, the Higgs sector of the NMSSM expands to a total of three CP-even scalars (h_1, h_2, h_3), two CP-odd scalars (a_1, a_2) and two charged scalars ($H^{\pm}$). If the mass of CP-odd Higgs boson a_1 is around 10 GeV, the NMSSM can account for the anomalous muon magnetic moment [24], and it also explains the discrepancy in the ratio of Υ widths $\Gamma(\Upsilon \to \tau\tau)/\Gamma(\Upsilon \to \mu\mu)$ between the value measured by the BaBar experiment and the value expected in the SM [25, 26, 27, 28].

Based on these facts, a direct search for the a_1 boson, produced via gluon fusion ($gg \to a_1$) and decaying to muon pair ($a_1 \to \mu\mu$) is performed in the mass regions 6-9 GeV and 11-12 GeV. Dimuon masses around the Υ resonances (9-11 GeV) are not included because uncertainties in the expected rate of Υ production make it difficult to distinguish an additional resonance in this mass range. Event selection for this search is quite simple. Events with two muons having opposite charge are selected. Each event is required to pass a trigger requiring two muons with $p_{\mathrm{T}} > 4$ GeV. As for muon identification, likelihood ratio-based selection is performed. The mass spectrum of the selected dimuon candidates is shown in figure 16. The signal search is performed by fitting the mass spectrum in the region of 6-12 GeV. The continuum background spectrum is modeled by using sidebands, while the spectrum shapes for signal and $\Upsilon(1/2/3s)$ are modeled by double-gaussian. No evidence for an excess is observed, and the obtained exclusion limits on the Higgs boson production are shown in figure 17.

5. Conclusions

A search for the Higgs boson at the Large Hadron Collider (LHC) running at a center-of-mass energy of 7 TeV is reported, based on early data collected by the ATLAS detector. For the Standard Model (SM) Higgs boson, the decay channels of $H \to \gamma\gamma$, $H \to WW^{(*)}$ and $H \to ZZ^{(*)}$ are analyzed, where the search for $H \to \gamma\gamma$ is performed by using a dataset with an integrated luminosity of 209 pb^{-1}, while $H \to WW^{(*)}$ and $H \to ZZ^{(*)}$ are searched by using

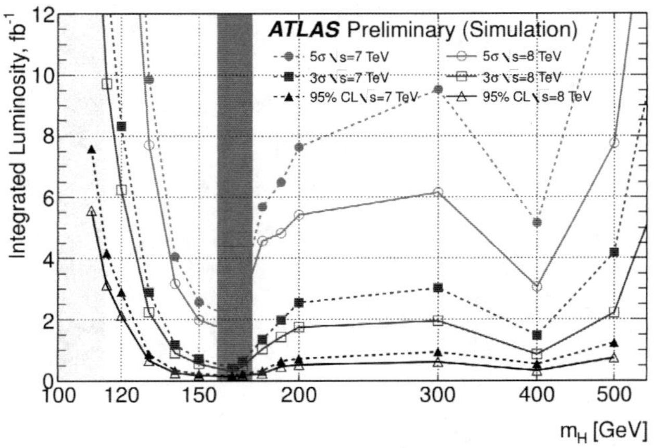

Figure 18. The luminosity required to give exclusion, evidence or discovery sensitivity for the Standard Model Higgs boson with data at $\sqrt{s}$ = 7 or 8 TeV, where the sensitivity is expected based on the simulation.

35-40 pb^{-1}. From the $H \to \gamma\gamma$ channel, the observed exclusion limits at 95% C.L. between 4.2 and 15.8 times the SM cross section are achieved in the Higgs boson mass range between 110 GeV and 140 GeV. With the $H \to WW^{(*)} \to l\nu l\nu$ channel, the Higgs boson in the mass range of $120 \leq m_H \leq 200$ GeV is searched, where the obtained limit of 1.2 times SM prediction is obtained at $m_H = 160$ GeV, coming closer to Tevatron limit and also the SM cross section. By using $H \to WW \to l\nu qq$, $H \to ZZ \to llll, ll\nu\nu$ and $llqq$, Higgs boson in the high mass region ($200 \leq m_H \leq 600$ GeV) is also searched, where the observed limits on the production rate over the SM prediction are 11.2 at $m_H = 400$ GeV ($H \to WW \to l\nu qq$), 24.0 at $m_H = 200$ GeV ($H \to ZZ \to llll$) and 3.5 at $m_H = 200$ GeV ($H \to ZZ \to ll\nu\nu/llqq$). A search for the Higgs boson in the extended models of the Standard Model is also performed. Neutral MSSM Higgs bosons are searched in the channel $A/H/h \to \tau^+\tau^-$, where the obtained exclusion limit supersedes the ones from previous experiments. By using the $a_1 \to \mu\mu$ channel, a light CP-odd Higgs boson generated by the next-to MSSM (NMSSM) model is searched, where the useful exclusion limits are obtained in the mass range of 6-9 GeV and 11-12 GeV.

The prospects for the Standard Model Higgs boson search at $\sqrt{s}$ = 7 TeV are finally considered in figure 18, where the projected sensitivity as a function of the Higgs boson mass is calculated based on the simulation. For the exclusion, it is predicted that the ATLAS result reaches the LEP limit with 4fb^{-1}. For the discovery of the signal events, 3σ observation and 5σ discovery are expected in a mass range of $m_H > 120$ GeV and $m_H > 135$ GeV with 10fb^{-1}, respectively.

References

[1] Evans L and Bryant P 2008 *JINST* **3** S08001
[2] Englert F and Brout R 1964 *Phys. Rev. Lett.* **13** 321
[3] Higgs P W 1964 *Phys. Lett.* **12** 132
[4] Higgs P W 1964 *Phys. Rev. Lett.* **13** 508
[5] Higgs P W 1966 *Phys. Rev.* **145** 1156
[6] Guralnik G S, Hagen C R and Kibble T W B 1964 *Phys. Rev. Lett.* **13** 585
[7] Abbiendi G *et al.* 2003 *Phys. Lett.* B **565** 61
[8] The TEVNPH Working Group for the CDF and D0 Collaborations 2010 Combined CDF and D0 upper limits on Standard Model Higgs-boson production with up to 6.7 fb^{-1} of data *Preprint* hep-ex/1007.4587

[9] The TEVNPH Working Group for the CDF and D0 Collaborations 2010 Combined CDF and D0 upper limits on Standard Model Higgs-boson production with up to 8.2 fb^{-1} of data *Phys. Rev. Lett.* **104** 061802 (*Preprint* hep-ex/1103.3233)
[10] Nilles N P 1984 *Phys. Rep.* **110** 1
[11] Haber H E and Kane G L 1985 *Phys. Rep.* **117** 75
[12] Maniatis M 2009 The NMSSM reviewed *Preprint* hep-ph/0906.0777
[13] ATLAS Collaboration 2009 Expected performance of the ATLAS experiment - detector, trigger and physics *Preprint* hep-ex/0901.0512 1480
[14] Cowan G, Cranmer K, Gross E and Vitels O 2011 Power-constrained limits *Preprint* hep-ph/1105.3166
[15] Cowan G, Cranmer K, Gross E and Vitels O 2011 *Eur. Phys. J.* C **71** 1
[16] Read A L 2002 *J. Phys.* G **28** 2693
[17] LHC Higgs Cross Section Working Group : Dittmaier S, Mariotti C, Passarino G, Tanaka R (Eds.) *et al.* 2011 Handbook of LHC Higgs cross sections: 1. Inclusive observables *Preprint* hep-ph/1101.0593
[18] Djouadi A, Kalinowski J and Spira M 1998 *Comput. Phys. Commun.* **108** 56
[19] ATLAS Collaboration 2011 *Phys. Rev.* D **83** 052005
[20] Barr A J, Gripaios B and Lester C G 2009 Measuring the Higgs boson mass in dileptonic W-boson decays at hadron colliders *J. High Energy Phys.* JHEP07(2009)072
[21] ATLAS Collaboration 2011 Limits on the production of the Standard Model Higgs Boson in pp collisions at $\sqrt{s} = 7$ TeV with the ATLAS detector *Preprint* hep-ex/1106.2748
[22] Schael S *et al.* 2006 *Eur. Phys. J.* C **47** 547
[23] The CDF Collaboration, the D0 Collaboration and Tevatron New Physics Higgs Working Group (TEVNPHWG) 2010 Combined CDF and D0 upper limits on MSSM Higgs boson production in tau-tau final states with up to 2.2fb^{-1} *Preprint* hep-ex/1003.3363
[24] Gunion J 2009 A light CP-odd Higgs boson and the muon anomalous magnetic moment *J. High Energy Phys.* JHEP08(2009)032 (*Preprint* hep-ph/0808.2509)
[25] Sanchis-Lozano M A 2002 Searching for new physics in leptonic decays of bottomonium *Mod. Phys. Lett.* A **17** 2265 (*Preprint* hep-ph/0206156)
[26] Sanchis-Lozano M A 2004 Leptonic universality breaking in Upsilon decays as a probe of new physics *Int. J. Mod. Phys.* A **19** 2183 (*Preprint* hep-ph/0307313)
[27] Zhang Y J and Shao H S 2009 Leptonic Decay of Upsilon, a Possible Signature of New Physics *Preprint* hep-ph/0911.1766
[28] BaBar Collaboration 2010 Test of lepton universality in Upsilon(1S) decays at BaBar *Phys. Rev. Lett.* **104** 191801 (*Preprint* hep-ex/1002.4358)
[29] Becerici Schmidt N, Cetin S A, Istin S and Sultansoy S 2010 The fourth Standard Model family and the competition in Standard Model Higgs boson search at Tevatoron and LHC *Eur. Phys. J. C* **66** 119

Journal of Physics: Conference Series **347** (2012) 012011 doi:10.1088/1742-6596/347/1/012011

Recent results from CMS on SUSY searches in leptonic final states

Robert Schöfbeck[1]
Institute of High Energy Physics, Nikolsdorfergasse 18, Vienna, Austria
E-mail: robert.schoefbeck@cern.ch

Abstract. We present the results of searches for new physics in various topologies that lead to one or more isolated leptons, jets, and missing transverse energy in the final state. The searches are performed using 35 pb^{-1} of data collected in 2010 by the CMS experiment at the LHC in pp-collisions at a center-of-mass energy of 7 TeV. No evidence for new physics is observed and limits are set on the production cross-section times the event acceptance for the searched topologies and in terms of the cMSSM.

1. Introduction

The vast majority of measurements with previous high-energy particle physics experiments below the TeV-scale can be successfully described by the Standard Model (SM) [1, 2]. It contains three generations of matter and is a gauge theory with a spontaneously broken electroweak sector where particles acquire their masses through the elusive Higgs boson. Searches for new physics beyond the Standard Model are motivated by several considerations, ranging from strong astrophysical evidence for "Dark Matter", to theoretical problems associated with explaining the observed particle masses and with maintaining the mass hierarchies in the presence of quantum corrections [3, 4]. Thus, in spite of its success in describing a vast range of phenomena, the SM is almost certainly incomplete as a description of fundamental particles and their interactions.

One of the most interesting candidates for physics beyond the Standard Model (BSM) is Supersymmetry (SUSY) [5, 6, 7, 8, 9, 10, 11]. This theoretical framework adresses - at least partially - many of the limitations of the SM. SUSY models predict a spectrum of new particles arising from a correspondence between SM fermions and SUSY partner bosons, and between SM bosons and SUSY partner fermions.

At the Large Hadron Collider (LHC) at CERN, supersymmetric particles, if they exist, are predicted to be produced dominantly via strong interactions, through the fusion of two gluons into a pair of gluinos (super-partners of the gluons), a pair of squarks (super-partners of the quarks), or a gluino and a squark. The LHC, with a proton-proton centre-of-mass energy $\sqrt{s}$ of 7 TeV, is a copious source of high-energy partons which allows experiments to probe supersymmetry beyond the limits previously set at LEP and at the Tevatron. Squarks and gluinos initiate a decay cascade in which quarks and subsequently other SM particles are produced, until the lightest supersymmetric particle (LSP) is created. The dynamics of the cascade depends on the SUSY model under consideration, and in particular on the masses of

[1] For the CMS Collaboration

Journal of Physics: Conference Series **347** (2012) 012011 doi:10.1088/1742-6596/347/1/012011

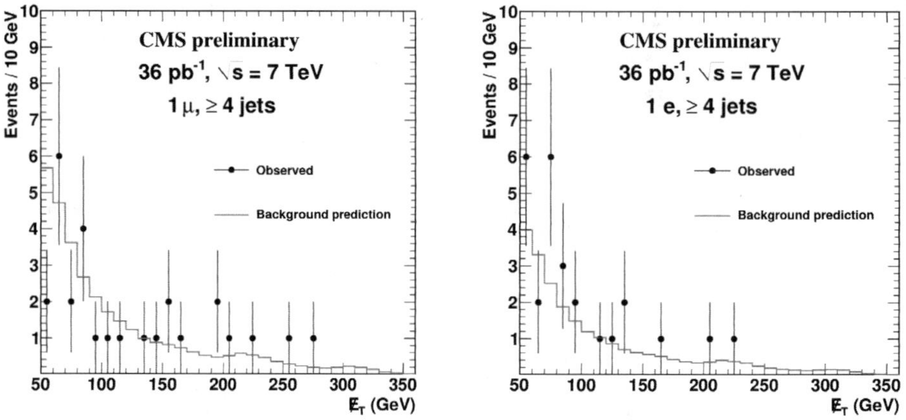

Figure 1. Measured vs. predicted E_T^{miss} distributions in muon and electron channels in the single lepton search channel using the lepton spectrum method. The data are shown as points with error bars, while the prediction from the lepton spectrum is shown as the red curve.

the supersymmetric particles. If R-parity is conserved, the LSP is unable to decay into SM particles and is therefore stable. If, in addition, the LSP is a neutralino, it is weakly interacting and thus escapes direct detection, generating missing transverse energy (E_T^{miss}) in the final state.

In these proceedings, the emphasis is put on the most recent leptonic analyses. Other important search fields such as hadronic signatures are also being covered by CMS [12, 13, 14].

Leptons can appear in the final state, for example if heavy neutralinos ($\tilde{\chi}_2^0 \rightarrow l^{\pm}\tilde{l}^{\mp} \rightarrow l^{\pm}l^{\mp}\chi_1^0$) or charginos ($\tilde{\chi}_1^{\pm} \rightarrow \chi_1^0 W^{\pm}$) are created in the decay cascades of the squark or gluino. All searches are performed using LHC data recorded with the CMS detector, corresponding to an integrated luminosity of 35 pb^{-1}. With an optimistic assumption of a typical SUSY cross-section of 2 pb this translates into approximately 70 signal events. This should be contrasted to the expected SM yield of e.g. 5500 events from $t\bar{t}$ when assuming the NLO value of this cross-section (157.5 pb).

2. Single lepton search

The single lepton analysis [15] selects events featuring jets, E_T^{miss}, and a single lepton in the final state. The presence of the lepton strongly reduces the contribution of the QCD multi-jet and Z $\rightarrow \nu\nu$+jets backgrounds, and provides several handles to build a data-driven prediction of the remaining background contribution from QCD, $t\bar{t}$, and W+jets. Events containing an additional lepton are vetoed, and handled by the di-lepton and multi-lepton analyses.

The event sample was collected using triggers based on the presence of a single electron or a single muon. The requirement of an H_T trigger was added when the peak luminosity increased beyond 2×10^{32} cm^{-2}s^{-1}. Here, H_T is the scalar sum of the transverse momenta of all jets as reconstructed at the trigger level[2]. The trigger selection is fully efficient with respect to the baseline selection applied offline, which consists of requiring (i) four jets with $p_T > 30$ GeV and $|\eta| < 2.4$ with $H_T > 500$ GeV; (ii) an isolated lepton, which can be either a muon with $p_T > 20$ GeV and $|\eta| < 2.1$, or an electron with $p_T > 20$ GeV and $|\eta| < 2.4$. H_T is the scalar sum over the transverse momenta of all jets above a threshold of 30 GeV. The search region is defined by an additional cut on the missing transverse energy, $E_T^{miss} > 250$ GeV. The contribution of the main background processes to the search region, $t\bar{t}$ and W+jets, is estimated using the lepton

[2] In the analysis context, H_T is calculated from jets as reconstructed offline.

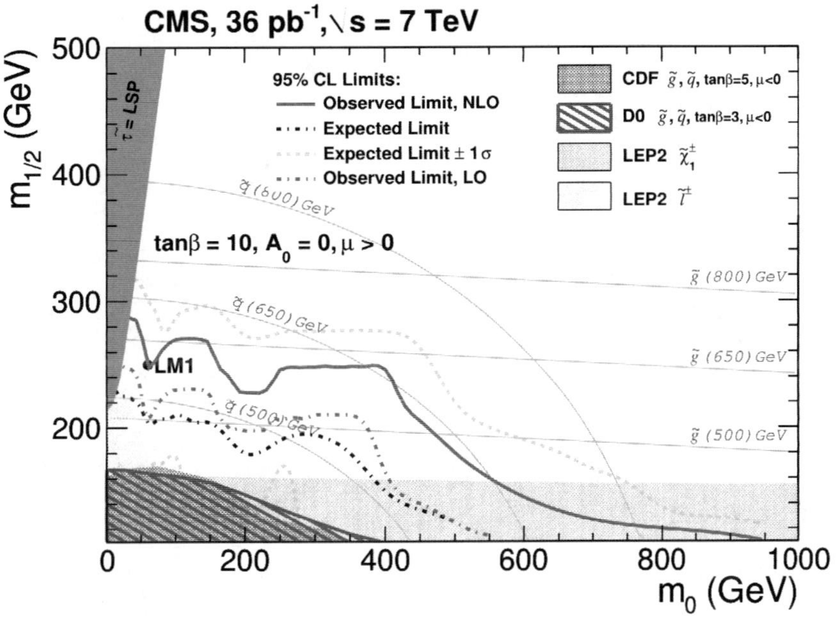

Figure 2. 95% CL exclusion contours for $\tan\beta = 10$ for the single lepton channel.

spectrum method. The foundation of this method is that, when the lepton and the neutrino are produced together in a W decay (either in $t\bar{t}$ or in W+jets events), the lepton p_{T} spectrum is directly related to the neutrino p_{T} spectrum. In contrast, no such correlation is present in SUSY events. Therefore, the lepton spectrum is used to predict the SM $E_{\mathrm{T}}^{\mathrm{miss}}$ distribution, after suitable corrections related to the effect of the W polarisation on the lepton and neutrino p_{T} spectra, to the lepton acceptance, to the reconstruction efficiency and to the difference in resolution. Figure 1 shows the prediction of the background together with the data for both the electron and the muon channels. Combining the electron and muon channels, 2 events are observed in the search region, while 3.8 ± 2.9 are expected from the background prediction. A 95% CL model independent upper limit of 4.1 signal events is calculated.

Figure 2 shows the 95% exclusion contour in the cMSSM plane for $A_0 = 0$, and $\mu > 0$ and $\tan\beta = 10$ which exceeds previous limits from Tevatron and LEP.

3. Same sign dilepton search

The same-sign di-lepton analysis [16] requires, in addition to jets and $E_{\mathrm{T}}^{\mathrm{miss}}$, exactly two isolated leptons of the same sign which can be electrons, muons or taus decaying hadronically. The event sample was collected using di-lepton and single-lepton triggers, but also H_{T} triggers, which provide sensitivity to events with low p_{T} electrons and muons. In the case of leptonic triggers, one of the electrons and muons must have $p_T > 20$ GeV and the second one must have $p_T > 10$ GeV. Both leptons must be isolated. The isolation requirement is based on the $RelIso$ variable, which is the sum of the energy deposits as measured in the tracker, the HCAL and the ECAL in a cone of $\Delta R < 0.3$ divided by the transverse momentum of the lepton[3]. We require $RelIso < 0.1$ for leptons of $p_T > 20$ GeV, and the isolation sum (i.e., the numerator of the $RelIso$ expression) to be less than 2 GeV for $p_T < 20$ GeV. We also require the presence of at least two reconstructed jets, implying $H_T > 60$ GeV. Finally, we require

[3] $\Delta R = \sqrt{\Delta\phi^2 + \Delta\eta^2}$ where ϕ is the azimuthal angle and η is the pseudo-rapidity.

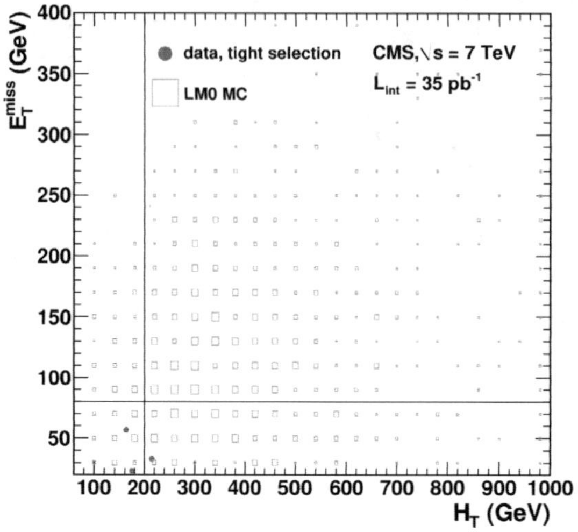

Figure 3. H_T versus E_T^{miss} scatter plots for the leptonically triggered same-sign di-lepton search. Overlayed are the three observed events (red dots) with the expected signal distribution for the LM0 benchmark point ($m_0 = 200\,\text{GeV}$, $m_{1/2} = 160\,\text{GeV}$, $\tan\beta = 10$, $\mu > 0$, and $A_0 = 400\,\text{GeV}$).

the missing transverse energy $E_T^{miss} > 30$ GeV (ee and $\mu\mu$) or $E_T^{miss} > 20$ GeV ($e\mu$). There are two kinematical search regions used in this analysis, the first has high E_T^{miss} ($E_T^{miss} > 80$ GeV); the second has high H_T ($H_T > 200$ GeV). These E_T^{miss} and H_T values were chosen to obtain an SM background expectation in simulation of less than one event in either of the two overlapping search regions. Figure 3 shows the scatter plot of the most sensitive discriminating variables H_T and E_T^{miss}. There is one event barely satisfying the H_T requirement but no event satisfying the E_T^{miss} requirement. For comparison, the LM0 benchmark point ($m_0 = 200\,\text{GeV}$, $m_{1/2} = 160\,\text{GeV}$, $\tan\beta = 10$, $\mu > 0$, and $A_0 = 400\,\text{GeV}$) is overlaid.

Hadronic triggers allow to explore the phase space with low-p_T electrons and muons, as well as final states with hadronic τ decays. We allow muons (electrons) with p_T as low as 5 (10) GeV, and restrict ourselves to τ_h with visible transverse momentum > 15 GeV, where τ_h refers to hadronic τ candidates only. All leptons must be isolated with $RelIso < 0.15$. For the ee, $e\mu$, and $\mu\mu$ final states, we require at least two jets, $H_T > 300$ GeV, and $E_T^{miss} > 30$ GeV. As backgrounds from QCD multijet production are significant for τ_h, we increase the E_T^{miss} and H_T requirements to $E_T^{miss} > 50$ GeV and $H_T > 350$ GeV in the $e\tau_h$, $\mu\tau_h$, and $\tau_h\tau_h$ final states.

The search selection and the data-driven background estimation techniques employed where chosen according to the trigger in use (lepton or hadron), and the channel ($l_i l_j$ where $l_{i,j} = e, \mu, \tau$) [16].

In all search regions, the predicted number of background events is compatible with zero, and no excess is observed. The results of the data-driven background estimations as described in [16] are summarized in Figure 4 and used to set limits in the cMSSM plane $m_0 - m_{1/2}$ for $\tan\beta = 3$, $A_0 = 0$ and $\mu > 0$. Again, previous limits from Tevatron and LEP are exceeded.

4. Opposite sign dilepton search

Requiring two opposite-sign leptons [17], the dominating SM background is $t\bar{t}$ and must be predicted from data. To this end, we select a data sample dominated by $t\bar{t}$ events using high-p_T lepton triggers and a pre-selection based on the $t\bar{t}$ cross-section measurement in the dilepton

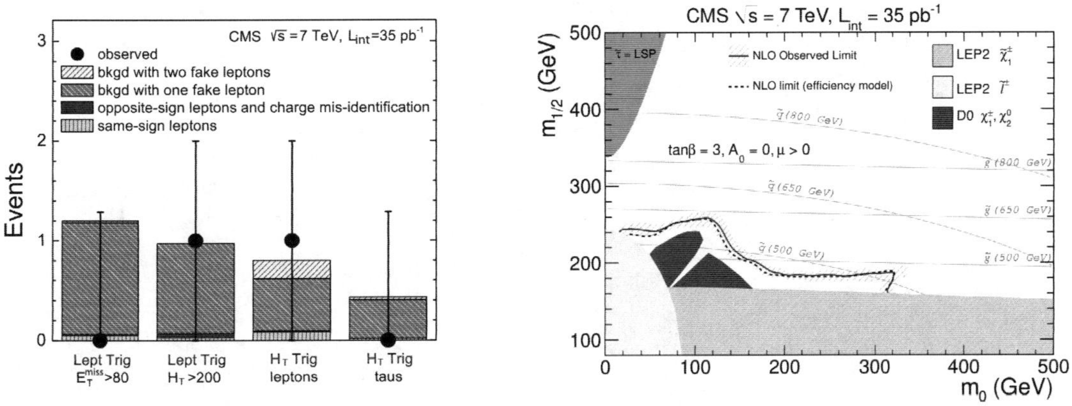

Figure 4. Left: A visual summary of the observed number of data events, the expected number of background events, and the composition of the background for the four search regions. Right: 95% CL exclusion contour in the $m_0 - m_{1/2}$ plane for CMSSM for the same-sign dilepton search.

channel [18]. We find good agreement between this data sample and predictions from Monte Carlo (MC) simulations in terms of the event yields and shapes of various kinematic distributions. The preselection requires two isolated opposite-sign leptons (e^+e^-, $e^\pm\mu^\mp$, or $\mu^+\mu^-$). At least one of the leptons must have $P_T > 20$ GeV and both must have $P_T > 10$ GeV. We require at least two jets of $P_T > 30$ GeV and $|\eta| < 2.5$, separated by $\Delta R > 0.4$ from leptons passing analysis selection with $P_T > 10$ GeV. We further require the event to have $H_T > 100$ GeV for the pre-selection and $H_T > 300$ GeV for the signal region. Finally, we remove events with a e^+e^- or $\mu^+\mu^-$ pair whose invariant mass is between 76 GeV and 106 GeV, or below 10 GeV. The signal region has an additional cut $y > 8.5$ where $y = E_{\mathrm{T}}^{\mathrm{miss}}/\sqrt{H_T}$. This selection preserves about 1% of the $t\bar{t}$ signal. The requirement is on y rather than $E_{\mathrm{T}}^{\mathrm{miss}}$ because the variables H_T and y are largely independent for the dominant $t\bar{t}$ background.

We estimate the $t\bar{t}$ background based on the idea [19] that in dilepton $t\bar{t}$ events the lepton and neutrinos from W decays have similar P_T spectra, reflecting the common boosts from the top and W decays (leaving aside W polarization effects). One can then use the observed $P_T(\ell\ell)$ distribution to model the distribution of the sum of neutrino P_T's, which is identified with $E_{\mathrm{T}}^{\mathrm{miss}}$. Then, in order to predict the $t\bar{t} \to$ dilepton contribution to a selection with $E_{\mathrm{T}}^{\mathrm{miss}}$ +X, one applies a cut on $P_T(\ell\ell)$+X instead. In practice one has to rescale the result of the $P_T(\ell\ell)$+X selection to account for the fact that any dilepton selection must include a moderate $E_{\mathrm{T}}^{\mathrm{miss}}$ cut in order to reduce Drell Yan backgrounds and to account for polarization effects. The result of these procedures is shown in Fig. 5. This method predicts $2.1 \pm 2.1(\mathrm{stat}) \pm 0.6(\mathrm{syst})$ while the SM MC yield is 1.3 events. One event is observed in data and other methods are used to cross-check that result [17]. No excess is seen beyond the SM prediction and we set limits in the cMSSM plane in Fig. 6 which substantially improve results from previous experiments.

5. Multi-lepton search

The multi-lepton analysis [20] selects events with three isolated leptons or more, acquired using single-lepton and di-lepton triggers. Candidate events in this search must have at least three leptons, of which at least one must be an electron or a muon. We classify multilepton events into search channels on the basis of the number of leptons, lepton flavour, and relative charges as well as charge and flavour combinations and other kinematic quantities described below. The

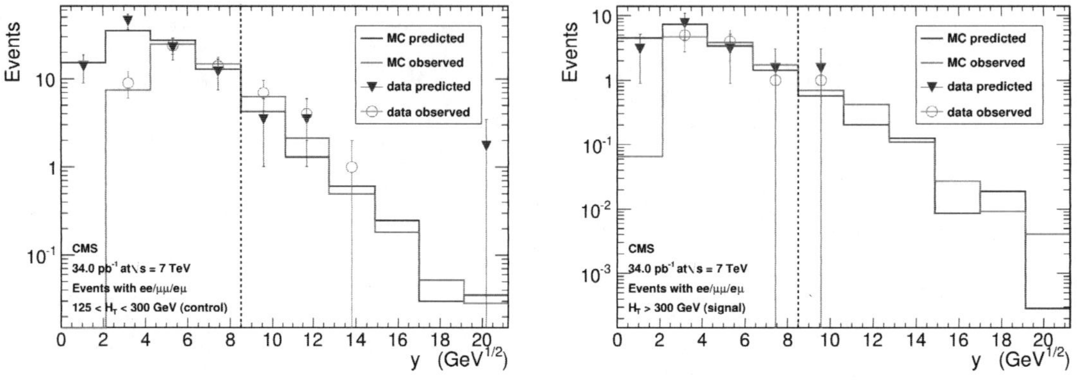

Figure 5. Distributions of y (observed) and $P_T(\ell\ell)/\sqrt{H_T}$ (predicted) for the control region (left) and signal region (right). We show the distributions in both MC and data.

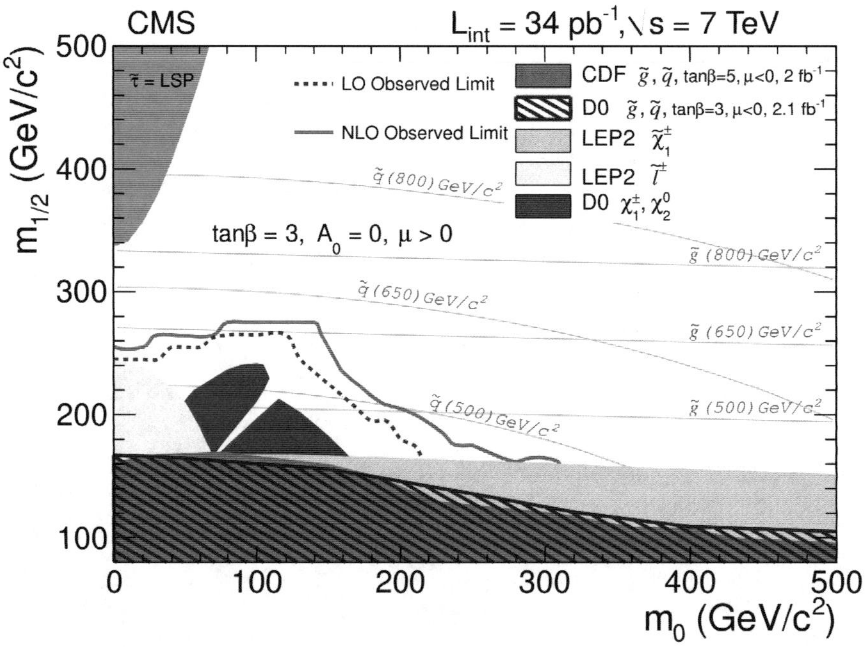

Figure 6. The observed 95% CL exclusion contour for the opposite-sign dilepton analysis at NLO (solid red line) and LO (dashed blue line) in the CMSSM $(m_0, m_{1/2})$ plane for $\tan\beta = 3$, $A_0 = 0$ and $\mu > 0$. The area below the curve is excluded by this measurement. Exclusion limits obtained from previous experiments are presented as filled areas in the plot. Thin grey lines correspond to constant squark and gluino masses.

channels are formed exclusively in that a candidate event can belong to one and only one channel. This avoids overlap between channels for limit-setting purposes. Each channel is treated as an independent search and the collective sensitivity of all channels to a model of new physics is obtained by combining exclusive sensitivities using standard statistical prescriptions.

There are 55 independent samples according to the relative charge of the leptons and their

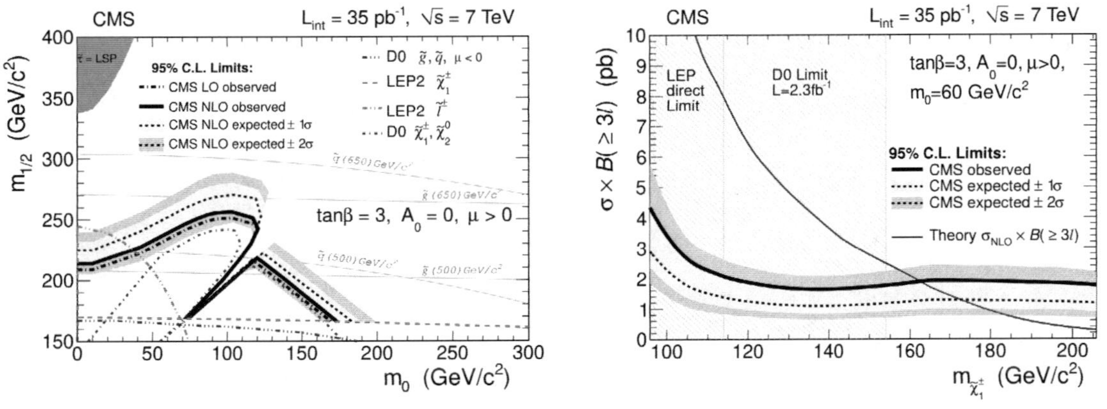

Figure 7. Left: 95% CL exclusion contour for the cMSSM scenario for the multi-lepton search along with the limits from the multilepton searches from the Tevatron [21] and the exclusion derived from slepton and chargino limits from LEP [22, 23, 24, 25, 26, 27]. Right: the expected and observed upper limits on the cross section times branching ratio $\sigma \times B(3\ell)$ as a function of the chargino mass.

flavour, which can be e, μ, and τ. The three-lepton requirement strongly reduces the Standard Model background, and the largest remaining background process is Z+jets, including Drell-Yan. The presence of hadronic activity in an event is characterised by the variable H_{T}. Jets used for the H_{T} determination must be well separated from any identified leptons; jets are required to have no lepton in a cone $\Delta R < 0.3$ around the jet axis. The remaining background is further suppressed by requiring $H_{\mathrm{T}} > 30$ GeV, $E_{\mathrm{T}}^{\mathrm{miss}} > 50$ GeV or a Z veto, depending on the considered final state. No excess is found with respect to the predicted background in search region, and limits are set in a variety of models. In particular, Fig. 7 shows the 95% CL exclusion in the cMSSM plane together with limits from Tevatron [21] and LEP searches [22, 23, 24, 25, 26, 27] for $A_0 = 0, \tan\beta = 3$, and $\mu > 0$. As can be seen, our results extend the excluded region in comparison with previous results from LEP and the Tevatron. For small values of m_0 the sleptons can become lighter than the gauginos, so the gauginos will decay into slepton and lepton (two-body decay), although for larger values of m_0 three-body decays will dominate.

While for two-body decays the branching fraction into leptons is 100%, it decreases rapidly for three-body decays. In the transition region from two- to three-body decays the leptons become soft and fail the p_{T} requirement [28]. Exclusion is therefore not possible, as shown by the non-excluded region between the two- and three-body decay regions. We exclude gluino masses up to 628 GeV for this choice of parameters.

The 95% CL upper limit on the cross section times branching fraction into 3ℓ varies from $\sigma_{95} = 0.8$ to 2 pb. The sensitivity to the chargino mass can be seen in the right panel of Fig. 7, where the NLO cross section for $m_0 = 60$ GeV equals the 95% CL experimental limit of $\sigma_{95} = 2$ pb for chargino mass of 163 GeV. Therefore, chargino masses above this value cannot be excluded.

6. Conclusions

We have presented results of searches for new physics in various topologies with one or more isolated leptons, jets, and missing transverse energy in the final state using the 35 pb^{-1} of the CMS 2010 dataset. No evidence for new physics is observed and limits are set on the production cross-section times the event acceptance for the searched topologies which exceed previous limits from LEP and the Tevatron.

References

[1] LEP2 Joint SUSY Working Group (ALEPH, DELPHI, L3 and OPAL Collaborations) 2004 *Notes LEPSUSYWG/01-03.1 and 04-01.1* URL `http://lepsusy.web.cern.ch/lepsusy`
[2] Alcaraz J (ALEPH and CDF and D0 and DELPHI and L3 and OPAL and SLD Collaborations) 2009 (*Preprint* `0911.2604`)
[3] Witten E 1981 *Nucl. Phys.* **B 188** 513
[4] Dimopoulos S and Georgi H 1981 *Nucl. Phys.* **B 193** 150
[5] Martin J 1997 *arXiv:hep-ph/9709356*
[6] Wess J and Zumino B 1974 *Nucl. Phys* **B 70** 39
[7] Nilles H P 1984 *Phys. Reports* **110** 1
[8] Haber H E and Kane G L 1987 *Phys. Reports* **117** 75
[9] Barbieri R, Ferrara S and Savoy C A 1982 *Phys. Lett.* **B119** 343
[10] Dawson S, Eichten E and Quigg C 1985 *Phys. Rev.* **D31** 1581
[11] Ellis J R, Hagelin J S, Nanopoulos D V, Olive K A and Srednicki M 1984 *Nucl. Phys.* **B238** 453–476
[12] Chatrchyan S *et al.* (CMS Collaboration) 2011 *arXiv:hep-ex/1107.1279*
[13] Chatrchyan S *et al.* (CMS Collaboration) 2011 *arXiv:hep-ex/1106.3272*
[14] Khachatryan V *et al.* (CMS Collaboration) 2011 *Phys.Lett.* **B698** 196–218 (*Preprint* `1101.1628`)
[15] Chatrchyan S *et al.* (CMS Collaboration) 2011 *arXiv:hep-ex/1107.1870*
[16] Chatrchyan S *et al.* (CMS Collaboration) 2011 *JHEP* **1106** 077 (*Preprint* `1104.3168`)
[17] Chatrchyan S *et al.* (CMS) 2011 *JHEP* **06** 026 (*Preprint* `1103.1348`)
[18] Khachatryan V *et al.* (CMS Collaboration) 2011 *Phys. Lett.* **B695** 424–443
[19] Pavlunin V 2010 *Phys. Rev.* **D81** 035005
[20] Chatrchyan S *et al.* (CMS Collaboration) 2011 *Submitted to JHEP* (*Preprint* `1106.0933`)
[21] Abazov V M *et al.* (D0 Collaboration) 2009 *Phys. Lett.* **B680** 34
[22] LEP2 Joint SUSY Working Group (ALEPH, DELPHI, L3 and OPAL Collaborations) 2004 *Combined LEP Selectron/Smuon/Stau Results, 183-208 GeV* URL `http://lepsusy.web.cern.ch/lepsusy/www/sleptons_summer04/slep_final.html`
[23] Heister A *et al.* (ALEPH Collaboration) 2002 *Phys. Lett.* **B526** 206
[24] Heister A *et al.* (ALEPH Collaboration) 2004 *Phys. Lett.* **B583** 247
[25] Abdallah J *et al.* (DELPHI Collaboration) 2003 *Eur. Phys. J.* **C31** 421
[26] Achard P *et al.* (L3 Collaboration) 2004 *Phys. Lett.* **B580** 37
[27] Abbiendi G *et al.* (OPAL Collaboration) 2004 *Eur. Phys. J.* **C32** 453
[28] Aaltonen T *et al.* (CDF Collaboration) 2008 *Phys. Rev. Lett.* **101** 251801

Study of short-lived resonances with the ALICE Experiment at the LHC

Ayben Karasu Uysal[1]

Department of Physics, Yildiz Technical University, Istanbul, Turkey

E-mail: akarasu@cern.ch

Abstract. The study of short-lived resonances allows the investigation of the collision dynamics and of the properties of the hot and dense medium created in high energy collisions. Moreover it is interesting to address the topics of the strangeness production by the analysis of strange resonances. First measurements of the $\phi(1020)$, $\Lambda^*(1520)$, $K^*(892)$, $\Xi^*(1530)$ and doubly charged $\Delta(1232)$ resonances in pp collisions at a center of mass energy of 7 TeV with the ALICE apparatus at the LHC are presented. Thermal model predictions of particle ratios in proton-proton collisions are shown.

1. Introduction

Resonances have unique characteristics to probe various properties of the hot and dense matter produced in relativistic heavy-ion collisions. The in-medium effects related to the high density and/or high temperature of the medium can modify various resonance properties, such as masses, widths and yields [1].

Since resonances may decay between chemical and thermal freeze-outs of the fireball, they can also provide information on hadronization and the time span between chemical and thermal freeze-out [2]. Comparison of the properties of the resonances obtained from heavy-ion collisions with the same obtained from elementary pp collisions is expected to provide evidences for possible in-medium effects.

In view of this, ALICE has attempted to detect and analyze resonances. Measurement of their properties such as mass, width, yield as a function of the transverse momentum can provide insight for understanding the dynamics of the medium created in the collision.

ALICE [3] is the LHC experiment mainly devoted to the study of ultrarelativistic heavy-ion collisions, in order to investigate the properties of the Quark-Gluon Plasma (QGP). Its excellent capabilities for particle identification in a wide transverse momentum (p_t) range (from ~0.1 to ~10 GeV/c), make it a very well suited device for resonances study.

In this contribution, results are presented for $K^*(892)^0$ (identified by charged $K\pi$ decay channel), $\phi(1020)$ (K^+K^- decay channel), $\Xi(1530)^0$ ($\Xi\pi$ decay channel), $\Lambda^*(1520)$ (pK decay channel) and doubly charged $\Delta(1232)$ ($p\pi$ decay channel) measurements done with a sample of data collected by ALICE in 2010. Section 2 shortly describes the analysis procedure and Section 3 shows some results. Section 4 presents the conclusions and an outlook.

[1] For the ALICE Collaboration

Journal of Physics: Conference Series **347** (2012) 012012 doi:10.1088/1742-6596/347/1/012012

2. Analysis

The Time Projection Chamber (TPC) and the Time-of-Flight (TOF) detectors are used to measure the resonance production through the detection of its hadronic decay products. The TPC provides identification and momentum information of the charged particles by measuring their ionization energy loss (dE/dx). Charged particles in the intermediate momentum range are identified in ALICE by the Time Of Flight (TOF) detector. The time measurement with the TOF, in conjunction with the momentum and track length measured by the tracking detectors is used to calculate the particle mass. The combination of these methods provides particle identification over the large transverse momentum range.

The online trigger selection requires a signal in either of the VZERO scintillator counters or at least one hit in either of the silicon pixel detector (SPD) layers. The primary vertex is reconstructed using either SPD tracklets or global tracks. Events are selected by requiring that the distance between the position of primary vertex and the geometrical centre of the apparatus along the beam axis be less than 10 cm.

The dataset corresponding to pp collisions at $\sqrt{s}$ = 7 TeV consisted of 25 to 385 millions of minimum-bias (MB) events, for the different resonances analyzed.

2.1. Reconstruction of resonances

Resonance signals are reconstructed by calculating the invariant mass of the pair of decay daughters. However, the pair invariant mass spectra have large underlying backgrounds due to many random combinatorial pairs, which have no correlation at all. Therefore, to get the real resonance invariant spectra one needs to subtract out the huge combinatorial background. This is done using event-mixing [4] or like-sign [5] techniques if the resonance signals are not clearly visible.

The result of the subtraction is fitted with a Breit-Wigner or a Voigtian (= convolution of Gaussian and Breit-Wigner) to describe the signal function plus a polynomial function to account for the remaining residual background. Background identification methods and fit functions used for each resonance species are summarized in the Table 1.

Table 1. Background identification methods and fit functions used for each resonance particle

Resonance	Background Technique	Signal Function	Residual Background Function
$K^*(892)$	Like-Sign	Breit-Wigner	Straight Line
$\phi(1020)$	Fit Function	Voigtian	2nd Order Polynomial
$\Xi(1530)^0$	Fit Function	Voigtian	2nd Order Polynomial
$\Lambda^*(1520)$	Event-Mixing	Breit-Wigner	2nd Order Polynomial
$\Delta(1232)$	Event-Mixing	p-wave Breit-Wigner	Straight Line

3. Results

$K^*(892)^0$, $\phi(1020)$ and $\Xi(1530)^0$ raw counts in pp collisions are corrected for reconstruction efficiency, acceptance and branching ratio. The obtained p_t spectra are compared with the PHOJET [6] and some widely used PYTHIA [7] Atlas-csc [8], Perugia-0 [9] and D6T [10] tunes. For $K^*(892)^0$, the best agreement is obtained for D6T, while PHOJET and Atlas-csc overestimate low p_t yields and underestimate high p_t ones. The $\phi(1020)$ is well reproduced by D6T below 2.5 GeV/c and by PHOJET and Atlas-csc above this limit. Both mesons are underestimated by Perugia-0. In the case of $\Xi(1530)^0$, none of the above Monte-Carlo (MC) models succeeds in describing the measured p_t spectrum. Figure 1 shows the comparison between corrected spectra and MC models.

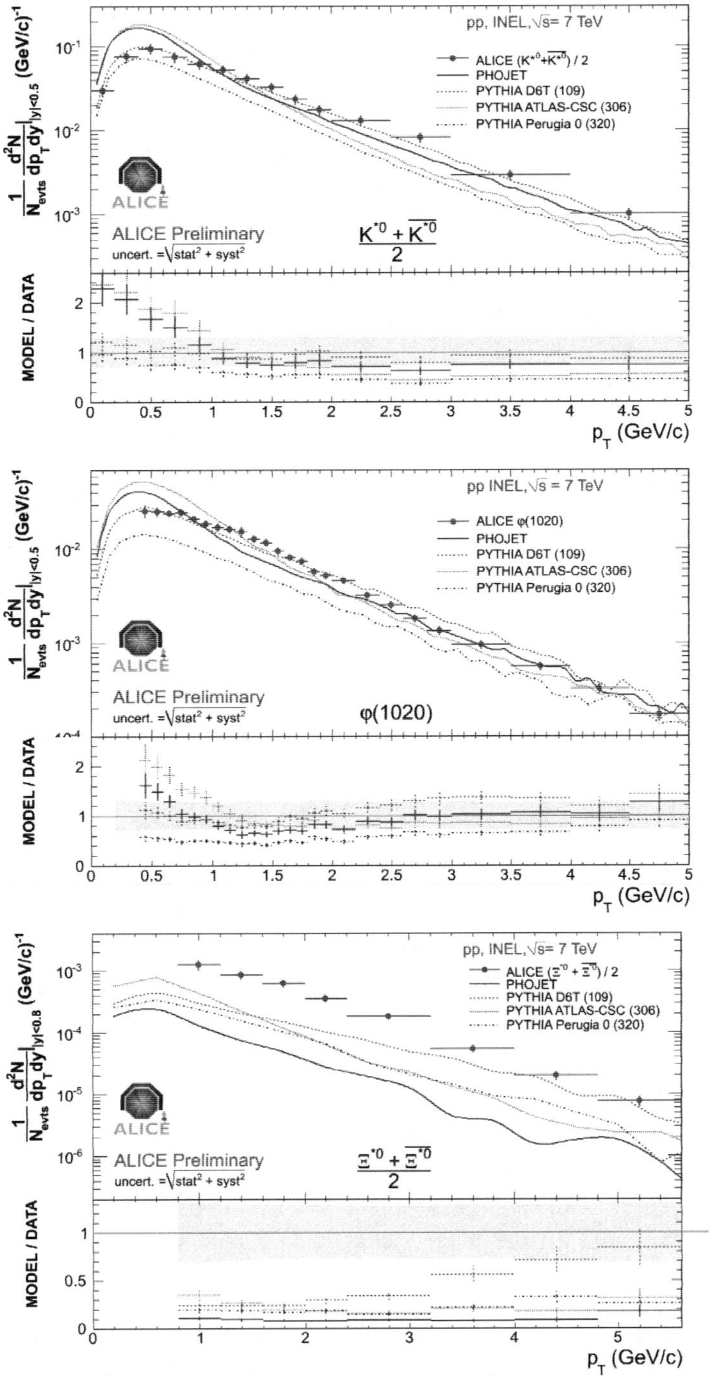

Figure 1. Transverse momentum spectra of (upper) $K^*(892)^0$, (middle) $\phi(1020)$ and (lower) $\Xi^*(1530)^0$ compared with different MC Models. Lower part of each plot shows the ratio between MC and data for each measured bin. The shadow band is the systematic error on data.

Figure 2 and Figure 3 show the extracted mass values for the doubly charged $\Delta(1232)$ and $\Lambda^*(1520)$ resonances as a function of p_t. The values are in agreement with the Particle Data Group

(PDG) [11] values within errors for $\Lambda^*(1520)$, while 30-50 MeV/c^2 mass shift is observed for doubly charged $\Delta(1232)$.

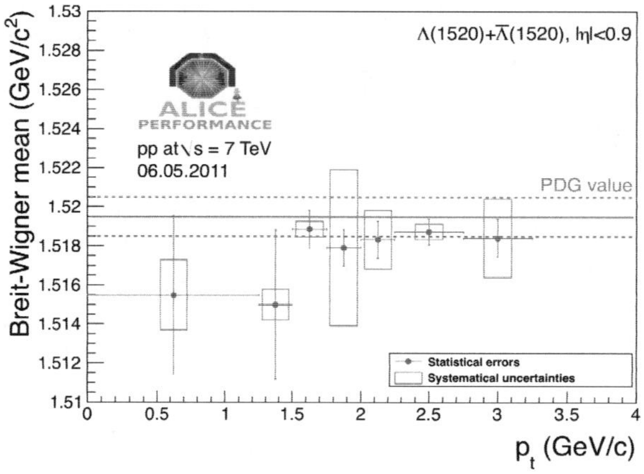

Figure 2. Extracted mass of $\Lambda^*(1520)$ + anti-$\Lambda^*(1520)$ as a function of p_t. Solid line is the PDG mass (1.5195 GeV/c^2) and the dashed lines are the PDG errors. Statistical errors are estimated from the fit function while systematical errors are calculated from the difference when $\Lambda^*(1520)$ and anti-$\Lambda^*(1520)$ are analyzed separately.

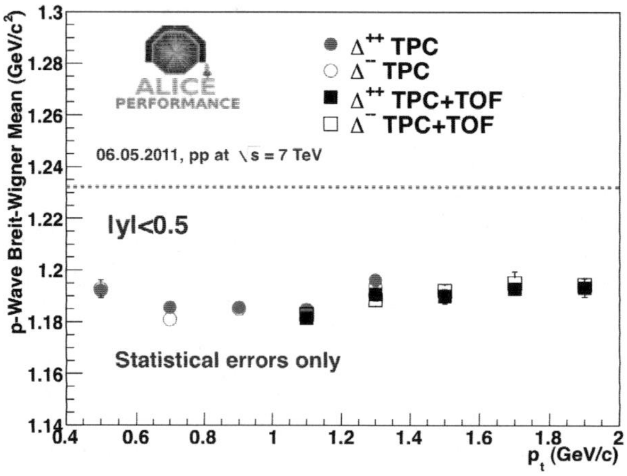

Figure 3. Extracted mass of (full symbols) Δ^{++} and (open symbols) Δ^{--} as a function of p_t. Solid line is the PDG mass (1.232 GeV/c^2). Errors are statistical only.

4. Conclusions

The resonance measurements have been performed using the dE/dx in the gas of the TPC and time-of-flight information from TOF. The combinations of these techniques allow us to cover a broad range of momentum.

$d^2N/dydp_t$ has been determined for $K^*(892)^0$, $\phi(1020)$ and $\Xi(1530)^0$ resonances through their hadronic decay channels. Spectra show different levels of agreement with various MC event generators: mesons are reasonably well reproduced by PYTHIA D6T, while PHOJET and PYTHIA Atlas-csc match $\phi(1020)$ above 2.5 GeV/c and Perugia-0 underestimates all spectra. The $\Xi(1530)^0$ is underestimated in all cases. The study of other resonances will complete the picture on pp collisions at $\sqrt{s}$= 7 TeV. The presented analysis in pp is done as a baseline and will be continued for the Pb-Pb data.

Journal of Physics: Conference Series **347** (2012) 012012 doi:10.1088/1742-6596/347/1/012012

The Δ^{++} and Δ^{--} masses are shifted found to be shifted by 30-50 MeV/c^2 and no p_t dependence of this shift is observed. Width values, except for the high momentum range, are in agreement within the error bars with the PDG value. The mass shift can be explained as a kinematical shift inside a thermal medium which is coupled to the $N\pi \leftrightarrow \Delta(1232)$ interaction [12].

In order to estimate the particle yield ratios, the statistical thermal model used. Ratios for pp collisions at $\sqrt{s}$ = 7 TeV for different temperature values have been calculated within strangeness canonical approach using the THERMUS code [13] and the results are presented in Figure 4. This shows that the particle ratios are very sensitive to the freeze-out temperature and they are inversely proportional to the mass difference.

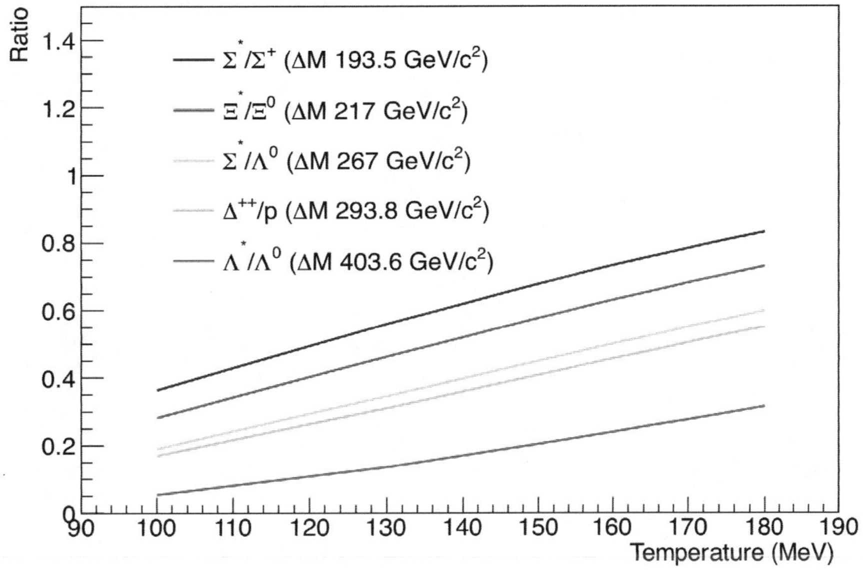

Figure 4. Ratios of particles with different temperatures predicted using the strangeness canonical THERMUS model. ΔM values show the mass differences between the two particles.

References

[1] Rapp R and Wambach J 2000 *Adv. Nucl. Phys.* **25** 1
[2] Torrieri G and Rafelski 2001 J *Phys. Lett.* B **509** 239
[3] Aamodt K et al. (ALICE Collaboration) 2008 *JINST* **3** S08001
[4] L'Hôte D 1994 *Nucl. Inst. Meth. in Phys. Res. Sec.* A **337** 544
[5] Adams J et al. 2003 *Phys. Rev. Lett.* **92** 092301
[6] Engel R et al. 1995 *Z. Phys.* C **66** 203
[7] Sjöstrand T, Mrenna S and Skands P 2006 *J. High Energy Phys.* **05** 026
[8] Field R 2008 *Acta Phys. Pol.* B **39** 2611
[9] Buttar C, Clements D, Dawson I and Moraes A 2004 *Acta Phys. Pol.* B **35** 433
[10] Skands P Z 2009 *Preprint* hep-ph/0905.3418 [arXiv]
[11] Nakamura K et al. (Particle Data Group) 2010 *J. Phys. G.* **37** 075021
[12] Weinhold W, Friman B and Nörenberg W 1998 *Phys. Lett.* B **433** 236
[13] Wheaton S and Cleymans J 2009 *Comput. Phys. Commun.* **180** 84

2nd International Conference on Particle Physics IOP Publishing
Journal of Physics: Conference Series **347** (2012) 012013 doi:10.1088/1742-6596/347/1/012013

Search for New Physics in rare heavy flavour decays at LHCb

Francesco Dettori[1]
INFN Sezione di Cagliari - S. P. Monserrato-Sestu Km 0.700 09042 Monserrato (CA) Italy
E-mail: Francesco.Dettori@cern.ch

Abstract. LHCb first searches for rare heavy flavour decays are reported in this contribution. 2010 dataset has been used to put bounds on $B^0_{d,s} \to \mu^+\mu^-$ decays which are comparable with present best limits. Searches for lepton number violating $B^+ \to h^-\mu^+\mu^+$decays have lead to tight limits on their branching ratios. Radiative and $B^0_d \to K^*\mu^+\mu^-$ decays have already been observed and will produce sensitive results in the very near future. Present results and future prospects are described in view of their potential for new physics searches.

1. Introduction

The study of rare decays is an effective method to search for physics beyond the Standard Model (SM) and is complementary to direct production searches; while the latter aim to the production of new real particles, the former explore the production of new virtual particles, which, by mediating these reactions, can modify their branching ratios or other sensitive parameters. For this reason higher energy ranges than directly accessible can be probed with rare decays. Furthermore, being suppressed or forbidden in the SM, the New Physics (NP) contributions can be at the same level or larger than the SM ones.

In this contribution we report the first searches for heavy flavour rare decays, in particular for b-hadron decays, obtained by the LHCb detector exploiting data taken in proton-proton collisions at $\sqrt{s} = 7$ TeV provided in 2010 and 2011 by the LHC. Most of the results presented here are based on the 2010 total integrated luminosity of 37 pb^{-1}, while some are updated with data from 2011 run. The LHCb detector is appositely designed to collect the forward $b\bar{b}$ pairs production in high energy pp collisions ($\sigma(pp \to b\bar{b}X) \sim 75.3\mu b$ at 7 TeV in the pseudorapidity range $2 < \eta < 6$[1]); it is built as a forward single arm spectrometer covering a rapidity range of $1.8 < y < 4.9$. LHCb is equipped as a standard multipurpose detector with a tracking system with excellent momentum, and therefore mass, resolution, RICH detectors and muon stations for particle identification and electromagnetic and hadronic calorimeters. A fully detailed description is available elsewhere [2].

In the following we describe different studies which are already or about to be published and prospects for important channels which will be subject of future publications.

[1] On behalf of the LHCb Collaboration

Published under licence by IOP Publishing Ltd

2. Search for $B^0_{d,s} \to \mu^+\mu^-$ decays

The $B^0_{d,s} \to \mu^+\mu^-$ decays are very rare decays in the Standard Model being flavour changing neutral currents (FCNC) and further suppressed for helicity reasons. Within SM their branching ratios are precisely predicted to be [3]:

$$\mathcal{B}^{SM}(B^0_s \to \mu^+\mu^-) = (3.35 \pm 0.32) \cdot 10^{-9} \tag{1}$$
$$\mathcal{B}^{SM}(B^0 \to \mu^+\mu^-) = (0.10 \pm 0.01) \cdot 10^{-9} \tag{2}$$

while in different NP models these rates can be highly enhanced and are therefore golden channels for the search of New Physics also for their clear experimental signature.

The search for $B^0_{d,s} \to \mu^+\mu^-$ decays presented here is based on about 37pb^{-1} of integrated luminosity collected in 2010. We provide here only an overview of this analysis, while for further details the reader should view Ref. [4]. This analysis starts with a soft preselection aiming at reducing the dataset to a manageable level and preserving high and similar signal and control channel efficiencies while reducing background. Good quality tracks displaced from the primary vertex are selected and fitted to form a vertex if the distance of closest approach between them is less than 3 mm; the secondary vertex is required to have good quality fit and to be displaced from the primary vertex. Tracks are identified as muons if they have a minimum of two hits in the last four muon stations. The dominant background after this selection is the combinatorial $b\bar{b} \to \mu^+\mu^- X$, while the double misidentification background from hadronic two-body decays is negligible.

After this selection a further discrimination is obtained exploiting two variables: the invariant mass of the muon pair and a multivariate likelihood discriminant (GL). The GL is built upon information from the distance of closest approach of the two muons, the B candidate lifetime, the impact parameter of the B and of the muons with respect to the primary vertex, the transverse momentum of the B and the isolation variable, which takes into account the number of tracks surrounding the candidate muons. The GL is by construction distributed uniformly between 0 and 1 for the signal, while background events accumulate at 0, as shown in Fig. 1(a) for LHCb Monte Carlo (MC) simulations as described in Ref. [5]. The GL probability distribution has been trained using MC simulations but calibrated using real data with hadronic two body decays ($B^0 \to h^+h^-$) in order to mimic the signal and di-muon invariant mass sidebands for the background. The result of this calibration is shown in Fig. 1(b) where the measured GL distributions are in agreement with what expected from MC simulations.

The signal invariant mass distribution has been parametrised using a Crystal Ball function. Its resolution, the σ of the gaussian portion, has been estimated with two methods. The first method interpolates the resolution obtained for charmonium and bottomonium resonances (J/ψ, $\psi(2S)$, $\Upsilon(1S)$, $\Upsilon(2S)$ and $\Upsilon(3S)$), while the second exploits $B \to h^+h^-$ decays without particle identification, which would modify the momentum spectra of the selected particles. The two methods agree and the estimated average resolution is $\sigma = 26.7 \pm 0.9$ MeV/c^2. The mean values of the B masses, obtained instead from fits to exclusive $B \to h^+h^-$ decays isolated using particle identification, are $M_{B^0} = 5275.0 \pm 1.0$ MeV/c^2 and $M_{B^0_s} = 5363.1 \pm 1.5$ MeV/c^2.

Three different channels, $B^+ \to J/\psi(\mu^+\mu^-)K^+$, $B^0_s \to J/\psi(\mu^+\mu^-)\phi(K^+K^-)$ and $B^0 \to K^+\pi^-$, are used to normalise the signal branching ratio in following way:

$$\mathcal{B}(B^0_q \to \mu^+\mu^-) = \mathcal{B}_{\text{norm}} \times \frac{\varepsilon_{\text{norm}}}{\varepsilon_{\text{sig}}} \times \frac{f_{\text{norm}}}{f_{B^0_q}} \times \frac{N_{B^0_q\to\mu^+\mu^-}}{N_{\text{norm}}} = \alpha_{B^0_q\to\mu^+\mu^-} \times N_{B^0_q\to\mu^+\mu^-} \tag{3}$$

where $\varepsilon_{\text{norm}}/\varepsilon_{\text{sig}}$ is the efficiency ratio, N and $\mathcal{B}$ are respectively the number of observed events and branching ratio of each channel, and $f_{\text{norm}}/f_{B^0_q}$ the ratio of the probabilities for a b quark to hadronise in the given B meson. This last factor gives the largest contribution to the

Journal of Physics: Conference Series **347** (2012) 012013 doi:10.1088/1742-6596/347/1/012013

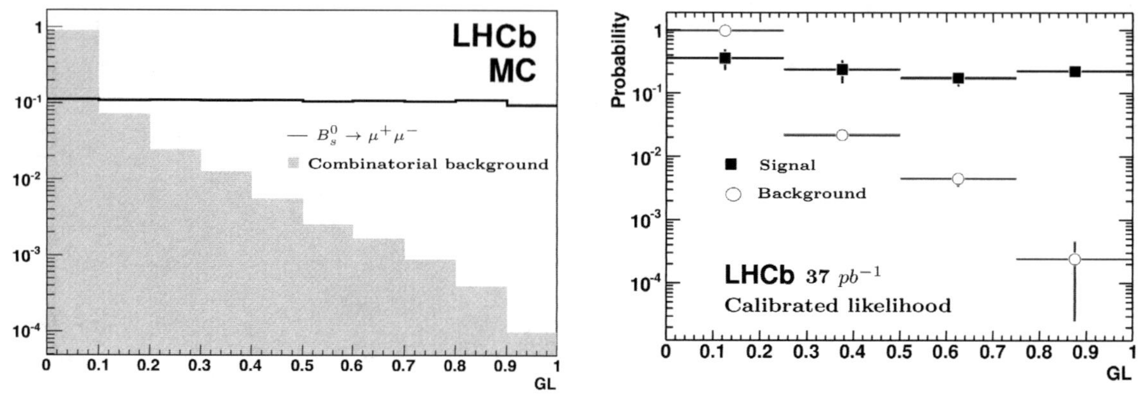

(a)

(b)

Figure 1. Distributions of the discriminant multivariate output GL for $B_s^0 \to \mu^+\mu^-$ signal and background in (a) MC simulations and (b) calibrated on real data (see text for details).

normalisation error in channels with different B meson with respect to the signal one; the currently used value [6] is $f_{B^0}/f_{B_s^0} = 3.71 \pm 0.47$, however the LHCb experiment is improving this observable [7] and consequently will reduce this error in the future. The single event sensitivity, α of Eq.(3), is calculated with each of the normalisation channels and the results are in agreement with each other, giving an average value of:

$$\alpha_{B_s^0\to\mu^+\mu^-} = (8.6 \pm 1.1) \cdot 10^{-9}$$

for the B_s^0 meson and

$$\alpha_{B^0\to\mu^+\mu^-} = (2.24 \pm 0.16) \cdot 10^{-9} \quad .$$

for the B^0 meson.

The search has been performed in 4×6 two dimensional bins of GL and invariant mass (in the signal region of ± 60 MeV/c^2 around the B masses). The two dimensional scatter plot of data events is shown in Fig. 2. For each of the 24 bins the compatibility of the number of observed events with the background only and the background plus signal hypotheses is calculated with the CL method [8] as a function of the signal branching ratio. Since no deviation is observed from the background-only hypothesis, the following upper limits on the signal branching ratios are put:

$$B^0 \to \mu^+\mu^- \quad < \quad 4.3(5.6) \cdot 10^{-8} \text{ at } 90\%(95\%)\text{C.L.} \tag{4}$$
$$B_s^0 \to \mu^+\mu^- \quad < \quad 1.2(1.5) \cdot 10^{-8} \text{ at } 90\%(95\%)\text{C.L.} \quad . \tag{5}$$

This analysis is going to be updated soon to include 2011 data. Currently, exploiting 58pb^{-1} of integrated luminosity collected in 2011, prospect for the sensitivity of this analysis has been produced. Yields are compatible with what have been observed in 2010 and resolutions and signal to background ratio are unchanged. Furthermore the analysis will take advantage of the use of full particle identification system and improved multivariate discriminants. Therefore it can be foreseen that the analysis of 2011 dataset will give results equal or better to the extrapolation with statistics of 2010 results presented here. In particular, in Fig. 3(a) the exclusion possibility for $B_s^0 \to \mu^+\mu^-$ branching ratio as a function of the integrated luminosity is shown, reaching the level of $5 \cdot 10^{-9}$ with 1fb^{-1}, *i.e.* 2011 year-end expected statistics. Conversely in Fig. 3(b) the discovery potential for a given branching ratio is shown, displaying that a $\mathcal{B}(B_s^0 \to \mu^+\mu^-) \sim 7 \cdot 10^{-9}$ could be measured at 3σ by the end of the year.

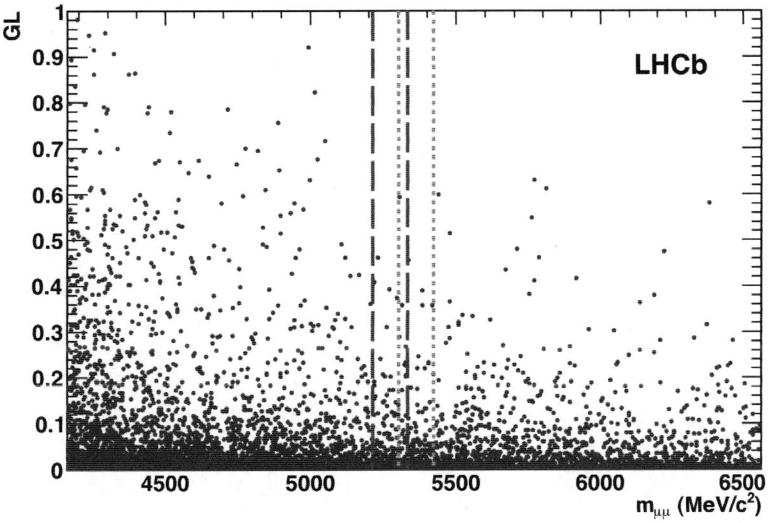

Figure 2. Scatter plot, GL versus di-muon invariant mass, of real data events observed by the LHCb experiment. The dashed green lines and the dotted orange lines indicate respectively the B^0 and B_s^0 signal regions.

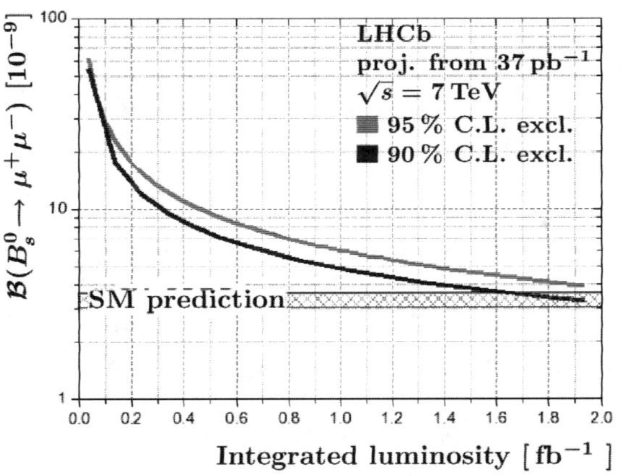

(a)

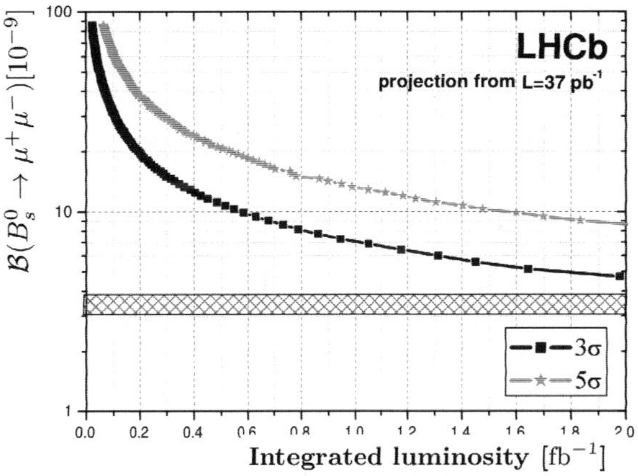

(b)

Figure 3. Prospects for the search of $B_s^0 \to \mu^+\mu^-$ decay in LHCb, as a function of integrated luminosity, obtained by extrapolating the 37pb^{-1} results. The exclusion possibility on branching ratio value at 90% (95%) C.L. is shown in blue (red) in (a) while the 3 (5) σ discovery potential is shown in blue (orange) in (b).

3. Angular observables in $B_d^0 \to K^* \mu^+ \mu^-$ decay

The $B_d^0 \to K^* \mu^+ \mu^-$ is a FCNC $b \to s\ell\ell$ reaction and is therefore at loop level in the SM. While its measured branching ratio ($1.05 \cdot 10^{-6}$ [9]) is in agreement with SM prediction within errors, NP can modify the differential decay rate as a function of the di-muon invariant mass (q^2). This translates into different observables which are sensitive to NP; for example the forward-backward asymmetry (A_{FB}) of the muons with respect to K^* in the di-muon rest frame and in particular its zero-crossing point [10] which is free of hadronic uncertainties.

Previous measurements [11, 12, 13] of A_{FB} are in fair agreement with SM predictions within the errors, which are still large. However the data points have opposite sign of A_{FB} with respect to SM in the region with q^2 lower than the J/ψ mass, favouring an opposite sign for the Wilson coefficient of the electromagnetic dipole coupling, which would be a symptom of NP models [14].

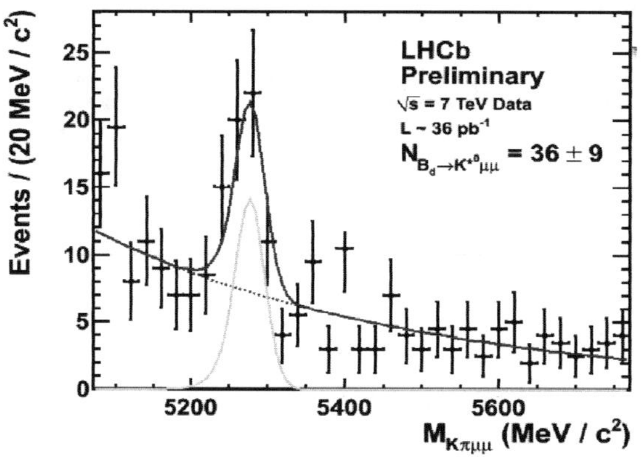

Figure 4. Invariant mass distribution of $K^* \mu^+ \mu^-$ in 36pb^{-1} of integrated luminosity from 2010 dataset.

LHCb is well suited for the measurement of angular observables in $B_d^0 \to K^* \mu^+ \mu^-$. Exploiting an analysis based on a Boosted Decision Tree (BDT) discriminant 36 signal events were observed in the 2010 dataset, as is shown in the invariant mass distribution in Fig. 4. With about 200pb^{-1} a dataset comparable to the previous measurements is expected at LHCb while about 1000 (600) events are expected in 1fb^{-1} with background to signal ratio $\sim 1(0.2)$, allowing precision measurements of $A_{FB}(q^2)$.

4. Photon polarisation in $b \to s\gamma$ transitions

Radiative $b \to s\gamma$ transitions are crucial to understand the possibility of NP couplings. While the branching ratios of these decays are in agreement with SM expectations, they offer the opportunity to measure the photon polarisation which in the SM should be dominantly leftly (rightly) polarised for b ($\bar{b}$) decays. On the other hand NP scenarios can host different couplings leading to a different photon polarisation.

As far as the $B_s^0 \to \phi\gamma$ process is concerned, since the dominant decays are $B_s^0 \to \phi\gamma_R$ and $\bar{B}_s^0 \to \phi\gamma_L$, the final state photon polarisation is related to the time dependent decay rate by means of the B mixing [15]. Firstly seen by Belle [16] the $B_s^0 \to \phi\gamma$ decay has been already observed at LHCb as demonstrated by Fig. 5(b) which shows the $\phi\gamma$ invariant mass for 88pb^{-1} of integrated luminosity which combine 2010 with part of 2011 dataset.

The photon polarisation measurement in the $B^0 \to K^*\gamma$ decay instead exploits the angular distribution of the e^+e^- from the photon conversion. Furthermore $B^0 \to K^*\gamma$ is also a

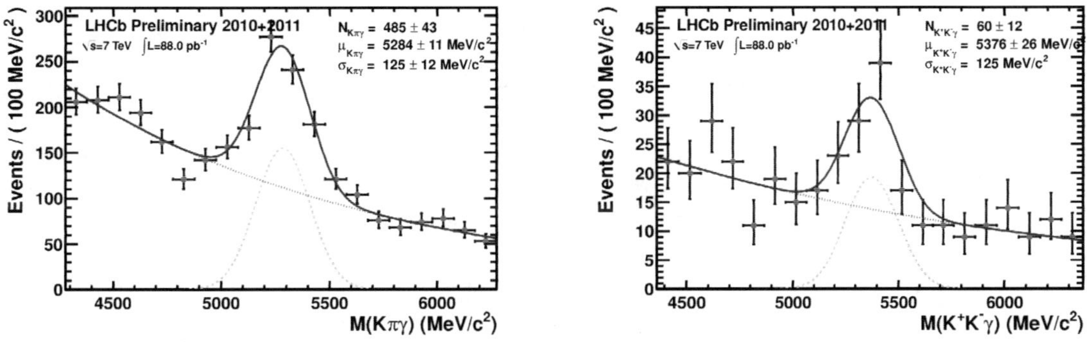

(a) (b)

Figure 5. Observation of radiative decays in the LHCb experiment: invariant mass distribution of (a) $B^0 \to K^*\gamma$ and (b) $B_s^0 \to \phi\gamma$ decays, obtained 88pb^{-1} of integrated luminosity collected in 2010 and part of 2011.

benchmark for all other radiative decays. The observation of this decay in LHCb is shown in Fig. 5(a).

First measurements of CP observables with radiative decays will be produced with a subset of 2011 data and are expected to be competitive with current results.

5. Search for $B^+ \to h^-\mu^+\mu^+$decays

The B meson decays $B^+ \to h^-\mu^+\mu^+(h = \pi, K)$, with lepton number violation $\Delta L = 2$, are forbidden in the Standard Model but could be present in new physics scenarios. In particular, allowing the existence of heavy Majorana neutrinos [17] leads naturally to diagrams as the one in Figure 6 for the $B^+ \to K^-\mu^+\mu^+$decay and similarly for $B^+ \to \pi^-\mu^+\mu^+$. The rate of these processes is highly suppressed by the mixing level between SM and heavy neutrinos but could be enhanced in case the latter is on its mass shell [18].

The search for $B^+ \to h^-\mu^+\mu^+$decays has been done in LHCb using 36 pb^{-1} of integrated luminosity collected in 2010. The analysis strategy was developed exploiting opposite sign muons and in particular, in order to optimise the selection, $B^+ \to h^+J/\psi(\to \mu^+\mu^-)$ was used as signal and the right invariant mass sideband of $B^+ \to K^+\mu^+\mu^-$ for the background. Same selection criteria were used for channels with a kaon or a pion in the final state, with the exception of the particle identification of the hadron.

Simple kinematic cuts exploiting transverse momentum, impact parameter with respect to primary vertex, secondary vertex fit χ^2 and flight distance were used in order to reject

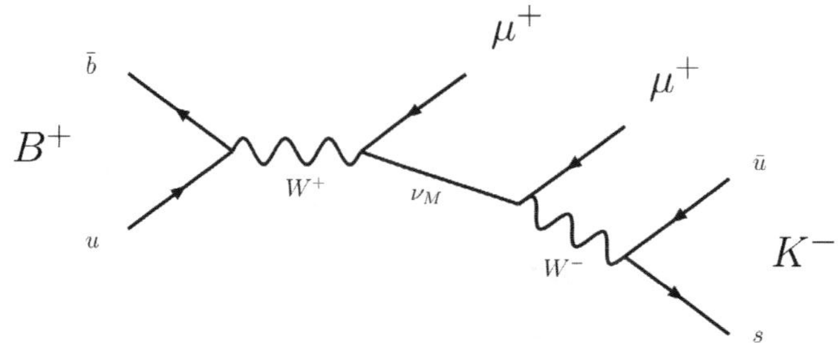

Figure 6. Feynman diagram of the $B^+ \to K^-\mu^+\mu^+$transition mediated by a Majorana neutrino.

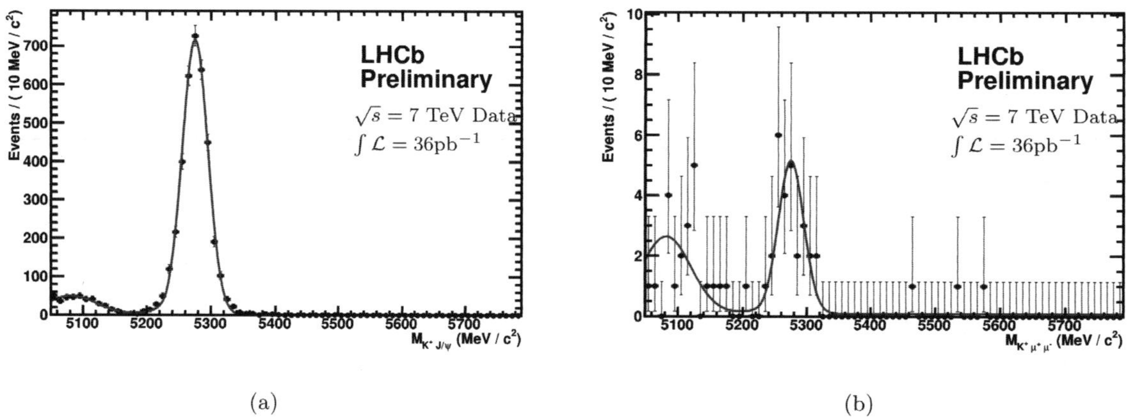

Figure 7. Invariant mass distributions of $B^+ \to K^-\mu^+\mu^+$decays with the two muons (a) in the J/ψ mass region and (b) not in it.

combinatorial background from primary vertex tracks. The invariant mass for $B^+ \to h^+\mu^+\mu^-$ with $|m_{\mu^+\mu^-} - m_{J/\psi}| < 50\text{MeV/c}^2$ is shown in Fig. 7(a) where it can be seen the peak of the $B^+ \to K^+J/\psi$ decay; conversely requiring $|m_{\mu^+\mu^-} - m_{J/\psi}| > 70\text{MeV/c}^2$ the plot in Fig. 7(b) is produced, showing the peak of the rare $B^+ \to K^+\mu^+\mu^-$ decay. From the latter distribution, the combinatorial background has been estimated to be 0.55 ± 0.31 (0.36 ± 0.25) events in the signal region of the $B^+ \to K^-(\pi^-)\mu^+\mu^+$ channel.

Different types of exclusive peaking backgrounds were studied using Monte Carlo simulations and the resulting invariant mass specta are shown in Fig. 8 where it can be seen that all the contributions are well below 0.1 events in the signal regions of both the $B^+ \to h^-\mu^+\mu^+$channels.

No events were observed in the signal mass region of both $B^+ \to K^-\mu^+\mu^+$and $B^+ \to \pi^-\mu^+\mu^+$channels leading to upper limits on the branching ratios of:

$$\mathcal{B}(B^+ \to K^-\mu^+\mu^+) < 4.3 \cdot 10^{-8} \tag{6}$$
$$\mathcal{B}(B^+ \to \pi^-\mu^+\mu^+) < 4.5 \cdot 10^{-8} \tag{7}$$

at 90% C.L. improving significantly previous upper limits [19].

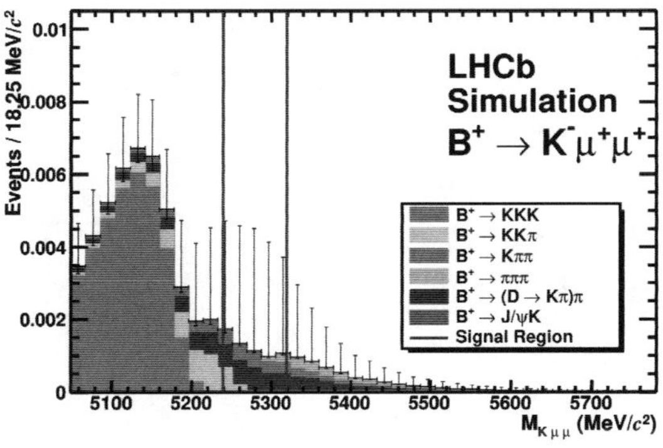

(a)

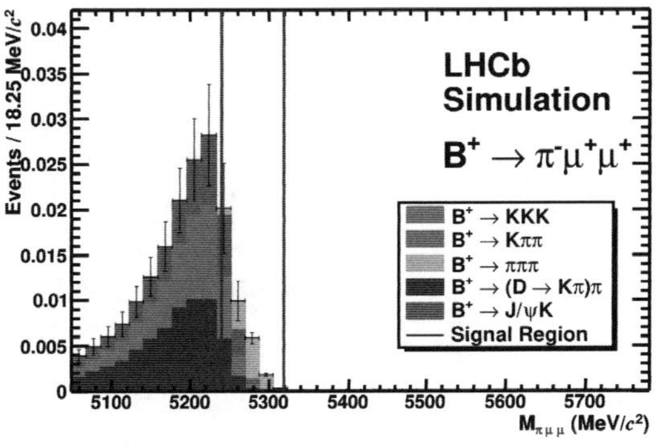

(b)

Figure 8. Invariant mass distributions of simulated peaking background that can fake the $B^+ \to h^-\mu^+\mu^+$decays.

6. Conclusions

LHCb experiment is starting to show its potential in the rare B decays sector after only one year of data taking and 36 pb^{-1} of integrated luminosity. Bounds on the $B^0_{d,s} \to \mu^+\mu^-$ decays branching ratios have been put which are comparable to current world best limits. Searches for $B^+ \to h^-\mu^+\mu^+$decays have put limits on their rate improving significantly previous measurements. Radiative decays and $B^0_d \to K^*\mu^+\mu^-$ decays have been observed and will produce very sensitive results in the near future.

Exploiting the increasing statistics, LHCb in 2011-2012 will be able to discover or to put tight limits on various New Physics models, confirming the importance of rare decays in NP searches.

References

[1] Aaij R *et al.* (LHCb Collaboration) 2010 *Phys. Lett.* **B694** 209–216 (*Preprint* [1009.2731])
[2] Alves A A *et al.* (LHCb Collaboration) 2008 *JINST* **3** S08005
[3] Gámiz E *et al.* (HPQCD Collaboration) 2009 *Phys. Rev. D* **80** 014503
[4] Aaij R *et al.* (LHCb Collaboration) 2011 *Phys. Lett.* **B699** 330–340 (*Preprint* [1103.2465])
[5] Beiyaev I *et al.* 2010 *Nuclear Science Symposium Conference Record (NSS/MIC), 2010 IEEE* pp 1155–1161 ISSN 1082-3654
[6] Asner D *et al.* (Heavy Flavor Averaging Group) 2010 (*Preprint* [1010.1589])
[7] Aaij R *et al.* (LHCb Collaboration) 2011 (*Preprint* [1106.4435])
[8] Junk T 1999 *Nucl.Instrum.Meth.* **A434** 435–443 (*Preprint* [hep-ex/9902006])
[9] Nakamura K *et al.* (Particle Data Group) 2010 *J.Phys.G* **G37** 075021
[10] Egede U, Hurth T, Matias J, Ramon M and Reece W 2010 *JHEP* **1010** 056 (*Preprint* [1005.0571])
[11] Wei J T *et al.* (Belle Collaboration) 2009 *Phys.Rev.Lett.* **103** 171801 (*Preprint* [0904.0770])
[12] Aubert B *et al.* (BABAR Collaboration) 2009 *Phys.Rev.* **D79** 031102 (*Preprint* [0804.4412])
[13] Aaltonen T *et al.* (CDF Collaboration) 2009 *Phys.Rev.* **D79** 011104 (*Preprint* [0804.3908])
[14] Bobeth C, Hiller G and van Dyk D 2010 *JHEP* **07** 098 (*Preprint* [1006.5013])
[15] Muheim F, Xie Y and Zwicky R 2008 *Phys. Lett.* **B664** 174–179 (*Preprint* [0802.0876])
[16] Wicht J *et al.* (Belle Collaboration) 2008 *Phys. Rev. Lett.* **100** 121801 (*Preprint* [0712.2659])
[17] Atre A, Han T, Pascoli S and Zhang B 2009 *JHEP* **05** 030 (*Preprint* [0901.3589])
[18] Cvetic G, Dib C, Kang S K and Kim C S 2010 *Phys. Rev.* **D82** 053010 (*Preprint* [1005.4282])
[19] Edwards K W *et al.* (CLEO Collaboration) 2002 *Phys. Rev.* **D65** 111102 (*Preprint* [hep-ex/0204017])

Journal of Physics: Conference Series **347** (2012) 012014 doi:10.1088/1742-6596/347/1/012014

Recent B-Physics results from CMS

Franco Simonetto[1]
c/o Dip. Fisica, via Marzolo 8, 35131, Padova, Italy
E-mail: franco.simonetto@pd.infn.it

Abstract. The CMS experiment has measured the production of heavy flavored hadrons and jets at 7 TeV. A large range of inclusive and exclusive cross-section measurements (Quarkonia, charged and neutral B hadrons) as well as studies of $b\bar{b}$ angular correlations were made and compared to theoretical predictions at Leading Order and Next to Leading Order precision. These results are here reviewed.

1. Introduction

There are several reasons to study heavy flavour Physics at the LHC. First the measurement of b-jets and b-hadrons production provides a valuable test of QCD predictions, complementing the measurements performed at lower centre of mass energies in $p\bar{p}$ collisions at the Tevatron. In addition, hints of new Physicis might show up in the evolution and decay of B-hadrons. While the B_d, B_u properties have been studied in detail at the B-factories, B_s properties, still largely unknown, can best be measured in a hadron collider. Finally, the top quark, the Higgs boson, and new particles expected in several New Physics models (SUSY Higgs, fourth generation quarks, etc.) decay preferentially to b-jets: a proper understanding of the production of b-quarks in the Standard Model is therefore mandatory to understand the SM background to these new Physics searches.

Out of the four detectors operating at the Large Hadron Collider at CERN, only LHCb was explicitly designed for Heavy Flavour Physics, while ATLAS and CMS are omni-purpose detectors, primarily devoted to the search for new Physics processes taking place at the highest possible energy. However, due to the refined self-triggering muon system, the high resolution silicon tracker, and the precise vertex reconstruction provided by the three layers of inner pixels, CMS provides competitive results also in the B-Physics sector. I review here some of these results. A detailed description of the CMS detector can be found in [1].

For the reader's ease, results are grouped in three sections: inclusive measurements, where b-jets are tagged by high p_T muons or displaced secondary vertices, semi-inclusive measurements of $b \to J/\psi X$ decays, and exclusive measurements of fully reconstructed $B \to J/\psi\ h$ decays, where the hadron h can be either a charged kaon, a neutral kaon or a ϕ, and the J/ψ is reconstructed in its $\mu^+\mu^-$ decay.

[1] On behalf of the CMS Collaboration

2. Measurements of b-jet production properties with inclusive tags

2.1. Inclusive $b \to \mu\bar{\nu}_\mu X$ decays

In the first stage of LHC operation at 7 TeV, the low luminosity allowed CMS to run a single muon trigger with loose requirements: the projection of the muon momentum in the direction perpendicular to the beam axis, p_T^μ, should exceed 3 GeV. Muons are identified by matching a track reconstructed in the silicon tracker to that reconstructed in the muon chambers. To improve track quality and background rejection, at the analysis stage tighter requirements are imposed on the muon transverse momentum, $p_T^\mu > 6$ GeV, and pseudorapidity $|\eta^\mu = -ln(tan(\theta/2))| < 2.1$ (θ is the polar angle between the track direction and the beam line).

Signal events are discriminated from muons produced in background processes (from charm decay, from the decays of charged pions and kaons, from hadron sailing through the detector to the muon chambers, or from the decays of $Q\bar{Q}$ bound states,"onia"), using jet information. Jets with $p_T^{jet} > 1$ GeV are reconstructed using all good quality charged tracks in the event (with $p_T^{track} > 0.3$ GeV), using the anti-k_T algorithm [2] with threshold R=0.5. Due to the large b-quark mass, the momentum of the muon, projected along the direction perpendicular to the jet axis ($p_T^{\mu,rel}$) is on average larger for signal events than for background. A fit to the $p_T^{\mu,rel}$ distribution determines the amount of signal and background in the data. Template shapes for the signal and for each background source are taken from the simulation. The fit result is displayed on Figure 1. Muon trigger, tracking and identification efficiencies are measured with a "Tag & Probe" method, using a sample of muons from $J/\psi \to \mu^+\mu^-$ decays [3]. One of the two muons (the "Tag") is identified. The Probe is then defined as a second track in the same event, with opposite charge to the Tag, which, combined to the Tag, has an invariant mass compatible with the J/ψ world average value [4]. The efficiency is determined in bins of the Probe p_T and η by comparing the number of probes passing and falling the selection criteria. After background subtraction and efficiency correction, with an integrated luminosity

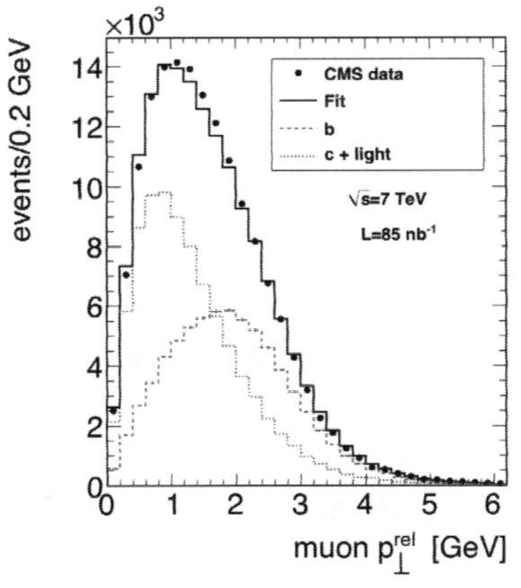

Figure 1. $p_T^{\mu,rel}$ distribution for data (points with error bars), best fit (black line), b and light quarks contributions (red and blue lines respectively).

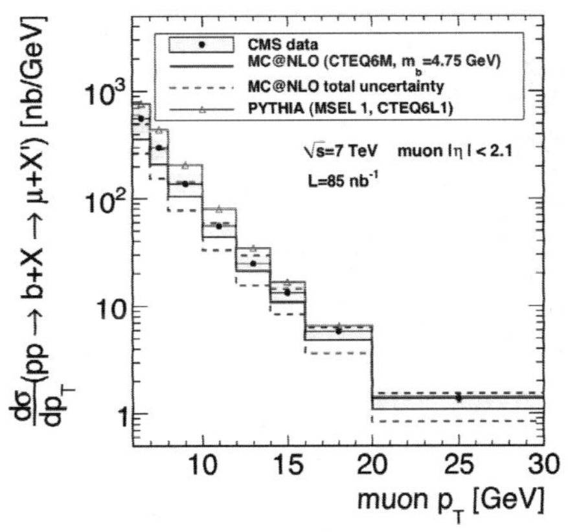

Figure 2. Differential b production cross section as a function of the muon transverse momentum.

of $\mathcal{L} = 0.85 \pm 0.09$ nb^{-1} CMS measures [5] :

$$\sigma(pp \to b\bar{b}X \to \mu Y) = 1.32 \pm 0.01_{\text{stat}} \pm 0.15_{\text{syst}} \pm 0.15_{\mathcal{L}}\ \mu\text{b}, \quad (1)$$

where the first error is statistical, the second is the quadratic sum of the systematic uncertainties (mostly due to the parameterisation of the $p_T^{\mu,rel}$ distributions), and the last term is the uncertainty on the luminosity, known, at the time of that measurement, with $\pm 11\%$ precision. The result is consistent with the prediction of NLO QCD, $0.95^{+0.42}_{-0.21}{}_{\text{scale}} \pm 0.09_{\text{mb}} \pm 0.05_{\text{pdf}}$ μb [6] while the PYTHIA [7] prediction is somewhat larger ($\sim 19\mu$b). Figure 2 shows the variation of the cross section, as a function of the muon transverse momentum, compared to the model predictions.

An independent inclusive cross section measurement is performed exploiting the single jets trigger. Jets with transverse momentum in the range $18 < p_T^{jet} < 300$ GeV are reconstructed combining informations from charged tracks, electromagnetic and hadronic calorimetric clusters with a Particle Flow tecnique [8]. While ensuring the same performance as that of the traditional calorimetric methods at high energy, the PF sizeably improves efficiency and resolution at low momentum ($p_T^{jet} < 100$ GeV). Jets from b are tagged reconstructing secondary vertices formed by at least three charged tracks and displaced by at least four standard deviations from the collision point. The vertex mass, computed from the tracks belonging to the secondary vertex, is used to separate b-jets from light quarks (see Figure 3). The efficiency for vertex reconstruction is measured as a function of p_t^{jet}, η^{jet} in a subset of muon tagged jets. The fraction of jets from b in the sample is measured in ranges of p_T^{jet} and η^{jet} with a fit to the vertex mass: an example is shown in Figure 3. Preliminary results [9] obtained from an integrated luminosity of 60 nb^{-1} are well consisted both with PYTHIA and with NLO calculations, which are however affected by large systematic uncertainties. Figure 4 shows the fraction of jets from $b\bar{b}$ events as a function of p_T^{jet}, compared to PYTHIA and NLO calculations.

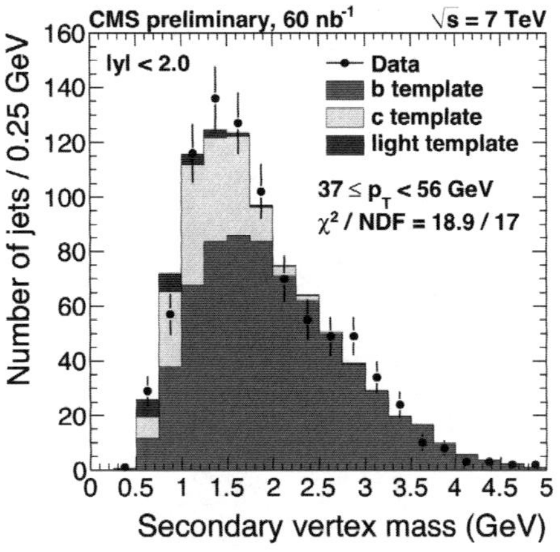

Figure 3. Invariant mass of the tracks at the secondary vertex. Points with error bands are data, overlayed to the contributions from b (red histogram), c (green) and light quarks (blue).

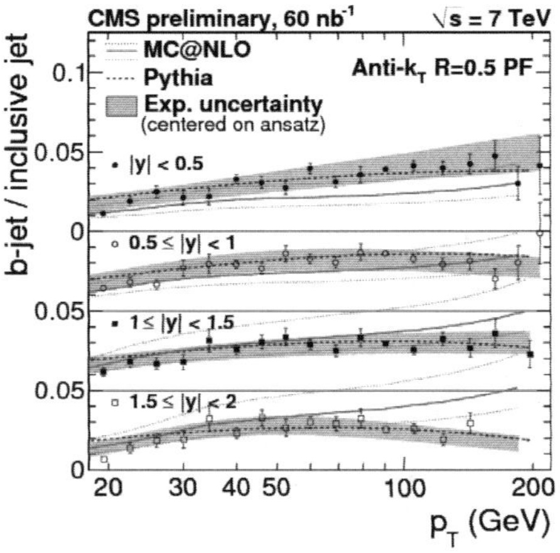

Figure 4. Measured b-jet cross section as a ratio to inclusive jet cross section. The NLO theory and Pythia MC predictions are shown for comparison.

Measuring the angular correlation between b-jets may shed light on the production of $b\bar{b}$ in proton proton interactions: gluon splitting (an $o(\alpha_s^3)$ process) favours the production of collinear b-jets, whereas jets from flavour creation ($o(\alpha_s^2)$) or flavour excitation ($o(\alpha_s^3)$), should be produced at larger angles. To improve the angular resolution, the b-jet direction is computed from the line connecting the primary to the secondary vertex. All events containing two secondary vertices, each composed by at least three charged tracks, an invariant mass exceeding

1.4 GeV, $p_T^{Vertex} > 8$ GeV, $|\eta^{vertex}| < 2$, and missing the primary vertex by at least five standard deviations, are employed. The angular distance between the two b vertices is measured by the quantity $\Delta R = \sqrt{(\Delta\phi^2 + (\Delta\eta)^2}$, where ϕ is the azimuthal angle. The data show a large excess of events at low ΔR as compared to the model expectation [10] (see Figure 5,6), hinting that gluon splitting contribution is significantly larger than expected.

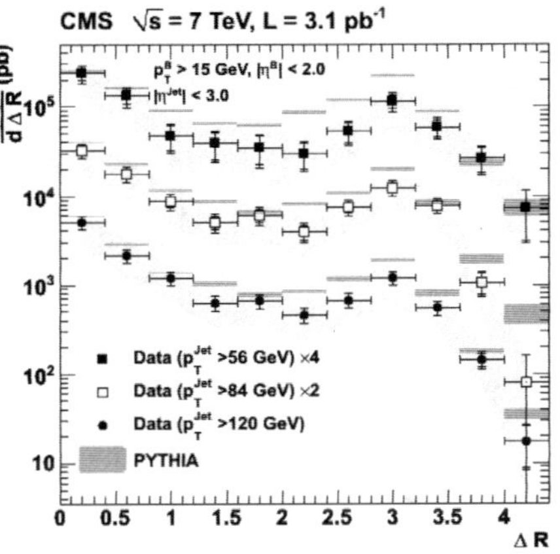

Figure 5. ΔR distribution for $b\bar{b}$ data, for three different thrsholds of the leading jet p_T. The green bands show Pythia predictions. The yellow band reflects the ±47% uncertainty in the absolute normalization.

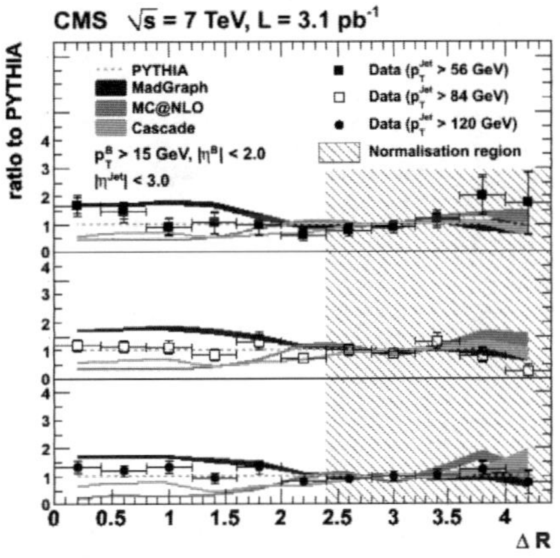

Figure 6. Ratio between data and Pythia predictions, normalized to unity in the region $\Delta R > 2.4$. A sizeable deviation from unity is observed at low ΔR values. None of the othem models considered reproduces the measured trend.

2.2. *Semi-Inclusive $b \to J/\psi(\to \mu^+\mu^-)X$ decays*

B hadrons decays to a final state containing a J/ψ have relatively large branching fractions and are easy to trigger. For these reasons, they are an interesting tool for studying B-Physics at hadron colliders. $J/\psi \to \mu^+\mu^-$ decays are reconstructed with efficiency ranging from 2% to 40% with good signal to noise ratio (see Figure 7). Particles from B decays are separated from prompt production with a fit to the distance of the J/ψ production point ($l_{J/\psi}$) from the primary vertex (see Figure 8). The contribution from combinatoric background is computed using opposite charge muon pairs with an invariant mass not compatible with the J/ψ ("side bands").

Efficiency is computed applying the Tag & Probe results to the two muons, accounting for correlations. The cross sections in the accepted region ($6.5 < p_T^{J/\psi} < 30$ GeV, and rapidity $|y| < 2.4$) are [11]:

$$\sigma(pp \to J/\psi X) = 70.9 \pm 2.1_{\rm stat} \pm 3.0_{\rm syst} \pm 7.8_{\mathcal{L}} \text{ nb} \quad (2)$$

$$\sigma(pp \to b\bar{b}X \to J/\psi Y) = 26.0 \pm 1.4_{\rm stat} \pm 1.6_{\rm syst} \pm 2.0_{\mathcal{L}}\text{nb} \quad (3)$$

The fraction of J/ψ produced in b decays increases with the transverse momentum in any rapidity range, as shown in Figure 9, where the CMS results in different rapidity ranges are compared to those of ATLAS, CDF, and LHCb.

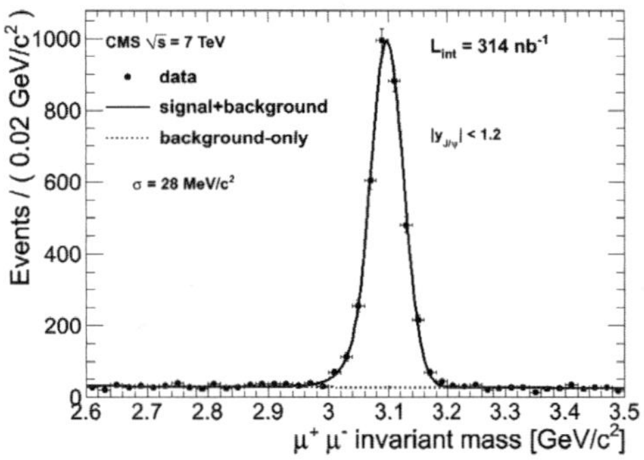

Figure 7. Di-muon invariant mass in the central rapidity region ($|y^{J/\psi}| < 1.2$)

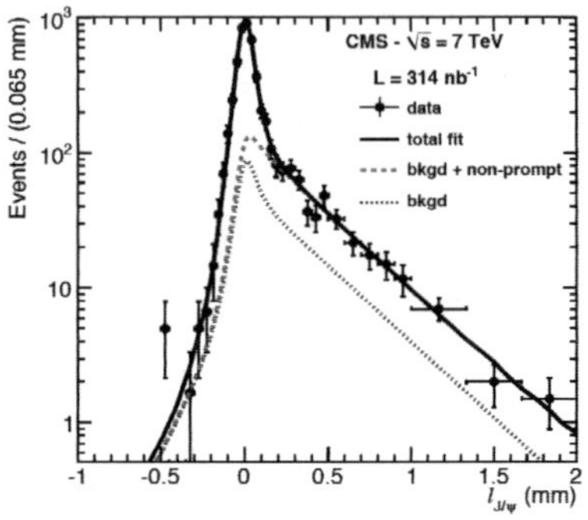

Figure 8. Decay lenght distribution, $l_{J/\psi}$, for J/ψ candidates

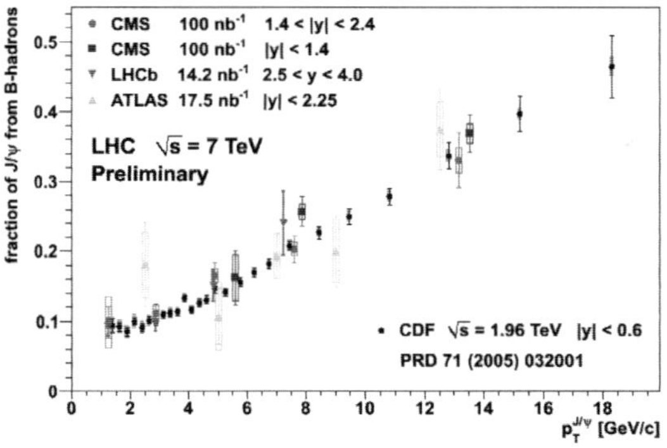

Figure 9. Fraction of J/ψ from b decays as a function of p_T. CMS results are compared to those of ATLAS, CDF, and LHCb.

2.3. *Fully reconstructed $B_q \to J/\psi(\to \mu^+\mu^-)h$ decays*

The large cross section for b hadrons production at the LHC allows the collection of large samples of fully reconstructed b-hadrons in $J/\psi h$ final states , where h stands for a low mass hadron. Only the most recent measurement of the cross section $\sigma(pp \to B_s X, B_s \to J/\psi\phi)$ is described here: the previos measurements of $B^+ \to J/\psi K^+$ and $B_d \to J/\psi K_s$ proceed through similar steps.

The $J/\psi \to \mu^+\mu^-$ decay triggers the event. Then pair of opposite charge particles are combined, assuming the Kaon mass hypothesys, to form the ϕ candidate. The four tracks are constrained to a common vertex, rejecting combinations with a probability smaller than 2%. Only combination with $p_T^{B_s} > 8GeV$ and rapidity $|y^{B_s} < 2.4$ are retained. In case of multiple candidates in the same event, that with the largest probability is chosen. The signal event yield is computed with a two-dimensional fit to the system invariant mass (M_{B_s}) and proper decay length (ct) (see Figure 10,11).

Besides the amount of signal, prompt and non-prompt background, the fit also determines some of the parameters of the *pdfs* employed to describe the background invariant mass distribution and signal and background proper time: the measured decay lenght for the signal ($ct = 478 \pm 26\mu m$) is consistent within 1.4 standard deviations with its expected value. The measurement is repeated in several p_T and $|y|$ ranges, with *pdf* parameters fixed to those obtained in the global fit. Using all the data collected during the first year of data taking at 7 TeV, for

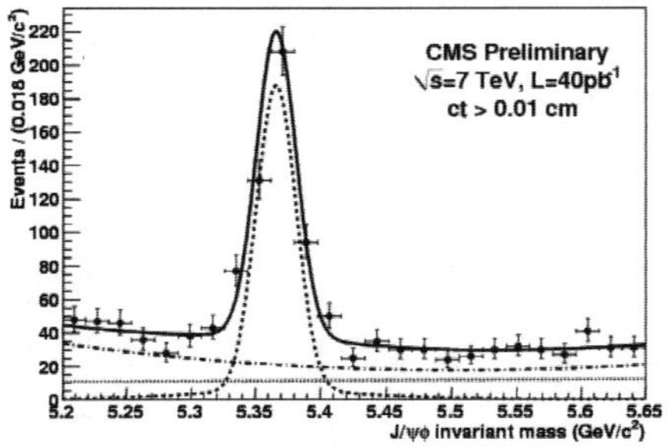

Figure 10. $J/\psi\phi$ invariant mass for data (point with error bars), best fit (continuos line), signal (dashed), prompt (dot-dashed) and non prompt(dotted) background. To enance the signal contribution, only events with $ct > 100\mu m$ are shown.

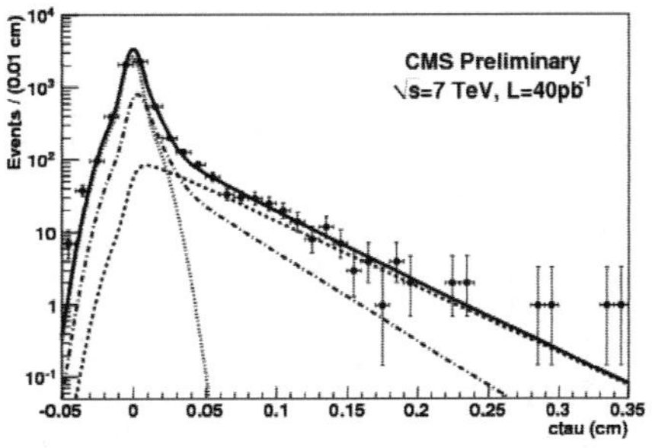

Figure 11. B_s decay lenght distribution for data (point with error bars), best fit (continuos line), signal (dashed), prompt (dot-dashed) and non prompt(dotted) background, for events with an invariant mass compatible with the B_s.

an integrated Luminosity of 40 nb^{-1}, CMS measures [12]

$$\sigma(pp \to B_s X) \cdot \mathcal{B}(B_s \to J/\psi\phi) \;=\; 6.9 \pm 0.6_{\text{stat}} \pm 0.5_{\text{syst}} \pm 0.3_{\mathcal{L}} \text{ nb} \qquad (4)$$

Figure 12 compares the cross sections measured by CMS for the three processess $B_u \to J/\psi K^+$ [13], $B_d \to J/\psi K_s$ [14], and $B_s \to J/\psi\phi$. The reader should note that the measurements have been performed in different p_T, y regions, as indicated in the figure.

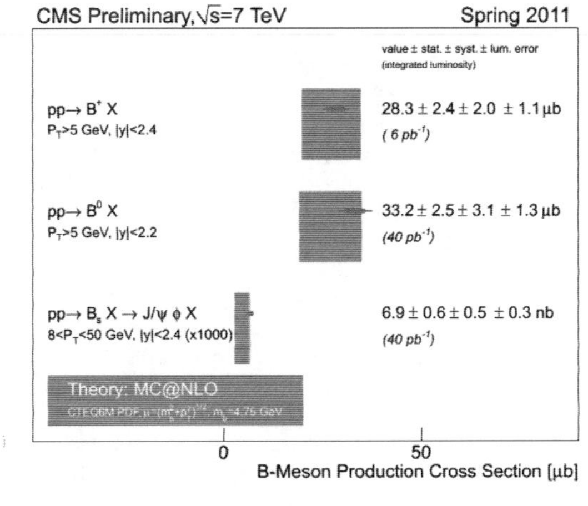

Figure 12. Summary of B meson cross section measurements performed by CMS with 7 TeV pp collisions at LHC. The inner error bars of the data points correspond to the statistical uncertainty, while the outer (thinner) error bars correspond to the quadratic sum of statistical and systematic uncertainties. The outermost brackets correspond to the total error, including a luminosity uncertainty which is also added in quadrature. Theory predictions at NLO are obtained using MC@NLO.

3. Conclusions

The CMS collaboration has provided an impressive amount of results on B-Physics using the data collected so far at the LHC collider, including eight published measurements and four preliminary public notes. Comparison to existing calculations of the cross section and angular correlations permits deeper understanding of the means of $b\bar{b}$ quarks production and hadronisation in hadron colliders. These high quality results proof the detector potential for the future programme of Heavy Flavour Physics, including searching for rare B decays, CP violation in the B_s sector, or using the b jets as a tool in new Physics search.

Journal of Physics: Conference Series **347** (2012) 012014 doi:10.1088/1742-6596/347/1/012014

References

[1] CMS collaboration, "*The CMS experiment at the CERN LHC*", 2008 JINST 3 S08004.
[2] M. Cacciari, G.P. Salam and G. Soyez, "*The anti-k_t jet clustering algorithm*", JHEP 04 (2008) 063.
[3] CMS Collaboration, *Performance of muon identification in pp collisions at* $\sqrt{s} = 7$ *TeV*, CMS Physics Analysis Summary CMS-PAS-MUO-10-002 (2010).
[4] K. Nakamura et al. (Particle Data Group)," *The Review of Particle Physics*", J. Phys. G 37, 075021 (2010) .
[5] CMS Collaboration," *Inclusive b-hadron production cross section with muons in pp collisions at* $\sqrt{s} = 7$ *TeV*", JHEP 1103 (2011) 090.
[6] S. Frixione, P. Nason and B.R. Webber, "*Matching NLO QCD and parton showers in heavy flavour production*", JHEP 08 (2003) 007 .
[7] T. Syöstrand, S. Mrenna and P.Z. Skands," *PYTHIA 6.4 Physics and Manual*", JHEP 05 (2006) 026.
[8] CMS Collaboration, *Particle-Flow Event Reconstruction in CMS and Performance for Jets, Taus, and Missing ET*, CMS PAS PFT-09-001 (2009).
[9] CMS Collaboration, "*Inclusive b-jet production in pp collisions at* $\sqrt{s} = 7$ *TeV* ",CMS-PAS-BPH-10-009,(2009).
[10] CMS Collaboration, "*Measurement of BB(bar) Angular Correlations based on Secondary Vertex Reconstruction at* $\sqrt{=}7$ *TeV*", JHEP 1103 (2011) 136.
[11] CMS Collaboration "*J/ψ prompt and non-prompt cross sections in pp collisions at* $\sqrt{s} = 7$ *TeV*,Eur.Phys.J. C71 (2011) 1575.
[12] CMS Collaboration "*Measurement of the B_s^0 Production Cross Sections in pp Collisions at* $\sqrt{s} = 7$ *TeV*" CERN-PH-EP-2011-06, Submitted to PRD-RC.
[13] CMS Collaboration, "*Measurement of the B^+ Production Cross Section in pp Collisions at* $\sqrt{s} = 7$ *TeV*", Phys.Rev.Lett.106:112001,(2011).
[14] CMS Collaboration, "*Measurement of the B_d^0 Production Cross Section in pp Collisions at* $\sqrt{s} = 7$ *TeV*", Phys. Rev. Lett. 106, 252001 (2011).

The Heavy Baryon Masses and Spin-Isospin Dependence

Z Ghalenovi, A A Rajabi and A Tavakolinezhad

Physics Department, Shahrood University of Technology, Shahrood, Iran

E-mail: z_ghalenovi@yahoo.com

Abstract. Heavy flavour baryons containing one charm (beauty) quark with light flavour combinations are studied using the hypercentral approach. The considered potential is a combination of Coulombic, linear confining and harmonic oscillator terms. An improved form of the hyperfine interaction and isospin dependent quark potential is introduced. The ground state masses of heavy baryons are computed. Also the magnetic moments of heavy baryons are computed using the effective mass of the confined quarks. The obtained results are very close to the ones obtained in experiments or in other works.

1. Introduction

Recently, there is renewed interest both experimentally and theoretically in the static properties of heavy flavor baryons such as mass and magnetic moments [1-5]. The purpose of this paper is to calculate the masses of the heavy baryons in a simple approximation, developed in [6-8]. In this work the considered potential is a combination of Coulombic, linear confining and harmonic oscillator terms. The non confining part of the potential is provided by the interaction with the Goldstone bosons, giving rise to a spin- and isospin-dependent interaction. More generally, one can expect that the quark-antiquark pair production can lead to an effective residual quark interaction containing an isospin (or flavor) dependent term and with these motivations in mind, we have introduced isospin dependent terms in the Hamiltonian [9, 10]. Thus the complete interaction used, is given by:

$$H(x) = V(x) + H_{\text{int}}(x) \tag{1}$$

Here V(x) is the confining potential:

$$V(x) = ax^2 + bx - c/x \tag{2}$$

where a, b and c are constants and x is the hyperradius. $H_{\text{int}}(x)$, the hyperfine interaction (nonconfining potential) is given by:

$$H_{\text{int}}(x) = H_S(x) + H_I(x) + H_{SI}(x) \tag{3}$$

where $H_S(x)$, $H_I(x)$ and $H_{SI}(x)$ are the spin, isospin and spin-isospin dependent interactions respectively, given by:

$$H_S = A_S \left(\frac{1}{\sqrt{\pi}\sigma_s}\right)^3 \exp(-x^2/\sigma_S^2)(\vec{s}_1.\vec{s}_2), \tag{4.a}$$

$$H_I = A_I\left(\frac{1}{\sqrt{\pi}\sigma_I}\right)^3 \exp(-x^2/\sigma_I^2)(\vec{t}_1\vec{t}_2), \tag{4.b}$$

$$H_{SI} = A_{SI}\left(\frac{1}{\sqrt{\pi}\sigma_{SI}}\right)^3 \exp(-x^2/\sigma_{SI}^2)(\vec{s}_1.\vec{s}_2)(\vec{t}_1\vec{t}_2)\ . \tag{4.c}$$

Here $\vec{s}_i$ and $\vec{t}_i$ are the spin and isospin operators of the i ›th quark respectively. The values of the hyperfine interaction constants are displayed in Table 2.

2. The Model

The model Hamiltonian for the baryon is expressed in terms of the Jacobian-coordinates (ρ, λ) as well as hypercentral coordinates (x) as be written as [10-12]

$$H = \frac{p_\rho^2}{2m_\rho} + \frac{p_\lambda^2}{2m_\lambda} + V(\rho,\lambda) = \frac{p_x^2}{2m} + V(x) \tag{5}$$

where m is the reduced mass:

$$m = \frac{2m_\rho m_\lambda}{m_\rho + m_\lambda} \tag{6}$$

Since $V = V(x)$ depends on x (the coordinates of all the three quarks) only, in the three-quark wave function one can factor out the hyperangular part, which is given by hyperspherical harmonics. The remaining hyperradial part of the wave function is determined by the hypercentral Schrödinger equation as follows [10-12]

$$\left[\frac{d^2}{dx^2} + \frac{5}{x}\frac{d}{dx} - \frac{\gamma(\gamma+4)}{x^2}\right]\psi_\gamma(x) = -2m\left[E - V(x)\right]\psi_\gamma(x) \tag{7}$$

where γ is the grand angular quantum number given by $\gamma = 2\nu + l_\rho + l_\lambda$; l_ρ and l_λ are the angular momenta associated with the ρ and λ variables and ν is a non-negative integer number. We investigate the ground state of baryons ($\gamma = 0$).

For the ground state of baryons and using the hyperradial approximation, where has been introduced in reference [8], the Schrödinger equation is given as:

$$\frac{d^2\chi(x)}{dx^2} + 2\mu[E - V(x) - \frac{15}{8\mu x^2})]\chi(x) = 0 \tag{8}$$

where μ is an arbitrary parameter with the dimension of mass, $\chi(x)$ is the reduced function:

$$\chi(x) = x^{5/2}\psi(x) \tag{9}$$

and $V(x)$ is the three-quark potential over the six-dimensional sphere that was defined in Eq. (2). We use a new variable $x' = \sqrt{\mu}x$, to eliminate an artificial dependence of equation (8) on μ, then equation (8) becomes

$$\chi''(x') + 2[E - V(x') - \frac{15}{8x'^2})]\chi(x') = 0 \tag{10}$$

We can solve equation (10) by the variational method. We introduce a simple variational ansatz for $\chi(x')$ as

$$\chi(x') = 2\sqrt{2}p^3 x'^{5/2} e^{-p^2x'^2} \tag{11}$$

where p is the variational parameter, and the numerical factor is chosen so that $\int \chi^2(x')dx' = 1$. The trial three-quark Hamiltonian admits explicit solutions for the wave function and the energy $E_0 = \min_p E(p)$, where

$$E(p) = \langle \chi | H | \chi \rangle \tag{12}$$

Now by using the condition $\frac{dE}{dp}\Big|_{p=p_0} = 0$, the value of p_0 is found.

3. The Heavy Baryon Masses and Magnetic Moments

The baryon masses are given by the three quark masses and the energy E_0 with the hyperfine interaction potential $< H_{\text{int}} >$ treated as a perturbation. The first order energy correction from the nonconfining potential $< H_{\text{int}} >$, as given in (3), can be obtained by using the unperturbed wave function (11) as follows

$$< H_{\text{int}} > = \int \chi H_{\text{int}} \chi \, dx' \tag{13}$$

Therefore, the Baryon mass becomes the sum of quark masses and energy of the perturbed system [10]:

$$M_{baryon} = m_{q1} + m_{q2} + m_{q3} + E_0 + < H_{\text{int}} > \tag{14}$$

This depends on the constituent quark masses m_q and potential parameters a , b and c that are listed in Tables 1-2. The quark masses are the QCD input and the potential parameters are obtained with a global fit to the experimental masses. We obtained the heavy baryon masses and compared with experimental data or other theoretical model predictions in Tables 3-4.

Within the baryons the masses of the quarks may get modified due to their binding interactions with the other two quarks. The effective quark mass is defined as

$$m_i^{eff} = m_i \left(1 + \frac{E_\gamma + < H_{\text{int}} >}{\sum_i m_i}\right) \tag{15}$$

such that the mass of the baryon is

$$M_B = \sum_i m_i^{eff} \, . \tag{16}$$

The magnetic moment of baryon is obtained in terms of its constituent quarks as [11,12]

$$\mu_B = \sum_i < \phi_{sf} | \mu_i \vec{\sigma}_i | \phi_{sf} > \tag{17}$$

where

$$\mu_i = \frac{e_i}{2m_i^{eff}} \tag{18}$$

Here, e_i and σ_i represent the charge and the spin of the quark $(s_i = \frac{\sigma_i}{2})$ constituting the baryonic state and $| \phi_{sf} >$ represents the spin-flavour wave function of the respective baryonic state. The computed magnetic moments of the single charm and beauty baryons with $J^P = 3/2^+$ are listed in table 5. As we see from Tables 3-5, our calculations are close to the ones obtained in experiments or in other works.

Table 1 The quark masses used in this model (in MeV)

m_u	320
m_d	325
m_s	440
m_c	1310
m_b	4670

Table 2 The fitted values of the parameters of potentials, obtained with a global fit to the experimental masses.

parameters	value
σ_S	2.37 fm
A_s	54.7 fm^2
σ_{SI}	2.15 fm
A_{SI}	90 fm^2
σ_I	1.12 fm
A_I	85 fm^2
a	0.23 fm^{-3}
b	0.61 fm^{-2}
c	0.52
μ	1 fm^{-1}

Table 3 Single charm baryon masses (masses are in GeV).

baryon	present work	Exp	[11]	[16]	[17]	[18]	|Theory-Exp|†
Σ_c^{++}	2.445	2.452	2.451	-	-	2.443	0.007
Σ_c^{*++}	2.532	2.518	2.516	-	-	2.506	0.014
Σ_c^{+}	2.450	2.451	2.461	2.448	2.455	2.460	0.001
Σ_c^{*+}	2.537	2.518	2.526	2.505	2.519	2.525	0.019
Σ_c^{0}	2.455	2.453	2.471	-	-	2.477	0.002
Σ_c^{*0}	2.542	2.518	2.536	-	-	2.544	0.024
Ξ_c^{+}	2.468	2.467	2.485	2.496	2.466	2.530	0.001
Ξ_c^{*+}	2.556	2.646	2.672	2.633	2.649	2.603	0.090
Ξ_c^{0}	2.473	2.471	2.494	-	-	2.548	0.002
Ξ_c^{*0}	2.561	2.646	2.680	-	-	2.623	0.085
Ω_c^{0}	2.588	2.699	2.696	2.701	2.718	2.620	0.111
Ω_c^{*0}	2.676	2.768	2.757	2.759	2.776	2.704	0.092

† The difference between our theoretical results and experimental data.

Table 4 Single beauty baryon masses (masses are in GeV).

baryon	present work	Exp	[11]	[14]	[16]	[17]	|Theory-Exp|†
Σ_b^+	5.805	5.807	5.801	-	-	-	0.002
Σ_b^{*+}	5.892	5.829	5.823	-	-	-	0.063
Σ_b^0	5.810	5.811	5.811	5.805	5.789	5.833	0.001
Σ_b^{*0}	5.897	5.832	5.834	5.834	5.844	5.858	0.065
Σ_b^-	5.815	5.815	5.821	-	-	-	0.000
Σ_b^{*-}	5.900	5.836	5.834	-	-	-	0.064
Ξ_b^0	5.828	-	5.872	5.937	5.879	-	-
Ξ_b^{*0}	5.916	-	5.936	5.963	5.967	-	-
Ξ_b^-	5.833	5.792	5.887	-	-	5.806	0.041
Ξ_b^{*-}	5.921	-	5.943	-	-	5.980	-
Ω_b^-	5.948	-	6.005	6.065	6.037	6.081	-
Ω_b^{*-}	6.036	-	6.065	6.088	6.090	6.102	-

† The difference between our theoretical results and experimental data.

Table 5 Magnetic moments of single charm and single beauty baryons with $J^P = 3/2^+$ in terms of nuclear magneton μ_N.

baryon	present work	[11]	[13]	[15]	[18]
Σ_c^{*++}	3.377	3.842	3.560	3.407	3.427
Σ_c^{*+}	1.097	1.252	1.170	1.130	1.158
Σ_c^{*0}	-1.115	-0.848	-1.230	-1.146	-1.101
Ξ_c^{*+}	1.390	1.513	1.430	1.264	1.242
Ξ_c^{*0}	-0.960	-0.688	-1.000	-0.986	-1.002
Ω_c^{*0}	-0.770	-0.865	-0.770	-0.833	-0.904
Σ_b^{*+}	3.461	3.234	-	3.082	-
Σ_b^{*0}	0.833	0.791	-	0.724	-
Σ_b^{*-}	-1.794	-1.655	-	-1.634	-
Ξ_b^{*0}	1.079	1.041	-	0.875	-
Ξ_b^{*-}	-1.596	-1.095	-	-1.477	-
Ω_b^{*-}	-1.368	-1.199	-	-1.292	-

* indicates $J^P = 3/2^+$ state.

4. Conclusions

In this paper we employed the hyperspherical formalism to study the masses and magnetic moments of baryons containing one heavy flavor quark in the ground state. We solved the Schrödinger equation by the variational method. We have shown that it is possible to find baryon masses by suitable confining and nonconfining interaction potentials. Using the theory of time-independent perturbation and the wave function, we could get the effects of spin and isospin potentials in the shift of baryon energy. Since the wave function of the system and the hyperfine interaction are independent of the baryon flavor, the value of the hyperfine interaction was identical for the baryons with the same spin- isospin and also the mass difference of baryons with $J^P = 3/2^+$ and $J^P = 1/2^+$ having same flavor is 88 MeV (For example $M(\Sigma_c^+) - M(\Sigma_c^{*+})$).Therefore the potential parameters have been chosen so that the errors of our results are minimized.

References

[1] C Albertus, J Nieves, J M Verde-Velasco, E Hernandez 2007 *Eur. Phys. J.* A **32** 183
[2] Faessler A et al. 2006 *Phy. Rev.* D **73** 094013
[3] Ebert D et al., 2002 *Phys. Rev.* D **66** 014008
[4] Matrasulov D U, Rakhimov Kh Yu, Butanov Kh T and Kanna F C 2002 *Eur. Phys. J.* A **14** 1
[5] Ebert D, Faustov R N and Galkin V O 2009 *Physics of Atomic Nuclei* **72** 1
[6] Fabre de la Ripelle M and Simonov Yu A 1991 *Ann. Phys.* (N.Y.) **212** 235
[7] Dubin A Yu, Kaidalov A B and Simonov Yu A 1994 *Phys Lett* . B **323** 41
[8] Narodetskii I M, Trusov M A 2001 *arXiv*; hep-ph/0104019v2
[9] Golzman L Ya and Riska D O 1996 *Phys. Rep.* C **268** 263
[10] Hassanabadi H, Rajabi A A and Zarrinkamar S 2008 *Modern Phys Lett* A **23** 527
[11] Patel B, Kumar Rai A and Vinodkumar P C 2008 *J. Phys.* G **35** 065001
[12] Patel B, Kumar Rai A and Vinodkumar P C 2008 *Pramana J. Phys* **70** 798
[13] Dhir R, Verma R C 2009 *Eur. Phys. J.* A **42** 243-49
[14] Ebert D, Faustov R N and Galkin V O 2005 *Phys. Rev.* D **72** 034026
[15] Majethiya A, Patel B and Vinodkumar P C 2008 *Eur. Phys. J.* A **38** 307-15
[16] Garcilazo H, Vijande J and Valcarce A 2007 *J. Phys.* G: *Nucl. Part. Phys* **34** 961-76
[17] Roberts E W and Muslema P 2008 *arXiv*:0711.2492v2 [nucl-th]
[18] Patel B, Kumar Rai A and Vinodkumar P C 2008 Arxiv: 0803.0221v1 [hep-ph]

Search for exotic new Physics at *BABAR*

Alberto Cervelli on behalf of the *BABAR* Collaboration
INFN sezione di Pisa and Universitá di Pisa, Largo Bruno Pontecorvo 3, 56124 Pisa, Italy
E-mail: alberto.cervelli@pi.infn.it

Abstract. We present the recent *BABAR* results for direct new physics searches. In particular we describe searches in the τ sector, with new results concerning lepton flavor violation searches, and in the Υ(nS) sector, where light Higgs and dark matter candidates are sought after and searches for lepton flavor violating decays are performed. We also present precise tests of lepton universality in Υ(nS) decays.

1. Introduction

BABAR operated for the last ten years, having been shut down in April 2008, during its long data-taking period it recorded events at different center of mass energies. The largest part of the *BABAR* data sample consists of 486 fb^{-1} taken around Υ(4S) resonance energy. To broaden its reach for new physics and in order to study Υ decays, the PeP-II asymmetric storage rings were operated also around Υ(3S) and Υ(2S) resonances, recording 30 fb^{-1} and 14 fb^{-1} around each resonance, respectively. A complete description of the *BABAR* detector can be found in [1].

2. Direct search for light Higgs and Dark Matter candidates

The Higgs mechanism, leading to the electroweak breaking, predicts an Higgs boson mass which is unstable after radiative corrections. Some new physics (NP) models solve this instability with the presence of a CP-odd Higgs singlet A^0, as predicted by Next to Minimal Supersymmetric Standard Model (NMSSM) [2], or an axion-like particle [3]. The former solution would solve the Higgs instability, while the latter would provide also a solution for the dark matter puzzle. Such a pseudoscalar particle can be searched in the decays of Υ(nS), which hence provides a great probe for direct NP searches and could be of paramount importance for the understanding of two of the most sought after answers in particle physics.

BABAR recorded one of the largest samples of Υ(3S) and Υ(2S), with seven times the statistics of the combined CLEO and Belle datasets for the former and half of the combined statistics for the latter. Searches were performed in three different channels: $\Upsilon(2S,3S) \to \gamma A^0, A^0 \to \mu^+\mu^-$ [4], $\Upsilon(3S) \to \gamma A^0, A^0 \to \tau^+\tau^-$ [5], and $\Upsilon(3S) \to \gamma A^0, A^0 \to$ invisible [6].

2.1. $\Upsilon(2S,3S) \to \gamma A^0, A^0 \to \mu^+\mu^-$

The $A^0 \to \mu^+\mu^-$ event signature is characterized by two charged tracks, of which at least one should be identified as a μ, forming the A^0 candidate, and an energetic γ which should have an energy $E_\gamma > 200$ MeV in the center of mass (CM) frame. A kinematic fit of the $\gamma\mu\mu$ vertex is performed in order to select the Υ candidate, the background shapes and yields are estimated on a subsample of the data which is then removed from the final selection and fit. A fit to the invariant mass of the two reconstructed μ is done in order to look for a peak produced by A^0, the fit is performed in the $0.212\ \mathrm{GeV/c^2} < m_{A^0} < 9.3\ \mathrm{GeV/c^2}$ range by looking at 300 MeV/c^2-wide windows that are moved by 2–5 MeV/c^2 steps, excluding the J/Ψ and $\Psi(2S)$ peak regions from the fit due to the large peaking backgrounds, resulting in 1951 fitting points. The possible bias in the fitted value of the signal yield is tested with a large number of pseudo-experiments, and is consistent with zero for all m_{A^0} values. The significance of any positive fluctuation is estimated through a likelihood ratio variable. Including the systematics the largest significances observed are 3.1σ ($\Upsilon(2S)$) and 2.8σ ($\Upsilon(3S)$), consistent with null hypothesis. The limits on $\mathcal{B}(\Upsilon(\mathrm{nS}) \to \gamma A^0) \times \mathcal{B}_{\mu\mu}$ are shown in Fig. 1 and Tab. 1.

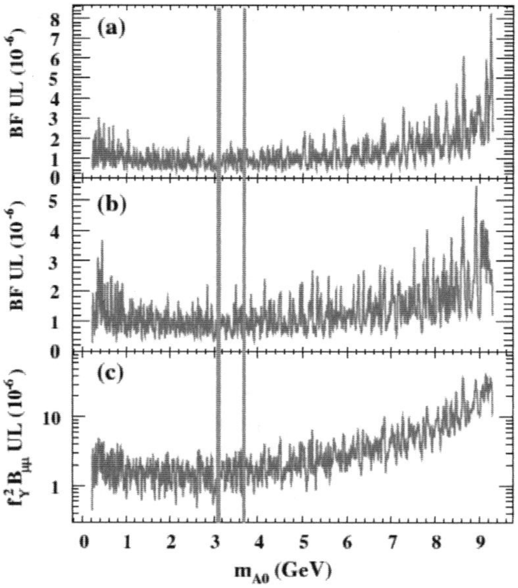

Figure 1. 90% Confidence Level (C.L.) upper limits on (a) $\mathcal{B}(\Upsilon(2S) \to \gamma A^0) \times \mathcal{B}_{\mu\mu}$, (b)$\mathcal{B}(\Upsilon(3S) \to \gamma A^0) \times \mathcal{B}_{\mu\mu}$, and (c) effective coupling $f^2_\Upsilon \times \mathcal{B}_{\mu\mu}$ as a function of m_{A^0}

The branching fractions (BFs) $\mathcal{B}(\Upsilon(\mathrm{nS}) \to \gamma A^0)$ are related to the effective coupling f_Υ of the bound b quark to the A^0 through [7]:

$$\frac{\mathcal{B}(\Upsilon(\mathrm{nS}) \to \gamma A^0)}{\mathcal{B}(\Upsilon(\mathrm{nS}) \to \ell^+\ell^-)} = \frac{f^2_\Upsilon}{2\pi\alpha}\left(1 - \frac{m^2_{A^0}}{m^2_{\Upsilon(nS)}}\right) \tag{1}$$

where $\ell = e, \mu$ and α is the fine structure constant. *BABAR* computed the experimentally accessible quantity $f^2_\Upsilon \mathcal{B}_{\mu\mu}$, averaging on the $\Upsilon(2S)$ and $\Upsilon(3S)$ samples, taking into account both correlated and uncorrelated uncertainties, the results are shown in Fig. 1(c).

2.2. $\Upsilon(3S) \to \gamma A^0, A^0 \to$ invisible

The decays of the light Higgs boson depend on its mass and couplings, as well as on the low-energy particle spectrum of the underlying theory. In certain NMSSM scenarios, particularly those in which the mass of the lightest supersymmetric particle (LSP) is larger than m_τ or if the Higgs mass is less than $2m_\tau$, the dominant decay of A^0 could be invisible: $A^0 \to \chi^0\chi^0$, where the neutralino χ^0 is the LSP. The cleanest experimental signature of such decays is production of monochromatic single photons in decays $\Upsilon \to \gamma A^0$, accompanied by a significant missing energy and momentum.

BABAR performed a search for a monochromatic peak in the missing mass distribution of events with a single high energy photon. The decay width of A^0 is assumed to be negligibly small compared to experimental resolution, and it is assumed that a single A^0 state exists in the range $0 < m_{A^0} \leq 7.8$ GeV/c^2; or if two or more states are present, they do not interfere.

Detection of the low-multiplicity single photon events requires dedicated trigger and filter lines. The *BABAR* detector offers two different trigger lines for such events: a high energy line requiring an isolated electromagnetic calorimeter (EMC) cluster with a CM energy $E^{\star}_{\gamma} > 2$GeV, and no tracks originating from the interaction region, which is refined at reconstruction level to have a photon energy $E^{\star}_{\gamma} > 3$GeV and no tracks reconstructed in the drift chamber (DCH) with $p^{\star} > 1$ GeV/c; and a low energy line which require $E^{\star}_{\gamma} > 1.5$GeV and no tracks reconstructed in the DCH with $p^{\star} > 0.1$ GeV/c. The two triggers present different backgrounds: the high energy line backgrounds are dominated by $e^+e^- \to \gamma\gamma$ events, especially in the energy region $E^{\star}_{\gamma} \sim E_{cm}/2$, where the photon energy shows a peak; on the other hand for the low energy line backgrounds are dominated by low angle radiative Bhabha events $e^+e^- \to \gamma e^+e^-$, where the two charged tracks are not reconstructed.

A limited number of variables can be used to select these very low multiplicity event samples, in particular the selection is focused on the photon quality requirements (number of crystals in the neutral clusters, and cluster shape), polar angle of the photon (which is distributed in the CM polar angle as $1 - \cos^2\theta^{\star}$ and peaked forward and backward for backgrounds), veto of extra particle in the event, and a Instrumented Flux Return (IFR) veto in the muon system, which suppress the $e^+e^- \to \gamma\gamma$ events where one of the photon is lost in the dead regions of the EMC but is reconstructed as an IFR cluster. The selection is optimized to maximize $\varepsilon_S/\sqrt{\varepsilon_B}$ where ε_S is the signal selection efficiency and ε_B is the background efficiency. MC events are used to estimate the signal efficiency, while the background is estimated using a data sub sample (about 10%) which is later included in the final fit.

The signal is extracted from an unbinned maximum likelihood fit to $m^2_{A^0}$, in steps of 0.1 GeV^2/c^4, the resolution on $m^2_{A^0}$ is estimated from simulated and reconstructed $e^+e^- \to \gamma\gamma$, and ranges between 1.5 GeV^2/c^4 (for m_{A^0} =0) and 0.7 GeV^2/c^4 (for m_{A^0} =8 GeV/c^2). The largest systematic uncertainties arise from the estimate of $e^+e^- \to \gamma\gamma$ peaking background yield (only in the high energy region) and its shape (affecting both regions). No signal excess was observed in the fit, and an Upper Limit (UL) was set for $\mathcal{B}(\Upsilon(3S) \to \gamma A^0) \times \mathcal{B}(A^0) \to$ invisible $<$ $(0.7 - 31) \times 10^{-6}$ (90% CL) . The limits as a function of m_{A^0} are shown in Fig.2.

2.3. $\Upsilon(3S) \to \gamma A^0, A^0 \to \tau^+\tau^-$

The $\Upsilon(3S) \to \gamma A^0, A^0 \to \tau^+\tau^-$ channel is reconstructed only in the case both τ leptons decay leptonically (i.e. $\tau \to e\nu_e\nu_\tau$ or $\tau \to \mu\nu_\mu\nu_\tau$). The selected events are required to contain at least one energetic photon candidate with $E_\gamma > 100$ MeV, and exactly two charged tracks, which should be identified as leptons (either an electron or μ). After this preliminary selection most of the background retained is due to radiative τ pair production and higher order QED processes, like two photon processes, leading to four tracks with low transverse momenta in the final state.

The backgrounds are reduced using 8 kinematic variables: the total CM energy calculated from the two tracks and the candidate photon; the squared missing mass; the aplanarity, which is the cosine of the angle between the photon direction and the plane identified by the two tracks momenta; the cosine of the angle between the photon and the tracks; the total transverse momentum of the event calculated in the CM frame; the polar angle of the missing momentum; and the cosine of the polar angle between the tracks in the photon recoil frame. The final selection is optimized to maximize $S/\sqrt{B}$, where S and B are the expected number of signal and of background events respectively. Since the background may vary as a function of the photon

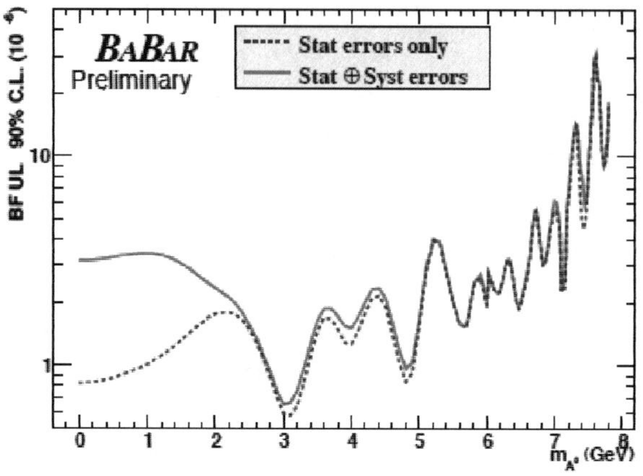

Figure 2. 90% confidence level ULs on the branching fraction $\mathcal{B}(\Upsilon(3S) \to \gamma A^0) \times \mathcal{B}(A^0) \to$ invisible. The dashed blue line shows the statistical uncertainties only, the solid red line includes the systematic uncertainties.

energy, the optimization is performed in five E_γ regions, which partially overlaps in order to reduce the discontinuity in the efficiency. The backgrounds after the selection are dominated by $e^+e^- \to \gamma\tau^+\tau^-$ events, with larger contribution observed for low E_γ values.

The analysis looks for an excess in a narrow region in the E_γ spectrum, since any peak in the recoil mass ($m_{\tau\tau}$) indicating the presence of a new particle decaying in two τ's, translates to a peak in the photon energy distribution. To extract the signal and background yields a two-step binned likelihood fit is made simultaneously for $\tau\tau \to ee, e\mu$, and $\mu\mu$ samples. In the first step of the fit a no signal hypothesis is assumed and the background function is fitted, in the second step a search for $\Upsilon(3\text{S}) \to \gamma A^0, A^0 \to \tau^+\tau^-$ is performed, assuming the A^0 to have a negligible width and the signal distribution is parametrized with a Crystal Ball function. The search for such signal is performed scanning for peaks in the E_γ distribution in steps equal to half the photon energy resolution. In total 307 scan points are examined and for each scan point the yield and its statistical uncertainty are obtained from the fit. The yield significance from the data is shown in Fig.3. The data points are consistent with the normal distribution and no significant evidence for any signal is found, and an UL is set: $\mathcal{B}(\Upsilon(3\text{S}) \to \gamma A^0, A^0 \to \tau^+\tau^-) = (1.5 - 16) \times 10^{-5}$.

A review for all Higgs and dark matter candidate searches can be found in Tab. 1.

Table 1. Results for Higgs and Axion-like particle searches performed by the BABAR Collaboration.

	Energy Range	Upper Limit	
$\Upsilon(2\text{S},3\text{S}) \to \gamma A^0, A^0 \to \mu^+\mu^-$	$0.212\text{GeV/c}^2 < m_{A^0} < 9.3\text{GeV/c}^2$	$(0.26 - 8.3) \times 10^{-6}$	[4]
$\Upsilon(3\text{S}) \to \gamma A^0, A^0 \to \tau^+\tau^-$	$0.212\text{GeV/c}^2 < m_{A^0} < 9.3\text{GeV/c}^2$	$(1.5 - 16) \times 10^{-5}$	[5]
$\Upsilon(3\text{S}) \to \gamma A^0, A^0 \to$ invisible	$m_{A^0} \leq 7.8\text{GeV/c}^2$	$(0.7 - 31) \times 10^{-6}$	[6]

3. Lepton Flavor Violation in τ decays

Lepton Flavor Violation (LFV) involving charged leptons has never been observed, and stringent experimental limits exist for both τ and μ decays [8, 9]. On the other hand, experimental observation of neutrino oscillations [10] implies that, within the Standard Model (SM), there are

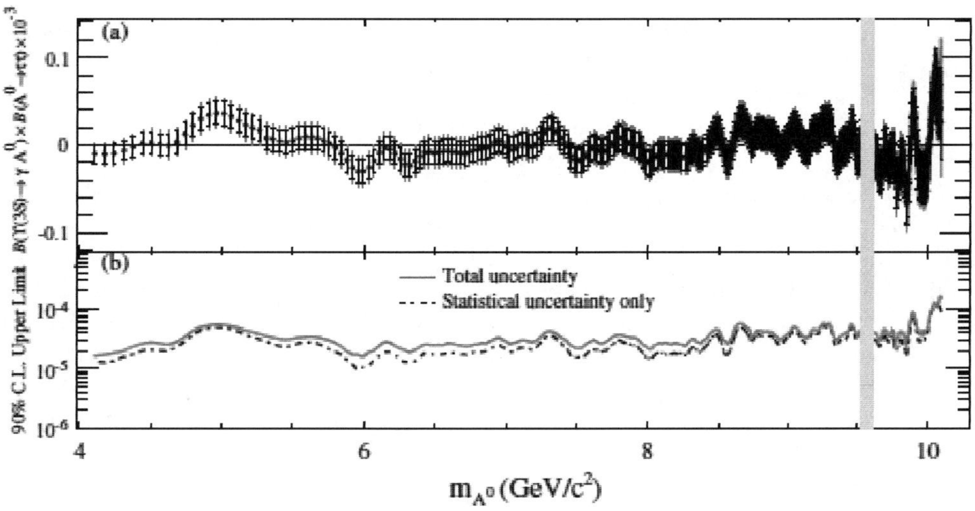

Figure 3. Results for $\Upsilon(3\mathrm{S}) \to \gamma A^0, A^0 \to \tau^+\tau^-$ (a) Product branching fractions as a function of the Higgs mass. For each point, both the statistical uncertainty (from the central value to the horizontal bar) and the total uncertainty (statistical and systematic added in quadrature) are shown (from the central value to the end of the error bar). In (b), the corresponding 90% C.L. upper limits on the product of the branching fractions versus the Higgs mass values are shown, with total uncertainty (solid line) and statistical uncertainty only (dashed line). The shaded vertical region represents the excluded mass range corresponding to the $\chi_{bJ}(2\mathrm{P}) \to \gamma\Upsilon(1\mathrm{S})$ states.

amplitudes contributing to LFV in the neutral sector, which would contribute to LFV in charged sector, albeit with effects well below the current experimental sensitivity [11]. Many descriptions of physics beyond the SM predict enhanced LFV in τ decays over μ decays, with branching fractions within or at least very close to the present experimental sensitivities [12, 13, 14]. An observation of LFV in τ decays would be a clear signature of NP, while improved limits would further constrain models.

We will present in the following results for $\tau^- \to \ell\gamma$ and $\tau^- \to \ell_1^-\ell_2^+\ell_3^-$ ($\ell_i = e, \mu$) searches: these two channels are two of the most studied τ LFV decays and have analogues in the $\mu \to eee$ and $\mu \to e\gamma$ decays. The observation of signal in more than one channel, among the eight decays we will present, would not only imply NP, but by looking at the ratios of the measured branching ratios (BR) it would be possible to obtain important information about the flavor structure of the NP leading to the LFV processes (a review of possible branching fraction ratios can be found in Ref. [15]).

In order to look for LFV in τ decays, events with low charged particle multiplicity are selected, the event space is divided in two non-overlapping hemispheres by a plane containing the interaction region and orthogonal to the thrust axis, calculated from both charged tracks and neutral energy deposit. Each track and reconstructed photon is associated to one of the two hemispheres. The two hemispheres are called the tag side (where a SM τ decay is observed) and the signal side (where the LFV decay candidate is observed). For $\tau^- \to \ell\gamma$ searches the signal candidate is expected to have an energetic photon with $E_\gamma > 1\mathrm{GeV}$ and only one charged track in the signal side, in the tag side the decay is categorized in five broad classes: e, μ, and π tags, where only one track is present, ρ tags where a candidate π^0 is observed along the tag side track, and 3h tag, where there are three tracks in the tag hemisphere. On the other hand in $\tau^- \to \ell_1^- \ell_2^+ \ell_3^-$ the signal hemisphere is required to contain exactly three tracks, while the tag side is required to have only one track and no neutral deposits corresponding to photons with $E_\gamma > 100$ MeV. Both analyses are performed in a blind way, in order to reduce biases introduced by the analysts, and the data are observed only after the selection is optimized and systematics are estimated.

The selection is optimized independently for each channel. For both searches the background is reduced by using particle identification (PID), kinematic information, and for $\tau^- \to \ell\gamma$ analysis also multivariate algorithms. The optimization is made in order to obtain the best expected UL, the number of expected background events in the blinded signal region is extrapolated from unblinded sidebands, using MC samples or data enriched control samples. The signal efficiency is estimated using MC. The UL is determined using a modified frequentist approach accounting for systematics errors.

The results for $\tau^- \to \ell^-\gamma$ are shown in Fig. 4, the main background contribution comes from radiative SM τ decays such as $\tau^- \to \ell^- \nu_\tau \bar{\nu}_\ell \gamma$, which constitute an irreducible background in the case the νs have low momenta.

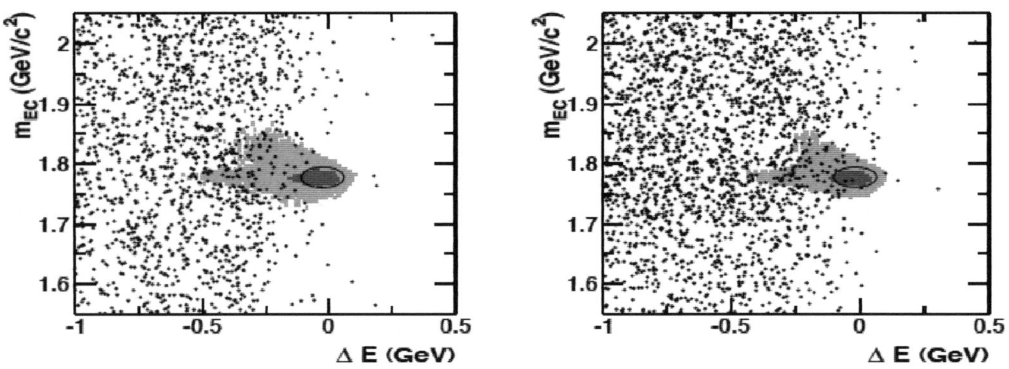

Figure 4. Sidebands and signal region for $e\gamma$ channel (left) and $\mu\gamma$ channel (right) in the $(m_{EC}, \Delta E)$ plane. Data events are shown as dots, and contours containing 90% (50%) of signal MC are shown as light- (dark-) shaded regions.

$\tau^- \to \ell_1^- \ell_2^+ \ell_3^-$ decays present smaller backgrounds due to the clear signature of the signal candidates, the main systematics associated to the analysis arise from the background model which have large errors, reflecting on large errors on background yields due to the lack of data events in the sidebands. Results are shown in Fig. 5, and a resume of the 90% CL ULs for both LFV searches in τ decays is reported in Tab. 2.

4. Lepton Flavor Violation in Υ decays

The large Υ(3S) and Υ(2S) samples available for *BABAR* made it possible to reach unprecedented sensitivities for the searches for rare decays of such resonances. Υ(nS) decays have been used to search for LFV in $\Upsilon(\text{nS}) \to \ell\ell'$ decay, which are as sensible as τ decays to NP. Four channels have been studied, looking at $e\tau$ and $\mu\tau$ final states for both Υ(3S) and Υ(2S). The LFV event signature is composed of one energetic lepton (either e or μ) and a τ reconstructed through leptonic or hadronic ($\tau \to \pi\pi^0(\pi^0)$) decays, the selection is partially common to the four channels with differences regarding mainly PID and τ-daughters kinematics.

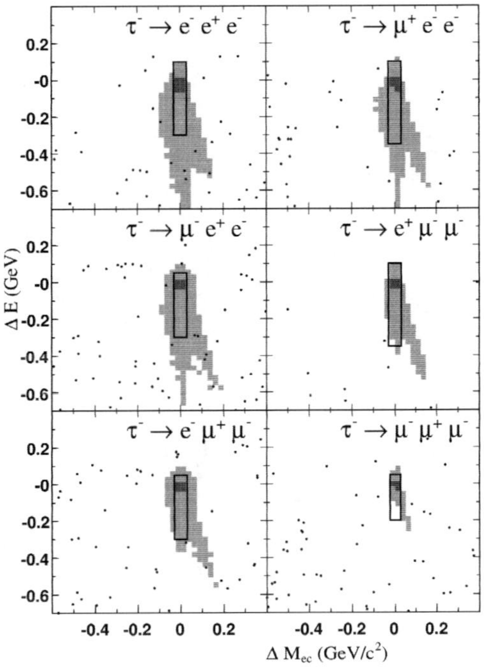

Figure 5. Data events are indicated with dots, boundaries of the signal region for each channel are denoted with the solid black line. The shading represents the MC signal distribution.

Table 2. Results for LFV searches in τ decays. Luminosity, efficiency and UL is shown for each channel.

$\tau \to$	$\mathcal{L}$ (fb^{-1})	$\varepsilon(\%)$	UL (10^{-8})	
$e^-e^+e^-$	468	8.6	297	[16]
$\mu^-e^+e^-$		8.8	2.2	
$\mu^+e^-e^-$		12.6	1.8	
$e^-\mu^+\mu^-$		6.4	3.2	
$e^+\mu^-\mu^-$		10.2	2.6	
$\mu^-\mu^+\mu^-$		6.6	3.3	
$e\gamma$	516	3.9	3.3	[17]
$\mu\gamma$		6.1	4.4	

The main discriminating variable x is defined as the ratio between the primary lepton momentum (i.e. the electron or muon produced in the Υ(nS) decay) and the CM beam energy. An unbinned maximum likelihood fit is performed on this variable to determine the signal and background yields. The PDF shapes of x are different for the signal and each background class: signal events are expected to peak at $x = x_{max} \sim 0.97$, τ-pair, which constitute the main background source, present a smooth distribution with an end-point at x_{max}, QED events peak at $x \sim 1$ since no missing energy is present, and hadrons have a shape similar to the τ pairs. the BF is calculated as $BF = N_{SIG}/(\varepsilon_{SIG} \times N_{\Upsilon(\mathrm{nS})})$. The main systematic uncertainties on BF measurement arise from the choice of the PDF shapes, due to the limited statistics in the selected sample. No signal evidence is found in any of the four channels under study, so ULs are set, the results are reported in Tab. 3 [18].

Table 3. Results for LFV searches in Υ decays, statistical error is noted first followed by systematic error, the last column shows the improvement with respect to previous measurements made by CLEO collaboration. UL for $\Upsilon(2S) \to e^{\pm}\tau^{\mp}$ and $\Upsilon(3S) \to e^{\pm}\tau^{\mp}$ were measured for the first time

	$\mathcal{B}(10^{-6})$	UL (10^{-6})	Improvement
$\mathcal{B}(\Upsilon(2S) \to e^{\pm}\tau^{\mp})$	$0.6^{+1.5+0.5}_{-1.4-0.6}$	< 3.2	First
$\mathcal{B}(\Upsilon(2S) \to \mu^{\pm}\tau^{\mp})$	$0.2^{+1.5+1.0}_{-1.3-1.2}$	< 3.3	$\times 5.5$
$\mathcal{B}(\Upsilon(3S) \to e^{\pm}\tau^{\mp})$	$1.8^{+1.7+0.8}_{-1.4-0.7}$	< 3.2	First
$\mathcal{B}(\Upsilon(3S) \to \mu^{\pm}\tau^{\mp})$	$-0.80.2^{+1.5+1.4}_{-1.5-1.3}$	< 3.3	$\times 3.7$

5. Lepton Flavor Universality

In the SM the couplings of the gauge bosons to leptons are independent of the lepton flavor, and, aside from small mass effects, the decay width for $\Upsilon(1S) \to \ell^+\ell^-$ should be identical for all leptons. Hence in the SM one expect the quantity:

$$R_{\ell\ell'}\Upsilon(1S) = \frac{\Gamma_{\Upsilon(1S)\to\ell\ell}}{\Gamma_{\Upsilon(1S)\to\ell'\ell'}} \tag{2}$$

with $\ell, \ell' = e, \mu, \tau$ and $\ell \neq \ell'$ to be close to one, in particular the value for $R_{\tau\mu}\Upsilon(1S)$ is predicted to be ~ 0.992 [19]. In the next-to-minimal supersymmetric extension of the SM, deviation from the SM expectation may arise due to a light CP-odd Higgs boson, A^0 [20], among other hypothetical particles, A^0 may mediate the following processes $\Upsilon(1S) \to \gamma A^0 \to \gamma\ell^+\ell^-$ or $\Upsilon(1S) \to \gamma\eta_b(1S)$, with $\eta_b(1S) \to A^0 \to \ell^+\ell^-$.

If the photon remains undetected, the lepton pair would be ascribed to the $\Upsilon(1S)$ and the proportionality of the coupling of the Higgs to the lepton mass would lead to an apparent violation of the lepton universality, and this effect should be larger for decays to τ pairs, and enhanced for higher mass resonances. The deviation of $R_{\tau\mu}(\Upsilon(1S))$ may be as large as $\sim 4\%$, depending on the A^0 mass.

BABAR performed a measurement of $R_{\tau\mu}(\Upsilon(1S))$, in the decays $\Upsilon(3S) \to \Upsilon(1S)\pi^+\pi^-$, with $\Upsilon(1S) \to \ell\ell$, and $\ell = e, \mu$. Only events with τ decaying in one charged particle are selected. One tenth of the data is used to validate the analysis method and the signal extraction procedure, the validation sample is then discarded from the final result to avoid biases. The selection requires exactly four charged tracks, and the $\Upsilon(1S) \to \ell\ell$ is formed by selecting two oppositely-charged tracks, constrained to come from a common vertex, while the other two tracks are used to reconstruct the $\Upsilon(3S) \to \Upsilon(1S)\pi^+\pi^-$ candidate. Different selection procedures are applied for μ and τ channel selection due to the presence of neutrinos in the latter case, which leads to larger background contributions arising from non-leptonic $\Upsilon(1S)$ decays and τ decays.

In the $\Upsilon(1S) \to \mu^+\mu^-$ channel PID is used to identify muons, and the $\Upsilon(3S)$ is reconstructed by selecting tracks compatible with being produced in the $\Upsilon(3S) \to \Upsilon(1S)\pi^+\pi^-$ decay. For the τ channel tighter selection criteria are applied, due to the presence of large missing energy, so only events with a visible energy in the CM frame larger than 5 GeV are selected, and further selection is applied on the di-pion momenta along with a multivariate algorithm taking as input event shape and kinematic variables. Finally in order to select $\Upsilon(3S) \to \Upsilon(1S)\pi^+\pi^-$ the invariant mass difference between the two resonance candidates is required to be less than 2.5 GeV/c^2, and the di-pion invariant mass is required to be between 0.28 and 0.90 GeV/c^2.

The final selection efficiency is estimated from MC simulated events and are measured to be $\varepsilon_{\mu\mu} = (44.57 \pm 0.04)\%$ and $\varepsilon_{\tau\tau} = (16.77 \pm 0.03)\%$. An extended unbinned maximum likelihood fit, applied to the two disjoint datasets is used to extract $R_{\tau\mu}$. The μ data sample is fitted with a two dimensional PDF, based on the invariant mass of the μ pair and on $M^{\text{reco}}_{\pi^+\pi^-}$, defined as the invariant mass of the system recoiling against the pion pair. The τ data sample is modeled with a one dimensional PDF of $M^{\text{reco}}_{\pi^+\pi^-}$. The data taken 30 MeV below the $\Upsilon(3S)$ resonance are used to model the background shapes. The result of the simultaneous fit is $R_{\tau\mu} = 1.006 \pm 0.013$, where the error is statistical only. Fig. 6 shows the projection of the fit for the three variables.

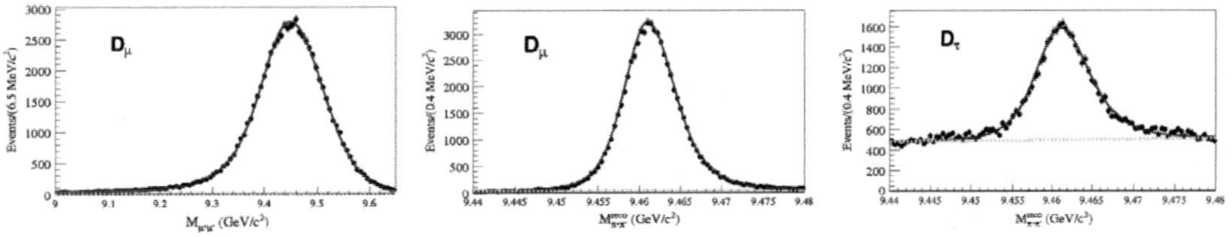

Figure 6. Fit projection for μ pair invariant mass (left) and $M^{\text{reco}}_{\pi^+\pi^-}$ (middle) in the μ sample, and for $M^{\text{reco}}_{\pi^+\pi^-}$ (right) in the τ sample. The dashed lines represent the background shape, while the solid line is the sum of signal and background contribution to the fit, and data are represented by the dots.

Several systematic errors cancel out in the ratio, however the residual systematics are related to the differences between data and simulation in the efficiency of event selection, PID, and the trigger efficiency. There is also a systematic uncertainty on signal and background yields due to the imperfect knowledge and the particular choice of the PDFs used in the fit. The total systematic uncertainties is estimated to be 2.2%.

Including all the systematic corrections, the ratio $R_{\tau\mu}$ is found to be [21]:

$$R_{\tau\mu} = 1.005 \pm 0.013 \pm 0.022 \tag{3}$$

where the first error is statistic only and the second is systematic. This result is the most precise test for this ratio and is consistent with the SM expectation, excluding the hypothesis of an A^0 with mass lower than 9 GeV/c^2 at 90% confidence level.

6. Conclusion

BABAR has proven to be a very versatile detector for the search for NP in rare and exotic decays, thanks to a very large data sample recorded, and the energy range exploited, with samples taken at Υ(4-3-2S) resonances, leading to a large production of $b\bar{b}$ bound states, τ pairs and hadronic decays. The high luminosity achieved and the constant development of new analysis tools and techniques made it possible to greatly improve the sensitivities for such rare processes over the years, with improvements far better than the ones provided by the increase in statistics only.

Many new results have been presented in this paper, and thanks to these improved limits many bounds to NP models could be set. The great number of searches for indirect NP effects of non SM particles demonstrate how a flavor factory can be a key infrastructure for seeking new answers even in the LHC era.

References

[1] B. Aubert *et al.* (BABAR Collaboration), Nucl. Instrum. Methods Phys. Res., Sect. A **479**, 1 (2002).
[2] R. Dermisek *et al.*, Phys. Rev. D **76**, 051105(R) (2007).
[3] Y. Nomura and J. Thaler, Phys. Rev. D **79**, 075008 (2009).
[4] B. Aubert *et al.* (BABAR Collaboration), Phys. Rev. Lett. **103**, 081803 (2009).
[5] B. Aubert *et al.* (BABAR Collaboration), Phys. Rev. Lett. **103**, 181801 (2009).
[6] B. Aubert *et al.* (BABAR Collaboration), arXiv:0808.0017.

[7] F. Wilczek, Phys. Rev. Lett. **39**, 1304 (1977).
[8] M.L. Brooks *et al.* (MEGA/LAMPF Collaboration), Phys. Rev. Lett. **83**, 1521 (1999)
[9] K. Inami, Nucl. Phys. B Proc. Suppl. **181-128**, 295 (2008).
[10] K2K Collaboration, M. H. Ahn *et al.*, Phys. Rev. Lett. **90**, 041801 (2003); KamLAND Collaboration, K. Eguchi *et al.*, Phys. Rev. Lett. **90**, 021802 (2003); SNO Collaboration, Q. R. Ahmad *et al.*, Phys. Rev. Lett. **89**, 011301 (2002).
[11] W.J. Marciano and A.I. Sanda, Phys. Lett. B **67** 303 (1977).
[12] P. Paradisi, JHEP **0510**, 006 (2005).
[13] K. S. Babu and C. Kolda, Phys. Rev. Lett. **89**, 241802 (2002).
[14] A. Brignole and A. Rossi, Phys. Lett. B **566**, 217 (2003).
[15] A.G. Akeroyd, M. Aoki, and Y. Okada, Phys. Rev. D **76**, 013004 (2007).
[16] J.P. Lees *et al.* (BABAR Collaboration), Phys. Rev. D **81**, 111101(R) (2010).
[17] B. Aubert *et al.* (BABAR Collaboration), Phys. Rev. Lett. **104**, 021802 (2010).
[18] J.P. Lees *et al.* Phys. Rev. Lett. **104**, 151802 (2010).
[19] C. Amsler et al. (Particle Data Group), Phys. Lett. B **667**, 1(2008).
[20] R. Dermisek and J. F. Gunion, Phys. Rev. Lett. **95**, 041801 (2005).
[21] P. del Amo Sanchez *et al.* (BABAR Collaboration), Phys. Rev. Lett. **104**, 191801 (2010).

Pion fragmentation functions from e^+e^- production

Maryam Soleymaninia[1,2], **Ali Khorramian**[1,2] **and Mohammad Moosavinejad**[2,3]

[1] Physics Department, Semnan University, Semnan, Iran
[2] School of Particles and Accelerators, Institute for Research in Fundamental Sciences (IPM), P.O.Box 19395-5531, Tehran, Iran
[3] Physics faculty, Yazd University, Yazd, Iran

E-mail: Maryam.Soleimaninia@gmail.com, khorramiana@theory.ipm.ac.ir, mmoosavi@yazduni.ac.ir

Abstract. We present the new sets of fragmentation functions for charged pions at leading order. They are fitted to data on inclusive charged-hadron production in e^+e^- annihilation taken by different collaborations, who discriminated between events with charm, bottom, and light-flavors fragmentation in their charged-hadron sample. They also agree with data on the production of neutral pions.

1. Introduction

In finding any physics signatures beyond the standard model and any new hadronic systems in high-energy hadron reactions it is necessary to have accurate Quantum Chromodynamics (QCD) predictions for cross sections or decay rates. As the new generation of high energy experiments with the detection of a final state hadron are taking place, further tests of QCD and the Standard Model require an accurate knowledge not only of the parton distribution functions (PDFs) and $\alpha_s(M_Z)$, but equally of the fragmentation functions (FFs) $D_i^h(x, Q^2)$. A fragmentation function indicates the probability of the transition of a parton i at scale Q into a hadron h carrying away a fraction x of the parton's momentum or energy in the center-of-mass (c.m.) frame.

The fragmentation functions are related to a nonperturbative aspect of QCD, hence they cannot be precisely calculated by theoretical methods. The situation is similar to the determination of the PDFs, where high-energy experimental data are used for their determination instead of theoretical calculations. In the data sample without discrimination of the hadron species, they distinguished between four cases, namely fragmentation of u, d, and s quarks; c quarks only; b quarks only; and all five quark flavors (u, d, s, c, and b).

Like $\alpha_s(M_Z)$, PDFs and FFs are important quantities because they are universal: according to the factorization theorem, once they are known at some suitably defined scale $Q = Q_0$, they can be calculated at any other scale Q and used in any type of process [1]. The most reliable way to determine them at a given scale is by fitting to inclusive single hadron production data in which the fraction x of available momentum or energy in the c.m. frame carried away by the hadron is measured.

Journal of Physics: Conference Series **347** (2012) 012017 doi:10.1088/1742-6596/347/1/012017

2. Single inclusive cross section in e^+e^- annihilation

The inclusive production of a hadron h by e^+e^- annihilation,

$$e^+e^- \to (\gamma, Z) \to h + X, \tag{1}$$

is described first by a partonic subprocess $e^+e^- \to q\bar{q}$ to produce a quark- antiquark pair and a gluon at NLO and second by a hadronization process in which a parton fragments to produce a hadron. The latter is a non-perturbative process that is described by a non-perturbative fragmentation function (non-pff). The hadron-production cross section for inclusive production of a hadron h is given by [2]

$$F^h(x, Q^2) = \frac{1}{\sigma_{tot}} \frac{d\sigma(e^+e^- \to hX)}{dx}. \tag{2}$$

To be able to compare our calculations with experimental data, we normalize the cross section to

$$\sigma_{tot} = N_c \sum_{i=1}^{N_f} e_{q_i}^2 \sigma_0, \tag{3}$$

where $\sigma_0 = (4\pi\alpha^2/3s)$ is the total cross section of $e^+e^- \to \mu^+\mu^-$ for massless leptons, $N_c = 3$, e_{q_i} is the electroweak charge of quark q_i in units of the positron charge, and N_f is the number of active flavors. The variable Q^2 is the virtual photon or Z^0 momentum squared in $e^+e^- \to \gamma$ (or Z^0) and it is expressed by the center-of-mass energy $\sqrt{s}$ as $Q^2 = s$. The variable x is the hadron energy E_h scaled to the beam energy $\sqrt{s}/2$, and it is defined by the fraction:

$$x \equiv \frac{E_h}{\sqrt{s}/2} = \frac{2E_h}{\sqrt{Q^2}}. \tag{4}$$

QCD factorization theory predicts that the fragmentation process is described by the summation of hadron productions from quarks, antiquarks, and gluons and this is similar to the structure functions in DIS [3, 4, 5]

$$F^h(x, Q^2) = \sum_i C_i(x, \alpha_s) \otimes D_i^h(x, Q^2). \tag{5}$$

According to the notation $\otimes$ that indicates a convolution integral definition

$$f(x) \otimes g(x) = \int_x^1 \frac{dy}{y} f(y) g\left(\frac{x}{y}\right), \tag{6}$$

and also coefficient functions $C_i(x, \alpha_s)$ describing partonic subprocess [6], the hadron-production cross section is given by

$$F^h(x, Q^2) = \sum_i \int_x^1 \frac{dz}{z} D_i^h(z, M_f^2) \frac{1}{\sigma_{tot}} \frac{d\sigma_i}{dy}\left(\frac{x}{z}, \mu^2, M_f^2\right). \tag{7}$$

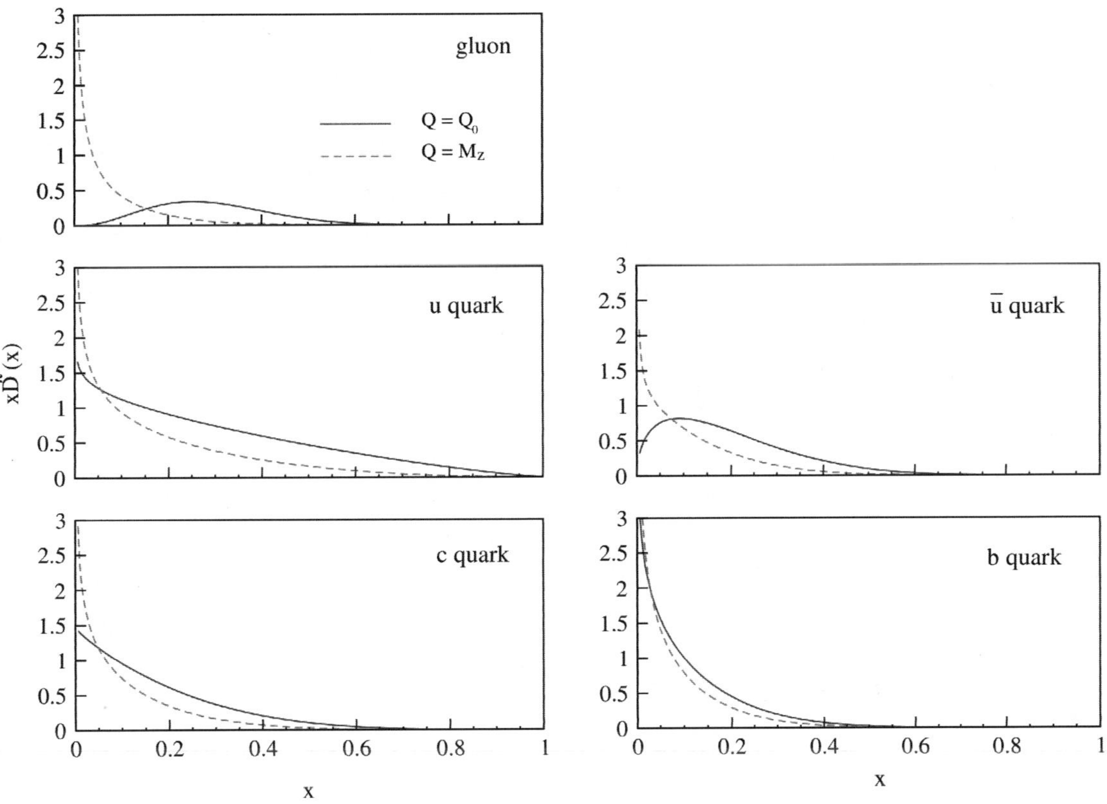

Figure 1. π^+ fragmentation functions at initial sale $Q = Q_0$ and $Q = M_Z$.

3. QCD analysis

The x dependence of the pion fragmentation functions are parameterized at the initial Q^2 ($\equiv Q_0^2$) in the same way as the PDF analysis [7, 8, 9, 10, 11, 12, 13]. Since they should vanish at x=1, we choose a simple polynomial form

$$D_i^\pi(x, Q_0^2) = N_i^\pi x^{\alpha_i^\pi}(1-x)^{\beta_i^\pi}, \tag{8}$$

where N_i^π, α_i^π, and β_i^π are parameters to be determined by a χ^2 analysis of $e^+e^- \to \pi X$ data. Since we have built in the $c\bar{c}$ and $b\bar{b}$ thresholds, we have three different starting scales Q_0 [14, 15]

$$Q_0 \;=\; \begin{cases} 1 \text{ GeV}, & \text{if} \quad i = u, d, s, g \\ m_c = 1.43 \text{ GeV}, & \text{if} \quad i = c \\ m_b = 4.3 \text{ GeV}, & \text{if} \quad i = b \end{cases} . \tag{9}$$

According to partonic structure of $\pi^+(u\bar{d})$ we take the same favored fragmentation functions for π^+ from u and $\bar{d}$ quarks. Also the pion productions from $\bar{u}$, d, s, and $\bar{s}$ are considered the same at the $Q_0^2 = 1$ GeV.

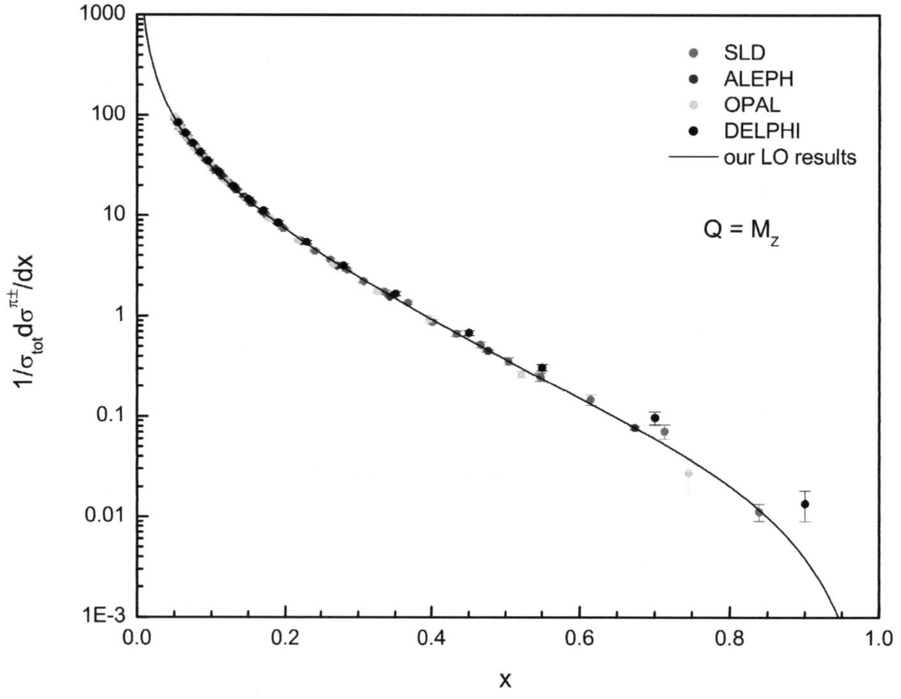

Figure 2. Comparison of our results with pion experimental data at $Q = M_Z$.

$$D_u^{\pi^+}(x,Q_0^2) = D_{\bar{d}}^{\pi^+}(x,Q_0^2) = N_u^{\pi^+} x^{\alpha_u^{\pi^+}} (1-x)^{\beta_u^{\pi^+}}, \tag{10}$$

$$\begin{aligned} D_{\bar{u}}^{\pi^+}(x,Q_0^2) &= D_d^{\pi^+}(x,Q_0^2) \\ &= D_s^{\pi^+}(x,Q_0^2) = D_{\bar{s}}^{\pi^+}(x,Q_0^2) \\ &= N_{\bar{u}}^{\pi^+} x^{\alpha_{\bar{u}}^{\pi^+}} (1-x)^{\beta_{\bar{u}}^{\pi^+}}. \end{aligned} \tag{11}$$

Gluon and heavy quarks fragmentation functions are defined separately.

$$D_g^{\pi^+}(x,Q_0^2) = N_g^{\pi^+} x^{\alpha_g^{\pi^+}} (1-x)^{\beta_g^{\pi^+}}, \tag{12}$$

$$D_c^{\pi^+}(x,m_c^2) = D_{\bar{c}}^{\pi^+}(x,m_c^2) = N_c^{\pi^+} x^{\alpha_c^{\pi^+}} (1-x)^{\beta_c^{\pi^+}}, \tag{13}$$

$$D_b^{\pi^+}(x,m_b^2) = D_{\bar{b}}^{\pi^+}(x,m_b^2) = N_b^{\pi^+} x^{\alpha_b^{\pi^+}} (1-x)^{\beta_b^{\pi^+}}. \tag{14}$$

These 15-dimensional parameters are determined by the charged-hadron production data of $e^+ + e^- \to h^\pm + X$ in which the deviation of the theoretical prediction from the data becomes minimal. We use the data obtained by the collaborations SLD [16], ALEPH [17], OPAL [18], DELPHI [19, 20], TASSO [21, 22, 23], TPC [24], and TOPAZ [25].

4. Conclusions

We presented new sets of FF for charged pions, at LO. Q^2 evolution results of FFs are shown in Fig. 1. The initial functions of π^+ are supplied at Q^2=1 GeV^2 for g and light quarks FFs and for heavy quarks at $Q^2 = m_c^2$ and m_b^2. They were fitted to data on inclusive charged-pion production in e^+e^- annihilation. In Fig. 2, our LO hadron-production cross section results are compared with the data at $Q = M_Z$ without separation on initial partons by the SLD, ALEPH, OPAL and DELPHI collaborations.

Acknowledgments

The authors would like to thank the ICPP-Istanbul II organizing committee for providing this opportunity and support us to participate in the conference.

References

[1] Hirai M and Kumano S 2011 *Preprint* arXiv:1106.1553 [hep-ph]
[2] Ellis R K, Stirling W J and Webber B R 1996 *QCD and Collider Physics* (Cambridge University Press)
[3] Soleymaninia M and Khorramian A N 2011 *Int. J. Mod. Phys.* A **26** 686
[4] Soleymaninia M, Khorramian A and Atashbar Tehrani S 2010 *Proceedings of the Conference in Honor of Murray Gell-Mann's 80th Birthday* (World Scientific) 451–458
[5] Khorramian A, Atashbar Tehrani S, Soleymaninia M and Batebi S 2009 *Hyperfine Interactions* **194** 341–345
[6] Nason P and Webber B R 1994 *Nucl. Phys.* B **421** 473; Erratum 1996 *ibid.* **480** 755. See also Altarelli G, Ellis R K, Martinelli G and Pi S-Y 1979 *Nucl. Phys.* B **160** 301; Furmanski W and Petronzio R 1982 *Z. Phys. C* **11** 293; Mitov A and Moch S 2006 *Nucl. Phys.* B **751** 18
[7] Khanpour H and Khorramian A N 2010 *Acta Phys. Polon.* B **41** 2929
[8] Khorramian A N, Atashbar Tehrani S, Taheri Monfared S, Arbabifar F and Olness F I 2011 *Phys. Rev.* D **83** 054017 (*Preprint* arXiv:1011.4873 [hep-ph])
[9] Taheri Monfared S, Khorramian A N, Atashbar Tehrani S and Haddadi Z 2011 *Nucl. Phys. Proc. Suppl.* **210-211** 125
[10] Taheri Monfared S and Khorramian A N 2011 *Int. J. Mod. Phys.* A **26** 692
[11] Arbabifar F, Khorramian A N, Taheri Monfared S and Atashbar Tehrani S 2011 *Int. J. Mod. Phys.* A **26** 625
[12] Khorramian A N, Khanpour H and Atashbar Tehrani S 2009 *PoS* EPS-HEP2009 393
[13] Khanpour H, Khorramian A N and Atashbar Tehrani S 2011 *Int. J. Mod. Phys.* A **26** 658
[14] Kniehl B A, Kramer G and Pötter B 2000 *Nucl. Phys.* B **582** 514 (2000) & 2001 *Nucl. Phys.* B **597** 337. There are also previous and recent analyses in Binnewies J, Kniehl B A and Kramer G 1995 *Z. Phys.* C **65** 471; 1995 *Phys. Rev.* D **52** 4947; 1996 *Phys. Rev.* D **53** 3573; Albino S, Kniehl B A and Kramer G 2005 *Nucl. Phys.* B **725** 181; 2006 *Nucl. Phys.* B **734** 50; Kniehl B A and Kramer G 2005 *Phys. Rev.* D **71** 094013; 2006 *Phys. Rev.* D **74** 037502
[15] Kretzer S 2000 *Phys. Rev.* D **62** 054001
[16] Abe K *et al.* (SLD collaboration) 2004 *Phys. Rev.* D **69** 072003
[17] Buskulic D *et al.* (ALEPH collaboration) 1995 *Z. Phys.* C **66** 355; Barate R *et al.* 1998 *Phys. Rep.* **294** 1
[18] Akers R *et al.* (OPAL collaboration) 1994 *Z. Phys.* C **63** 181
[19] Abreu P *et al.* (DELPHI collaboration) 1998 *Eur. Phys. J.* C **5** 585
[20] Abreu P *et al.* (DELPHI collaboration) 1995 *Nucl. Phys.* B **444** 3
[21] Brandelik R *et al.* (TASSO collaboration) 1980 *Phys. Lett.* B **94** 444
[22] Althoff M *et al.* (TASSO collaboration) 1983 *Z. Phys.* C **17** 5
[23] Braunschweig W *et al.* (TASSO collaboration) 1989 *Z. Phys.* C **42** 189
[24] Aihara H *et al.* (TPC collaboration) 1984 *Phys. Rev. Lett.* **52** 577; 1988 *Phys. Rev. Lett.* **61** 1263
[25] Itoh R *et al.* (TOPAZ collaboration) 1995 *Phys. Lett.* B **345** 335

Journal of Physics: Conference Series **347** (2012) 012018 doi:10.1088/1742-6596/347/1/012018

The study of heavy contribution in diffractive parton distribution functions

Sara Taheri Monfared[1,2], Ali Khorramian[1,2] and Shahin Atashbar Tehrani[2]

[1] Physics Department, Semnan University, Semnan, Iran

[2] School of Particles and Accelerators, Institute for Research in Fundamental Sciences (IPM), P.O.Box 19395-5531, Tehran, Iran

E-mail: sara.taherimonfared@gmail.com, khorramiana@theory.ipm.ac.ir, Atashbar@ipm.ir

Abstract. In this paper we give a brief introduction to the description of QCD analysis in order to determine the diffractive parton distribution functions (DPDFs) at the next-to-leading order (NLO) approximation. We focus on key features of analysis for all available diffractive HERA data and outline challenges for considering heavy quark contributions within the framework of the fixed flavour number scheme (FFNS) which is in full agreement with the diffractive heavy quarks experimental data.

1. Introduction

Diffraction process in high energy physics is very important to complete the understanding of strong interactions. This can give us a better understanding of the mechanisms of production and make the eventual corrections on the calculations. Significant progress has been made in understanding diffraction in terms of QCD by studying virtual photon dissociation in deep inelastic scattering (DIS) at HERA [1]. As well as being sensitive to novel features of parton dynamics in the high density, low x regime, diffractive DIS cross sections are used to extract DPDFs, an essential ingredient in predicting many diffractive processes at the LHC [2, 3].

We have previously extracted DPDFs from H1 2006 data [4] in Refs. [5, 6]. Here, we perform a global fit on all available HERA data [4, 7, 8, 9, 10, 11, 12]. We believe that the full HERA data sample analysis is a powerful technique to achieve the best precision possible in extracting DPDFs. In this paper, the extracted parton distributions from the current analysis have been used to predict heavy structure functions and the comparison of their predictions with experimental data are presented.

2. Theoretical background of the QCD analysis

Global fits to determine parton distributions [13, 14, 15, 16, 17, 18, 19, 20, 21, 22, 23, 5, 24, 25] use available data and the most up-to-date QCD calculations to best determine the parton distributions. Consequently, we consider all three methods of selecting observables at H1 and ZEUS to achieve the best result. At HERA, diffractive events were selected either by the detection of the final state proton (the Leading Proton Spectrometer, LPS, in the ZEUS case and the Forward Proton Spectrometer, FPS, in the H1 case) [7, 9, 11, 12] or on the basis of a large rapidity gap between the system X and the outgoing proton (LRG method) [4, 11]. The diffractive contribution was also identified by the M_X method [8, 10] based on the shape of the mass distribution of the system X. In order to keep the compatibility between all data sets, cross section measurements from FPS/LPS and M_X methods are corrected to H1 LRG domain, by multiplying them with the appropriate global factors.

We summarize the key features of our QCD analysis here. In our calculation, we consider pomeron as an object with parton distribution function. These DPDFs are parametrized at a starting scale and are evolved to higher factorization scales using a numerical solution of the DGLAP evolution equations [26] in FFNS with three massless quarks and heavy charm treated as massive particles. In other word, the number of active (light) flavors n_f appearing in the splitting functions and the corresponding Wilson coefficients will be fixed, n_f=3. Furthermore, the evaluation of the running strong coupling $\alpha_s(Q^2)$ requires to match $\alpha_s^{n_f}$ at $Q = m_h$ (m_c=1.41 GeV, m_b=4.50 GeV, m_t=175 GeV), i.e. $\alpha_s^{n_f}(m_h) = \alpha_s^{n_f-1}(m_h)$. we have performed all Q^2-evolutions in Mellin n-moment space and used the QCD-PEGASUS program [27] for the NLO evolution.

Finally, we extract DPDFs from the global χ^2 fit on all HERA diffractive DIS data obtained by various methods with very large uncertainties associated with the treatment of proton dissociation processes using the CERN subroutine MINUIT [28].

3. Open charm production in diffraction

The data from the HERA running period 1998-2000 were analyzed by both H1 [29] and ZEUS [30] collaborations. The charm quark was tagged by the reconstruction of $D^{*\pm}(2010)$ meson in diffractive DIS and PHP regimes. H1 used also another method based on the measurement of the displacement of tracks from primary vertex to identify the D^* production in the sample of DIS events only. In figures 1 and 2 we show our results for heavy diffractive cross section and

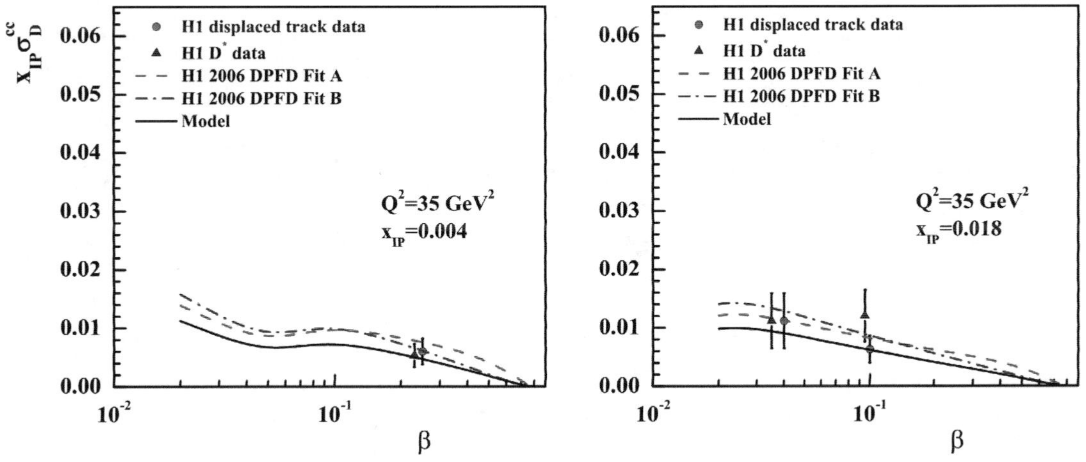

Figure 1. Comparison of our result for the contribution of the charm quarks to the diffractive cross section with H1 DPDF Fit A and Fit B [4] shown as a function of β for two different values of $x_{I\!P}$. The data are obtained from the H1 displaced track method and D^* production in DIS [29]. The error bars of the data points represent the statistical and systematic error in quadrature.

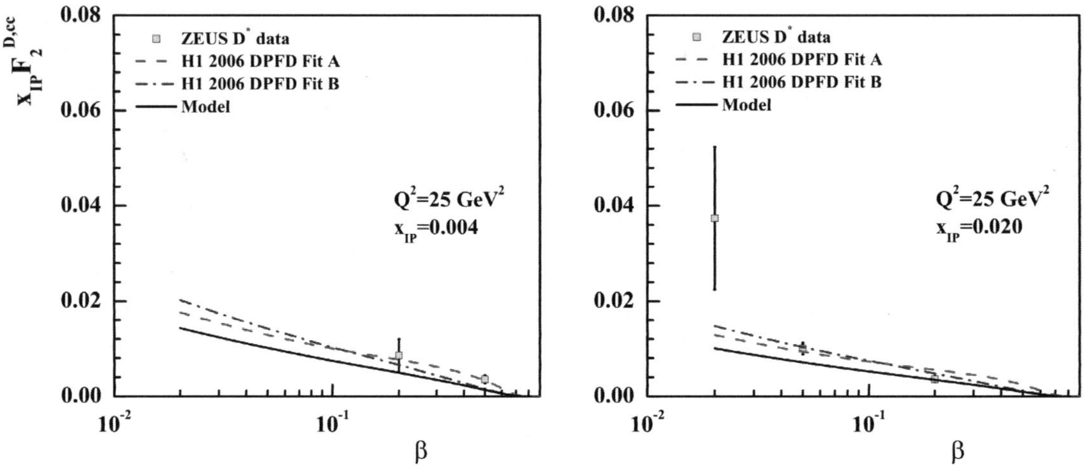

Figure 2. Comparison of our result for the contribution of the charm quarks to the diffractive structure function with H1 DPDF Fit A and Fit B [4] shown as a function of β for two different values of $x_{I\!P}$. The data are obtained from the ZEUS D^* production in DIS [30]. The error bars of the data points represent the statistical and systematic error in quadrature.

structure function together with H1 2006 DPDF [4] Fit A and Fit B and we compare them with H1 and ZEUS data.

4. Conclusion

In the present paper, all the most up-to-date data on diffractive deep inelastic scattering processes for the structure functions and reduced cross-section which allow us to be quite confident about the reliability of our predictions, have been analyzed in the standard parton model approach of perturbative QCD. Within large errors a good agreement is observed in figures 1 and 2, which supports the validity of QCD factorization in diffractive DIS.

Acknowledgments

S. T. M. gratefully acknowledges the generous support of the organizing committee of ICPP-Istanbul II for providing the opportunity to attend the conference.

References

[1] Arneodo M and Diehl M 2005 *Preprint* arXiv:hep-ph/0511047
[2] Albrow M G *et al.* (FP420 R and D Collaboration) 2009 *JINST* **4** T10001 (*Preprint* arXiv:0806.0302 [hep-ex])
[3] De Roeck A and Thorne R S 2011 *Preprint* arXiv:1103.0555 [hep-ph]
[4] Aktas A *et al.* (H1 Collaboration) 2006 *Eur. Phys. J.* C **48** 715 (*Preprint* arXiv:hep-ex/0606004)
[5] Taheri Monfared S and Khorramian A N 2011 *Int. J. Mod. Phys.* A **26** 692
[6] Taheri Monfared S, Khorramian A N and Atashbar Tehrani S 2010 *Proceedings of the Conference in Honor of Murray Gell-Mann's 80th Birthday* (World Scientific) 503–510
[7] Aktas A *et al.* (H1 Collaboration) 2006 *Eur. Phys. J.* C **48** 749 (*Preprint* arXiv:hep-ex/0606003)
[8] Chekanov S *et al.* (ZEUS Collaboration) 2004 *Eur. Phys. J.* C **38** 43 (*Preprint* arXiv:hep-ex/0408009)
[9] Chekanov S *et al.* (ZEUS Collaboration) 2005 *Nucl. Phys.* B **713** 3 (*Preprint* arXiv:hep-ex/0501060)
[10] Chekanov S *et al.* (ZEUS Collaboration) 2009 *Nucl. Phys.* B **816** 1 (*Preprint* arXiv:0812.2003 [hep-ex])
[11] Chekanov S *et al.* (ZEUS Collaboration) 2008 *Nucl. Phys.* B **800** 1 (*Preprint* arXiv:0802.3017 [hep-ex])
[12] H. Collaboration 2010 *Preprint* arXiv:1010.1476 [hep-ex]

[13] Khorramian A N, Atashbar Tehrani S, Taheri Monfared S, Arbabifar F and Olness F I 2011 *Phys. Rev.* D **83** 054017 (*Preprint* arXiv:1011.4873 [hep-ph])
[14] Khorramian A N, Khanpour H and Atashbar Tehrani S 2010 *Phys. Rev.* D **81** 014013 (*Preprint* arXiv:0909.2665 [hep-ph])
[15] Khorramian A N and Atashbar Tehrani S 2008 *Phys. Rev.* D **78** 074019 (*Preprint* arXiv:0805.3063 [hep-ph])
[16] Atashbar Tehrani S and Khorramian A N 2007 *JHEP* **0707** 048 (*Preprint* arXiv:0705.2647 [hep-ph])
[17] Khorramian A N and Atashbar Tehrani S 2007 *JHEP* **0703** 051 (*Preprint* arXiv:hep-ph/0610136)
[18] Khorramian A N, Mirjalili A and Atashbar Tehrani S 2004 *JHEP* **0410** 062 (*Preprint* arXiv:hep-ph/0411390)
[19] Soleymaninia M and Khorramian A N 2011 *Int. J. Mod. Phys.* A **26** 686
[20] Soleymaninia M, Khorramian A and Atashbar Tehrani S 2010 *Proceedings of the Conference in Honor of Murray Gell-Mann's 80th Birthday* (World Scientific) 451–458
[21] Khorramian A, Atashbar Tehrani S, Soleymaninia M and Batebi S 2009 *Hyperfine Interactions* **194** 341–345
[22] Khanpour H and Khorramian A N 2010 *Acta Phys. Polon.* B **41** 2929
[23] Taheri Monfared S, Khorramian A N, Atashbar Tehrani S and Haddadi Z 2011 *Nucl. Phys. Proc. Suppl.* **210-211** 125
[24] Arbabifar F, Khorramian A N, Taheri Monfared S and Atashbar Tehrani S 2011 *Int. J. Mod. Phys.* A **26** 625
[25] Khanpour H, Khorramian A N and Atashbar Tehrani S 2011 *Int. J. Mod. Phys.* A **26** 658
[26] Gribov V and Lipatov L 1972 *Sov. J. Nucl. Phys.* **15** 438 (1972 *Yad. Fiz.* **15** 781);
Gribov V and Lipatov L 1972 *Sov. J. Nucl. Phys.* **15** 675 (1972 *Yad. Fiz.* **15** 1218);
Dokshitzer Y 1977 *Sov. Phys. JETP* **46** 641 (1977 *Zh. Eksp. Teor. Fiz.* **73** 1216);
Altarelli G and Parisi G 1977 *Nucl. Phys.* B **126** 298
[27] Vogt A 2005 *Comput. Phys. Commun.* **170** 65 (*Preprint* arXiv:hep-ph/0408244)
[28] James F and Roos M 1975 *Comput. Phys. Commun.* **10** 343
[29] Aktas A *et al.* (H1 Collaboration) 2007 *Eur. Phys. J.* C **50** 1 (*Preprint* arXiv:hep-ex/0610076)
[30] Chekanov S *et al.* (ZEUS Collaboration) 2003 *Nucl. Phys.* B **672** 3 (*Preprint* arXiv:hep-ex/0307068)

Journal of Physics: Conference Series **347** (2012) 012019 doi:10.1088/1742-6596/347/1/012019

Top quark as a gate to new physics: Z' production in top events at an e^+e^- collider

Ambroise Espargilière
LAPP, Annecy, France
Laboratoire d'Annecy le Vieux de Physique des Particules (LAPP)
CNRS : UMR5814 - IN2P3 - Université de Savoie

E-mail: ambroise.espargiliere@lapp.in2p3.fr

Abstract. In Randall-Sundrum theories a dark matter candidate can arise consequently to baryon number conservation in the form of a massive right handed Dirac neutrino interacting through an additional $SU(2)_{\mathrm{R}}$ gauge group. The neutral $SU(2)_{\mathrm{R}}$ gauge boson, called Z', is supposed to couple only to the Standard Model right handed top quark. In e^+e^- collisions it will therefore be produced preferentially in top events. Its production is studied in the channel $e^+e^- \to t\bar{t}Z'$ and a method to measure its mass is proposed. The generator level measurement shows a precision better than 10% for a light Z' ($m_{Z'} \in [200, 500]$ GeV).

1. Introduction

Many theories have been built to fill in the Standard Model (SM) deficiencies (super-symmetry, Kalusa-Klein, ...). The so-called Grand Unified Theories (GUTs) have to cope with many questions to be relevant. The dark matter puzzle is one of those. Dark matter is constituted of yet unknown massive particles. Such particles account for the major part of the galaxy masses and can be detected thanks to gravitational effects (galaxy dynamics, gravitational lensing). Dark matter accounts for $\approx 25\%$ of the universe's energy budget. In many GUTs, a dark matter particle candidate arises as a spin-off of the baryon number conservation (*e.g.* the neutralino being stabilised thanks to the R-parity in super symmetry). In the case of Randall-Sundrum theories, the baryon number conservation leads to the stability of a particle having the properties of a right handed neutrino ν' [1, 2]. This ν' interact *via* an additional $SU(2)_{\mathrm{R}}$ gauge group involving, among others, a neutral gauge boson called Z'.

The expected main properties of these new particles (details in [3] and [4]) are the following:

- Small coupling to the SM through a Z^0/Z' mixing at the percent level. Too high mixing would conflict with direct dark matter detection constraints.
- The right handed top quark is the only SM particle that directly couples to the Z'
- The Z' mass $m_{Z'}$ must be in the TeV range but as light as possible.
- The ν' mass $m_{\nu'}$ must be a bit below $m_{Z'}/2$ to comply with the relic density constraints.

As a consequence, this Z' can not be directly produced in e^+e^- collisions in a resonant process such as $e^+e^- \to Z'$ and the benefit must be taken from its strong coupling to the top quark. In the next section, the Z' production in e^+e^- collisions is described and the choice of the studied Z' decay channel is discussed. In section 3 the selection of the events showing a Z' is drawn and in section 4 the cross section of the production process is measured using the event selection method and a technique for the measurement of the Z' mass $m_{Z'}$ is proposed. The analysis is done at the generator level. Data are generated using the software CalcHEP 2.5.5 [5] and Pythia 8 [6] is used to propagate and decay the particles. Some inputs from a full detector simulation are considered in section 4.3.

2. Z' production in 3 TeV e^+e^- collisions

Since the Z' is supposed to couple in the SM mainly (or uniquely) to the top quark, a favoured production process in e^+e^- collisions is the emission by an out-going $t\bar{t}$ pair. The corresponding diagram is displayed in figure 1.

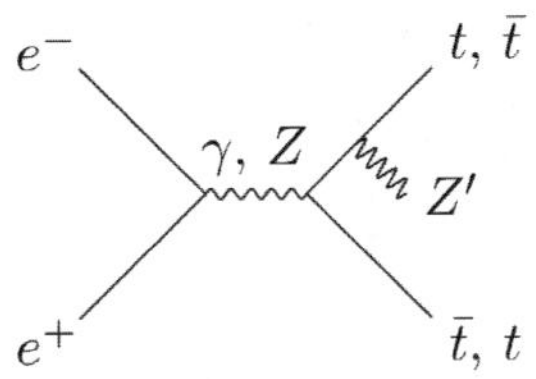

Figure 1. Z' production by emission from an out-going $t\bar{t}$ pair.

In the present set-up, if the small Z^0/Z' mixing is neglected, two decay channels are possible for the Z': the decay into a $t\bar{t}$ pair, allowed if its mass is sufficient, and the decay into $\nu'\bar{\nu}'$ pair, always allowed since the ν' has been taken at a mass below $m_{Z'}/2$.

Recent LHC results (elsewhere in those proceedings) have shown that a leptophobic Z', that would show up as a resonance in di-jet mass or $t\bar{t}$ pair invariant mass spectra, are very unlikely for Z' masses below ≈ 1.5 TeV. On the other hand, in the process discussed above, if the Z' decayed into a $t\bar{t}$ pair, a four top event would be obtained, giving therefore a striking signal at a 3 TeV e^+e^- collider (the SM cross section for four top events is ≈ 20 ab). Therefore the present study investigate the possibility of a light Z' decaying into invisible particles, hence avoiding the present experimental limits. Z' masses above the $t\bar{t}$ threshold are also considered. The cross sections of this process for different masses of the Z' are shown in table 1 at a centre of mass energy of 3 TeV and includes the effect of Initial State Radiations (ISR).

Table 1. Total process cross sections and Z' decay branching ratios times cross section.

$m_{Z'}$	Cross sect.	$\sigma_{Z'\to\text{inv.}}$	$\sigma_{Z'\to t\bar{t}}$
200 GeV	14.9 fb	14.9 fb	—
300 GeV	8.7 fb	8.7 fb	—
400 GeV	5.6 fb	2.4 fb	3.2 fb
500 GeV	3.7 fb	1.2 fb	2.5 fb
600 GeV	2.6 fb	0.8 fb	1.8 fb
700 GeV	1.8 fb	0.5 fb	1.3 fb

3. Z' event selection

The main characteristics of signal and backgrounds are summarised below. A method to discriminate the signal events is proposed.

3.1. Signal and background characteristics

The signal consists of multi-jet events with a large number of particles and a certain amount of missing energy. It is expected to be very close to SM s-channel $t\bar{t}$ events which constitute the main background. In a first step, the tagging of the top events is assumed to have a full efficiency and purity. Therefore, the backgrounds considered are only the SM inclusive $t\bar{t}$ events and the following background channels are leftover for this study (cross sections given at 3 TeV for boson hadronic decays only):

- $e^+e^- \rightarrow W^+W^-$ (263 fb)
- $e^+e^- \rightarrow Z^0Z^0$ (16 fb)
- $e^+e^- \rightarrow W^+W^-Z^0$ (11 fb)
- $e^+e^- \rightarrow Z^0Z^0Z^0$ (< 1 fb)

The present results shall then be scaled according to the real top tagging performance.

Both s and t channel $t\bar{t}$ production are considered, their respective Feynman diagrams are displayed in figure 2 (note that, for background 2, only the main diagram is represented) and their respective cross sections are summed up in table 2.

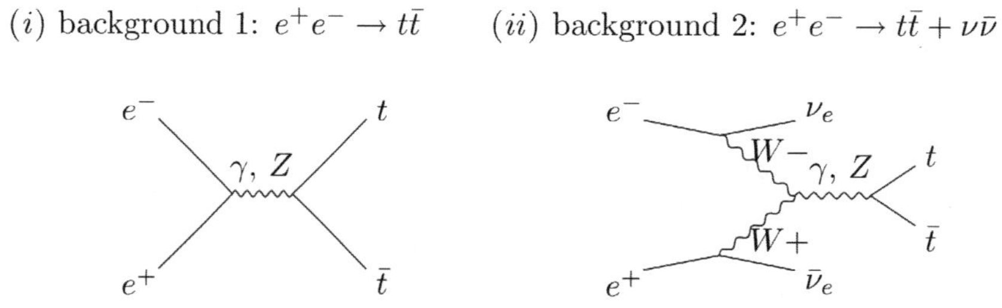

Figure 2. Diagrams of the considered background channels

Table 2. background channel cross sections.

Channel	Cross sect.
$e^+e^- \rightarrow t\bar{t}$	19.87 fb
$e^+e^- \rightarrow t\bar{t} + \nu\bar{\nu}$	5 fb

3.2. Selection results

The event selection is performed with two cuts on the output of a pair of Boosted Decision Trees (BDT) trained separately for the rejection of each background. The discriminative input variables are listed in table A1. They are all related to the event dynamics, event shape and to missing energy. The BDT output distributions for signal and background is displayed on figure 3 and the performance of the separation are summarised in table 3. S_{max} denotes the maximum significance and therefore corresponds to the optimal cut. The global efficiency of the event selection is simply the product of the efficiencies of the two single cuts : $(0.908 \pm 0.3) \times (0.974 \pm 0.2) = 0.884 \pm 0.004$.

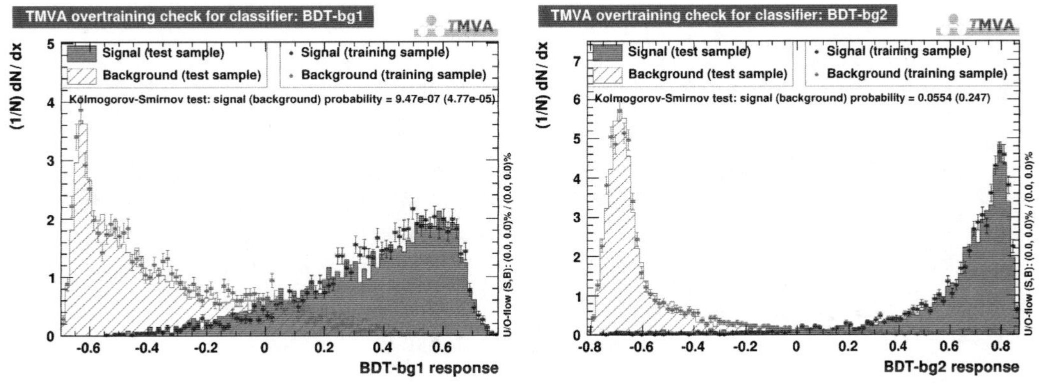

Figure 3. BDT output distributions for signal and both backgrounds.

Table 3. BDT performance. The third column gives for comparison the performance of a simple cut on the missing transverse momentum for the Bg1 rejection only. $\int$ROC is the integral of the Reciever Operating Characteristic (ROC) curve of the BDT.

	Sig. *vs* Bg1	Sig. *vs* Bg2	$\not{p}_T$ cut
$\int$ROC	0.955	0.971	N/A
$S_{\max}$	99.6	113.8	88.6
$\epsilon @ S_{\max}$	90.8%	97.4%	85%
$B @ S_{\max}$	18.0%	14.6%	40.3%
$S_{\max}$ cut	-0.0378	-0.3391	175

4. Measurement of process cross section and Z' mass

4.1. Method for Z' mass measurement

The Z' mass can be deduced from the event total invariant mass spectrum of the selected events. This spectrum should ideally show an upper end-point at 3 TeV- $m_{Z'}$.

Nevertheless, the ISR induces a smearing of the collision energy, as shown in figure 4. It therefore induces a smearing of the spectrum and then a smoothing of the spectrum upper edge. A function, called "smooth gate" in the following, defined in eq. 1, models well the shape of the spectrum. It is fitted to the event total invariant mass spectrum in the absence of the signal as shown in figure 5(c). This way the background contribution is modelled and could be subtracted from the total spectrum. The area of the remaining spectrum would then be proportional to the cross section of the $e^+e^- \to t\bar{t}Z'$ process. To measure the Z' mass, the "smooth gate" is fitted to the background subtracted spectrum as shown in figure 5(d) and the s_2 parameter, corresponding to the upper slope inflexion point, is related to $m_{Z'}$ through the relation 3 established in the following.

The "smooth gate" is defined by:

$$f(x) \quad = \quad \frac{A}{(e^{\frac{s_1 - x}{p_1}} + 1)(e^{\frac{x - s_2}{p_2}} + 1)}, \tag{1}$$

where s_i, $i \in \{1, 2\}$ are the inflection point abscissae and p_i, $i \in \{1, 2\}$ are the slope steepnesses, see graphical representation with arbitrary values of the parameters in figure 5(b).

The procedure described above is applied for $m_{Z'}$ from 200 to 700 GeV with a constant number of events. The use of a constant number of events is meant to probe the s_2 sensitivity

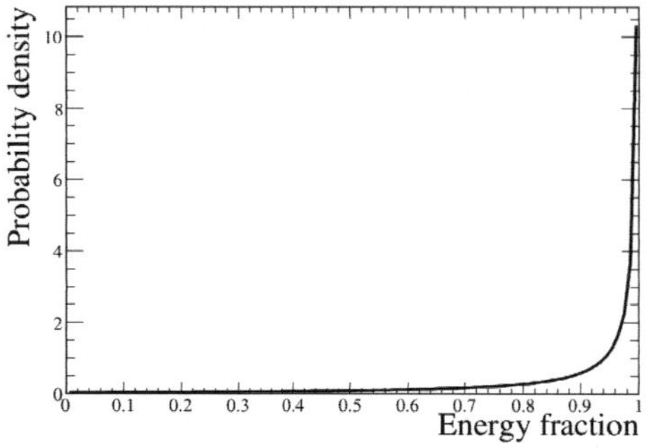

Figure 4. Distribution of the fraction of nominal collision energy due to ISR.

to $m_{Z'}$ only. The resulting event invariant mass spectra are displayed in figure 5(e). Their fitted "smooth gate" upper inflexion point abscissa are summed up in table 4 and plotted in figure 5(f). This figure shows as well how a straight line fits to the data points and the residuals are shown in the histogram placed in the corner of the figure. The fit information allows to define a relation between the s_2 parameter and $m_{Z'}$ which reads:

$$s_2 = (2757 \pm 8)\,\mathrm{GeV} - (1.05 \pm 0.02) \times m_{Z'}\,, \tag{2}$$

which gives a direct formula for $m_{Z'}$:

$$m_{Z'} = \frac{(2757 \pm 8)\,\mathrm{GeV} - s_2}{1.05 \pm 0.02}\,. \tag{3}$$

Table 4. Summary of the "smooth gate" upper inflexion point abscissae (s_2)

$m_{Z'}$ (GeV)	s_2 (GeV)
200	2530 ± 6
300	2458 ± 6
400	2340 ± 7
500	2235 ± 7
600	2119 ± 8
700	2005 ± 10

4.2. Model dependent results

The model predicted statistics for $1\,\mathrm{ab}^{-1}$ of data are used in order to estimate the expected cross section measurement precision and the expected resolution on $m_{Z'}$.

4.2.1. Event selection and cross section measurement Table 5 summarizes the main features of the analysis using the model predicted statistics: for each mass, the table shows the number of events passing the BDT cuts and surviving the background subtraction and the significance of the new physics signal. The significance S is defined as $\frac{s}{\sqrt{s+b}}$, where s and b are respectively the number of signal and background events. The measured cross sections are shown as well and compared to the predicted ones. The present method tend to underestimate the process cross section for low $m_{Z'}$ and slightly over-estimate it at higher values. The displayed uncertainties are given by $\Delta\sigma = \sigma(\sqrt{N_{\mathrm{yield}}}/N_{\mathrm{yield}} + \Delta\epsilon/\epsilon)$, where N_{yield} is the number of events passig the selection cut and ϵ is the selection efficiency.

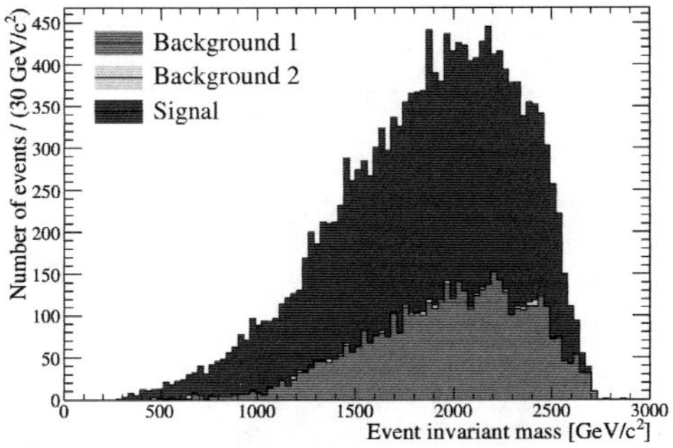

(a) Event total invariant mass spectrum for signal and both backgrounds ($m_{Z'}$ =200 GeV).

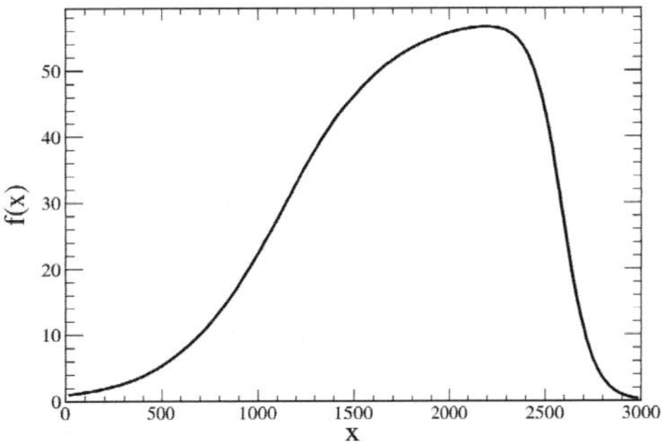

(b) "Smooth gate" function with arbitrary parameter values (A=58, s_1=1127, p_1=274, s_2=2589, p_2=76).

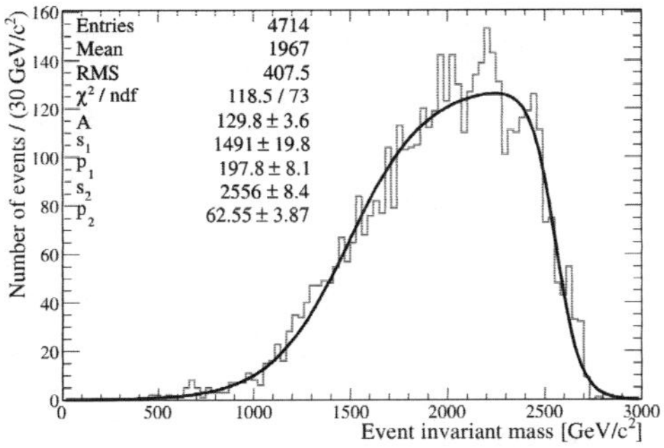

(c) Event total invariant mass spectrum after event selection, for background only, fitted with the "smooth gate" function.

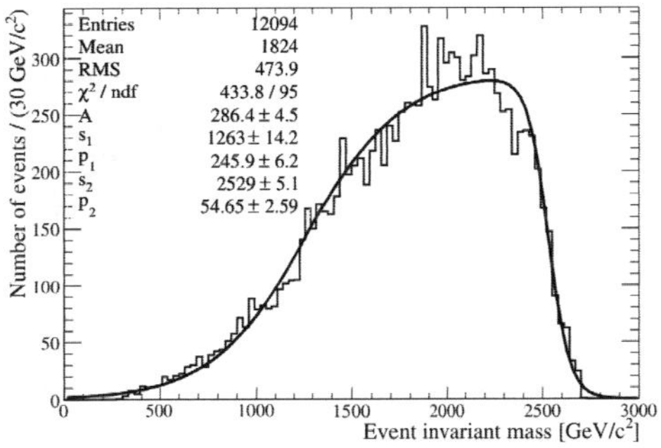

(d) Event total invariant mass spectrum after event selection and background subtraction fitted with the "smooth gate" function for $m_{Z'}$ = 200 GeV.

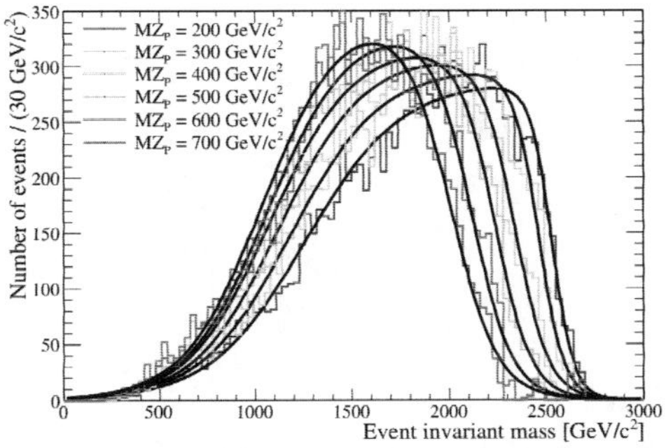

(e) Event total invariant mass spectrum after event selection and background subtraction fitted with the "smooth gate" function for various Z' masses.

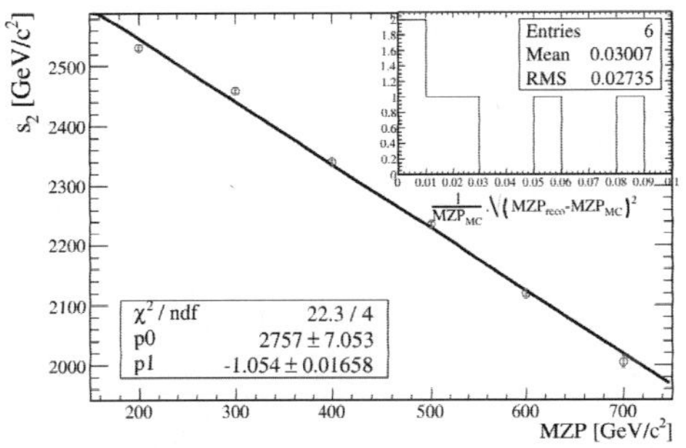

(f) s_2 parameter *versus* $m_{Z'}$. The corner histogram shows the residuals of the data points as measured from the fit.

Figure 5.

Table 5. Z' events using model predicted statistics and cross section measurement results (data for 1 ab^{-1} integrated luminosity)

$m_{Z'}$	Events yield	Significance	True cross sect.	Measured cross sect. (fb)
200	12094	92.7	14.9 fb	13.68 ± 0.19
300	7411	66.6	8.7 fb	8.38 ± 0.15
400	2150	24.6	2.4 fb	2.43 ± 0.07
500	1192	13.5	1.2 fb	1.35 ± 0.05
600	815	8.4	0.8 fb	0.92 ± 0.04
700	642	5.6	0.5 fb	0.73 ± 0.04

4.2.2. Z' mass measurement and expected resolution Table 6 summarizes the value of the s_2 parameter with its error for each Z' mass and the corresponding measured mass deduced from relation 3. It appears that above about 500 GeV, the mass measurement is less precise and underestimates the true value. This effect even increases with $m_{Z'}$. The origin of this effect is that the lower signal statistics at higher $m_{Z'}$ makes the fit of the "smooth gate" more sensitive to background fluctuations and introduces a bias not taken into account by the $m_{Z'}$ uncertainty formula (eq. 4). A bin to bin background subtraction may reduce this effect but the accumulation of more data would be the best solution to reduce it. However, the conclusion is that additional care must be taken for background subtraction to make this method reliable in the case of fewer than about 1000 events passing the selection cuts.

Thanks to eq. 3, it is now possible to predict the Z' mass from a $t\bar{t}$ event invariant mass spectrum to which a "smooth gate" function is fitted (see definition in eq. 1). As $m_{Z'} = (k - s_2)/\alpha$, the uncertainty on this prediction reads:

$$\begin{aligned} \Delta m_{Z'} &= \underbrace{\frac{\Delta k}{\alpha}}_{\substack{\text{constant}\\ \text{term}}} \oplus \underbrace{\frac{\Delta s_2}{\alpha}}_{\substack{\text{fit}\\ \text{term}}} \oplus \underbrace{\frac{m_{Z'}\Delta\alpha}{\alpha}}_{\substack{\text{linear}\\ \text{term}}} . \\ &= 7.8\,\text{GeV} \oplus 0.95\Delta s_2 \oplus 1.9\% m_{Z'} \end{aligned} \tag{4}$$

The errors Δk and $\Delta\alpha$ are inherent to the measurement method itself. Hence, the constant and the linear term of eq. 4 describe systematic uncertainties. They may be reduced by refining the analysis on the s_2 dependency on $m_{Z'}$. For instance, using more statistics for each value of $m_{Z'}$ and/or using more values of $m_{Z'}$ are possible refinements. Whereas the error Δs_2 comes from the event invariant mass spectra fitted with the "smooth gate" function and can only be improved by the accumulation of more statistics. Thus, the so-called "fit term" reflects the statistical uncertainty on $m_{Z'}$. The total resolution is predicted using eq. 4 and values from table 5 and 6. The result is plotted in figure 6.

4.3. Inputs from full simulation

The full simulation has been performed in the framework of the preparation of CLIC CDR [7]. The detector model is the CLIC version of the ILD detector. CLIC beamstrahlung has been simulated in details and the $\gamma\gamma \rightarrow$ *hadrons* machine background ($\approx$ 3 interaction per bunch crossing) has been included. Pandora PFA [8] has been used for data reconstruction.

Standard Model data sample have been available to evaluate the detector resolution on the variables and test top tagging performance. Figure 7 focuses on the resolution on the event total invariant mass and illustrates the fact that this variable is not robust against the finite detector

Table 6. Z' mass measurement results.

$m_{Z'}$ (GeV)	s_2 (GeV)	Measured $m_{Z'}$ (GeV)
200	2529 ± 6	217 ± 11
300	2448 ± 8	294 ± 13
400	2342 ± 20	395 ± 22
500	2282 ± 40	454 ± 40
600	2374 ± 69	365 ± 67
700	2527 ± 77	219 ± 74

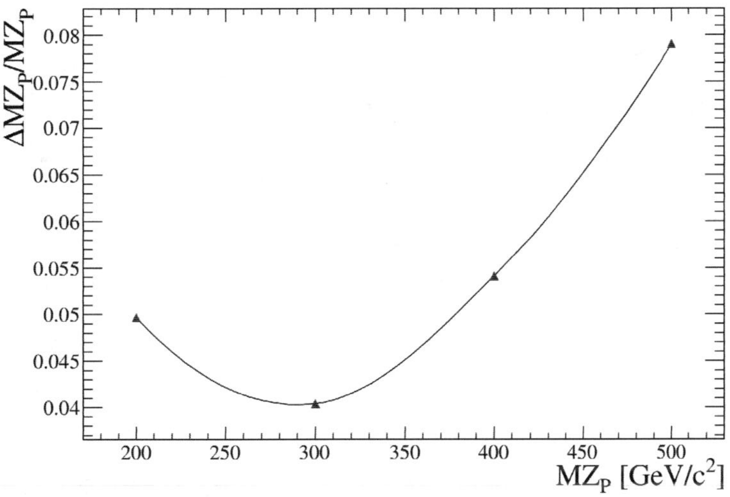

Figure 6. Expected error on $m_{Z'}$ *vs.* $m_{Z'}$.

resolution and the machine background. So far, the event total invariant mass was similar to the $t\bar{t}$ pair invariant mass which is the key variable for the $m_{Z'}$ measurement, therefore, a careful reconstruction of the top quarks four-vector will allow a reliable calculation of the $t\bar{t}$ pair invariant mass. The selection of boosted hadronic top events is an example of strategy to perform a precise reconstruction of top quark four-vectors without messing up with the problems of missing energy.

In figure 8, the separation between s-channel $t\bar{t}$ events and inclusive hadronic $W^+ W^-$ events is illustrated. A boosted decision tree is used as well. The preliminary performance sketched here indicate that $t\bar{t}$ events may be rather well extracted and that no dramatic degradation of the results presented is expected.

5. Conclusion

The Dark Matter candidate emerging from Randall-Sundrum theories (and possibly others) has the quantum numbers of a right handed Dirac neutrino ν'. This neutrino interacts *via* an additional gauge group $SU(2)_{\rm R}$ as a right handed analogy of the weak interaction. A Z' boson, being the neutral gauge boson of $SU(2)_{\rm R}$, couples only to the right handed top quark in the Standard Model and therefore can be produced in e^+e^- collision only *via* emission by an out-going $t\bar{t}$ pair.

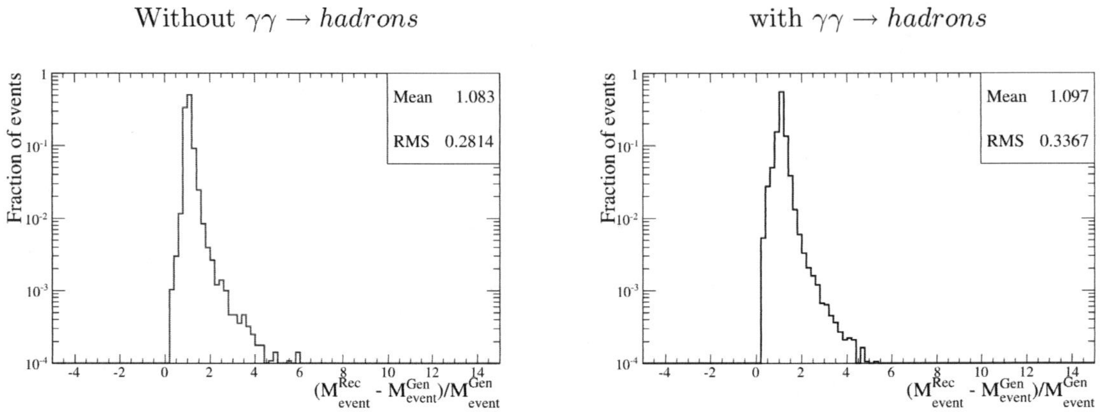

Figure 7. Resolution on the event total invariant mass after full simulation.

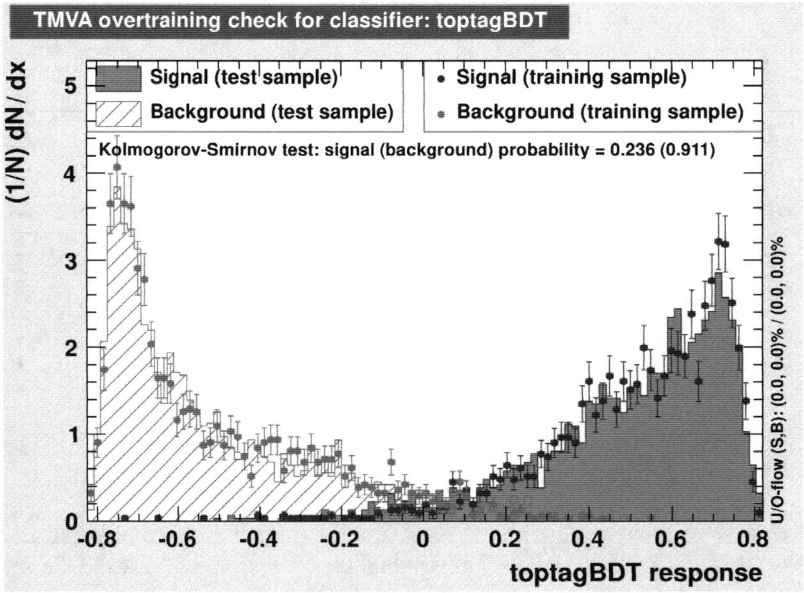

Figure 8. Separation between $t\bar{t}$ and $W^+ W^-$ events (preliminary).

This Z', if too light, can only decay into $\nu'\bar{\nu}'$ and avoids the present day LHC exclusions of a Z' resonance decaying into $t\bar{t}$ pairs. In this case the Z' can be discovered in $t\bar{t}$ event samples and its mass can be measured with a precision better than 10% with $1\,\mathrm{ab}^{-1}$ of data collected at a 3 TeV e^+e^- collider. The analysis has been carried out at the generator level, preliminary results from full simulation indicates that the total event invariant mass is not robust against a finite detector resolution and special care must be given to precise reconstruction of the top quark four-vectors (*e.g.* by selecting boosted hadronic tops) and the $t\bar{t}$ pair invariant mass will replace the event invariant mass in the analysis. The top tagging performance seems promissing and no critical degradation of those results is expected. A complementary note on the generator level study can be found in [9].

Acknowledgement

The first acknowledgement goes to the conference organisers for giving me the opportunity to present this work. The author would like to thank Geneviève Bélanger, Marco Battaglia and Géraldine Servant for their help in guiding this study at the beginning (the choice of the channel, the model parameter values to use...). Many thanks as well for Sasha Pukhov and Geneviève Bélanger again, for answering all my questions about the RHNM and CalcHEP and to Peter Skands for helping me fix the interface between CalcHEP and Pythia 8. The author would like especially to thank Jean-Jacques Blaising for advice, useful discussions and precious help.

Appendix A. BDT input variables

Table A1. Summary of the analysis variables

Variable name	Description	Discriminating power
`missPT`	Missing transverse momentum	3.462e-01
`missE`	Missing energy	3.114e-01
`jet1E`	First hemisph. energy	3.012e-01
`jet2E`	Second hemisph. energy	2.496e-01
`jet2P`	Second hemisph. momentum	2.365e-01
`missP`	Missing momentum	2.250e-01
`jet1P`	First hemisph. momentum	2.138e-01
`jetPTdiff:=jet1P-jet2P`	hemisph. PT diffrence	2.122e-01
`missA`	Angle of missing momentum	1.134e-01
`jet2PT`	Second hemisph. transverse momentum	1.078e-01
`sphericity`	Sphericity	1.024e-01
`aplanarity`	Aplanarity	9.443e-02
`jetEratio:=jet2E/jet1E`	Hemisph. energy ratio	9.302e-02
`Jet1PT`	First hemisph. transverse momentum	7.763e-02
`jetEdiff:=jet1E-jet2E`	Hemisph. energy difference	6.089e-02

Hemisphere The event hemispheres are defined as the half-spaces separated by the plane containing the interaction point and perpendicular to the event axis. The event axis is defined as the eigenvector of the event sphericity tensor S corresponding to its largest eigenvalue. The sphericity tensor S is given by :

$$S^{ab} = \frac{\sum_i p_i^a p_i^b}{\sum_i ||\vec{p_i}||^2} .$$

References

[1] Agashe K and Servant G 2004 *Phys. Rev. Lett.* **93** 231805
[2] Agashe K and Servant G 2005 *J. Cosmol. Astropart. Phys.* JCAP02(2005)002 *(Preprint hep-ph/0411254)*
[3] Bélanger G, Pukhov A and Servant G 2008 *J. Cosmol. Astropart. Phys.* JCAP01(2008)009 *(Preprint 0706.0526 [hep-ph])*
[4] Jackson C B *et al* 2010 *J. Cosmol. Astropart. Phys.* JCAP04(2010)004 *(Preprint 0912.0004 [hep-ph])*
[5] Pukhov A 2004 *Preprint hep-ph/0412191*
[6] PYTHIA 8 home page: `http://home.thep.lu.se/~torbjorn/Pythia.html`
[7] CLIC collaboration: `http://clic-study.org`
[8] Thomson M 2009 *Nucl. Instrum. Meth.* A **611** 25–40 (*Preprint arXiv:0907.3577 [physics.ins-det]*)
[9] Espargilière A 2011 *HAL : in2p3-00609168*

2nd International Conference on Particle Physics IOP Publishing
Journal of Physics: Conference Series **347** (2012) 012020 doi:10.1088/1742-6596/347/1/012020

The electromagnetic calorimeter of the SuperB Experiment

Davide Pinci[1]
INFN Sezione di Roma, Italy
E-mail: davide.pinci@roma1.infn.it

Abstract.
SuperB will run on a very high luminosity asymmetric e^+e^- flavour factory. Being a natural partner of hadron colliders, it will provide unique information about the details of the physics discoveries in the coming decade. The SuperB detector is based on the BABAR apparatus, with those modifications required to operate at a luminosity of $10^{36}\text{cm}^{-2}\text{s}^{-1}$ or above. In this presentation a summary of the R&D work performed on the new electromagnetic calorimeter will be shown.

1. Introduction

SuperB [1] is a project aiming to study with great precision CP violating and rare processes of heavy quarks (c and b) and leptons (τ). It will be based on an asymmetric electron (4.2 GeV) positron (6.7 GeV) machine designed to provide a luminosity of $10^{36}\text{cm}^{-2}\text{s}^{-1}$. The SuperB detector concept is based on the BABAR apparatus [2], shown in Fig. 1, with those modifications required to operate at a higher luminosity and with a reduced center-of-mass boost.

The current BABAR detector consists of a tracking system with a five layer double-sided silicon strip vertex tracker (SVT) and a 40 layer drift chamber (DCH) inside a 1.5T magnetic field, a Cherenkov detector with fused silica bar radiators (DIRC), an electromagnetic calorimeter (EMC) consisting of 6580 CsI(Tl) crystals and an instrumented flux-return (IFR) for K^0_L detection and μ identification. Further improvements are needed to cope with higher beam-beam and other beam-related backgrounds, as well as to improve detector hermeticity and performance. The rates of the main sources of background are reported in Table 1 together with the expected rates of interesting events.

Several R&D are required to implement the needed upgrades on the various subdetectors.

2. The Electromagnetic Calorimeter

The SuperB Electromagnetic Calorimeter (EMC) provides energy and direction measurement of photons and electrons and is an important component in the identification of electrons versus other charged particles. The system contains three components, shown in Fig. 1:

- the barrel calorimeter, reused from BABAR, that will cover 85% of the total solid angle;
- the forward endcap calorimeter, replacing the BABAR forward endcap, that will cover 5.6% of the total solid angle;

[1] On behalf of the SuperB-EMC group

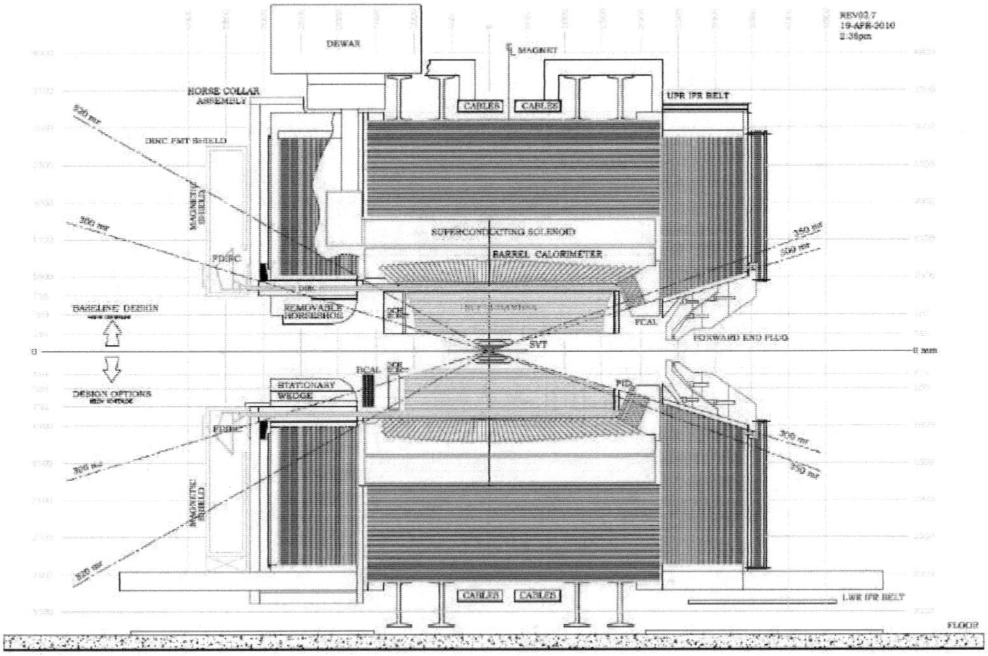

Figure 1. The SuperB Detector.

Table 1. Parameters of the main background and signal processes expected.

Process	Cross Section (mbarn)	Events per bunch crossing	Rate (kHz)
Radiative Bhabha	340	850	300 10^6
e^+e^- production	7.3	18	7 10^6
e^+e^- production seen by L0	0.3	1	0.3 10^6
Elastic Bhabha	10^{-4}	2.5 10^{-4}	100
Υ(4S)	10^{-6}	2.5 10^{-6}	1

- the backward endcap calorimeter, a new device improving the backward solid angle coverage (3.1 % of total solid angle covered).

The total solid angle covered for a massless particle in the center-of-mass is 94.1% of 4π.

2.1. The Barrel Calorimeter

The barrel calorimeter for SuperB is the existing BABAR CsI(Tl) crystal calorimeter. Estimated rates and radiation levels indicate that this system will continue to survive and function in the SuperB environment. There are 48 rings of crystals in polar angle, with 120 crystals in each azimuthal ring, for a total of 5,760 crystals. The crystal length ranges from 16 X_0 to 17.5 X_0. They are read out by two redundant PIN diodes connected to a multi- range amplifier. In

Journal of Physics: Conference Series **347** (2012) 012020 doi:10.1088/1742-6596/347/1/012020

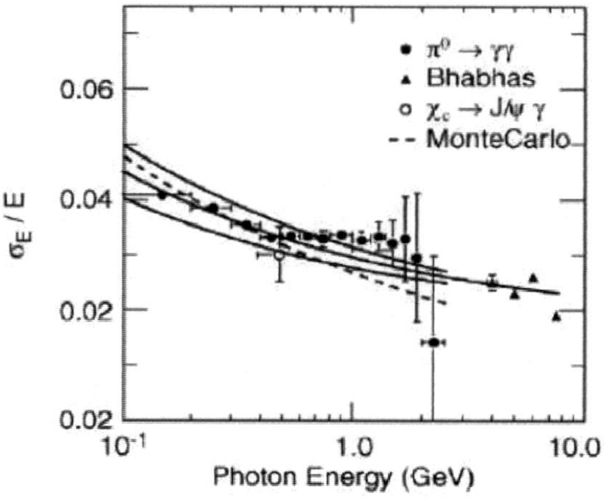

Figure 2. The BABAR energy resolution on photons.

BABAR this detector was able to provide a very good energy resolution described as:

$$\frac{\sigma_E}{E} = \frac{2.3\%}{\sqrt[4]{E(GeV)}} \oplus 1.35\%$$

and shown in Fig. 2. This behaviour can be taken as a reference for the performance required to the SuperB calorimeter: a resolution of the order of 4% in the 100 MeV region.

2.2. The Forward Endcap Calorimeter

The background level in the whole calorimeter system will be quite higher than in BABAR. The worst increase is expected in the forward endcap part (FWD). According to the last simulations, a rate of more than 2 MHz photons with an energy of few MeV, produced in radiative Bhabha processes, is expected per crystal in the FWD. These will produce a continous and fluctuating level below the relevant signals (Fig. 3) spoiling their energy measurement. In order to cope

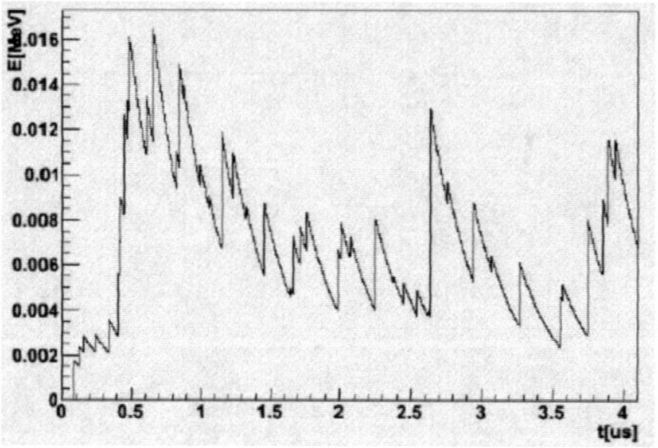

Figure 3. Behaviour of the light emitted by a FWD crystal due the radiative Bhabha photons.

with this background a fast enough crystal is needed to replace the CsI(Tl) ones that have a light yield decay time of more than 1 μs.

Journal of Physics: Conference Series **347** (2012) 012020 doi:10.1088/1742-6596/347/1/012020

2.2.1. The LYSO prototype The baseline solution is given by the LYSO(Ce). It is a high density crystal (ρ =7.4 g/cm^3), very fast (40 ns of decay time) and very bright (1 p.e. per keV released) [3], really promising for applications in high energy physics. In order to test the properties of this material a matrix of 25 crystals, each readout by one APD, was assembled and tested with electron beam at CERN-T10 with particles of energy between 1 GeV and 3 GeV and at the Frascati-BTF [4] in the 100 MeV - 500 MeV range. Among other measurements, in both cases the energy resolution was evaluated. The spectrum of the response of the detector to a beam of 500 MeV electrons is shown in Fig. 4. The distribution was fitted with a "Crystal-Ball" function

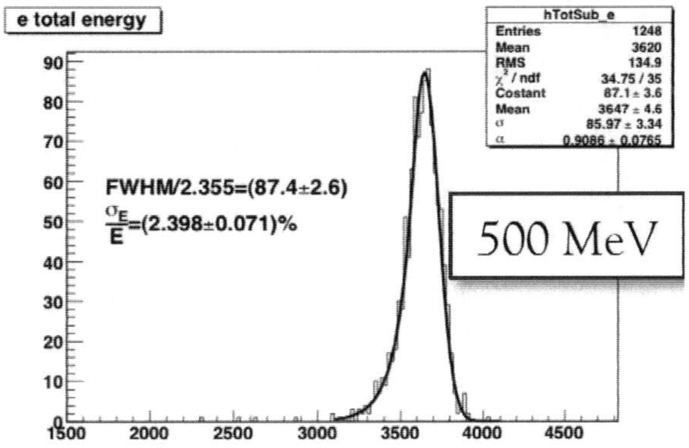

Figure 4. The response spectrum of the prototype for 500 MeV electrons.

able to describe the asymmetry due the non-perfect energy containment. A resolution of 2.4% was found. Fig. 5 shows the behavior of the relative resolution obtained at CERN-T10 and at the BTF as a function of the electron energy. From this plot it appears evident that a constant

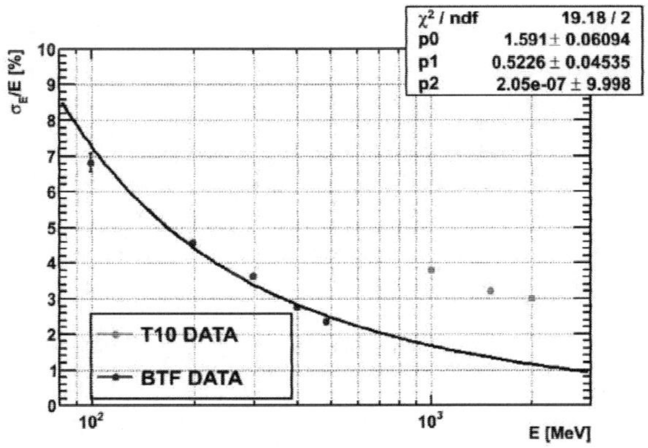

Figure 5. The relative resolution of the prototype as measured at CERN-T10 (1 GeV¡E¡2 GeV) and at the BTF (below 1 GeV).

effect, very likely due to an energy spread of the beam, was present at CERN downgrading the response of the detector. The BTF results were fitted with a curve described as:

$$\frac{\sigma_E}{E} = \frac{1.6\%}{\sqrt{E(\text{GeV})}} \oplus \frac{0.5\%}{E(\text{GeV})}$$

It is anyway important to outline that, for lower values, the energy spread of the beam becomes important also at the BTF. The tests performed have shown that LYSO is a very good solution to equip the SuperB FWD calorimeter.

2.2.2. The BGO alternative LYSO is a quite expensive material and some possible alternatives are taken into a account. One of these is represented by the BGO, the same material used, for example, in the L3 experiment [5]. BGO is as dense as the LYSO, less brighter and its light output has a decay constant of about 300 ns [3]. From measurements with a 18 cm long crystal, readout with a PMT, a resolution of 10% was evaluated for the 660 keV photons of a ^{137}Cs source (Fig. 6). From random trigger runs the pedestal width was evaluated to be 10 times smaller than the energy resolution and thus completely negligible. This means at least 150 p.e./MeV and thus a 1.1% of statistical term is expected at 50 MeV.

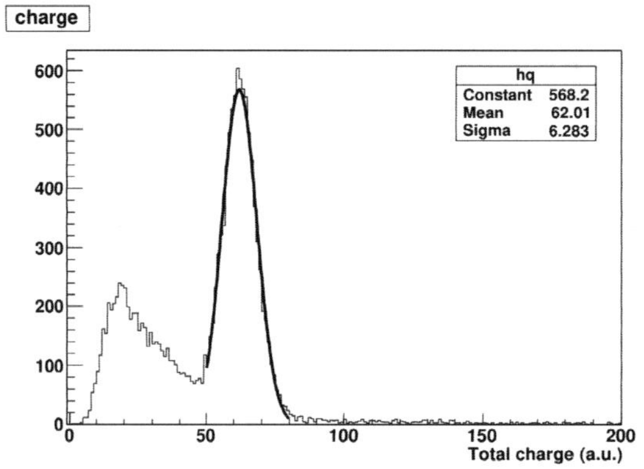

Figure 6. Response spectrum of a single BGO crystal to the 660 keV photons of a ^{137}Cs source.

Moreover, in order to reduce the effect of the background, the integration time of the pre-amplifier in the front-end electronics can be reduced. In this case, only a fraction f of the total charge will be integrated. The effect of this reduction on the energy resolution and its behavior as a function of f was studied. The results obtained are shown in black in Fig. 7. The relative energy resolution behavior as a function of f was fitted with a curve described by the function:

$$\frac{\sigma_E}{E} = \frac{\sigma_0}{\sqrt{f}}$$

Since the pedestal contribution is negligible, no other terms are needed. The energy resolution is found to upgrade as the square root of the total charge and thus to be dominated by the number of collected photo-electrons. From the fit, σ_0 is determined to be $(9.1 \pm 0.1)\%$. For $f = 1$ a resolution of 9.1% is expected, in agreement with what is found from the total charge measurements.

In particular, for an integration time of 100 ns a resolution of 12.7% was measured which allows to extrapolate a statistical contribution to the energy resolution at 50 MeV of about 1.5%.

2.2.3. Background effects With the latest results on the expected background at SuperB, the effects of soft photons produced by the radiative Bhabha on the energy resolution were simulated. The total charge collected in 5x5 crystal matrices and its fluactuations for different integration times were evaluated. The results show that in the barrel, the use of an integration time of

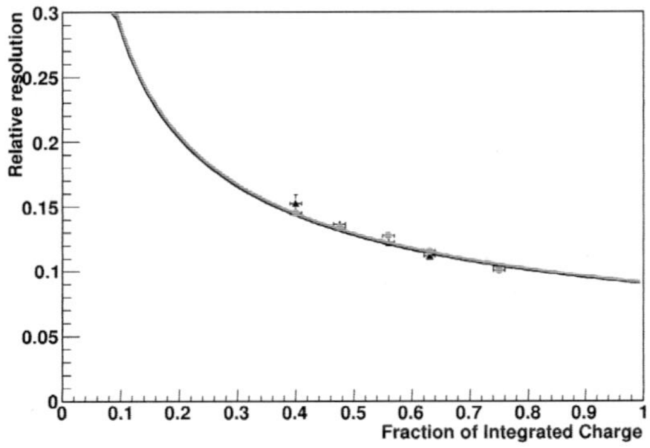

Figure 7. Behaviour of the resolution σ_E/E as a function of the fraction f for two different sets of measurements.

300 ns instead of 700 ns (as it was in BABAR) can reduce the baseline fluctuation form 2.4 MeV to 0.6 MeV in the central part and from 7.3 MeV to 2 MeV in the worst case. In the FWD, the use of LYSO crystal would result in a baseline fluctuation of 0.5 MeV in the external part and 2.5 MeV in the internal part. The BGO with an integration time of 100 ns would give 0.5 MeV and 1.5 MeV respectively.

3. Conclusion

A new forward endcap calorimeter is being developed for SuperB. LYSO has already shown to fulfill the requirements to equip that region with a resolution of 2.3% at 500 MeV. However, it represents a quite expensive solution.

A possible alternative are BGO crystals with an optimized readout electronics to reduce the background effect without spoiling the energy resolution.

References

[1] Bona M et al. 2007 SuperB: a high-luminosity heavy flavour factory. Conceptual design report *Preprint* 0709.0451v2 (INFN/AE07/2, SLAC-R-856, LAL 07-15, also available at `http://www.pi.infn.it/SuperB/CDR`)
[2] Aubert B et al. (BABAR Collaboration) 2002 The BABAR detector *Nucl. Instrum. Meth.* A **479** 1 (*Preprint* hep-ex/0105044)
[3] Nakamura K et al. (Particle Data Group) 2010 *J. Phys.* G **37** 075021
[4] Mazzitelli G, Ghigo A, Sannibale F, Valente P and Vignola G 2003 Commissioning of the DAΦNE beam test facility *Nucl. Instrum. Meth.* A **515** 524
[5] Adeva B et al. (L3 Collaboration) 1990 *Nucl. Instrum. Meth.* A **289** 35

2nd International Conference on Particle Physics IOP Publishing
Journal of Physics: Conference Series **347** (2012) 012021 doi:10.1088/1742-6596/347/1/012021

Performance of the ATLAS Inner Detector at the LHC

Thijs Cornelissen[1]
Bergische Universität Wuppertal, Gaußstraße 20, 42097 Wuppertal, Germany
E-mail: `thijs.cornelissen@cern.ch`

Abstract. Since the LHC startup in 2009, the ATLAS inner tracking system has played a central role in many ATLAS physics analyses. Rapid improvements in the calibration and alignment of the detector allowed it to reach nearly the nominal performance in the timespan of a few months. The tracking performance proved to be stable as the LHC luminosity increased by five orders of magnitude during the 2010 proton run, while the performance was only slightly degraded in the extremely dense heavy ion collisions.

1. Introduction

The ATLAS Inner Detector [1] started recording collisions from the LHC [2] beams in November 2009. Since then, approximately 600 million events have been recorded, reconstructed offline, and used to understand and improve the performance of the tracking system in great detail. The Inner Detector consists of three components: a pixel detector with very high granularity and precision, surrounded by a silicon strip tracker (SemiConductor Tracker, SCT), and finally the Transition Radiation Tracker (TRT), a tracker based on drift tubes (straws). The Pixel and SCT cover the pseudorapidity[2] range up to $|\eta| < 2.5$, while the TRT extends up to $|\eta| < 2.0$. The Inner Detector is immersed in a solenoid field of about 2 T, which allows a measurement of the momenta of the particles. Table 1 summarizes the specifications of the subcomponents in the tracking system.

Table 1. Specifications of the Inner Detector.

Detector	#channels	#hits per track	R-ϕ resolution	R-z resolution
Pixel	80M	~3	$< 10\ \mu$m	$\sim 100\ \mu$m
SCT	6.3M	~4 (modules)	17 μm	$\sim 800\ \mu$m
TRT	320k	~30	130 μm	-

[1] On behalf of the ATLAS Collaboration
[2] Pseudorapidity is defined as $\eta = -\ln \tan(\theta/2)$, where θ is the polar angle

2. Operation and performance

The Inner Detector is fully operational since September 2008, when the first cosmic tracks were detected in the Pixel Detector. A very large fraction of the detector is functional: approximately 97% of the Pixel Detector, 99% of the SCT Detector, 98% of the TRT Detector. The remaining dead channels are stored and corrected for in the offline reconstruction. The efficiencies were found to be nearly 100% for the Pixel and SCT Detectors, about 95% for the TRT. The noise rates are 10^{-10} for the Pixel Detector, $< 5 \cdot 10^{-4}$ for the SCT, around 2% for the TRT [3]. These values are equal or better than the design specifications [4]. The data is continuously monitored offline, e.g. to assess the impact of occasional operational issues on the tracking performance. Only a small fraction of data had to be discarded due to holes in the tracking acceptance.

Apart from the efficiencies at the hit level, the track reconstruction and vertex reconstruction efficiencies were also studied [5]. The track reconstruction efficiency was taken from the simulation, with the systematic errors estimated from data/MC disagreements. In the initial phase the agreements were generally very good, except in the forward region $|\eta| > 2.2$ were some material was found to be missing in the detector description. The tracking efficiency as a function of p_T is shown in Figure 1. The highest efficiency is obtained for high-p_T, central tracks. The vertex efficiency was obtained in a data driven way, using an event sample enriched with collisions. In Figure 2 it can be seen that the vertexing efficiency is practically 100% for vertices with three or more tracks.

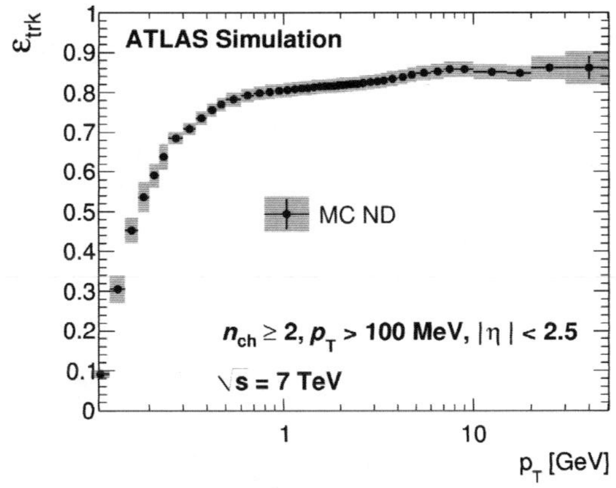

Figure 1. Tracking efficiency as a function of p_T [5].

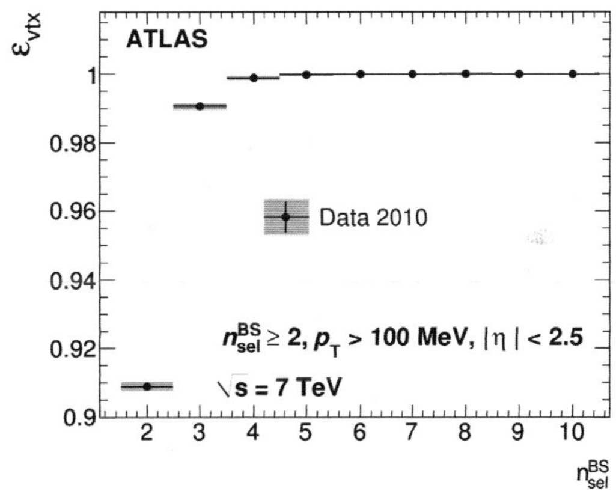

Figure 2. Vertex reconstruction efficiency as a function of the number of tracks [5].

3. Alignment

To be able to complete the ATLAS physics program, the relative positions of the sensors in the Inner Detector need to be known within 10 micrometers or better. The tolerances used during the assembly of the detector were much larger than that, therefore the full alignment needs to be obtained by studying the data [6]. The starting point during the LHC startup was the aligment obtained using cosmic-ray data in 2008 and 2009. The performance using this alignment was reasonable in the barrel part, but clearly more limited in the endcaps where statistics with the cosmic-ray data were low. The first collision runs at 7 TeV greatly increased the number of available tracks, allowing a more detailed alignment to be performed, eventually down to the level of individual sensors. Figure 3 shows that the measurement precision of the TRT was practically at the level of the simulation (which assumes a perfect alignment and calibration), after just a few months of data taking. In addition to the improved alignment, the precision of the Pixel Detector also benefitted greatly from the corrections for distorted sensor surfaces, which were enabled in time for the autumn 2010 reprocessing of the data. Figure 4 shows the improvement in the pixel resolution in this reprocessing, compared to the initial processing in spring 2010.

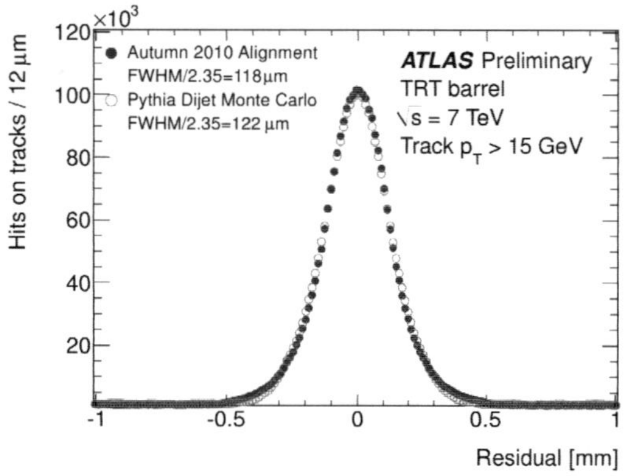

Figure 3. Hit residuals in the TRT barrel, compared between data and simulation [6].

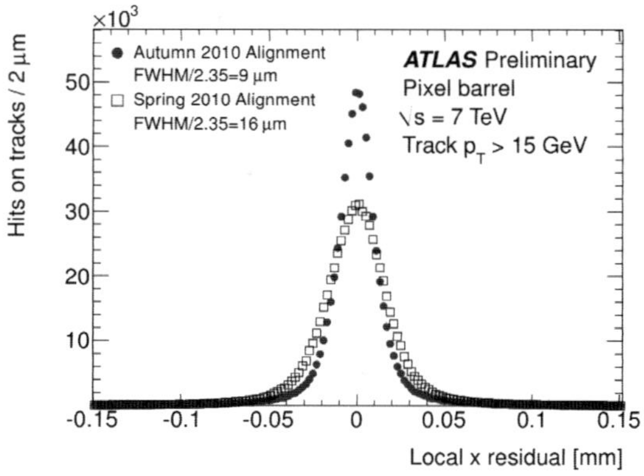

Figure 4. Hit residuals in the Pixel barrel, using the initial alignment (spring 2010) and an updated alignment (autumn 2010) [6].

4. Particle identification

The Inner Detector can distinguish between several particle types, depending on the momentum of the particle. In particular the separation between electrons and pions plays a very important role in many physics analyses that require electrons. Compared to pions, electrons are much more likely to produce transition radiation in the TRT radiator material. This radiation often produces large energy deposits in the straw ($\sim$10 keV, compared to 2 keV for a MIP). By counting the number of straws on track with a large energy deposit ('high threshold hits'), the probability of this particle being an electron can be determined. The large number of straws per track makes for a powerful discriminant. Figure 5 shows the probability for a straw on track to pass the high threshold as a function of momentum, for electrons and pions. More information about the performance of the TRT may be found in Ref. [7].

In addition to detecting electrons, both the TRT and Pixel Detector can also distinguish between pions, kaons and protons at low momentum using their specific energy loss $\mathrm{d}E/\mathrm{d}x$, as shown in Figure 6. The deuteron band is also visible in the right half of the plot. The $\mathrm{d}E/\mathrm{d}x$ based particle identification is used mainly to tag decays of Standard Model hadrons (e.g. the Λ and ϕ particles) [8], but also to search for stable, highly ionizing, supersymmetric particles [9].

5. Material studies

The material budget of the Inner Detector was studied using hadronic interactions and photon conversions. These are sensitive to the thickness of the material in interaction lengths and radiation lengths, respectively. Figure 7 shows the positions of the reconstructed hadronic interactions. The beam pipe can be seen near $R = 30$ mm, while the structure near $R = 50$ mm is the first layer of the Pixel Detector. The radius of the interactions in the beam pipe varies as a function of ϕ, due to the beam pipe being displaced with respect to the nominal beam line. This shift has since been implemented in the detector description used in the simulation and reconstruction. Figure 8 shows the positions of reconstructed photon conversions in the central

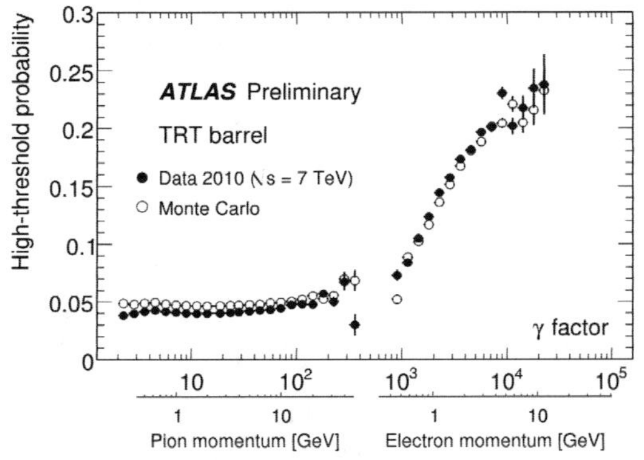

Figure 5. High threshold hit probability in the TRT barrel.

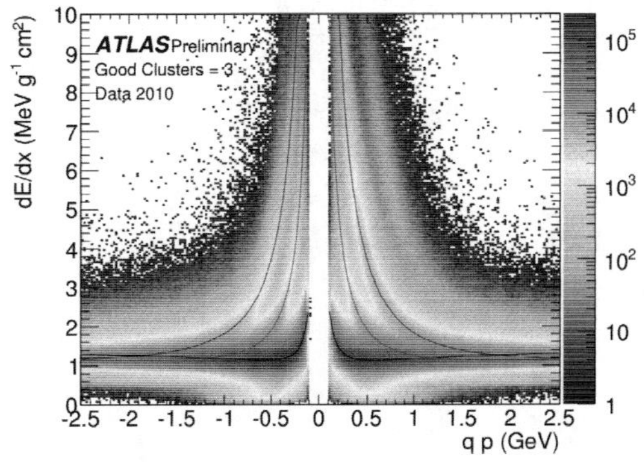

Figure 6. Charge deposited per unit length in the Pixel Detector [10].

part of the tracking system, compared with the simulation. It can be seen that the material distribution is well understood.

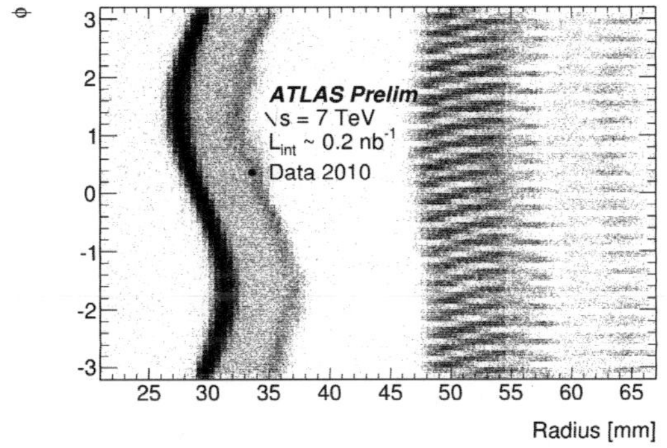

Figure 7. Reconstructed positions of hadronic interactions [11].

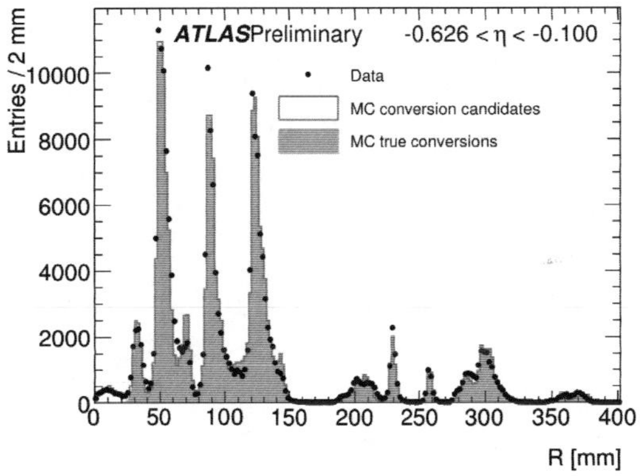

Figure 8. Reconstructed positions of photon conversions, comparing data and simulation.

6. Conclusions

The ATLAS Inner Detector has met or exceeded the expectations since the LHC startup in 2009. For the most part the performance (efficiency, resolutions) is in very good agreement with the simulation. The performance stayed almost constant as the LHC luminosity increased by six orders of magnitude. The Inner Detector will continue to play a central role in the fulfillment of the ATLAS physics program.

7. References

[1] The ATLAS Collaboration 2008 *JINST* **3** S08003.
[2] Evans L (ed.) and Bryant P (ed.) 2008 *JINST* **3** S08001.
[3] The ATLAS Collaboration 2010 *Eur.Phys.J.* **C70** 787.
[4] The ATLAS Collaboration 1997 *CERN/LHCC/97-16.*
[5] The ATLAS Collaboration 2011 *New J.Phys.* **13** 053033.
[6] The ATLAS Collaboration 2011 *ATLAS-CONF-2011-012.*

Journal of Physics: Conference Series **347** (2012) 012021 doi:10.1088/1742-6596/347/1/012021

[7] Bingul A, These proceedings.
[8] The ATLAS Collaboration 2010 *ATLAS-CONF-2010-023.*
[9] The ATLAS Collaboration 2011 *Phys. Lett. B* **701** 1.
[10] The ATLAS Collaboration 2011 *ATLAS-CONF-2011-016.*
[11] The ATLAS Collaboration 2011 *ATLAS-CONF-2010-058.*

Journal of Physics: Conference Series **347** (2012) 012022 doi:10.1088/1742-6596/347/1/012022

Electroweak and QCD measurements at LHCb

Victor Coco[1]
NIKHEF
E-mail: cocov@nikhef.nl

Abstract. The LHCb experiment at the LHC is fully instrumented over a unique pseudo-rapidity range in the forward region. Although it has been designed for b-physics, LHCb is able to provide valuable informations on particle production in this region of phase space. We summarize in this paper the electroweak bosons production measurement performed at $\sqrt{s} = 7$ TeV during the first half of 2010 data taking and the $\bar{\Lambda}/\Lambda$, $\bar{\Lambda}/K_s^0$ and $\bar{p}/p$ production ratios measurements performed at $\sqrt{s} = 900$ GeV and $\sqrt{s} = 7$ TeV.

1. Introduction

The LHCb experiment is dedicated to CP violation measurements and rare decays involving b and c-hadrons at the LHC [1]. Since most of the correlated $b\bar{b}$ pairs are produced at low angle with respect to the beam line, the LHCb detector has been designed as a single arm spectrometer instrumented over the $2 < \eta < 5$ pseudo-rapidity range. This unique coverage of the forward region, together with excellent tracking and particle identification performances allows the probing of particle production at the LHC in an uncharted region of phase space.

Theoretical predictions on W and Z production have uncertainties of between 3% and 10%, depending on rapidity, where the dominant uncertainty is due to knowledge of the parton distribution functions (PDFs). A precise measurement of their production at the LHCb would provide input to constrain the PDFs, both in the unique forward region ($\eta > 2.5$), and in a rapidity region which is in common to ATLAS and CMS ($2 \leq \eta \leq 2.5$). W and Z production cross sections, as well as cross section ratios and W charge asymmetry, have been measured in LHCb using approximately 16.5 pb^{-1} recorded in 2010 at $\sqrt{s} = 7$ TeV [2] and are summarised in this paper.

Hadron production measurements in this region also give important inputs to tune the hadronisation models in the LHCb acceptance, since these models have been extrapolated from measurements performed in a different energy and rapidity regime. Production cross-section of several particles have been measured at the LHCb. K_s^0 cross section measurement at $\sqrt{s} = 900$ GeV [3] and ϕ cross section measurement at $\sqrt{s} = 7$ TeV [4] probe strangeness production. J/ψ [5], $\psi(2S)$ [6], prompt charm [7] and b [8] cross section measurements at $\sqrt{s} = 7$ TeV probe heavy flavour production. We present here the measurements of light particle production ratios: $\bar{\Lambda}/\Lambda$,$\bar{\Lambda}/K_s^0$ [9] and $\bar{p}/p$ [10] both at $\sqrt{s} = 900$ GeV and $\sqrt{s} = 7$ TeV.

[1] On behalf of the LHCb Collaboration

Journal of Physics: Conference Series **347** (2012) 012022 doi:10.1088/1742-6596/347/1/012022

2. Electroweak measurements

2.1. Production cross sections

Muons are selected in the pseudo-rapidity range $2 < \eta < 4.5$. They are required to have $p_T > 20$ GeV/c and good track quality and momentum measurement. $Z \to \mu\mu$ candidates are made out of two such muons with the di-muon mass in the range $81 \leq M_{\mu\mu} \leq 101$ GeV/c^2, while $W \to \mu\nu$ candidates require the muon to be prompt and isolated to reduce heavy flavour background [2]. All efficiencies are determined on data with tag and probe methods.

Considered backgrounds for $Z \to \mu\mu$ candidates are $Z \to \tau\tau$, b and c events with two semi-leptonic decays, and generic QCD events with misidentified muons. For $W \to \mu\nu$, the considered backgrounds are $Z \to \mu\mu$ with one muon outside acceptance, $Z \to \tau\tau$, $W \to \tau\nu$, b and c events with a semi-leptonic decay and generic QCD events. The fits of the candidates' distributions shown in Figures 1 and 2, allow to extract the number of signal candidates and backgrounds which are reported on Table 1. $Z \to \mu\mu$ backgrounds are found to be negligible after the selection.

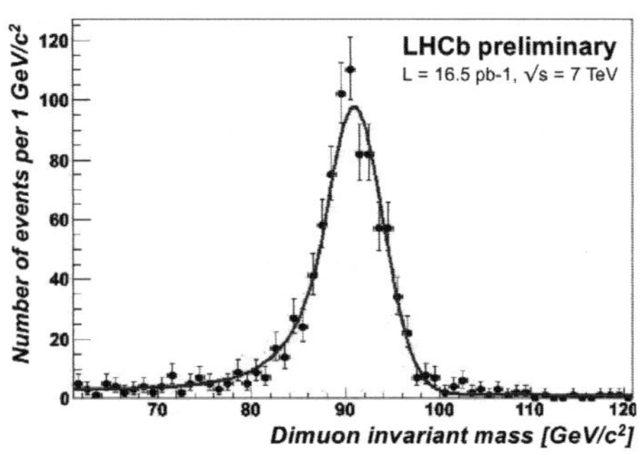

Figure 1. Invariant mass of the selected Z candidates.

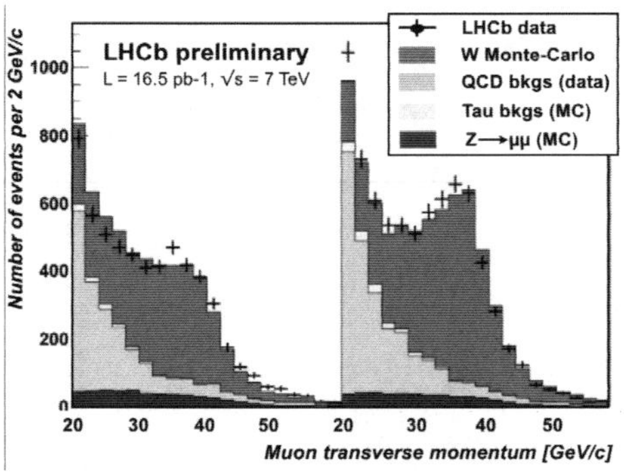

Figure 2. Transverse momentum of the muon of selected W candidates. Left for negative muons and right for positive muons.

Table 1. Number of signal and background candidates extracted from the fit of candidates distributions in Figure 1 and 2.

	$Z \to \mu\mu$	$W^+ \to \mu^+\nu$	$W^+ \to \mu^-\bar{\nu}$
Number of signal candidates	883	7624	5732
Number of background candidates	1.2 ± 1.2	2194 ± 150	1654 ± 150

We find:

$$\sigma_Z(2. < \eta_\mu < 4.5, p_T^\mu > 20, 81 < m_Z < 101) = 73 \pm 4 \pm 7.3\,\text{pb},$$
$$\sigma_{W^+}(2. < \eta_\mu < 4.5, p_T^\mu > 20) = 1007 \pm 48 \pm 101\,\text{pb},$$
$$\sigma_{W^-}(2. < \eta_\mu < 4.5, p_T^\mu > 20) = 680 \pm 40 \pm 68\,\text{pb}.$$

where the first uncertainty is statistical and systematic (except luminosity) combined and the second is systematic due to absolute luminosity measurement.

It is interesting to note that all systematic uncertainties related to efficiencies are limited by the amount of statistics available to the tag and probe method. The uncertainty from background estimation is related to the knowledge of the background shapes, while the dominant 10% uncertainty comes from the absolute luminosity measurement [11]. All systematic uncertainties are expected to improve significantly in the update of the analysis.

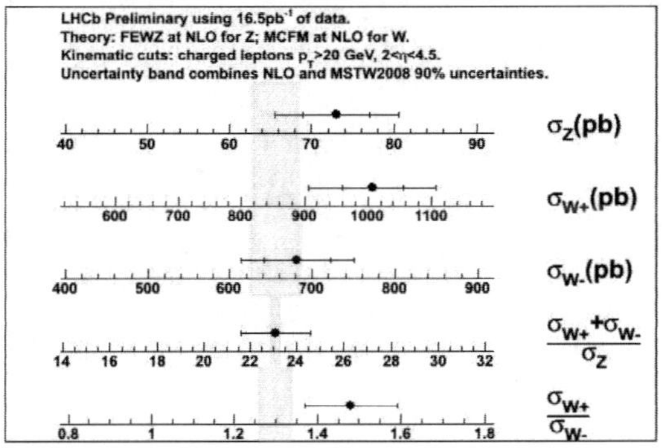

Figure 3. Z , W^+ and W^- cross-section measurements and ratios compared to the MCFM NLO prediction. The inner uncertainty combines the statistical and systematic experimental uncertainties. The outer uncertainty bar includes the luminosity uncertainty. The yellow bar is the theoretical uncertainty. Predicted and measured cross-sections are in good agreement.

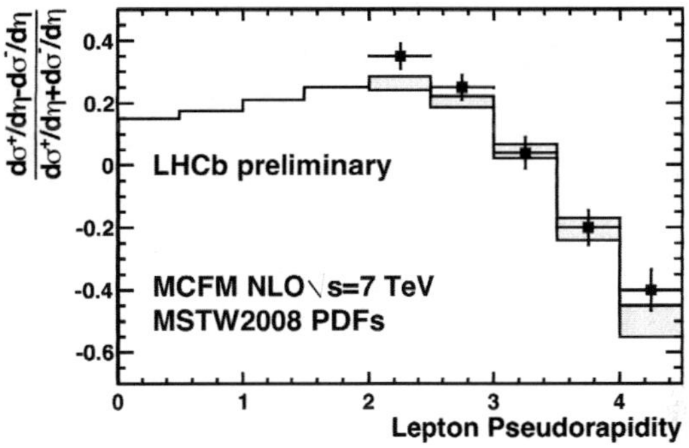

Figure 4. W charge asymmetry in bins of lepton pseudo-rapidity compared to the MCFM prediction. Measurements and prediction are in good agreement. The shaded area is the uncertainty arising from the MSTW08NLO PDF set.

2.2. Cross section ratios

Several cross-section ratios have been computed, as shown in Figure 3. In this case the systematics are reduced since the contribution of the absolute luminosity measurement cancels. The W charge asymmetry has also been evaluated:

$$A_W = \frac{\sigma_{W^+} - \sigma_{W^-}}{\sigma_{W^+} + \sigma_{W^-}}$$

.

It is shown in five bins of rapidity in Figure 4. The main theory uncertainties on the predictions arise from the knowledge of u and d quark PDFs. It has been shown that this measurement already reduces the uncertainty at high x on the PDFs for NNPDF2.1 [12]. Improved constraints are expected from the update of the analysis.

3. Soft QCD measurements

3.1. $\bar{\Lambda}/\Lambda$ and $\bar{\Lambda}/K^0_s$ ratios

The ratio of the production cross-sections of baryon/anti-baryon, such as $\bar{\Lambda}/\Lambda$, allow to study the baryon-number transport from the beam particles to the final state. $\bar{\Lambda}/K^0_s$ ratio on the other hand is a measure of the baryon to meson suppression, which is a good test for different fragmentation models.

High purity sample of prompt K_s^0 decaying into $\pi^+\pi^-$ and Λ, $\bar{\Lambda}$ decaying into $p\pi$ are selected based on a Fisher discriminant combining the impact parameter of the V^0 particles and their daughters. Only events with one primary vertex are selected to avoid diffractive event contribution [9]. No particle identification is used here.

Systematics are reduced since the major uncertainty, coming from absolute luminosity measurement, cancels through the ratio. Remaining source of systematics comes from kinematic correction of the Monte Carlo simulation used to evaluate selection efficiency, uncertainty in the interaction with material and diffractive events pollution. Depending on the bins in η and p_T, they add up to 0.02-0.06 for $\bar{\Lambda}/\Lambda$ and 0.02-0.03 for $\bar{\Lambda}/K_s^0$.

The measurement is performed with $0.3\,\mathrm{nb}^{-1}$ at $\sqrt{s} = 900\,\mathrm{GeV}$ and $1.38\,\mathrm{nb}^{-1}$ at $\sqrt{s} = 7\,\mathrm{TeV}$. $\bar{\Lambda}/\Lambda$ is in rather good agreement with Perugia 0 tune for Pythia at low rapidity while at high rapidity, extreme models of baryon transport seams to be favoured [13], see Figure 5. This behaviour is observable both at $\sqrt{s} = 900$ GeV and $\sqrt{s} = 7$ TeV.

$\bar{\Lambda}/K_s^0$ shows an excess over the whole range of rapidity and at both center of mass energy, see Figure 6.

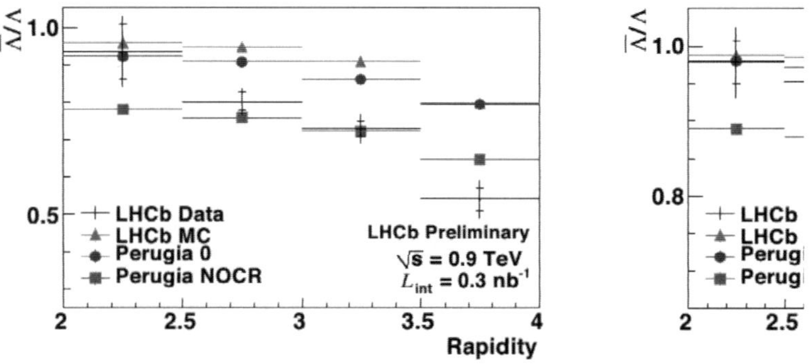

Figure 5. The production cross-section ratio $\bar{\Lambda}/\Lambda$ at $\sqrt{s} = 900$ GeV and $\sqrt{s} = 7$ TeV as a function of rapidity compared with the predictions of the LHCb MC tune [14], Perugia 0 and Perugia NOCR [13]. Vertical lines show the combined statistical and systematic uncertainties and the short horizontal bars (where visible) show the statistical component.

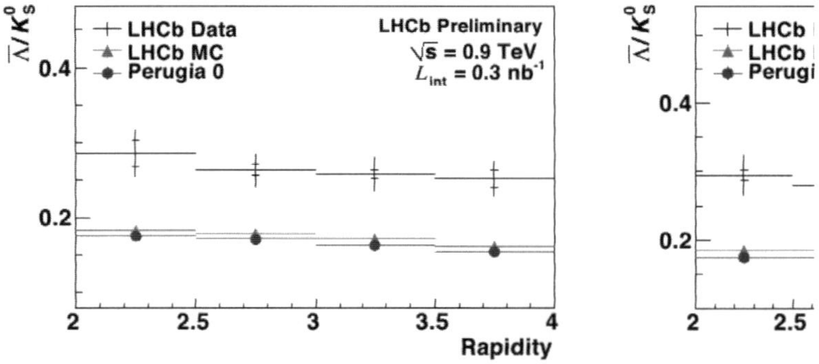

Figure 6. The production cross-section ratio $\bar{\Lambda}/K_s^0$ at $\sqrt{s} = 900$ GeV and $\sqrt{s} = 7$ TeV as a function of rapidity compared with the predictions of the LHCb MC tune [14], Perugia 0 and Perugia NOCR [13]. Vertical lines show the combined statistical and systematic uncertainties and the short horizontal bars (where visible) show the statistical component.

3.2. $\bar{p}/p$ *ratio*

Further tests of the baryon number transport have been performed by measuring the $\bar{p}/p$ production ratio. The measurement is done with 0.3 nb^{-1} at $\sqrt{s}$ = 900 GeV and 0.2 nb^{-1} at $\sqrt{s}$ = 7 TeV, [10].

Prompt protons with $p > 5$ GeV/c are selected, exploiting the powerful hadronic separation capabilities of the LHCb RICH system, in order to prevent K and π contamination. The performance of the particle identification is calibrated on kinematically isolated samples of $\phi \to K^+K^-$, $K_s \to \pi^+\pi^-$ and $\Lambda \to \pi p$.

The main systematic comes from cross contamination between p, K and π to the level of 2% to 10% at high rapidity. Results are consistent with Monte Carlo models at $\sqrt{s}$ = 7 TeV but differ significantly at $\sqrt{s}$ = 900 GeV, especially at low p_T, as shown in Figure 7. They show a similar behaviour when compared to fragmentation models as the $\bar{\Lambda}/\Lambda$ results.

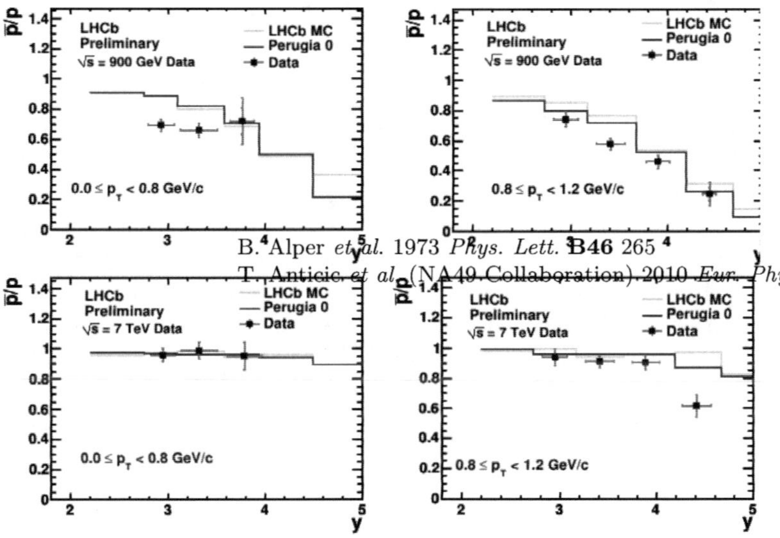

B. Alper *et al.* 1973 *Phys. Lett.* **B46** 265
T. Anticic *et al* (NA49 Collaboration) 2010 *Eur. Phys. J.* **C65** 9

Figure 7. Distribution of the ratio $\bar{p}/p$ against rapidity, for different p_T ranges. Up for $\sqrt{s} = 900\,\text{GeV}$, down for $\sqrt{s} = 7\text{TeV}$. Data are compared with the predictions of the LHCb MC tune [14] and Perugia 0 [13].

3.3. Comparison with other experiments

Another way to present the results of baryon transport number is to show the ratio of anti-baryon to baryon as a function of the rapidity loss, $\Delta y = y_{beam} - y$ with y_{beam} the rapidity of the incoming beam. It allows the comparison of the LHCb results with those of the previous experiments. Figure 8 shows that the LHCb measurement is in good agreement with the previous measurements.

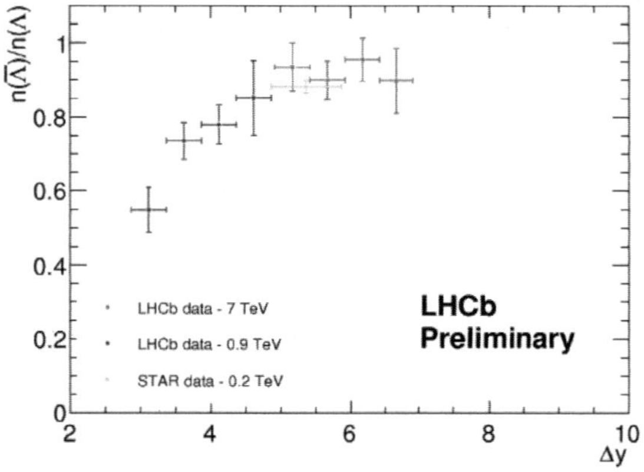

Figure 8. Left: The LHCb $\bar{\Lambda}/\Lambda$ ratio versus Δy is compared to a single data point from the STAR Collaboration [15]. Right: The LHCb $\bar{p}/p$ ratio versus Δy is compared to measurements performed by other experiments [16]. The error bars represent the total uncertainty, apart from the points marked ISR, where they signify the statistical uncertainty alone.

4. Conclusions

The cross-sections and ratios of W and Z bosons have been measured using 16.5 pb^{-1} of data. The luminosity uncertainty of 10% dominates the precision of the cross-section measurements. This luminosity uncertainty cancels in the cross-section ratios, which provide a more precise test of Standard Model predictions and probe of parton density functions. All results are consistent with NLO predictions. Preliminary measurements for ratios several neutral hadrons and protons suggest lower baryon suppression and higher baryon transport in data than in the Monte Carlo models investigated. The LHCb data are consistent with data from lower energy experiments.

The first studies with data from the 2009 and 2010 runs showed that the LHCb experiment is ready for its core physics program. Early data were used to measure particle production in an unexplored region of phase space, producing valuable inputs for the tuning of Monte-Carlo models and for constraining PDFs.

References

[1] A. A. Alves *et al.* (LHCb Collaboration) 2008 *JINST* 3 S08005
[2] R. Aaij *et al.* (LHCb Collaboration) LHCb-CONF-2011-012
[3] R. Aaij *et al.* (LHCb Collaboration) 2010 *Phys. Lett.* **B693** 69 (*Preprint* 1008.3105)
[4] R. Aaij *et al.* (LHCb Collaboration) 2011 *Preprint* 1107.3935 (submitted to Phys. Lett. B)
[5] R. Aaij *et al.* (LHCb Collaboration) 2011 *Eur. Phys. J.* **C71** 1645
[6] R. Aaij *et al.* (LHCb Collaboration) 2011 LHCb-CONF-2011-026
[7] R. Aaij *et al.* (LHCb Collaboration) 2011 LHCb-CONF-2010-013
[8] R. Aaij *et al.* (LHCb Collaboration) 2010 *Phys. Lett.* **B694** 209 (*Preprint* 1009.2731)
[9] R. Aaij *et al.* (LHCb Collaboration) 2011 *Preprint* 1107.0882 (submitted to JHEP)
[10] R. Aaij *et al.* (LHCb Collaboration) 2010 LHCb-CONF-2010-009
[11] P. Hopchev 2010 LHCb-PROC-2010-015 (*Preprint* 1005.4398)
[12] M. Ubali, LPC electroweak workshop, April 2011
[13] P. Z. Skands 2010 *Phys. Rev.* **D82** 074018
[14] I. Belyaev *et al.* 2010 LHCb-PROC-2010-056
[15] B. I. Abelev *et al.* (STAR Collaboration) 2007 *Phys. Rev.* **C75** 064901
[16] A. K. Aamodt *et al.* (ALICE Collaboration) 2010 *Phys. Rev. Lett.* **105** 072002
I. G. Bearden *et al.* (BRAHMS Collaboration) 2005 *Phys. Lett.* **B607** 42
B. I. Abelev *et al.* (STAR Collaboration) 2009 *Phys. Rev.* **C79** 034909
A. M. Rossi, G. Vannini, A. Bussiere, E. Albini, D. D'Alessandro and G. Giacomelli 1975 *Nucl. Phys.* **B84** 269
M. Banner *et al.* 1972 *Phys. Lett.* **B41** 547
B. Alper *et al.* 1973 *Phys. Lett.* **B46** 265
T. Anticic *et al.* (NA49 Collaboration) 2010 *Eur. Phys. J.* **C65** 9

2nd International Conference on Particle Physics IOP Publishing
Journal of Physics: Conference Series **347** (2012) 012023 doi:10.1088/1742-6596/347/1/012023

Electroweak results with the ATLAS 2010 data

Lidia Dell'Asta[1]
Universitá and INFN Milano
E-mail: `dellasta@cern.ch`

Abstract. The ATLAS Experiment is one of the two multi-purpose detectors at the LHC. During the year 2010 it has collected 45 pb^{-1} of pp collisions at a center-of-mass energy of 7 TeV. In this paper the ATLAS electroweak results with 2010 data are described. Measurements of total inclusive W and Z production cross sections are presented, as well as differential cross sections and W charge asymmetry. As the production of pairs of bosons is an important process for investigating the electroweak sector of the Standard Model, the measurements of the cross section of such processes are also presented.

1. Introduction

Electroweak measurements are among the first studies performed at the LHC with early 2010 pp collision data at a center-of-mass energy of 7 TeV.

W and Z final states are interesting because they are the "standard candle" used to get to know the detector, i.e. understand and calibrate it. They are used, for example, to validate the different object reconstructions and identifications, such as electron and muon reconstruction, to define the energy scale of missing energy and trigger efficiencies. Moreover, channels with W and Z bosons in the final state are important backgrounds to searches for new particles.

Diboson production plays an important role in electroweak physics. Their production rates and the kinematic distributions of the final states are sensitive to triple gauge boson couplings. In Figure 1 the Standard Model s-channel Feynman diagram for W^+W^- production through the quark-antiquark annihilation is shown. This channel contains the WWZ and $WW\gamma$ triple gauge boson coupling (TGC) vertices. Moreover, diboson channels are an important background to Higgs boson and new physics searches.

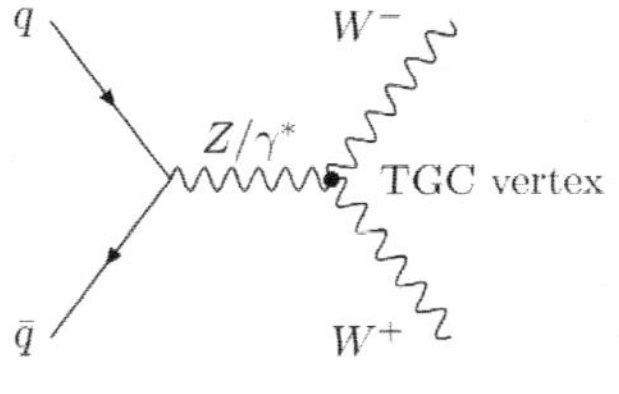

Figure 1. Standard Model s-channel Feynman diagram for W^+W^- production through the quark-antiquark annihilation.

ATLAS (A Toroidal LHC ApparatuS) is one of the two multi-purpose experiments at the LHC. It consists of an inner tracking detector surrounded by a superconducting solenoid which provides a 2 T magnetic field, electromagnetic and hadronic calorimeters and a muon

[1] On behalf of the ATLAS Collaboration

spectrometer with a toroidal magnetic field. The inner detector provides precision tracking of charged particles for $|\eta| < 2.5$. It consists of a silicon pixel detector, a silicon strip detector and a straw tube tracker that also provides transition radiation measurements for electron identification. The calorimeter system covers the pseudorapidity range $|\eta| < 4.9$. It is composed of sampling calorimeters with either liquid argon (LAr) or scintillating tiles as active media. In the region $|\eta| < 2.5$ the electromagnetic LAr calorimeter is finely segmented and plays an important role in electron identification. The muon spectrometer has separate trigger and high-precision tracking chambers which provide muon identification in $|\eta| < 2.7$ [1].

In the year 2010 the ATLAS Experiment has collected 45 pb^{-1} of data from pp collisions at a center-of-mass energy of 7 TeV. It corresponds to 93.6% of the luminosity delivered by the LHC. This high fraction of data collected was achieved thanks to the good performance of all sub-detectors. In Figure 2 the cumulative luminosity during the 2010 pp run at 7 TeV as function of time is shown. Considering data quality requirements, the luminosity available then for analyses amounts to $\sim$35 pb^{-1}. During the running of the first year, the luminosity profile changed considerably, due to an intense machine commissioning. Therefore the amount of multiple interactions per beam crossing increased throughout the data taking. This is shown in Figure 3, where the maximum mean number of interactions per beam crossing is plotted as a function of time.

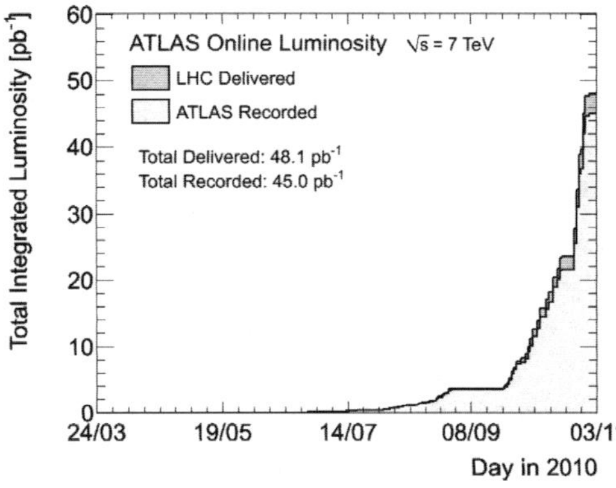

Figure 2. Online plot of the cumulative luminosity versus day delivered to (green), and recorded by ATLAS (yellow) during the 2010 pp data taking.

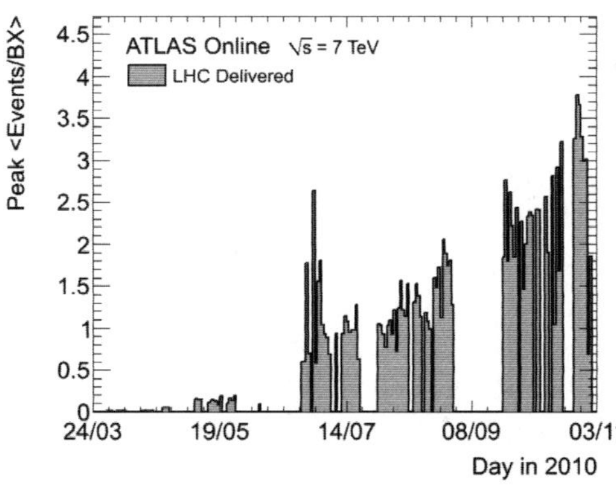

Figure 3. The maximum mean number of events per beam crossing versus day during the 2010 pp data taking.

In this paper, several cross section measurements are presented. In ATLAS, the total cross section (here for the $W \to \ell\nu_\ell$ decay) is calculated as:

$$\sigma_W^{\text{tot}} \times BR(W \to \ell\nu_\ell) = \frac{N_{obs} - N_{bkg}}{A_W C_W \mathcal{L}} \tag{1}$$

where N_{obs} is the number of observed events in data, N_{bkg} is the number of background events (extracted from data or from Monte Carlo, depending on the analysis), $\mathcal{L}$ is the integrated luminosity corresponding to the selected runs and trigger choice, A_W denotes the kinematic and geometric acceptance (fiducial acceptance) for the signal process and is determined from generator level Monte Carlo, and C_W is the ratio between the number of reconstructed signal events passing the selections of the analysis and the number of events within the fiducial acceptance.

The cross section is also measured in the fiducial region as:

$$\sigma_W^{\text{fid}} \times BR(W \to \ell\nu_\ell) = \frac{N_{obs} - N_{bkg}}{C_W \mathcal{L}}. \tag{2}$$

This cross section is not affected by significant theoretical uncertainties. Therefore, future improvements on the prediction of the acceptance can be used to extract improved total cross section measurements.

2. W and Z boson production

In the following, the W and Z/γ^* cross section measurements are presented in the electron and muon channels (Section 2.1). The cross section measurement for the production of W and Z bosons in association with jets is described in Section 2.2. Finally, in Section 2.3 the measurement of the W charge asymmetry in the muon channel is presented.

2.1. W and Z/γ^* cross section measurements

The W and Z/γ^* cross sections have been measured by ATLAS [2] in their leptonic decay channels. In particular, the electron and muon channels have been studied.

For the $W \to e\nu_e/\mu\nu_\mu$ channels, the signature is one lepton and missing energy from the neutrino. Therefore the event selection requires the presence of one well reconstructed lepton with $p_T > 20$ GeV, sizable missing energy, $E_T^{\text{miss}} > 25$ GeV, and transverse mass $m_T > 40$ GeV. In Figures 4 and 5 the distributions of the transverse mass of the selected W^+ to electron/muon candidates are shown.

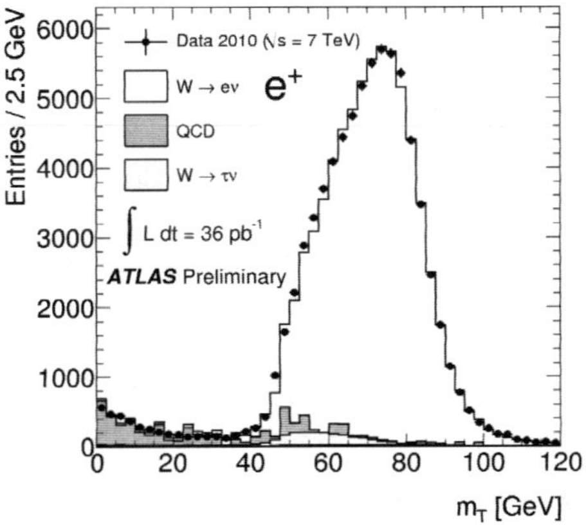

Figure 4. Distribution of the transverse mass of the selected $W \to e\nu_e$ candidates after all cuts, for positive charge [2].

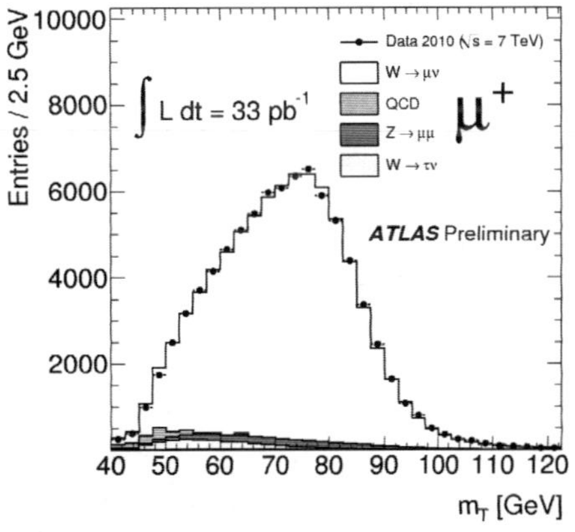

Figure 5. Distribution of the transverse mass of the selected $W \to \mu\nu_\mu$ candidates after all cuts, for positive charge [2].

For the $Z \to ee/\mu\mu$ channels, the signature is two opposite charged leptons. Therefore in this case the event selection requires the presence of two reconstructed leptons with opposite charge and the invariant mass of the lepton pair in the range $66 < m_{\ell\ell} < 116$ GeV. In Figures 6 and 7 the distributions of the invariant mass of the Z boson candidates are shown, both in the electron and in the muon channels.

For the two analyses the main backgrounds come from QCD jets and other electroweak processes. The estimation of the electroweak background is taken from Monte Carlo, while for the QCD background the estimation is extracted directly from data. As seen in Figures 4-7, the

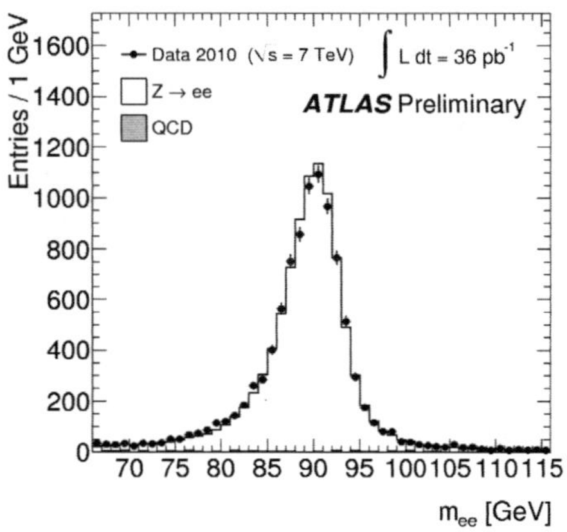

Figure 6. Invariant mass distribution of candidate $Z \to ee$ events [2].

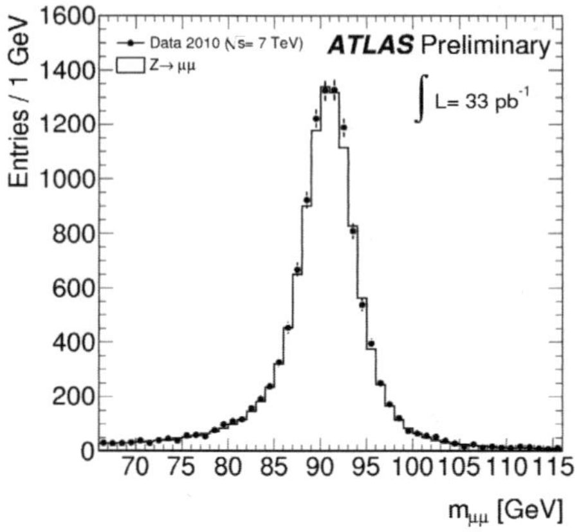

Figure 7. Invariant mass distribution of candidate $Z \to \mu\mu$ events [2]. The QCD background is found to be negligible.

amount of residual background, after the event selections described before, is small both for W and Z channels.

The main systematics affecting the cross section measurement come from the uncertainty on the electron reconstruction and identification ($\sim$1.5/3%) and the uncertainty on the missing energy scale (2% for W channels). It has to be noticed that the resulting systematic uncertainty is already smaller than the systematic on luminosity (3.4%) [3].

The systematic on the acceptance ($\sim$3/4%) is dominated by the uncertainties on the PDFs and on showering models. It is evaluated taking into account three contributions:

- uncertainties within one PDF set, which are derived using the `CTEQ 6.6` PDF [4],
- uncertainties between different PDF sets (the maximal difference between the `MRST LO*`[5], `CTEQ 6.6` and `HERAPDF 1.0` [6] sets is taken as a systematic),
- difference between the `PYTHIA` [7] and `MC@NLO` [8] simulations, using the same PDF set, `CTEQ 6.6`.

The results of the measurement of the W and Z/γ^* cross sections are reported in Table 1.

Table 1. Results of the W and Z/γ^* cross section measurements.

	$\sigma^{\text{tot}}_{W^\pm} \cdot \text{BR}(W \to \ell\nu_\ell)$ [nb]
W^+	$6.257 \pm 0.017(\text{stat.}) \pm 0.152(\text{syst.}) \pm 0.213(\text{lumi.}) \pm 0.188(\text{acc.})$
W^-	$4.149 \pm 0.014(\text{stat.}) \pm 0.102(\text{syst.}) \pm 0.141(\text{lumi.}) \pm 0.124(\text{acc.})$
W	$10.391 \pm 0.022(\text{stat.}) \pm 0.238(\text{syst.}) \pm 0.353(\text{lumi.}) \pm 0.312(\text{acc.})$

	$\sigma^{\text{tot}}_{Z/\gamma^*} \cdot \text{BR}(Z/\gamma^* \to \ell\ell)$ [nb], $66 < \text{m}_{\ell\ell} < 116$ GeV
Z/γ^*	$0.945 \pm 0.006(\text{stat.}) \pm 0.011(\text{syst.}) \pm 0.032(\text{lumi.}) \pm 0.038(\text{acc.})$

Figure 8 summarizes the results on the cross section measurements showing the measured and predicted W/Z cross section ratio. Good agreement with predictions is observed. Figure 9

Journal of Physics: Conference Series **347** (2012) 012023 doi:10.1088/1742-6596/347/1/012023

shows the measured and predicted W^- vs. W^+ cross sections times leptonic branching ratios: already with the 2010 amount of data, the W cross section measurement has sensitivity to PDFs and some constraints can be set.

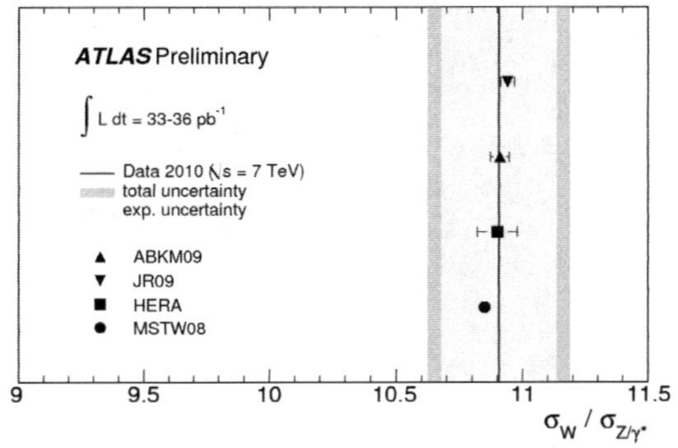

Figure 8. Measured and predicted W/Z cross section ratio. The experimental uncertainty of the measurement includes statistical and experimental systematic uncertainties [2].

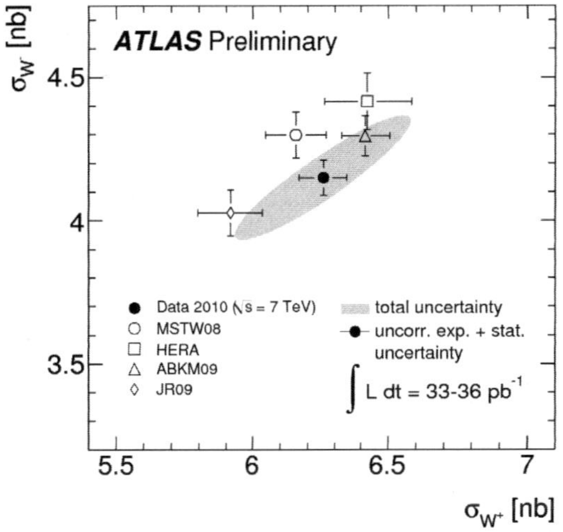

Figure 9. Measured and predicted W^- vs. W^+ cross sections times leptonic branching ratios. The projections of the ellipse to the axes correspond to one standard deviation uncertainty of the cross sections. The uncertainties of the predictions are the PDF uncertainties [2].

2.2. W+jets and Z/γ^*+jets cross sections

The study of massive vector boson production in association with one or more jets is an important test of QCD. Moreover, W/Z+jets processes are a significant background to studies of Standard Model processes such as $t\bar{t}$ or single-top production, as well as searches for the Higgs boson and for physics beyond the Standard Model. Therefore the measurements of the cross section and kinematic properties of W/Z+jets processes and comparisons to theoretical predictions are of significant interest.

ATLAS has measured the cross sections for the W+jets production [9] and Z/γ^*+jets production [10] with the boson decaying either in the electron or the muon channel.

Jets are reconstructed with the anti-k_t algorithm [11] with a radius parameter $R = 0.4$. All jets considered in the analysis for the W+jets (Z/γ^*+jets) cross section measurement are required to have a transverse momentum $p_T > 20$ (30) GeV and a pseudorapidity in the range $|\eta| < 2.8$. In addition, a lepton-jet overlap removal is applied: if a jet and a lepton passing some identification requirements are within $\Delta R = \sqrt{\Delta\eta^2 + \Delta\phi^2} < 0.5$, the jet is removed, regardless of the jet p_T or η.

Both in the W+jets and Z/γ^*+jets analyses, events are required to fire a single lepton trigger. Then, for the W+jets channel, the event selection is based on the presence of one isolated lepton (with $E_T > 20$ GeV). The events must not have any additional identified lepton. Moreover, events are required to have $E_T^{\text{miss}} > 25$ GeV, and transverse mass $m_T > 40$ GeV, as in the analysis for the W inclusive cross section measurement. For the Z/γ^*+jets analysis instead, the presence of two opposite charge leptons with $E_T > 20$ GeV is required. Furthermore, the invariant mass of the lepton pair must be $66 < m_{\ell\ell} < 116$ GeV.

The backgrounds are estimated with different techniques for the different analyses and also depending on the channel. For example, in the W+jets analysis, the number of QCD background events was estimated by fitting, in each jet multiplicity bin, the E_T^{miss} distribution in the data (without the E_T^{miss} cut) to a sum of two templates: one for the QCD background and another which included signal and the leptonic backgrounds. In both muon and electron channels, the shapes for the latter template were obtained from simulation. While in the muon channel the template for the QCD background was obtained from simulation, in the electron channel it was obtained from the data because the mechanisms by which a jet fakes an electron are difficult to simulate.

The main systematics affecting the measurements come from the jet energy scale, which is smaller in the W+jets channel (~9%) while is more important for the Z/γ^*+jets (~10 - 20%, depending on the p_T and η of the jet). Another important systematic for the W+jets channel comes from the treatment of pile-up: it amounts to ~7%.

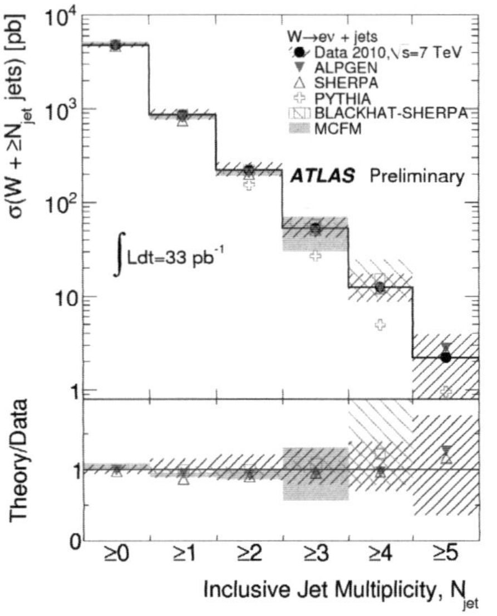

Figure 10. W+jets cross section as a function of the inclusive jet multiplicity for electron channel. Also shown are predictions from ALPGEN, SHERPA, PYTHIA, MCFM and BLACKHAT-SHERPA, and the ratio of theoretical predictions to data (PYTHIA is not shown in the ratio) [9].

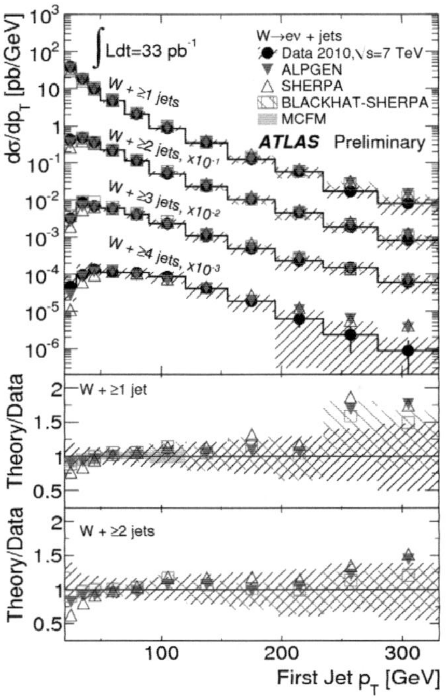

Figure 11. W+jets cross section as a function of the p_T of the first jet in the event for the electron channel. The p_T of the first jet is shown separately for events with ≥ 1 jet to ≥ 4 jet. The ≥ 2 jet, ≥ 3 jet and ≥ 4 jet distributions have been scaled down by factors of 10 and 100, 1000 respectively. Also shown are predictions from ALPGEN, SHERPA, MCFM and BLACKHAT-SHERPA, and the ratio of theoretical predictions to data for ≥ 1 jet and ≥ 2 jet events [9].

The results of the measurements are shown in Figures 10 to 13. In Figure 10 the W+jets cross section as a function of jet multiplicity for the electron channel is shown, while Figure 11 shows the W+jets cross section as a function of the p_{T} of the leading jet in the event for the electron channel, separately for events with $\geq$ 1 jet to $\geq$ 4 jet. The cross sections are quoted in the limited kinematic region: $E_{\mathrm{T}}^{\mathrm{j}} > 20$ GeV, $|\eta^{\mathrm{j}}| < 2.8$, $E_{\mathrm{T}}^{\ell} > 20$ GeV, $|\eta^{\mathrm{e}}| < 2.47$ (excluding $1.37 < |\eta^{\mathrm{e}}| < 1.52$), $|\eta^{\mu}| < 2.4$, $p_{\mathrm{T}}^{\nu} > 25$ GeV, $M_{\mathrm{T}} > 40$ GeV, $\Delta R^{\ell \mathrm{j}} > 0.5$, where ℓ, j and ν denote lepton, jet and neutrino, respectively. Along with data, the predictions from ALPGEN [12], SHERPA [13], PYTHIA, MCFM [14] and BLACKHAT-SHERPA [15], and the ratio of theoretical predictions to data are also displayed. Good agreement is observed with the predictions of the multi-parton matrix element generators ALPGEN and SHERPA. Calculations based on NLO matrix elements in MCFM and in BLACKHAT-SHERPA are also in good agreement with the data.

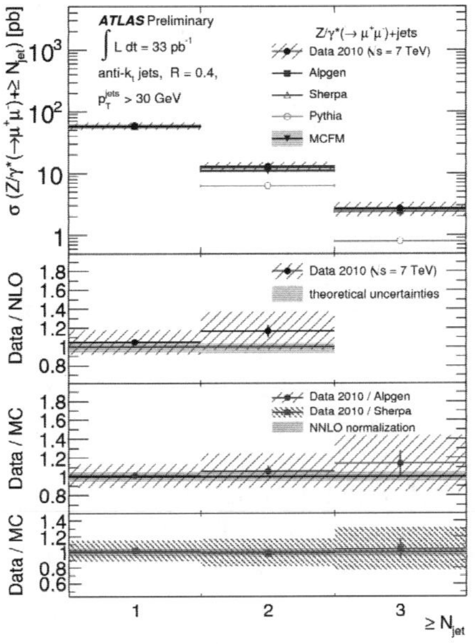

Figure 12. Z/γ^*+jets cross section as a function of the inclusive jet multiplicity for the muon channel. The measurements are compared to NLO pQCD predictions from MCFM (only available in the first two bins and including uncertainties), as well as the predictions from ALPGEN, SHERPA, and PYTHIA ($\times$1.17) [10].

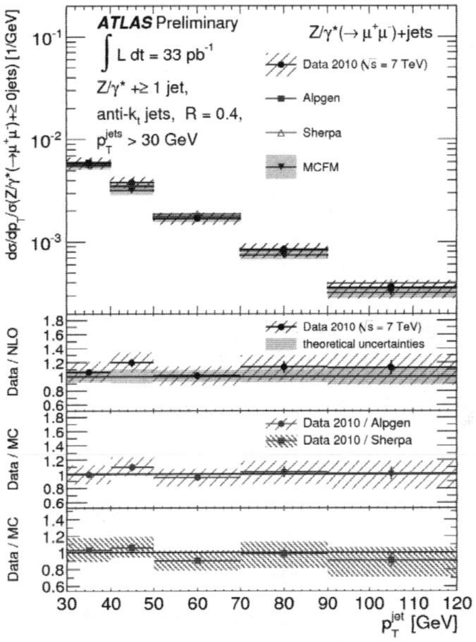

Figure 13. Z/γ^*+jets cross section as a function of the p_{T} of the jet for the muon channel. The measurements are compared to NLO pQCD predictions from MCFM, as well as the predictions from ALPGEN and SHERPA [10].

Figure 12 shows the measured Z/γ^*+jets production cross section in the muon channel as a function of the inclusive jet multiplicity, while Figure 13 shows the measured inclusive cross section as a function of the p_{T} of the jet, in events with at least one jet with $p_{\mathrm{T}}^{\mathrm{j}} > 30$ GeV and $|\eta^{\mathrm{j}}| < 2.8$ in the final state, and normalized to Z/γ^* Drell-Yan cross section. The results are defined in a limited kinematic range for the Z/γ^* decay products. In the muon channel the measurements are presented in the region: $66 < m_{\mu\mu} < 116$ GeV, $p_{\mathrm{T}}^{\mu} > 30$ GeV, $|\eta^{\mu}| < 2.4$, and $\Delta R^{\mathrm{j}\mu} > 0.5$. The measurements are compared to NLO pQCD predictions from MCFM, as well as the predictions from ALPGEN and SHERPA. The measured cross sections are described by the NLO pQCD predictions, which include non-perturbative corrections, as well as by the predictions from LO matrix elements supplemented by parton showers, as implemented in the ALPGEN and SHERPA Monte Carlo generators.

2.3. W charge asymmentry

The measurement of the W boson charge asymmetry is sensitive to quark distribution via the dominant production process $u\bar{d}(\bar{u}d) \to W^{+(-)}$. It provides complementary information to that obtained from measurements of inclusive deep inelastic scattering cross sections at the HERA ep collider, as the HERA data do not strongly constrain the ratio between u and d valence quarks in the kinematic regime of low x, where x is the proton momentum fraction carried by the parton. In particular the measurement of the W charge asymmetry at the LHC can contribute to the understanding of PDFs in the parton momentum fraction range $10^{-3} \lesssim x \lesssim 10^{-1}$.

The charge asymmetry has been studied in ATLAS in the muon channel [16]. The event selection is very similar to the selection used in the $W \to \mu\nu$ cross section measurement described before.

The asymmetry varies significantly as a function of the pseudorapidity η_μ of the charged decay lepton due to its strong correlation with the momentum fraction x of the partons producing the W boson. It is defined from the cross sections for $W \to \mu\nu_\mu$ production $d\sigma_{W\mu^\pm}/d\eta_\mu$ as:

$$A_\mu = \frac{d\sigma_{W\mu^+}/d\eta_\mu - d\sigma_{W\mu^-}/d\eta_\mu}{d\sigma_{W\mu^+}/d\eta_\mu + d\sigma_{W\mu^-}/d\eta_\mu} \tag{3}$$

where the cross sections include the event kinematical cuts used to select $W \to \mu\nu_\mu$ events.

Systematic effects on the W production cross-section measurements are typically the same for positive and negative muons, mostly canceling in the asymmetry. All systematic uncertainties on the asymmetry measurement are determined in each $|\eta_\mu|$ bin accounting for correlations between the charges. The dominant sources of systematic uncertainties on the asymmetry come from trigger and reconstruction efficiencies.

The measured differential charge asymmetry in eleven bins of muon absolute pseudorapidity is shown in Figure 14. Also shown are expectations from W predictions at NLO with different PDF sets: CTEQ 6.6, HERA 1.0 and MSTW 2008 [17]; all predictions are presented with 90% confidence-level error bands.

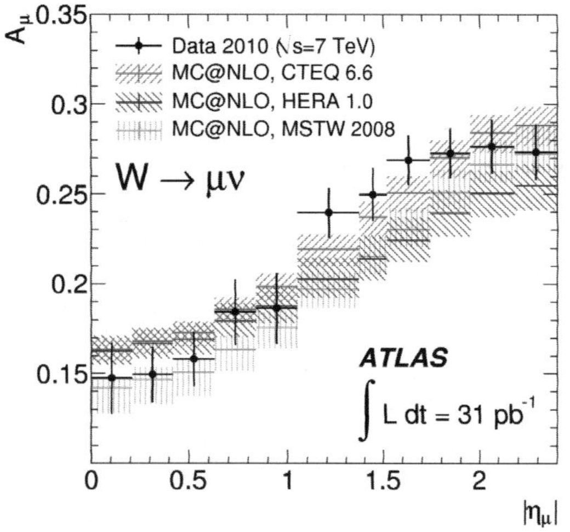

Figure 14. The muon charge asymmetry from W boson decays in bins of absolute pseudorapidity. The data (shown with error bars including the statistical and systematic uncertainties) are compared to MC@NLO predictions with different PDF sets [16].

A χ^2 comparison using the measurement uncertainty and the central value of the PDF predictions yields values per number of degrees of freedom of 9.16/11 for the CTEQ 6.6 PDF set, 35.81/11 for the `HERA 1.0` PDF set and 27.31/11 for the `MSTW 2008` PDF set. Whereas none of the predictions are inconsistent with these data, the predictions are not fully consistent with each other since they are all phenomenological extrapolations in x. The input of the data presented here is therefore expected to contribute to the determination of the next generation of PDF sets, helping to reduce PDF uncertainties, particularly the shapes of the valence quark distributions in the low-x region.

3. Diboson production

In the following, some measurements of diboson production cross sections are presented. In Section 3.1 the measurements of $W\gamma$ and $Z\gamma$ cross sections are presented. The analysis of W^+W^- production is described in Section 3.2 while in Section 3.3 the measurement of $W^{\pm}Z$ production is shown.

3.1. $W\gamma$ and $Z\gamma$ cross section measurements

The measurement of W and Z bosons in association with high energy photons provides important tests of the Standard Model. In fact, physics beyond the Standard Model, such as composite structure of W and Z bosons, new vector bosons, and techni-mesons would enhance those production cross sections and alter the event kinematics.

ATLAS studies [18] use measurements of $pp \to \ell^{\pm}\nu_{\ell}\gamma + X$ and $pp \to \ell^+\ell^-\gamma + X$ with an integrated luminosity of approximately 35 pb^{-1}. Events are selected by requiring the presence of a W or Z boson candidate along with an associated isolated photon having a transverse energy $E_{\mathrm{T}} > 15$ GeV and being separated from the closest electron or muon ℓ by $\Delta R^{\ell\gamma} > 0.7$.

The event selection is based on the presence of one high p_{T} lepton and one high E_{T} photon. In addition, for the $W\gamma$ channel the event is required to have $E_{\mathrm{T}}^{\mathrm{miss}} > 25$ GeV and $m_{\mathrm{T}} > 40$ GeV, while for the $Z\gamma$ channel events with $m_{\ell\ell} > 40$ GeV are selected. The suppression of photons from final state radiation FSR is done by means of an isolation cut.

A total of 192 $W\gamma$ candidates (95 in the electron and 97 in the muon channel) and 48 $Z\gamma$ candidates (25 in the electron and 23 in the muon channel) pass all the requirements. In Figure 15 the three body transverse mass[2] of those $W\gamma$ candidate events, considering both the electron and muon channels, is shown. In Figure 16 the three body invariant mass for the $Z\gamma$ candidate events is shown, again for electron and muon channels together. In both distributions a good agreement between data and predictions can be observed.

The summary of the measurements of the $W\gamma$ and $Z\gamma$ cross section is reported in Table 2, along with the NLO theoretical predictions. The main systematic uncertainties affecting the measurements come from photon reconstruction and identification efficiency ($\sim$11%). While the current measurements are not strongly sensitive to possible new physics, the distributions of kinematic variables determined from the leptons and photons are consistent with the predictions from the Standard Model in a new kinematic regime.

The ratio of the $W\gamma$ to $Z\gamma$ cross sections, defined as

$$R = \frac{\sigma_{pp\to\ell^{\pm}\nu_{\ell}\gamma}}{\sigma_{pp\to\ell^+\ell^-\gamma}} \tag{4}$$

can be measured with a higher relative precision than the individual cross sections since both experimental and theoretical uncertainties partially cancel. This ratio is a test of the $WW\gamma$ triple gauge coupling predicted by the Standard Model. In Figure 17 the measured ratio of the production cross sections of $W\gamma$ and $Z\gamma$, together with the Standard Model prediction are reported.

[2] defined as $m^2{}_{\mathrm{T}}(\ell,\nu,\gamma) = (\sqrt{M^2_{\ell\gamma} + |\vec{p}_{\mathrm{T}}(\gamma) + \vec{p}_{\mathrm{T}}(\ell)|^2} + \vec{E}_{\mathrm{T}}^{\mathrm{miss}})^2 - |\vec{p}_{\mathrm{T}}(\gamma) + \vec{p}_{\mathrm{T}}(\ell) + \vec{E}_{\mathrm{T}}^{\mathrm{miss}}|^2$.

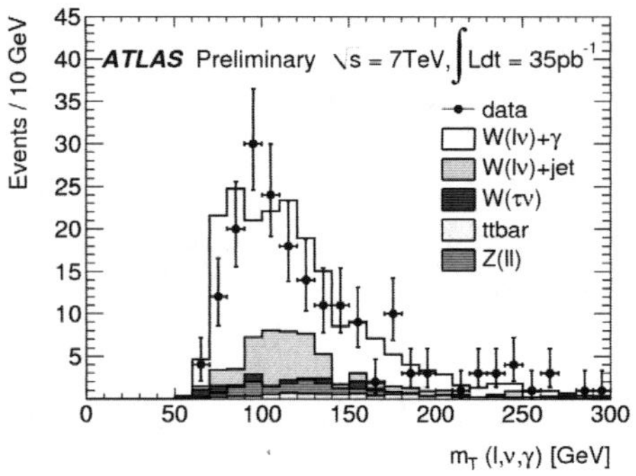

Figure 15. Distribution for the combined electron and muon decay channels of the three body transverse mass $m_T(\ell, \nu, \gamma)$ of the $W\gamma$ candidate events. Monte Carlo predictions for signal and backgrounds are also shown [18].

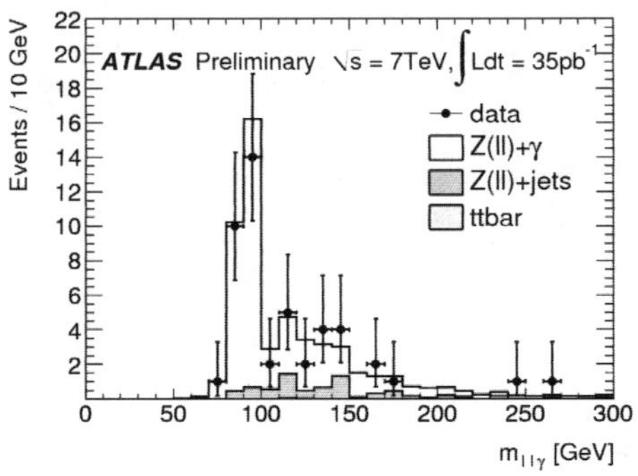

Figure 16. Three body invariant mass $m(\ell^+, \ell^-, \gamma)$ distribution for $Z\gamma$ data candidate events. Monte Carlo predictions for signal and backgrounds are also shown. Both the electron and muon decay channels are included [18].

Table 2. Production cross sections of the $pp \to \ell^{\pm}\nu_\ell\gamma + X$ and $pp \to \ell^+\ell^-\gamma + X$ processes at $\sqrt{s} = 7$ TeV. Both the experimental measurements and the Standard Model NLO predictions are given. The uncertainty in the Standard Model prediction is the systematic uncertainty coming from PDFs [18].

	σ [pb]	NLO prediction [pb]
$pp \to \ell^{\pm}\nu_\ell\gamma$	$42.5 \pm 4.2(\text{stat.}) \pm 7.2(\text{syst.}) \pm 1.4(\text{lumi.})$	42.1 ± 2.7
$pp \to \ell^+\ell^-\gamma$	$6.4 \pm 1.2(\text{stat.}) \pm 1.6(\text{syst.}) \pm 0.2(\text{lumi.})$	6.9 ± 0.5

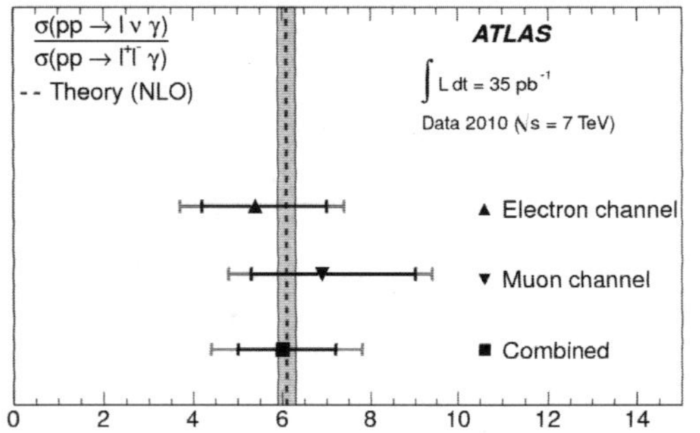

Figure 17. The measured ratio of the production cross sections of $W\gamma$ and $Z\gamma$, together with the Standard Model prediction. Results are shown for the electron and muon final states as well as for their combination. The error bars represent the statistical and the statistical plus systematic uncertainties. The one standard deviation uncertainty in the Standard Model prediction is represented by the yellow band [18].

3.2. W^+W^- production

The W^+W^- process plays an important role in electroweak physics. The production rate and kinematic distributions of W^+W^- are sensitive to the triple gauge couplings of the W boson and W^+W^- production is an important background to Standard Model Higgs boson searches.

ATLAS has measured the W^+W^- production cross section [19]. Candidate W^+W^- events are reconstructed in the fully leptonic decay channel (including the channels with taus decaying leptonically), looking for $\ell^+\nu_\ell\ell^-\nu_\ell$ events. This final state has a better signal to background ratio than the semi-leptonic or hadronic channels. In Figure 18 the event display of a W^+W^- candidate is shown: in this case one W decays in $e\nu_e$ and the other in $\mu\nu_\mu$. In the event display the electron and muon can be seen together with the direction of missing energy.

Figure 18. Event display of a $WW \rightarrow e\nu_e\mu\nu_\mu$ candidate.

The event selection starts from the requirement that the event has fired a single lepton trigger. The signal selection requires two opposite sign leptons with a $p_\mathrm{T} > 20$ GeV and missing energy.

The main backgrounds for the fully leptonic channel are W+jets, Drell-Yan production, top production (both $t\bar{t}$ and Wt) and other diboson processes. The main sources of background have been evaluated from Monte Carlo, except from the background from W+jets. This background was estimated directly from data, as the rate at which hadronic jets are mis-identified as leptons may not be accurately described in the Monte Carlo. The W+jets background is derived by defining a control region, similar to W^+W^- signal selection, that is enriched in W+jets events due to the use of an alternative lepton definition. The selected events are then required to pass the full W^+W^- event selection, where the jet is treated as if it was a fully identified lepton. The W+jets background is then estimated by scaling this control sample by a measured fake factor.

To reject the background from Z bosons, a cut on the dilepton invariant mass is applied, removing events with $|m_{\ell\ell} - \mathrm{m_Z}| < 10$ GeV and $m_{\ell\ell} < 15$ GeV. Events are required to have a large amount of missing energy, where the missing energy in this case is defined as:

$$E_{\mathrm{T,rel}}^{\mathrm{miss}} = \begin{cases} E_\mathrm{T}^{\mathrm{miss}} \times \sin(\Delta\phi) & \text{if} \quad \Delta\phi < \pi/2 \\ E_\mathrm{T}^{\mathrm{miss}} & \text{if} \quad \Delta\phi \geq \pi/2 \end{cases}$$

where $\Delta\phi$ is the difference in the azimuthal angle ϕ between the $E_\mathrm{T}^{\mathrm{miss}}$ and the nearest lepton or jet. This definition of $E_{\mathrm{T,rel}}^{\mathrm{miss}}$ helps in rejecting more efficiently those events for which missing energy is likely to have been mis-measured.

For the ee and $\mu\mu$ channels the requirement is $E_{\mathrm{T,rel}}^{\mathrm{miss}} > 40$ GeV, while for the $e\mu$ channel it is $E_{\mathrm{T,rel}}^{\mathrm{miss}} > 20$ GeV. Another important requirement to suppress background, in particular the $t\bar{t}$ contribution, is to veto events with reconstructed jets. The effect of this selection can be seen by comparing Figures 19 and 20: the distribution of $E_{\mathrm{T,rel}}^{\mathrm{miss}}$ for events passing the full event selection apart from the cut on $E_{\mathrm{T,rel}}^{\mathrm{miss}}$ are reported. In Figure 19 the jet veto is not applied, while in Figure 20 it is: the $t\bar{t}$ background is highly suppressed.

After the selection, 8 candidate events are selected in data: 1 in the e^+e^- channel, 2 in the $\mu^+\mu^-$ channel and 5 in the $e^\pm\mu^\mp$ channel. In Figure 21 the distribution of the dilepton system p_T for these W^+W^- candidates is shown.

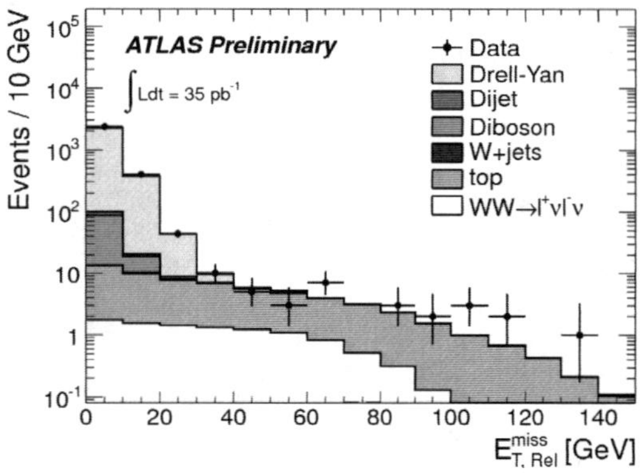

Figure 19. Relative E_T^{miss} distribution for the ee and $\mu\mu$ sample without the jet-veto requirement [19].

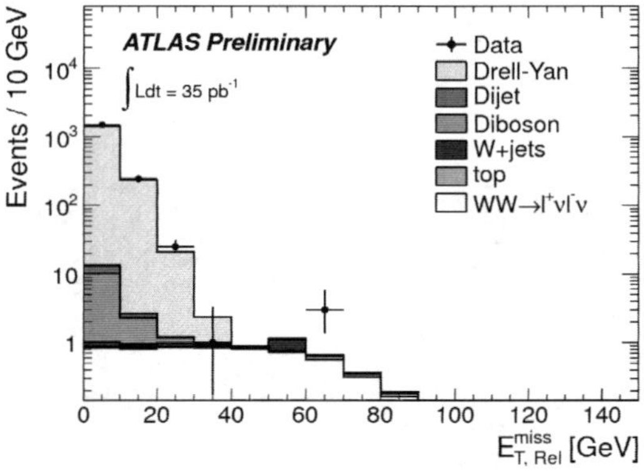

Figure 20. Relative E_T^{miss} distribution for the selected ee and $\mu\mu$ events. The distributions show events with all selection criteria applied except for relative E_T^{miss} [19].

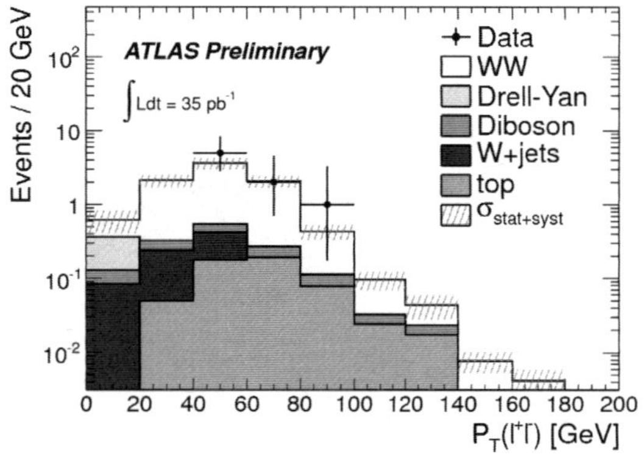

Figure 21. Distributions of the dilepton system p_T for W^+W^- candidates. The points are the data and the stacked histograms are from Monte Carlo predictions except the W+jets background, which is obtained from data-driven methods [19].

ATLAS measures a cross section of $\sigma_{W^+W^-}^{tot} = 41^{+20}_{-16}(\text{stat.}) \pm 5(\text{syst.}) \pm 1(\text{lumi.})\text{pb}$. This is in agreement with the NNLO theoretical prediction, which is 44.3 ± 3pb. Note that the systematic uncertainity on the measurement is far smaller than the statistical uncertainty. This measurement will therefore profit from the high integrated luminosity collected in the year 2011.

3.3. $W^{\pm}Z$ production

ATLAS has also measured the $W^{\pm}Z$ production cross section [20]. The analysis uses four channels with leptonic decays ($W^{\pm}Z \to \ell\nu_\ell\ell\ell$) involving electrons and muons: eee, $ee\mu$, $\mu\mu e$ or $\mu\mu\mu$ (including secondary e or μ leptons from the decay of τ leptons) plus missing transverse energy, E_T^{miss}. The results are based on an integrated luminosity of 205 pb^{-1} collected in early 2011. The main sources of background to the leptonic $W^{\pm}Z$ signal are ZZ, $Z\gamma$, Z/γ^*+jets, and top events. The signal and background contributions are mainly modeled with Monte Carlo simulation and validated with data-driven techniques.

At least one single lepton trigger is required to fire in order to select the event. Events with two leptons of the same flavour and opposite charge with an invariant mass within 10 GeV of the Z boson mass are selected. This reduces much of the background from QCD and top production, and some diboson backgrounds.

Events are then required to have at least 3 reconstructed leptons originating from the same primary vertex, two leptons from a $Z \rightarrow \ell\ell$ decay and an additional third lepton. This requirement reduces the Z/γ^*+jets, top, and some diboson backgrounds. Next, the $E_{\mathrm{T}}^{\mathrm{miss}}$ in the event is required to be greater than 25 GeV and the transverse mass of the system formed from the third lepton and the $E_{\mathrm{T}}^{\mathrm{miss}}$ is required to be greater than 20 GeV. These cuts suppress the remaining backgrounds from Z and diboson production.

Figures 22 and 23 show the transverse momentum of W and Z bosons for the selected events.

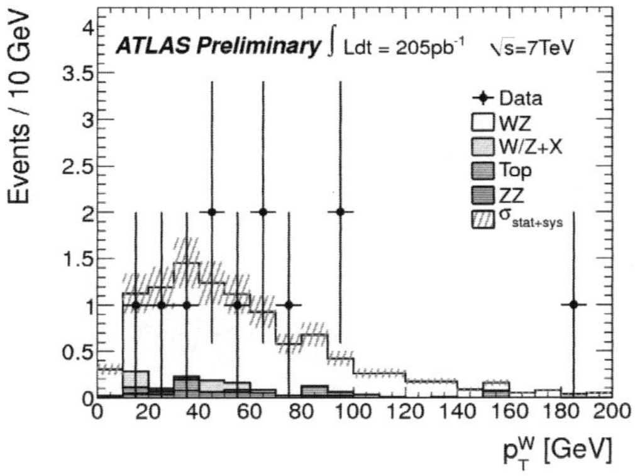

Figure 22. Transverse momentum of the W in $W^{\pm}Z$ candidate events [20].

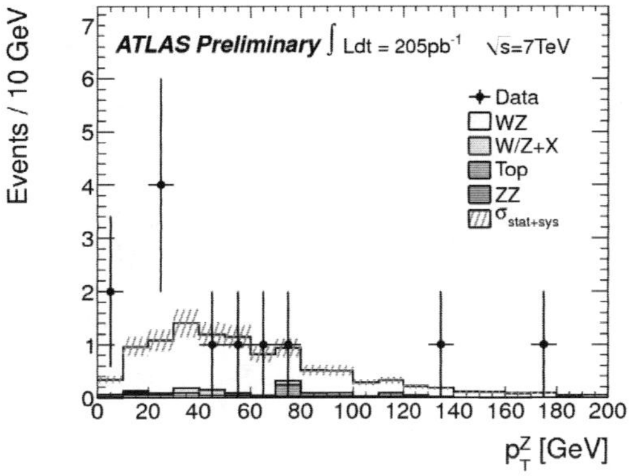

Figure 23. Transverse momentum of the Z in $W^{\pm}Z$ candidate events [20].

ATLAS observes 12 $W^{\pm}Z$ candidates in data, with $9.1 \pm 0.2(\mathrm{stat}) \pm 1.3(\mathrm{sys})$ signal and $2.0 \pm 0.3(\mathrm{stat}) \pm 0.7(\mathrm{sys})$ background events expected. The signal definition includes the contribution from tau decays into electrons or muons, which accounts for about 0.5 events.

The final results for the combined total inclusive cross section measurement for the $W^{\pm}Z$ bosons decaying directly into electrons and muons excluding contributions from tau decays is: $\sigma_{WZ}^{\mathrm{tot}} = 18^{+7}_{-6}(\mathrm{stat.})^{+3}_{-3}(\mathrm{syst.})^{+1}_{-1}(\mathrm{lumi.})\mathrm{pb}$. This measurement is found to be in agreement with the NNLO theoretical prediction, which is 16.9 pb. Also in this case, the main uncertainty on the measurement comes from the amount of data available for the measurement. Therefore also this measurement will profit from the high statistic collected during the full 2011 period.

4. Conclusions

To conclude, in this paper the recent electroweak results of the ATLAS Experiment with 2010 data have been presented. They are summarized in Figure 24.

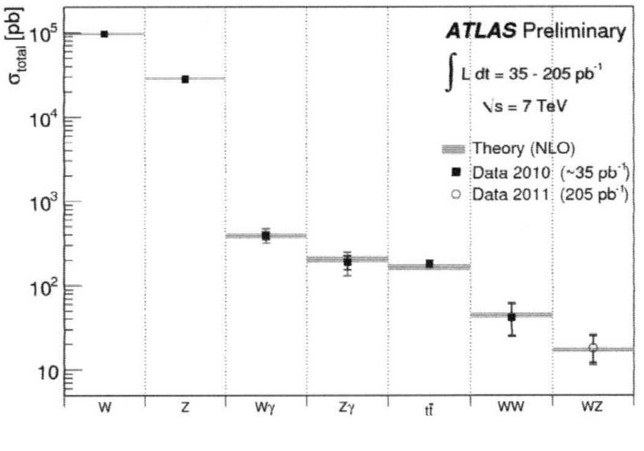

Figure 24. Summary of the ATLAS cross-section measurements from the 2010 and early 2011 datasets, including inclusive W and Z, diboson $W\gamma$ and $Z\gamma$, $W^{+}W^{-}$, $W^{\pm}Z$ and $t\bar{t}$ production. The dark error bar represents the statistical uncertainty. The red error bar represents the full uncertainty, including systematics and luminosity uncertainties.

In 2011, already more than 1 fb^{-1} of pp collisions has been collected. Such amount of data will give the possibility to improve those measurements for which the statistical error is the dominating uncertainty, like the measurements of the diboson production. With these data it will also be possible to measure differential cross sections. Moreover new analyses will be possible with the increased statistics, for example the measurement of the $W/Z+b$ in W/Z+jets events and the ZZ production cross section.

References

[1] The ATLAS Collaboration, *The ATLAS Experiment at the CERN Large Hadron Collider*, JINST **3** S08003 (2008).
[2] The ATLAS Collaboration, *A measurement of the total $W^{\pm}$ and Z/γ^{*} cross sections in the e and μ decay channels and of their ratios in pp collisions at $\sqrt{s} = 7$ TeV with the ATLAS detector*, ATLAS-CONF-2011-041, `https://atlas.web.cern.ch/Atlas/GROUPS/PHYSICS/CONFNOTES/ATLAS-CONF-2011-041/`.
[3] The ATLAS Collaboration, *Updated Luminosity Determination in pp Collisions at $\sqrt{s} = 7$ TeV using the ATLAS Detector*, ATLAS-CONF-2011-011, `https://atlas.web.cern.ch/Atlas/GROUPS/PHYSICS/CONFNOTES/ATLAS-CONF-2011-011/`.
[4] Nadolsky PM, et al., *Implications of* `CTEQ` *global analysis for collider observables*, Phys. Rev. **D78** (2008) 013004.
[5] Sherstnev A and Thorne RS, *Parton Distributions for LO Generators*, Eur. Phys. J. **C55** (2008), 553.
[6] H1 and ZEUS Collaborations, *Combined Measurement and QCD Analysis of the Inclusive ep Scattering Cross Sections at HERA*, JHEP **01** (2010) 109.
[7] Sjoestrand T, Mrenna S and Skands P, `PYTHIA 6.4` *physics and manual*, JHEP **05** (2006), 026.
[8] Frixione S and Webber BR, *Matching NLO QCD computations and parton shower simulations*, JHEP **06** (2002), 029.
[9] The ATLAS Collaboration, *Measurement of the production cross section for W-bosons in association with jets in pp collisions using 33 pb^{-1} of data at $\sqrt{s} = 7$ TeV with the ATLAS detector*, ATLAS-CONF-2011-060, `https://atlas.web.cern.ch/Atlas/GROUPS/PHYSICS/CONFNOTES/ATLAS-CONF-2011-060/`.
[10] The ATLAS Collaboration, *Measurement of the production cross section for Z/γ^{*} in association with jets in pp collisions at $\sqrt{s} = 7$ TeV with the ATLAS Detector*, ATLAS-CONF-2011-042, `https://atlas.web.cern.ch/Atlas/GROUPS/PHYSICS/CONFNOTES/ATLAS-CONF-2011-042/`.
[11] Cacciari M, Salam G P and Soyez G, *The anti-k_t jet clustering algorithm*, JHEP **04** (2008), 063.
[12] Mangano ML, et al., `ALPGEN`, *a generator for hard multi-parton processes in hadronic collisions*, JHEP **0307** (2003) 001.
[13] Gleisberg T, et al., *Event generation with* `SHERPA 1.1`, JHEP **0902** (2009) 007.
[14] Campbell JM, Ellis RK, Rainwater DL, *Next-to-leading order QCD predictions for W+2jet and Z+2jet production at the CERN LHC*, Phys. Rev. **D68** (2003) 094021.
[15] Berger CF, et al., *Precise predictions for W+4 jet production at the Large Hadron Collider*, arXiv:1009.2338 [hep-ph] (2010).
[16] The ATLAS Collaboration, *Measurement of the W charge asymmetry in the $W \to \mu\nu$ decay mode in pp collisions at $\sqrt{s} = 7$ TeV with the ATLAS detector*, Phys. Lett B 701 (2011) 31.
[17] Martin AD, Stirling WJ, Thorne RS, and Watt G, *Parton distributions for the LHC*, Eur. Phys. J. **C63** (2009) 189.
[18] The ATLAS Collaboration, *Measurement of $W\gamma$ and $Z\gamma$ production at $\sqrt{s} = 7$ TeV with the ATLAS Detector*, ATLAS-CONF-2011-013, arXiv:1106.1592, submitted to JHEP.
[19] The ATLAS Collaboration, *Measurement of the $W^{+}W$ cross section in $\sqrt{s} = 7$ TeV pp collisions with ATLAS*, ATLAS-CONF-2011-015, arXiv:1104.5225, submitted to PRL.
[20] The ATLAS Collaboration, *Measurement of the $W^{\pm}Z$ Production Cross-Section in Proton-Proton Collisions at $\sqrt{s} = 7$ TeV with the ATLAS Detector*, ATLAS-CONF-2011-084, `https://atlas.web.cern.ch/Atlas/GROUPS/PHYSICS/CONFNOTES/ATLAS-CONF-2011-084/`.

Journal of Physics: Conference Series **347** (2012) 012024 doi:10.1088/1742-6596/347/1/012024

Collective behaviour of partons could be a source of energetic hadrons

M K Suleymanov

COMSATS Institute of Information Technology, Park Road Chak Shahzad Town, Islamabad, Pakistan; Institute of Physics ANAS, 33, H.Javid ave., Baku, Azerbaijan; Joint Institute for Nuclear Research, Dubna, Moscow region, Russia

E-mail: mais_suleymanov@comsats.edu.pk

Abstract. We discuss the idea that collective behaviour of the quarks/partons, which has been intensely discussed for the last 40 years in relativistic hadron-nuclear and nuclear-nuclear interactions and confirmed by new data coming from the ultrarelativistic heavy ion collisions, can lead to energetic particle production. Created from hadronization of the quark/parton (or quarks/partons), energetic particles could get the energy of grouped partons from coherent interactions. Therefore, we think that in the centre of some massive stars, a medium with high density, close to Quantum Chromodynamic one could be a source of the super high-energy cosmic rays.

1. Introduction

Cosmic rays are one of the main components of the Universe. They could provide crucial information about the beginning of the Universe and its evolution. Figure 1 demonstrates the variety of cosmic rays in a diagram displaying the flux of cosmic particles as a function of their energy (the figure was taken from [1]). Even though there have been a lot of research conveyed on cosmic rays and their applications, there is a lot about them that we don't know. One of these still unknown aspects is the source of cosmic rays with super high energies (especially in the range over 10^{15} *eV*) [2]; we don't even know for sure whether their origin is galactic or extragalactic. It was proposed that massive stars could be the source of these rays and their super high energies could have resulted from further acceleration by very strong magnetic fields generated by these stars [3]. However, theoretical predictions show that these magnetic fields could be too weak to accelerate particles to energies of order 10^{15} *eV*. So it is natural that we look for an alternative source of these super high-energy cosmic rays, which does not involve acceleration [4]. The presented paper focuses on one of the possible alternative sources. We propose that these particles could be produced by the same massive stars, however without any acceleration, as a result of a collective behavior of quarks/partons through coherent interactions in the extreme (high density and/or high temperature) states (possibly from the Quark Gluon Plasma) of strongly interacting matter.

2. Phases of strongly interacting matter

It is considered, that during first $\sim 10^{-(22\text{-}23)}$ *sec* of nuclear-nuclear and hadron-nuclear interactions at relativistic and ultrarelativistic energies, the matter produced exists as a strongly interacting one, and

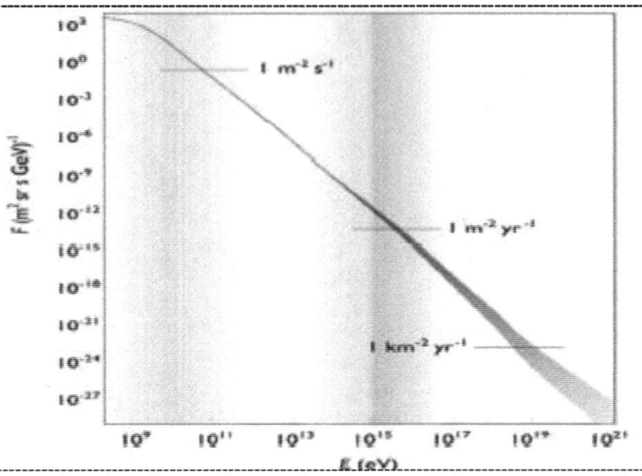

Figure 1. The flux of cosmic ray particles as a function of their energy. The flux for the lowest energies (yellow zone) are mainly attributed to solar cosmic rays, intermediate energies (blue) to galactic cosmic rays, and highest energies (purple) to extragalactic cosmic rays.

this new system contains only hadrons. Figure 2 shows the Quantum Chromodynamics phase diagram for strongly interacting matter (the figure was taken from [5]). Expectations are such, that with increase in density the nuclear medium will transform into extreme dense states of the strongly interacting matter, and then it will pass a mixed phase around critical density (here we expect the appearance of the quark-parton degree of freedom). After that we expect matter to be in the onset states of deconfinement [5], which could occur because of percolation effect [6].

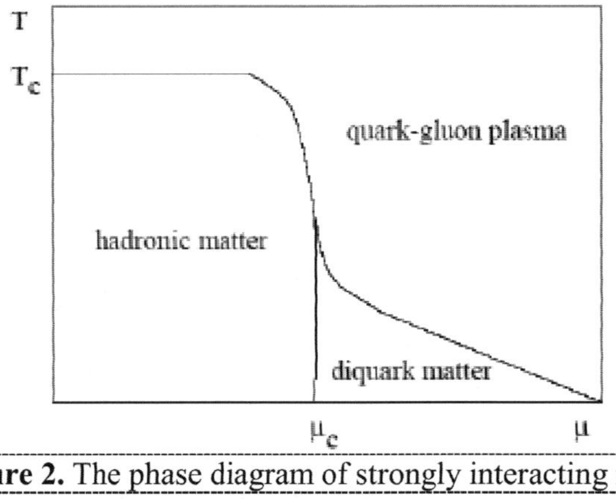

Figure 2. The phase diagram of strongly interacting matter.

3. Some evidence from the experiments.

The experimental data on relativistic and ultrarelativistic hadron-nuclear and nuclear-nuclear interactions have been showing the collective behavior for the quarks/partons in hot and dense matter [7-12]. We believe that as a result of coherent interactions in the system, quark(s)/parton(s) can get the energy of the grouped quark/parton system, therefore the only limiting factor on the final maximum energy of these quark(s)/parton(s) is the energy of the grouped quarks/partons in the system. The quarks/partons can hadronize and become super high-energy hadron cosmic rays.

Journal of Physics: Conference Series **347** (2012) 012024 doi:10.1088/1742-6596/347/1/012024

The following effects we consider as some experimental evidence on quarks/partons collective phenomenon: *JINR* Cumulative effect [7] ; *CERN EMC* effect [8]; recent *RHIC* [9-11] and LHC [12] effects on saturation of v_2 as a function of the p_T.

3.1. JINR Cumulative effect.

When we say "energetic" (or "cumulative") hadrons we think about hadrons formed as a result of hadronization of quark(s)/parton(s), that can get the grouped quark/parton energy in the high density (and/or temperature) medium. A.M.Baldin first predicted the formation of energetic hadrons through the cumulative effect [7]. In the experiment at JINR Dubna it was shown that at energies of several *GeV*, particle production in nuclear collisions is set up in the asymptotic regime [7]. This observation was described as the achievement of invariability of physical picture of the secondary particle production in nuclear fragmentation with increasing collision energy or achievement of so-called limiting fragmentation of nuclei. Figure 3 [13] shows the energy dependence (*E*) for the invariant cross section of cumulative protons (with momentum in the range 0,4 – 1,0 *GeV/c* and with angles: 160^0 - 164^0 in lab frame) produced at different targets by pions and protons (blacked points). The σ_{in} is the total cross section for inelastic πp and *pp* interactions.

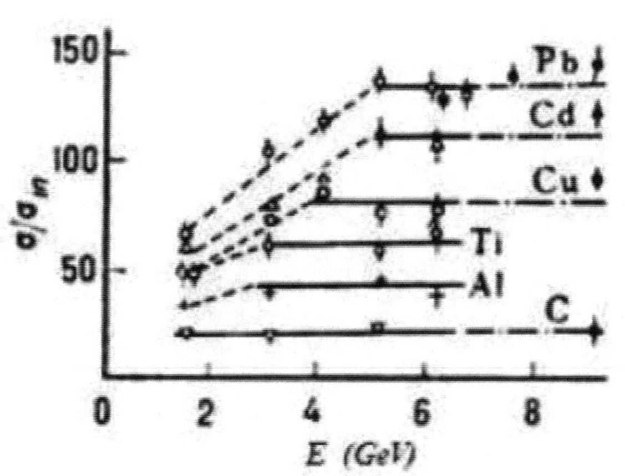

Figure 3. The energy dependence for the invariant cross section of cumulative protons (with momentum in the range 0.4 – 1.0 *GeV/c* and with angles: 160^0 - 164^0 in the lab frame) produced at different targets by pions and protons (blacked points). The σ_{in} is the total cross section for inelastic πp and *pp* interactions.

Cumulative processes involve production of secondary particles with energies beyond the kinematic limit of free nucleon collisions at relativistic hadron-nuclear and nuclear-nuclear interactions. This means that the secondary cumulative particles in these processes can be produced from systems of grouped nucleons, as a result of their collective behavior. Moreover, they get the total energies of the systems. In order to characterize these particles the cumulative parameter *x* was introduced, so that a secondary particles with $x > 1$ are called cumulative ones. The following expression defines the value of *x*:

$$x = \frac{(E_i - P_L)}{m_N} , \qquad (1)$$

where E_i , P_L and m_N are the energy, longitudinal momentum of a particle and m_N is the mass of the nucleon respectively.

The value of x is always less than 1 for particles produced from free nucleon collisions. Figures 4 [14] and 5[15] show the x and momentum spectra, respectively, for secondary charged particles produced in the hadron-nuclear reactions at relativistic energies. Appearance and existence of particles with $x>1$ can be observed from figure 4. In figure 5 the arrow points on the value of $x=1$.

Ref. [16] points to some theoretical explanation for the appearance of cumulative particles in hadron-nuclear and nuclear-nuclear interactions at relativistic energies. The authors propose that the effect is a result of nucleon collective phenomena.

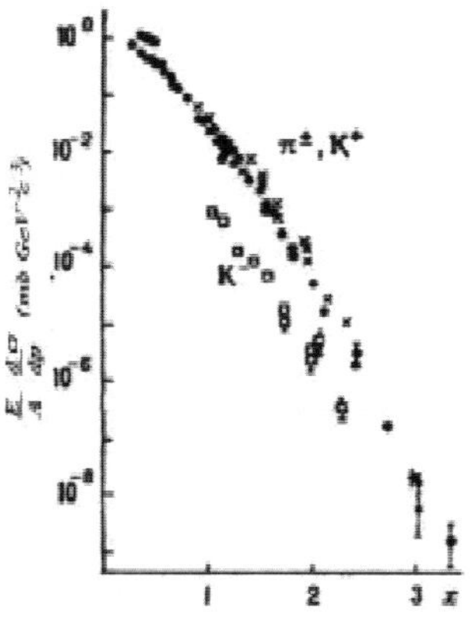

Figure 4. The x spectra for secondary charged particles produced in the hadron-nuclear reactions at relativistic energies.

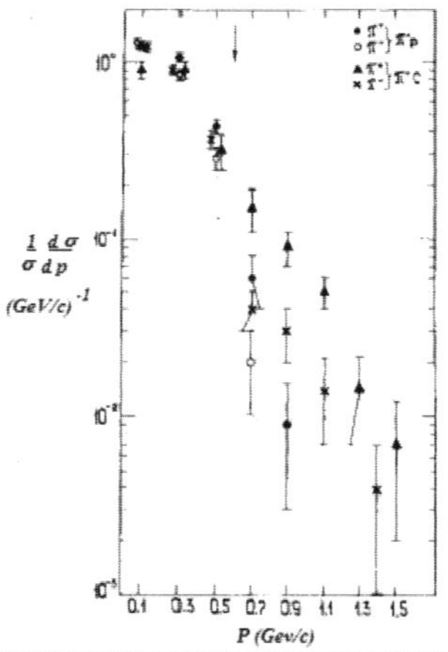

Figure 5. The momentum spectra for secondary charged particles produced in the hadron-nuclear reactions at relativistic energies.

The Coherent Tube Model (CTM) [17] can give us even a clearer explanation for the energetic (cumulative) particle production. Here the interaction of a hadron with a target nucleus results from its simultaneous collision with the tube of nucleons of cross section σ that lie along its path in the target nucleus. For an interaction of projectile with momentum p_{lab} the cumulative square of the center-of-mass energy is $\sim s_i \cong 2imp_{\ lab}$ (i is a number of nucleons , m - a nucleon mass). In paper [18] unusually strong A dependence (stronger than the commonly assumed A or $A^{2/3}$) of the cross section for $p + A \rightarrow J/\Psi + X$ reaction at incident energies below $30\ GeV$ was described quantitatively using cumulative effects (via energy rescaling).

Ref. [19] discusses CTM for high energy nucleus-nucleus collisions. In this case two tubes are considered : i_1 nucleons in incident tube and i_2 nucleons in the target tube. The c.m. energy squared for this tube-tube collision is approximately given by $\sim s_{i1i2} \cong 2i_1i_2mp_{\ lab.}$

3.2. CERN EMC-effect

About 25 years ago, deep-inelastic muon-nucleus scattering [8] at the European Muon Collaboration (*EMC*) at *CERN* revealed that the structure function F_2 and, hence, quark and gluon distribution of a nucleon bound in a nucleus differ for nuclei with different masses. Figure 6 shows the ratios of structure functions for different nuclei [20]. The paper [20] discusses that none of the popular models explaining the *EMC* effect seem satisfactory and presents a new point of view on the effect as a simple relativistic phenomenon.

Journal of Physics: Conference Series **347** (2012) 012024 doi:10.1088/1742-6596/347/1/012024

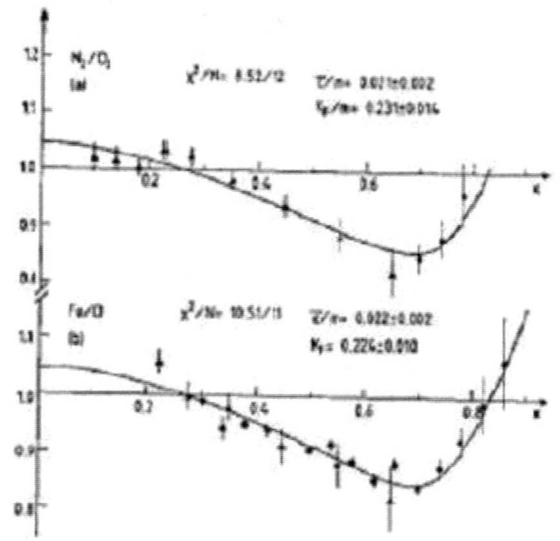

Figure 6. The ratios of structure functions for different nuclei

3.3. Scaling behavior of v_2 at ultrarelativistic heavy ion collisions.

Experimental results from studies of the azimuthal anisotropy of final states in nuclear-nuclear collisions at the energies of relativistic and ultrarelativistic heavy-ion colliders (RHIC, LHC) exhibit a collective behavior, which is likely to be formed at an early, parton, stage of the space-time evolution of produced hot and dense matter [9-11]. The data on azimuthal anisotropy indicates that strongly interacting matter produced in the final state under extreme conditions behaves as a nearly ideal liquid rather than ideal gas made of quarks and gluons.

Scaling behavior of v_2 (see figure 7 [10] and figure 8 [11]) provides an opportunity to assume that the collective behavior of the partons defines the dynamics of an expansion in the longitudinal plane; evolution of the collective flow occurs on the parton state.

The first measurement of elliptic flow of charged particles in *Pb-Pb* collisions at the center of mass energy per nucleon pair $\sqrt{s_{NN}} = 2.76\ TeV$ with the *ALICE LHC* detector [12], shows that the p_t – dependence of the v_2 does not change within uncertainties from *200 GeV* to *2.76 TeV*.

4. Discussion and summary

In the language of parton models, the facts pointed out in previous sections can indicate that high density (and/or temperature) nuclear matter can contain strongly correlated multiparton/multiquark formations and involve collective behavior in the system. The collective behavior could lead to formation of coherent groups of quarks/partons (for example coherent tube). As a result of coherent interactions quark(s)/parton(s) can get the energy of grouped quarks/partons (coherent nucleons or quark/parton tubes for example) and become super high-energy quark(s)/parton(s), since in this case the resulting energy will depend on the parameters of the system of grouped quark(s)/parton(s) (tube(s)) too. In the framework of the Coherent Tube Model the cumulative square of the center-of-mass energy s will depend on a number i of the quarks/partons, grouped in the tube(s), and increase with i (section 3.1). So if we consider a system with a temperature around *150-170 MeV* (hot matter) and density 7-10 times greater than that of a normal nucleus (dense matter), then in this system the particle(s) with energy ~ 10^{15} *eV* could be produced as a result of coherent interactions with system of grouped quarks/partons (or tube(s)) containing ~ 10^7 quarks/partons and more. In this case the collective behavior which was observed at relativistic and ultrarelativistic energies is a necessary condition for coherent system (or tube(s)) formation. The quark(s)/parton(s) can leave the hot and/or dense medium with minimum energy loss and transform into a super high-energy hadron(s), where their energies are restricted only by the total energy of the quarks/partons system.

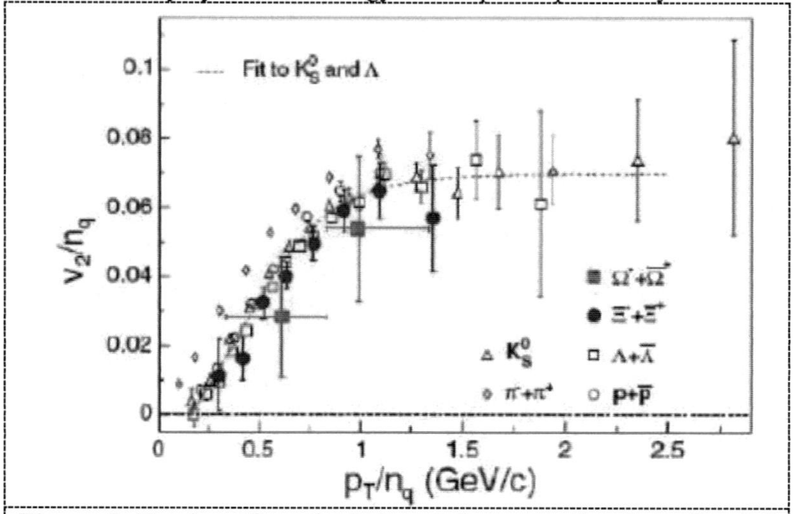

Figure 7. Number of quarks (n_q) scaled v_2 as a function of scaled p_T for $\Xi+\Xi^{+}bar$ (filled circles) and $\Omega+\Omega^{+}bar$ (filled squares). Same distributions are also shown for $\pi\pi$ (open diamonds), $p+\ pbar$ (open triangles) [21], K_S^0 (open circles), $\Lambda+\Lambda bar$ (open squares) [22]. All data are from 200 GeV $Au + Au$ minimum bias collisions. A dot-dashed-line is the scaled result of a fit to K_S^0 and Λ [23].

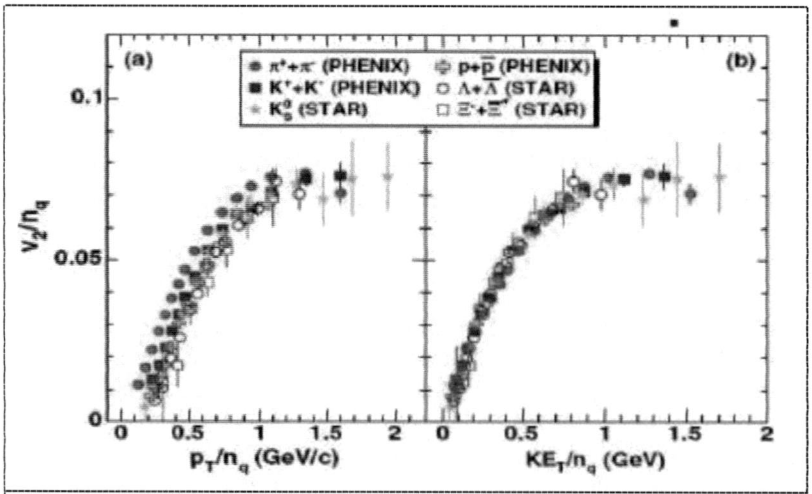

Figure 8. (a) $v_2=n_q$ vs $p_T=n_q$ and (b) $v_2=n_q$ vs $K\ E_T=n_q$ for identified particle species obtained in minimum bias $Au + Au$ collisions. The *STAR RHIC* data is from Refs. [24-25]

Now if we turn to cosmic rays, we can see that, it is possible for some medium with high density (and/or temperature) close to the *QCD* critical one to be a source of cosmic particles with super high energies. This medium can evolve in the center of some massive star (e.g. neutron star [26]). High density (and/or temperature) of this medium can lead to the deconfinement. Quark(s)/parton(s) with big values of x or energy can be formed in this system as a result of collective phenomenon and coherent interactions, where they hadronize and appear as super high-energy cosmic particles. As we have mentioned above, here the maximum energy of cosmic particles is limited only by the values of the total energy of the grouped quarks/partons.

Final notes:

1. Experimental results on relativistic and ultrarelativistic hadron-nuclear and nuclear-nuclear collisions point out the collective behavior of the quarks/partons;
2. Energetic quarks/partons can be produced from coherent interactions of the grouped quarks/partons;
3. In the center of some massive stars where the conditions are similar to ones suitable for deconfinement, the produced energetic quarks/partons can transform into super high-energy cosmic rays, where their energies are limited only by the values of total energy of the grouped quarks/partons.

Acknowledgments
The author would like to thank Dr. S. M. Junaid Zaidi (Rector of CIIT, Islamabad) for his support for author's participation in the ICPP2011. Also thanks to A.M. Suleymanzade for help in preparing the paper.

References

[1] Swordy S 2001 *Space Science Reviews* **99** 85.
[2] Ginzburg V L 1993 *Phys.Usp.* **163** 45
[3] Ptitsina K V and Troitsky S V 2010 *Phys.Usp.* **187** 723
[4] Suleymanov M et al. 2009 *PoS EPS-HEP2009* 406
[5] Satz H 2000 hep-ph/0007069
[6] Satz H 2002 hep-ph/0212046; Brzychczy J 2004 nucl-th/0407008; Pajar C 2005 hep-ph/0501125
[7] Baldin A M et al. 1974 *Sov.J. Nucl.Phys.* **18** 41; Baldin A M 1977 *PEPAN* **8** 429
[8] Aubert J J et al. 1983 *Phys. Lett.* B **123** 275
[9] Okorokov V A 2009 *Physics of Atomic Nuclei* **72** 147
[10] Adamset J et al. 2005 *Phys.Rev.Lett.* **95** 122301
[11] Adareet A et al. 2007 *Phys.Rev.Lett.* **98** 162301
[12] Aamodt K et al. 2010 arXiv[nucl-ex]:1011.3914v1
[13] Stavinski V S 1979 *PEPAN* **10** 949
[14] Stavinski V S 1988 Proceedings of the IX International Seminar on problems of high energy physics 14-19 July Dubna p 190
[15] Angelov N 1979 *Sov. Journal of Nuc.Phys.* **29** 5
[16] Stricman M I and Francfurt L L 1980 *PEPAN* **11** 571; Efremov AV 1982 *PEPAN* **13**
[17] Berlad G, Dar A and Eilam G 1976 *Phys. Rev.* D **13** 161; Afek Y, Berlad G, Eilam G and Dar A 1976 Technion Report No. PH-76-12
[18] Afek Y, Berlad G, Eilam G and Darf A *Phys. Rev.Lett.* **37** 947
[19] Afek Y, Berlad G, Dar A and Eilam G 1976 Technion –PH-76-87
[20] Efremov A V 1986 *Phys. Lett.* B **174** 219
[21] Adler S S et al. 2003 *Phys. Rev. Lett.* **91** 182301
[22] Adams J et al. 2004 *Phys. Rev. Lett.* **92** 052302
[23] Dong X, Esumi S, Sorensen P, Xu N, Xu Z 2004 *Phys. Lett.* B **597** 328
[24] Adams J et al. 2004 *Phys. Rev. Lett.* **92** 052302
[25] Adams J. et al. 2005 *Phys. Rev. Lett.* **95** 122301
[26] Lyne A G and Smith F G 1990 *Pulsar Astronomy* (Cambridge University Press)

The ATLAS TRT and its Performance at LHC

Ahmet Bingül[1]
Department of Engineering Physics, University of Gaziantep, Gaziantep 27310, Turkey

E-mail: bingul@gantep.edu.tr

Abstract. The ATLAS Transition Radiation Tracker (TRT) is the outermost of the three sub-systems of the ATLAS Inner Detector at the Large Hadron Collider at CERN. It consists of almost 300000 thin-wall drift tubes (straws) providing on average 30 two-dimensional space points with 0.12-0.15 mm resolution for charged particle tracks with $|\eta| < 2$ and $p_T > 0.5$ GeV/c. Along with continuous tracking, it provides particle identification capability through the detection of transition radiation X-ray photons generated by high velocity particles in the many polymer fibers or films that fill the spaces between the straws. Custom-built analog and digital electronics is optimized to operate as luminosity increases to the LHC design. In this talk, a review of the commissioning and first and current operational experience of the TRT detector and its performance at LHC will be presented. Emphasis will be given to performance studies based on the reconstruction and analysis of LHC collisions. A comparison of the TRT response and the particle identification in pp and Pb-Pb collisions will be presented. The results are also compared with the expected performance.

1. Introduction

The ATLAS experiment[1] is one of the two general purpose detectors at the Large Hadron Collider (LHC) at CERN. The LHC is a hadron storage ring with a 27 km circumference allowing the collision of protons and lead ions in four interaction points. For proton collisions, designed center of mass enegy is E_{CM} = 14 TeV, interaction rate is f = 40 MHz which corresponds T = 25 ns bunch spacing and instantaneous luminosity $L = 10^{34}$ cm^{-2}s^{-1}. Beams were first injected at energies of 450 GeV on September 2008 followed by a 14 month shutdown period for repairs. As of November 2009 there had been collisions at E_{CM} = 900 GeV for a short time before the commissioning for 3.5 TeV beam energy began in 2010. First events at this energy were seen on March 30, 2010, and have been provided ever since with constantly increasing luminosity reaching values as high as 2×10^{33} cm^{-2}s^{-1} so far. At the end of 2010, the LHC delivered a short run of Pb-Pb collisions at centre-of-mass energy of nucleons at E_{CM} = 2.76 TeV. In Pb-Pb collisions, 10 μb^{-1} data have been delivered by LHC.

Currently, the LHC is running at E_{CM} = 7 TeV with T = 50 ns bunch spacing and has delivered a total of 1.05 fb^{-1}, and ATLAS has recorded 1.01 fb^{-1} in the 2010 and 2011 7 TeV proton collisions. The E_{CM} = 7 TeV program is scheduled to be continued throughout 2011 and 2012. In 2013 there will be a shutdown allowing the LHC to be repaired and reach its design parameters afterwards.

[1] On behalf of ATLAS TRT Collaboration

Journal of Physics: Conference Series **347** (2012) 012025 doi:10.1088/1742-6596/347/1/012025

2. The ATLAS Detector

The ATLAS detector is a hermetic 4π multipurpose detector located 100 m underground at one of the four LHC interaction points. It employs appropriate detection techniques for studying the remnants of high energy proton collisions up to TeV energies. Viewed from the outside, its most striking feature is a muon system consisting of eight air-core toroid coils providing a 1.4T magnetic field for muon detection and measurement. Further inside is a multilayer calorimetric system. It employs various energy measurement techniques depending on the actual location inside the detector (and hence the projected energy and particle density during operation). The calorimeter encloses a superconducting solenoid magnet which provides a 2T magnetic field for the Inner Detector.

The Inner detector covers the range of $|\eta| < 2.5$ and provides tracking and particle identification for charged particles (e-π) with a transverse momentum $p_T > 0.5$ GeV/c. The inner detector consists of pixel and silicon micro-strip detectors surrounded by a Transition Radiation Tracker (TRT). A schematic view of the inner detector is shown in Figure 1.

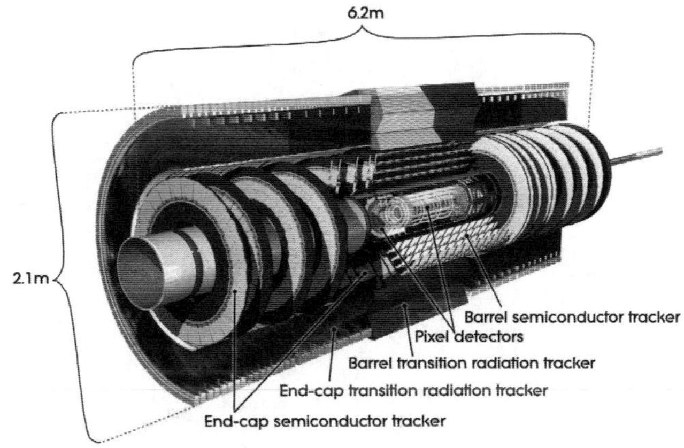

Figure 1. Cut-away image of the ATLAS Inner Detector.

The calorimeter system, covers the pseudorapidity range $|\eta| < 4.9$ and uses a variety of techniques. The electromagnetic calorimeter is a lead liquid argon (LAr) detector in the barrel region and in the forward region copper LAr technology is used; it provides e-γ trigger, identification and measurement. The hadronic calorimeter with scintillator tiles or LAr as the active material and with either steel, copper or tungsten as the absorber material, provides trigger, jet measurements and missing transverse energy measurement. Finally the Muon Spectrometer is based on the magnetic deflection of muon tracks in the large superconducting air-core toroid magnets, instrumented with separate trigger and high-precision tracking chambers covering the pseudorapidity range up to $|\eta| < 2.7$.

The trigger system of ATLAS comprises of three levels: the first level (L1), which is fully hardware, is implemented in the calorimeters and the Muon Spectrometer. The other two levels, the level two (L2) and the Event Filter (EF) are software based: the L2 accepts data from defined Regions Of Interests (ROI) of L1 and the EF which provides a full event reconstruction on computer farms [2].

3. The Transition Radiation Tracker

The ATLAS Transition Radiation Tracker (TRT) is the outermost of the three sub-systems of the ATLAS inner detector. It extends to a radius of 1082 mm from the interaction region. Its design was driven by the need to get continuous tracking with long lever arm and electron identification in the pseudorapidity range $|\eta| < 2$. It was also designed to provide a point resolution of 130 μm and withstand the challenging LHC environment with a high particle density and accumulated radiation dose [3].

3.1. *TRT design*

TRT is a gaseous detector. It consists of 4 mm diameter straw tubes wound from a multilayer film reinforced with carbon fibers and containing a 30 μm gold plated tungsten wire in the center [4]. Keeping the wire at ground, the potential difference between the straw wall and the wire is 1.5 kV. Therefore the charge clusters created through primary ionization, as depicted in Figure 2, undergo avalanche multiplication near the wire with a amplification factor of $2.5x10^4$. The straw is filled with a gas mixture of 70% Xe, 27% CO_2 and 3% O_2. Gas composition and high voltage are chosen for stability and transition radiation performance.

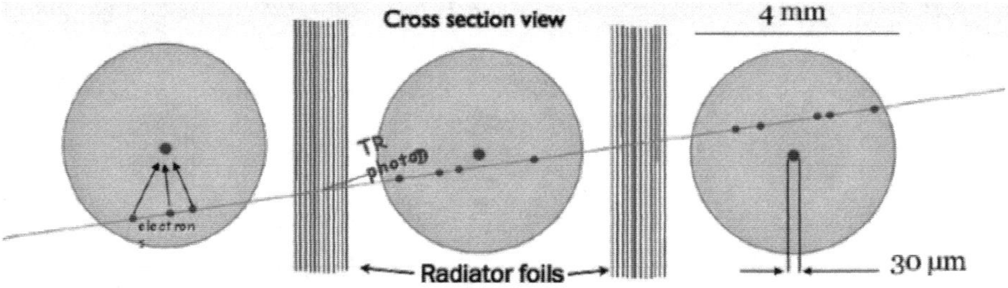

Figure 2. Working principle of the TRT. Primary ionization clusters are generated by traversing charged particles and absorbed TR photons.

For electron identification the TRT exploits Transition Radiation (TR), soft X-ray photon emitted by a charged particle when traversing the boundary between materials with different dielectric constants (ε_1, ε_2)[2]. The emitted photons are detected in the straw tubes by absorption on xenon atoms and subsequent ionization. In this way TR deposits a much higher energy in a single straw than an ionizing particle usually does. In ATLAS, normally only electrons reach a velocity which is needed to generate Transition Radiation thus the detection of a TR photon indicates the traversing of an electron. The emission of TR happens in dedicated radiators which are interleaved with the straws. By introducing multiple layers of radiator material and choosing their spacing in the right way, the emission of TR can be stimulated coherently amplifying the yield of emitted transition radiation.

The TRT barrel covers $|\eta| < 1$ and comprises totally 105088 readout channels parallel to beam axis arranged in 73 layers. Each straw has a length of 144 cm. The design was chosen to ensure the hit rate of a single readout channel is kept below 20 MHz at LHC design luminosity[5]. The end-cap regions cover the range $1 < |\eta| < 2$. Each end-cap contains 122880 39 cm long straws oriented radially in 160 layers[6].

Each charged track traversing the detector volume crosses on average 30 straws with a point resolution of 130 μm over a distance of roughly half a meter. With these measurements the TRT significantly improves the momentum resolution for charged tracks as can be seen in Figure 3.

3.2. *Signal formation and digitization*

A particle crossing a straw causes some primary ionization centers to create electrons which drift to the anode wire where electrons are multiplied in an electric avalanche. Hence a measureable signal can be generated. A Transition Radiation photon, created in one of the radiator layers, is absorbed in the

[2] Note that the emitted TR energy and intensity are proportional to $\varepsilon_1 - \varepsilon_2$ and particle Lorentz factor, $\gamma = E/m$, respectively. In addition, TR emission angle with respect to the particle direction is proportional to $1/\gamma$.

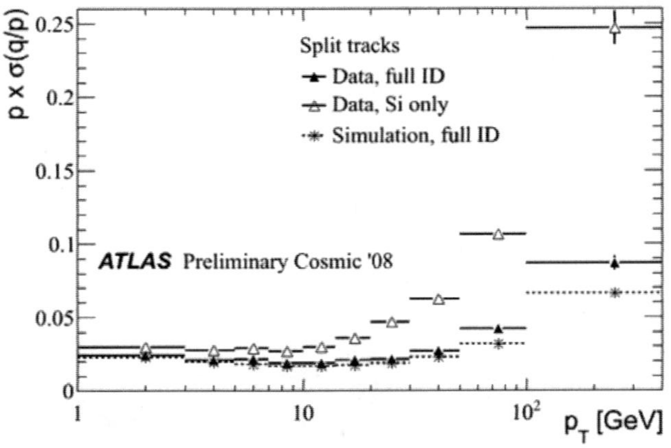

Figure 3. Momentum resolution of the ATLAS tracking system with and without using the TRT information. Contribution of TRT at high energy can be seen clearly

Xe gas creating primary ionization as well. This additional ionization from TR increases the electric pulse height obtained in the straw tube. To measure the signal, the individual straws are directly coupled to the ASDBLR(Amplifier-Shaper-Discriminator with Baseline Restoration) frontend chip. Here electric pulses are amplified, shaped and the slow ion drift tail of the signal is suppressed[7]. The ASDBLR moreover contains two independent discriminators set to different thresholds: a low threshold (300 eV) for registering the passage of minimum ionizing particles and a high threshold (6 keV) to flag the absorption of transition radiation X-rays.

The ASDBLRs are coupled to DTMROC(Digital Time Measurement Read-Out Chip) digitization chips. For each triggered event, 75 ns (i.e. three bunch crossings) of the signal are digitized with different timing characteristics. Note that the time resolution of TRT is about 3 ns. Leading edge of the pulse gives information about the drift time of the electrons to the wire and hence the distance of closest approach of the track to the wire. So, this information is used for tracking of the charged particle passing though the straws.

An example of digitization and timing of TRT pulse is shown in Figure 4.

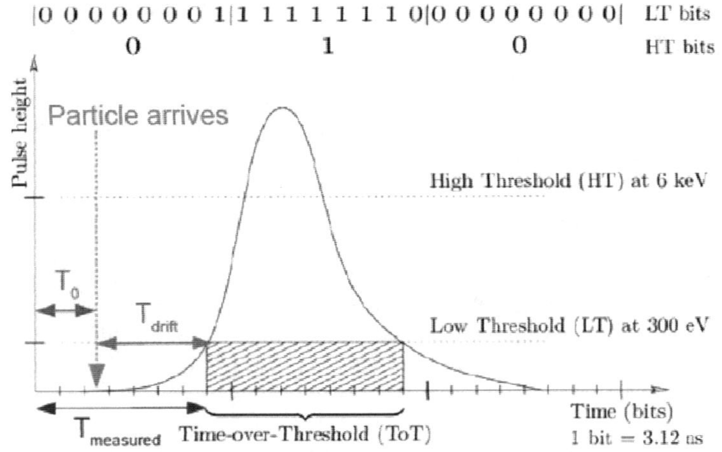

Figure 4. Digitization and timing of a TRT pulse.

Journal of Physics: Conference Series **347** (2012) 012025 doi:10.1088/1742-6596/347/1/012025

3.3. *TRT calibration*

By design each straw measures a time which has to be converted into spatial coordinates. We need a calibration function which relates the measured time and track's closest distance to the wire. In TRT, to do that two calibration procedures are applied. One is the R-t relation which converts drift time to a measured drift radius. For all TRT straws, one R-t relation is used for Barrel and one for end-cap. The other one is the T_0 calibration the time between the beginning of the readout window and the physical arrival of a particle originating from the bunch crossing associated with that window. T_0 calibration is performed in hardware level. Each calibration is performed regularly after every run.

An example R-t curve can be seen in Figure 5.

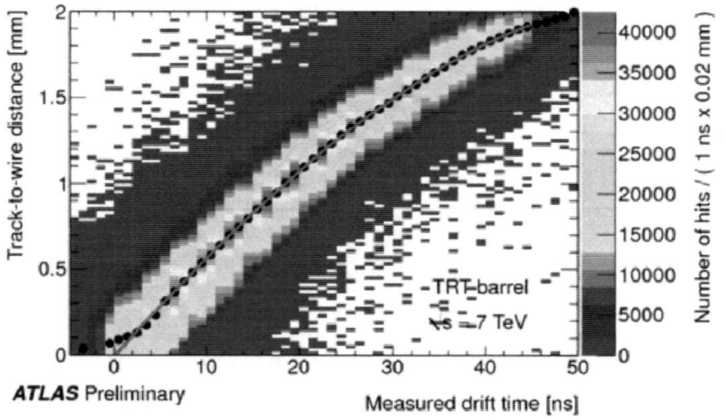

Figure 5. Example for a R-t calibration curve.

3.4. *Particle identification with TRT*

The TRT offers two different ways of identifying particles.

For separating electrons from charged pions, the fraction of high threshold hits on a track can be used. As indicated in Figure 6, the probability (which depends on particle γ value) for generating a transition radiation photon and a subsequent high threshold hit is much higher for electrons than for pions[3]. With this way, TRT separates electrons from pions over a momentum range between 1 GeV/c and 150 GeV/c.

TRT also provides particle identification information in the form of dE/dx measurements for charged particles. As an estimator for this quantity, the measured time over threshold, ToT, (i.e. the time between leading and trailing edge) is taken as a starting point. After applying various corrections to accommodate for physical variation (e.g. the actual length the track was passing through the straw) a good estimator for the specific energy loss can be achieved as demonstrated in Figure 7. Proton and kaon bands are clearly visible and distinguishable. Using ToT technique, particle identification with TRT can be used for the charged particles whose momentum is less than 10 GeV/c[3].

3.5. *TRT performance in proton and heavy ion collisions*

For quantifying the tracking performance of the TRT there are two key quantities: The spatial resolution and the average straw efficiency. The alignment and calibration procedures yield the resolutions of 118 μm for barrel and 132 μm for end-cap hit resolutions in 7 TeV proton collisions.

[3] Consider a particle with momentum p = 10 GeV/c crossing a total of 30 straws. Using binomial theorem and assuming the probability is constant for all straws one can calculate expected number of HT hits. For example, for the electron hypothesis expected number of high threshold hits is 6 while for the pion hypothesis it is 1.2.

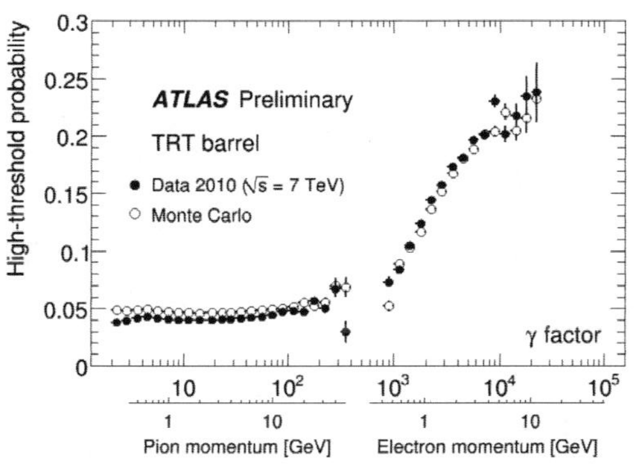

Figure 6. Probability of a TRT high-threshold hit as a function γ factor as measured in 7 TeV collision events.

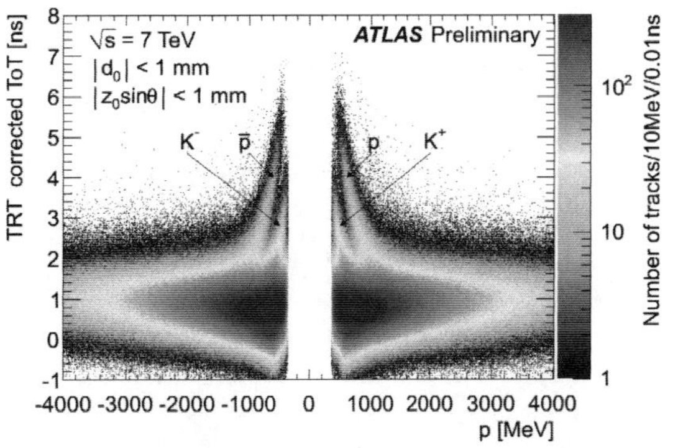

Figure 7. Shown are the TRT corrected ToT versus track momentum. One can see the different populations from different particles.

An example resolution distributions are shown in Figure 8 for the barrel and in Figure 9 for the end-cap regions for the data and Monte Carlo studies which assume the perfectly aligned detector.

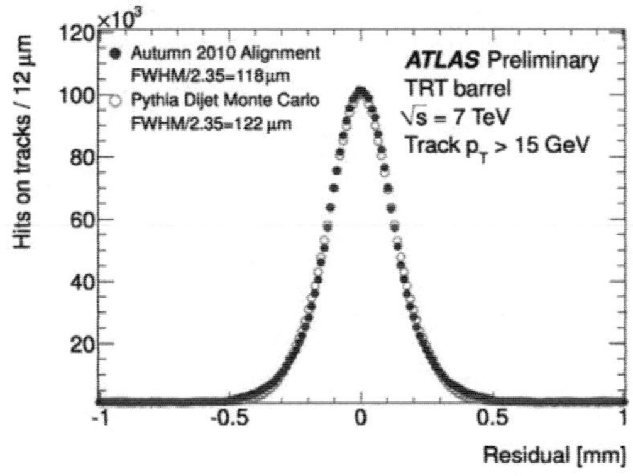

Figure 8. Residual distribution for the barrel.

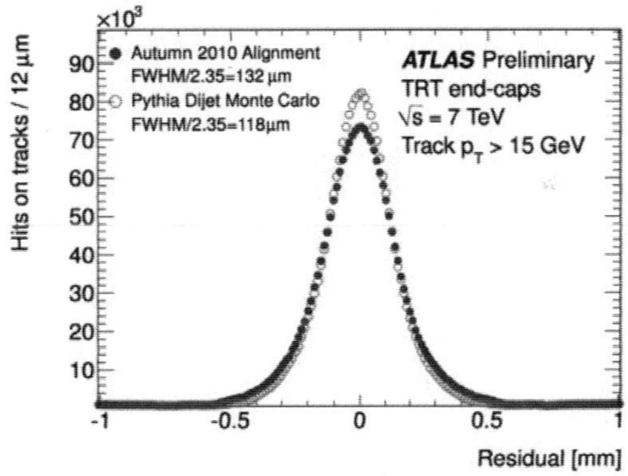

Figure 9. Residual distribution for the end-cap.

The straw hit efficiency is defined as number of hits on straw divided by number of tracks passing that straw. As an example the efficiency curve in the barrel for 7 TeV running is shown in Figure 10. Inefficiency at outer straw radius is due to reconstruction effects and smaller electric signal pulse size.

Although ATLAS was not designed with heavy ion physics in mind, the whole detector was operated and data was being recorded and analyzed. The TRT was especially challenged by the harsh conditions of high energy ion-ion collisions. Depending on centrality of the events, several thousand charged tracks resulting in detector occupancies as high as 95% could be observed. However, even in high occupancy, TRT contributes to the tracking information[3].

4. Summary

The TRT is an excellent detector improving ATLAS tracking with its long lever arm and the high number of hits per track. It also provides discrimination between electrons and charged pions. During LHC stable beams in 2010 and 2011 TRT achieved 100% data taking efficiency. The alignment and calibration procedures resulted in ~120(130) μm hit resolution in Barrel(End-cap). Barrel hit resolution is better than the design expectations.

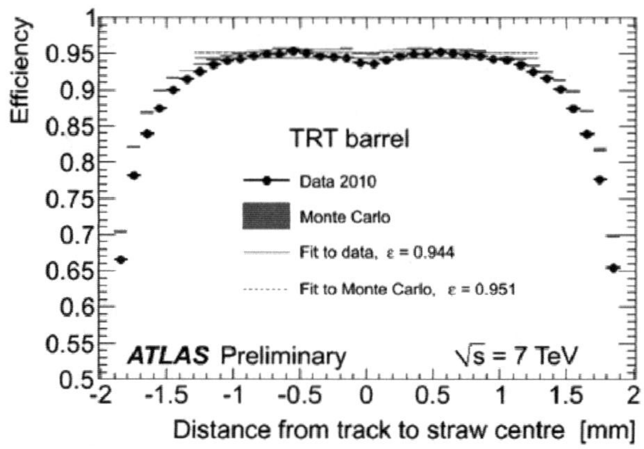

Figure 10. Example efficiency curve for the barrel.

The results obtained from heavy ion collisions indicate that the TRT provides useful tracking information even in high occupancy. Thus giving us confidence that the TRT is ready for at least ten more years of LHC running to be recorded.

5. Acknowledgement

The ATLAS collaboration wishes to thank CERN for the efficient commissioning and operation of the LHC during this initial high-energy data-taking period as well as the support staff from our institutions without whom ATLAS could not be operated efficiently.

We acknowledge the TRT colleagues making the detector perfect.

References

[1] ATLAS Collaboration, G. Aad et al. 2008, *The ATLAS Experiment at the CERN Large Hadron Collider*, JINST 3 S08003.

[2] Nicolaidou R 2011, *The 13th International Conference on B-Physics at Hadron Machines - Beauty2011*, Amsterdam, The Netherlands, ATL-COM-PHYS-2011-568

[3] Schaepe S 2011, *Commissioning and performance of the ATLAS Transition Radiation Tracker with first high energy pp and Pb-Pb collisions at LHC*, ATL-INDET-PROC-2011-001.

[4] Abat E et al. 2008, *The ATLAS Transition Radiation Tracker (TRT) proportional drift-tube: Design and Performance*, JINST 3 P02013

[5] Abat E et al. 2008, *The ATLAS TRT Barrel Detector*, JINST 3 P02014.

[6] Abat E et al. 2008, *The ATLAS TRT end-cap detectors*, JINST 3 P10003.

[7] Abat E et al. 2008, *The ATLAS TRT electronics*, JINST 3 P06007.

Journal of Physics: Conference Series **347** (2012) 012026 doi:10.1088/1742-6596/347/1/012026

Recent charm physics results from *BABAR*

Ryan M. White[1]
University of South Carolina, Dept. of Physics and Astronomy, Columbia, SC, USA
E-mail: `white@physics.sc.edu`

Abstract. In this proceeding, recent charm physics results from the *BABAR* experiment are discussed. The studies include a Dalitz plot and partial-wave analysis of the decay $D_s \to K^+K^-\pi^+$, a recent measurement of the leptonic decay constant f_{D_s}, the precision mass and width measurements of the $D_{s1}(2535)$ meson, searches for new resonances in inclusive e^+e^- collisions, searches for flavor changing neutral currents, lepton-flavor and lepton-number violating decays, and searches for CP violation.

1. Introduction

The Charm physics studies discussed in this proceeding make use of the $\Upsilon(4S)$ data set with an integrated luminosity of $\sim$480 fb^{-1} collected by the *BABAR* detector. The *BABAR* detector is described in detail elsewhere [1]. The components that are important to the analyses described here follows. Charged particle tracks are detected, and their momenta measured, with a cylindrical drift chamber (DCH) and a silicon vertex tracker (SVT), both operating inside a 1.5 T solenoidal magnetic field. Photon energies are measured with a CsI(Tl) electromagnetic calorimeter (EMC). Identification of charged particle species is determined from information collected from a ring-imaging Cherenkov detector (DIRC) and specific energy-loss measurements in the SVT and DCH. Additionally, muons are identified with information collected from a instrumented flux return (IFR) which is made of layers of iron, brass, limited streamer tubes (LSTs), and resistive plate chambers (RPCs).

2. Dalitz plot and Partial Wave Analysis of the Decay $\mathbf{D_s \to K^+K^-\pi^+}$ [2]

Scalar mesons are still a puzzle in light meson spectroscopy. New claims for the existence of broad states close to threshold such as $\kappa(800)$ [3] and $f_0(600)$ [4], have reopened discussion about the composition of the ground state $J^{PC} = 0^{++}$ nonet, and about the possibility that states such as the $a_0(980)$ or $f_0(980)$ may be 4-quark states, due to their proximity to the $K\overline{K}$ threshold [5]. This hypothesis can be tested only through accurate measurements of the branching fractions and the couplings to different final states. It is therefore important to have precise information on the structure of the $\pi\pi$ and $K\overline{K}$ $\mathcal{S}$-waves. In this context, D_s^+ mesons can shed light on the structure of the scalar amplitude coupled to $s\bar{s}$. The understanding of the $K\overline{K}$ $\mathcal{S}$-wave is also of great importance for the precise measurement of CP-violation in B_s oscillations using $B_s \to J/\psi\phi$ [6, 7].

[1] On behalf of the *BABAR* Collaboration

$m_{K^+K^-}$ range(MeV/c^2)	$f_{\mathcal{S}-wave}(\%)$	$f_{\mathcal{P}-wave}(\%)$
1019.456 ± 5	3.5 ± 1.0	96.5 ± 1.0
1019.456 ± 10	5.6 ± 0.9	94.4 ± 0.9
1019.456 ± 15	7.9 ± 0.9	92.1 ± 0.9

Table 1. $\mathcal{S}$-wave and $\mathcal{P}$-wave fractions computed in three K^+K^- mass ranges around the $\phi(1020)$ peak. Errors are statistical only.

The D_s^+ meson decay to $K^+K^-\pi^+$ is studied to determine the $\mathcal{S}$-wave structure in the vicinity of the $\phi(1020)$ resonance using a partial wave analysis technique. The results of this study are used in the Dalitz plot analysis to determine the resonant structure of this decay.

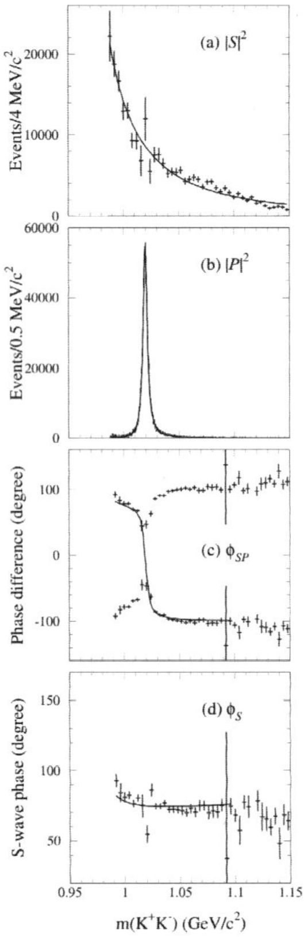

Figure 1. Squared (a) $\mathcal{S}$- and (b) $\mathcal{P}$-wave amplitudes; (c) phase difference $\phi_{\mathcal{SP}}$; (d) $\phi_{\mathcal{S}}$. The curves result from the fit.

Model-independent information of the K^+K^- $\mathcal{S}$-wave in the vicinity of the $\phi(1020)$ resonance is obtained by performing a partial wave analysis in the threshold region. We measure the fraction of $\mathcal{S}$ and $\mathcal{P}$-wave in various K^+K^- mass regions, shown in table 1. We then extract a phenomenological description of the $\mathcal{S}$-wave, assuming that it is dominated by the $f_0(980)$. We fit the $|\mathcal{S}|$, $|\mathcal{P}|$, and $\phi_{\mathcal{SP}}$ distributions (figure 1), where we assume empirically a Breit-Wigner (BW) to parameterize the $f_0(980)$ and the spin-1 Relativistic Breit-Wigner (RBW) for the $\phi(1020)$. The resulting mass and width, $m_0 = (0.922 \pm 0.003_{stat})$ GeV/c^2 and $\Gamma_0 = (0.24 \pm 0.08_{stat})$ GeV/c^2, are used to parameterize the $f_0(980)$ in the Dalitz plot analysis, where an unbinned maximum likelihood (ML) fit of the events is performed to determine the resonant contributions of this decay.

The decay is found to be dominated by the $\phi(1020)$ resonance in K^+K^- and the $\bar{K}^{*0}(892)$ resonance in $K^-\pi^+$, although $\mathcal{S}$-wave contributions exist as well. The magnitudes and phases of each resonance are measured relative to the $\bar{K}^{*0}(892)$. In addition to the magnitudes and phases as parameters in the fit, the $\bar{K}^{*0}(892)$ and the $f_0(1370)$ masses and widths are free to float. The results of the fit are shown in figure 2. The magnitudes, phases, and fit fractions of the resonances in the decay model are shown in table 2.

The fitted parameters for the $\bar{K}^{*0}(892)$ are:

$$m_{\bar{K}^{*0}(892)} = (895.6 \pm 0.2_{stat} \pm 0.3_{sys})\ \text{MeV}/c^2$$
$$\Gamma_{\bar{K}^{*0}(892)} = (45.1 \pm 0.4_{stat} \pm 0.4_{sys})\ \text{MeV}$$

and for the $f_0(1370)$

$$m_{f_0(1370)} = (1.22 \pm 0.01_{stat} \pm 0.04_{sys})\ \text{GeV}/c^2$$
$$\Gamma_{f_0(1370)} = (0.21 \pm 0.01_{stat} \pm 0.03_{sys})\ \text{GeV}.$$

3. Absolute Branching Fractions of Leptonic Decays of the D_s and a Measurement of the Leptonic Decay Constant f_{D_s} [8]

The D_s^- meson can decay purely leptonically via annihilation of the $\bar{c}$ and s quarks into a W^-. In the Standard Model (SM), the leptonic partial width $\Gamma(D_s^- \to \ell^-\bar{\nu}_\ell)$ is given by

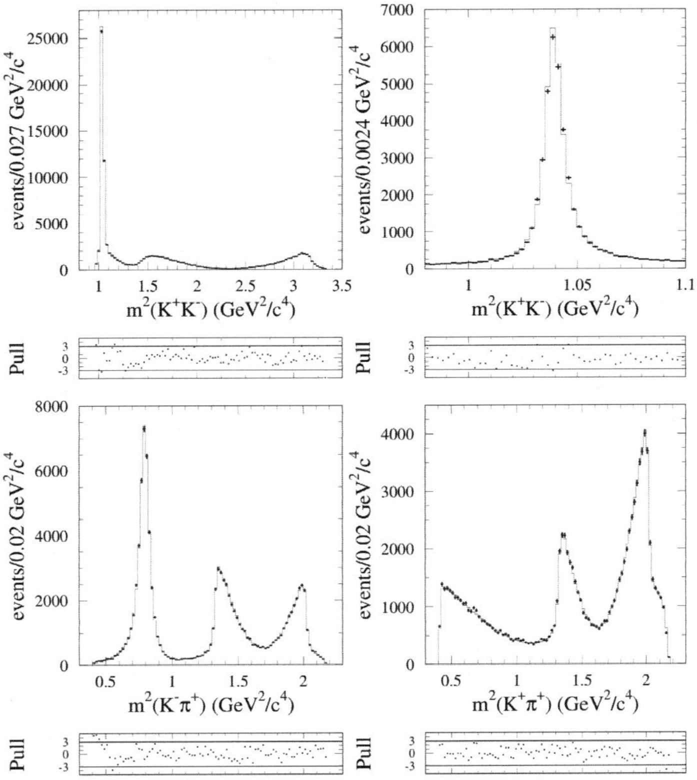

Figure 2. $D_s^+ \to K^+K^-\pi^+$: Dalitz plot projections. The data are represented by points with error bars, the fit results by the histograms.

Decay Mode	Decay fraction(%)	Amplitude	Phase(radians)
$\bar{K}^*(892)^0K^+$	$47.9 \pm 0.5 \pm 0.5$	$1.(Fixed)$	$0.(Fixed)$
$\phi(1020)\pi^+$	$41.4 \pm 0.8 \pm 0.5$	$1.15 \pm 0.01 \pm 0.26$	$2.89 \pm 0.02 \pm 0.04$
$f_0(980)\pi^+$	$16.4 \pm 0.7 \pm 2.0$	$2.67 \pm 0.05 \pm 0.20$	$1.56 \pm 0.02 \pm 0.09$
$\bar{K}_0^*(1430)^0K^+$	$2.4 \pm 0.3 \pm 1.$	$1.14 \pm 0.06 \pm 0.36$	$2.55 \pm 0.04 \pm 0.22$
$f_0(1710)\pi^+$	$1.1 \pm 0.1 \pm 0.1$	$0.65 \pm 0.02 \pm 0.06$	$1.36 \pm 0.05 \pm 0.20$
$f_0(1370)\pi^+$	$1.1 \pm 0.1 \pm 0.2$	$0.46 \pm 0.03 \pm 0.09$	$-0.45 \pm 0.11 \pm 0.52$
χ^2/NDF	1.24		

Table 2. Results from the $D_s^+ \to K^+K^-\pi^+$ Dalitz plot analysis. Quoted uncertainties are statistical and systematic, respectively. The error on each fit fraction is obtained by propagating the full covariance matrix from the fit.

$$\Gamma = \frac{G_F^2 M_{D_s}^3}{8\pi}\left(\frac{m_\ell}{M_{D_s}}\right)^2\left(1-\frac{m_\ell^2}{M_{D_s}^2}\right)^2 |V_{cs}|^2 f_{D_s}^2, \tag{1}$$

where M_{D_s} and m_ℓ are the D_s^- and lepton masses, respectively, G_F is the Fermi coupling constant, and V_{cs} is an element of the Cabibbo-Kobayashi-Maskawa (CKM) quark mixing matrix. These decays provide a clean probe of the pseudoscalar meson decay constant f_{D_s}.

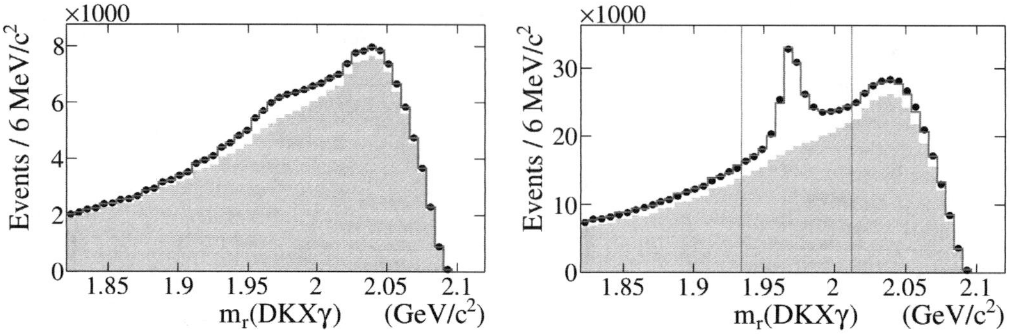

Figure 3. $m_r(DKX\gamma)$ distribution for the total Wrong-Sign (left) and Right-Sign (right) samples. The Right-Sign sample requires that the charm and strange quark content of the DKX be consistent with recoiling from a D_s^-, while the Wrong-Sign sample requires the charm and strange quark content to be consistent with recoiling from a D_s^+.

Within the SM, f_{D_s} has been predicted using several methods [9, 10]; the most precise method uses unquenched LQCD calculations. As of 2008, experimental averages [11] were significantly larger than the theoretical predictions at that time. Models of new physics (NP), including a two-Higgs doublet [12] and leptoquarks [13], may have explained this difference. In addition, f_{D_s} measurements provide a cross-check of QCD calculations which predict the impact of NP on B and B_s meson decay rates and mixing. High precision determinations of f_{D_s}, both from experiment and theory, are necessary in order to discover or constrain effects of NP.

We measure the absolute branching fractions of leptonic D_s^+ decays with a method similar to the one used by the Belle Collaboration [14, 15]. An inclusive sample of D_s^+'s is obtained by reconstructing the rest of the event in reactions of the kind $e^+e^-\to c\bar{c}\to DKXD_s^{*-}$, where $D_s^{*-}\to D_s^-\gamma$. Here, D represents a charmed hadron (D^0, D^+, D^*, or Λ_c^+), K represents the K_S^0 or K^+ required to balance strangeness in the event, and X represents additional pions produced in the $c\bar{c}$ fragmentation process. When the charmed hadron is a Λ_c^+ an additional anti-proton is required to assure baryon number conservation. No requirements are placed on the decay products of the D_s^+ so that the selected events correspond to an inclusive sample. The 4-momentum of each D_s^- candidate, p_r, is measured as the difference between the momenta of the colliding beam particles and the fully reconstructed $DKX\gamma$ system: $p_r = p_{e^+} + p_{e^-} - p_D - p_K - p_X - p_\gamma$. The inclusive D_s^- yield is obtained from a binned fit to the distribution in the recoil mass $m_r(DKX\gamma) \equiv \sqrt{p_r^2}$ shown in figure 3. Within this inclusive sample, we determine the fraction of events corresponding to $D_s^-\to\mu^-\bar{\nu}_\mu$, $D_s^-\to e^-\bar{\nu}_e$, and $D_s^-\to\tau^-\bar{\nu}_\tau$ decays (fits shown in figure 4) by extracting the yield from a binned likelihood fit. In the $D_s^-\to\mu^-\bar{\nu}_\mu$ and $D_s^-\to e^-\bar{\nu}_e$ channels, the signal yield is extracted from $m_r^2(DKX\gamma\mu(e))$. Since the $D_s^-\to\tau^-\bar{\nu}_\tau$ events contain more than one neutrino E_{extra} is used to extract the signal yield, where E_{extra} is defined as the total energy of the EMC clusters with individual energy greater than 30 MeV and not overlapping with the $DKX\gamma$ candidates. In the SM, ratios of the branching fractions for these decays are $e^-\bar{\nu}_e$:$\mu^-\bar{\nu}_\mu$:$\tau^-\bar{\nu}_\tau$=2×10^{-5}: 1 : 10, due to helicity and phase-space suppression.

Using the measured branching fractions, we determine the D_s^- decay constant using Eq. (1) and the known values for m_ℓ, m_{D_s}, $|V_{ud}|$ (we assume $|V_{cs}| = |V_{ud}|$), and the D_s^+ lifetime [16]. The f_{D_s} values are listed in table 3; the systematic uncertainty includes the uncertainties on these parameters (1.9 MeV). Finally, we obtain the error-weighted average $f_{D_s} = (258.6 \pm 6.4(stat) \pm 7.5(syst))$ MeV. The precise determination of f_{D_s} from *BABAR* updates the HFAG average of $f_{D_s} = 257.3 \pm 5.3$ MeV [11], which is 1.6σ larger than the current theoretical prediction [9].

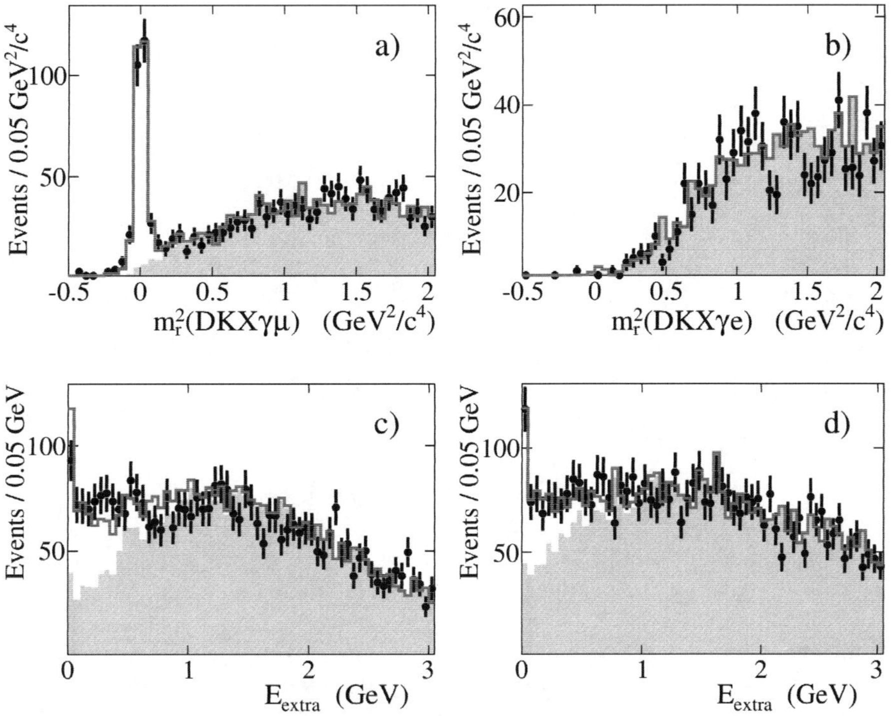

Figure 4. Fitted distributions of (a) $m_r^2(DKX\gamma\mu)$, (b) $m_r^2(DKX\gamma e)$, (c) E_{extra} for $D_s^- \to \tau_{e\nu\nu}^-\bar{\nu}_\tau$, (d) E_{extra} for $D_s^- \to \tau_{\mu\nu\nu}^-\bar{\nu}_\tau$ candidates. In each figure, the points represent the data with statistical error bars, the open histogram is from the fit described in the text, and the solid histogram is the background component from the fit.

Decay	$\bar{\varepsilon}$	Signal Yield	$\mathcal{B}(D_s^- \to \ell^-\bar{\nu}_\ell)$	f_{D_s} (MeV)
$D_s^- \to e^-\bar{\nu}_e$	70.5%	6.1 ± 2.2 ± 5.2	$< 2.3 \times 10^{-4}$ at 90% C.L.	
$D_s^- \to \mu^-\bar{\nu}_\mu$	67.7%	275 ± 17	$(6.02 \pm 0.38 \pm 0.34)\times 10^{-3}$	265.7 ± 8.4 ± 7.7
$D_s^- \to \tau^-\bar{\nu}_\tau$ $(\tau^- \to e^-\bar{\nu}_e\nu_\tau)$	61.6%	408 ± 42	$(5.07 \pm 0.52 \pm 0.68)\times 10^{-2}$	247 ± 13 ± 17
$D_s^- \to \tau^-\bar{\nu}_\tau$ $(\tau^- \to \mu^-\bar{\nu}_\mu\nu_\tau)$	59.5%	340 ± 32	$(4.91 \pm 0.47 \pm 0.54)\times 10^{-2}$	243 ± 12 ± 14

Table 3. Average efficiency ratios, signal yields, branching fractions, and decay constants for the leptonic D_s decays. The first uncertainty is statistical and the second is systematic.

4. Precision Mass and Width Measurement of the $D_{s1}(2535)$ Meson [17]

Unlike D mesons, the theoretical description of D_s^+ mesons is problematic because the masses and widths of the $D_{s0}^*(2317)^+$ and $D_{s1}(2460)^+$ states [18, 19, 20, 21, 22, 23] are not in agreement with potential model calculations based on HQET [24]. Theoretical explanations for the discrepancy invoke $D^{(*)}K$ molecules [25], chiral partners [26, 27], unitarized chiral models [28, 29], tetraquarks [30, 31], and lattice calculations [32, 33], but a satisfactory description is still lacking (see [34, 35] for more details). Improved measurements of the $D_{s1}(2536)^+$ meson parameters can lead to a better understanding of these states.

In this analysis a precise measurement of the $D_{s1}(2536)^+$ mass and decay width is performed based on a high statistics data sample. The D_{s1}^+ is reconstructed via its decay mode $D^{*+}K_S^0$, with $K_S^0 \to \pi^+\pi^-$ and $D^{*+} \to D^0\pi^+$. The D^0 is reconstructed through two decay modes, $K^-\pi^+$ and $K^-\pi^+\pi^+\pi^-$, which will be labeled $K4\pi$ and $K6\pi$, respectively, in the following. To improve

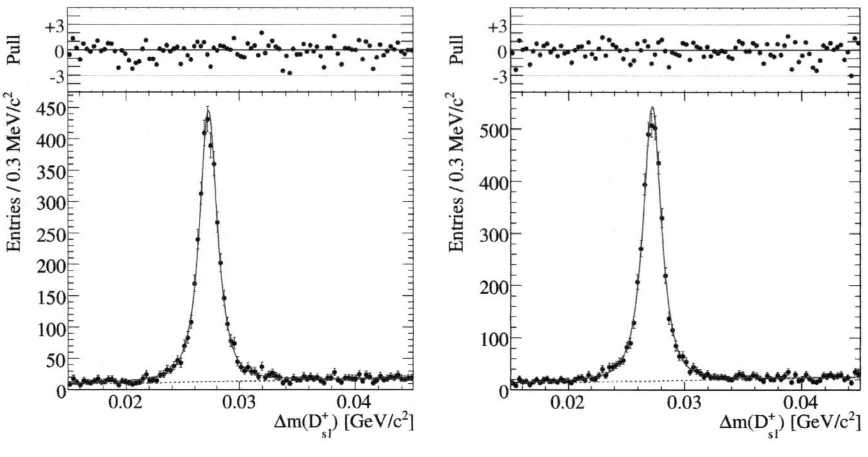

Figure 5. Fit of a RBW convolved with the resolution function to the D_{s1}^+ candidate mass difference spectra in data, for the (left) $K4\pi$ and (right) $K6\pi$ modes. The dotted line indicates the background line shape. The upper parts of the figures show the normalized fit residuals.

the mass resolution, the mass difference $\Delta m(D_{s1}^+) = m(D_{s1}^+) - m(D^{*+}) - m(K_S^0)$ is examined. The fit to $\Delta m(D_{s1}^+)$ (figure 5), is parameterized with a signal function that consists of a RBW shape numerically convolved with a $p^*(D_{s1}^+)$-dependent Gaussian resolution function, where p^* is the momentum in the CM, and a linear function is used to describe the background. The resolution model is derived from D_{s1}^+ signal Monte Carlo (MC) by studying the difference Δm_{res} between the reconstructed and generated D_{s1}^+ mass values. The detector resolution is found to have a half-width-at-half-max (HWHM) of about 0.55 MeV/c^2. We determine from the fit of $\Delta m(D_{s1}^+)$ the mean mass differences to be:

$$\begin{aligned} \Delta m(D_{s1}^+)_0 &= 27.231 \pm 0.020 \text{ MeV}/c^2 \quad (K4\pi), \\ \Delta m(D_{s1}^+)_0 &= 27.205 \pm 0.018 \text{ MeV}/c^2 \quad (K6\pi), \end{aligned}$$

and total width values

$$\begin{aligned} \Gamma(D_{s1}^+) &= 1.000 \pm 0.049 \text{ MeV} \quad (K4\pi), \\ \Gamma(D_{s1}^+) &= 0.941 \pm 0.045 \text{ MeV} \quad (K6\pi). \end{aligned}$$

The fitted values for the two D^0 decay modes agree within the statistical errors. The signal yield is 3704 ± 71 for $K4\pi$ and 4334 ± 78 for $K6\pi$. Combining these results, and adding the D^{*+} and K_S^0 masses, the final value for the $m(D_{s1}(2536)^+) = (2535.08 \pm 0.01 \pm 0.15)$ MeV/c^2 and for the width, $\Gamma(D_{s1}^+) = (0.92 \pm 0.03 \pm 0.04)$ MeV.

5. Search for New Resonances Decaying to Dπ and D$^*\pi$ in Inclusive e$^+$e$^-$ Collisions near $\sqrt{s} = 10.58$ GeV [36]

The spectrum of quark-antiquark systems was predicted using a relativistic chromodynamic potential model [37], yet experimentally little is known of the spectrum of mesons consisting of a charm quark and an up or down quark. The low mass spectrum of the $c\bar{u}$ or $c\bar{d}$ systems consists of the ground states (1S), orbital excitations with angular momentum $L = 1, 2$ (1P,1D), and the first radial excitations (2S). Besides the ground state mesons D and D^*, only two 1P states have been well-established experimentally [16], known as the $D_1(2420)$ and the $D_2^*(2460)$.

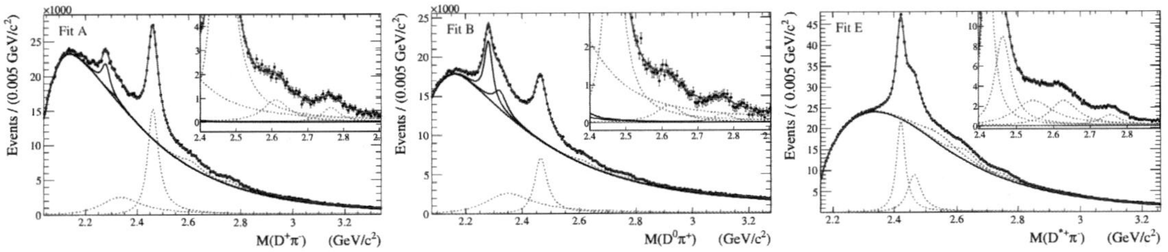

Figure 6. Mass distribution for $D^+\pi^-$ (left), $D^0\pi^+$ (middle), and $D^{*+}\pi^-$ (right) candidates. Points correspond to data, with the total fit overlaid as a solid curve. The lower solid curves are the backgrounds, and the dotted curves are the signal components. The inset plots show the background-subtracted distributions.

The measurement of these states is possible due to their narrow widths ($\sim$ 30 MeV), while other 1P states have broad widths on the order of $\sim$ 300 MeV, making them difficult to detect.

To search for additional states that have not been observed, the *inclusive* production of $D^+\pi^-$, $D^0\pi^+$, and $D^{*+}\pi^-$ final states in the reaction $e^+e^- \rightarrow c\bar{c} \rightarrow D^{(*)}\pi X$, where X is any additional system, is analyzed. In order to measure the resonance parameters, the mass difference is examined, where $M(D^+\pi^-) = m(K^-\pi^+\pi^+\pi^+) - m(K^-\pi^+\pi^+) + m_{D^+}$, $M(D^0\pi^+) = m(K^-\pi^+\pi^+) - m(K^-\pi^+) + m_{D^0}$, and $M(D^{*+}) = m(K^-\pi^+(\pi^+\pi^-)\pi_s^+\pi^-) - m(K^-\pi^+(\pi^+\pi^-)\pi_s^+) + m_{D^{*+}}$, where m_{D^+}, m_{D^0}, and $m_{D^{*+}}$ are the nominal mass values [16]. The mass spectrum is modeled with a smooth background function, BW functions convolved with resolution and bias functions for peaking backgrounds, and RBW functions to describe the signal resonances. The signal RBW shapes are corrected for the mass resolution by convolving them with the detector resolution. The detector resolution using the mass variables is about 3 MeV/c^2. The masses and widths of resonances determined from fits to the various final states are found in table 5, and the fits to the data are shown in figure 6.

Due to the vector nature of the D^{*+}, the $D^{*+}\pi^-$ final states contain resonances with different spin-parity (J^P) which have different helicity-angle distributions. The helicity angle θ_H is defined as the angle between the primary pion and the slow pion from the D^{*+} decay. To extract the signal resonances in the $D^{*+}\pi^-$ distribution, the fit is performed for $|\cos\theta_H| > 0.75$ and for $|\cos\theta_H| < 0.5$ separately. Using the signal parameters determined from these two fits, the final parameters for the $D^{*+}\pi^-$ are determined by fitting the entire sample, fixing the parameters from the previous fit (indicated in table 5).

Further information as to the spin-parity nature of some of these states is accesible by studying the helicity distributions (figure 7) of the $D^{*+}\pi^-$ final state. The $D^{*+}\pi^-$ data is divided into 10 sub-samples corresponding to $\cos\theta_H$ intervals of 0.2 from -1 to +1. Each sample is fit with the shape parameters fixed to the values determined in the final fit of the entire sample. The yields are extracted from these fits and the resulting distributions are fit to various Y functions, indicated in the plots.

We observe for the first time four signal peaks, which we denote $D(2550)^0$, $D^*(2600)^0$, $D(2750)^0$, and $D^*(2760)^0$. We also observe the isospin partners $D^*(2600)^+$ and $D^*(2760)^+$. The $D(2550)^0$ and $D^*(2600)^0$ have mass values and helicity-angle distributions that are consistent with the predicted radial excitations $D_0^1(2S)$ and $D_1^3(2S)$. The $D^*(2760)^0$ signal observed in $D^+\pi^-$ is very close in mass to the $D(2750)^0$ signal observed in $D^{*+}\pi^-$; however, their mass and width values differ by 2.6σ and 1.5σ, respectively. Four $L = 2$ states are predicted to lie in this region [37], but only two are expected to decay to $D^+\pi^-$. This may explain the observed features.

Resonance	Channel	Yield (x10^3)	Mass MeV/c^2	Width MeV
$D_1(2420)^0$	$D^{*+}\pi^-$	102.8±1.3±2.3	2420.1±0.1±0.8	31.4±0.5±1.3
	$D^{*+}\pi^-$	214.6±1.2±6.4	2420.1(fixed)	31.4(fixed)
$D_2^*(2460)^0$	$D^+\pi^-$	242.8±1.8±3.4	2462.2±0.1±0.8	50.5±0.6± 0.7
	$D^{*+}\pi^-$	136±2±13	2462.2(fixed)	50.5(fixed)
$D(2550)^0$	$D^{*+}\pi^-$	34.3±6.7±9.2	2539.4±4.5±6.8	130±12±13
	$D^{*+}\pi^-$	98.4±8.2±38	2539.4(fixed)	130(fixed)
$D^*(2600)^0$	$D^+\pi^-$	26.0±1.4± 6.6	2608.7±2.4±2.5	93±6±13
	$D^{*+}\pi^-$	50.2±3.0±6.7	2608.7(fixed)	93(fixed)
	$D^{*+}\pi^-$	71.4±1.7±7.3	2608.7(fixed)	93(fixed)
$D(2750)^0$	$D^{*+}\pi^-$	23.5±2.1±5.2	2752.4±1.7±2.7	71±6±11
$D^*(2760)^0$	$D^+\pi^-$	11.3±0.8±1.0	2763.3±2.3±2.3	60.9±5.1±3.6
$D_2^*(2460)^+$	$D^0\pi^+$	110.8±1.3±7.5	2465.4±0.2±1.1	50.5(fixed)
$D^*(2600)^+$	$D^0\pi^+$	13.0±1.3±4.5	2621.3±3.7±4.2	93(fixed)
$D^*(2760)^+$	$D^0\pi^+$	5.7±0.7±1.5	2769.7±3.8±1.5	60.9(fixed)

Table 4. Summary of the results. The first error is statistical and the second is systematic; "fixed" indicates the parameters were fixed to the values from other fits

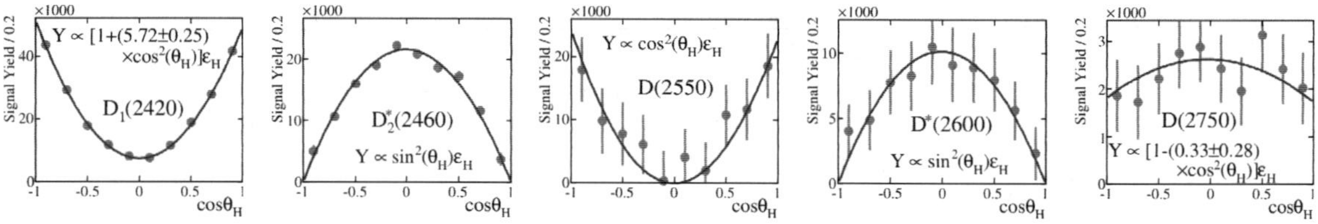

Figure 7. Distribution in $\cos\theta_H$ for each signal in $D^{*+}\pi^-$. The error bars include statistical and correlated systematic uncertainties. The curve is a fit using the function Y shown in the plot; ε_H is the efficiency as a function of $\cos\theta_H$.

6. Search for Rare and Forbidden Charm Decays

Flavor changing neutral current (FCNC) processes have been studied extensively for K and B mesons, in $K^0 - \bar{K}^0$ and $B^0 - \bar{B}^0$ mixing, and in rare FCNC decays such as $s \to d\ell^+\ell^-$, $b \to s\gamma$ and $b \to s\ell^+\ell^-$ decays. The results agree with expectations within the framework of the SM. There are ongoing efforts to improve the measurements and the theoretical predictions, and to measure new effects, such as CP violation, in FCNC processes.

The recent observation of $D^0 - \overline{D}^0$ mixing [44, 45] has increased interest in FCNC processes in the charm sector. Of particular interest is the source of $D^0 - \overline{D}^0$ mixing. If the mixing is due to physics beyond the SM, it could also give rise to measurable effects in FCNC charm decays. In the SM, the FCNC decays $X_c^+ \to h^+\ell^+\ell^-$ are expected to be heavily suppressed due to cancellations of amplitudes through the Glashow-Iliopoulos-Maiani (GIM) mechanism. For example, the $c \to u\ell^+\ell^-$ transitions illustrated in figure 8 yield branching fractions for $D \to X_u\ell^+\ell^-$ of $\mathcal{O}(10^{-8})$ [46, 47]. These decays are masked by the presence of long-distance contributions from intermediate vector resonances such as $D \to X_uV$, $V \to \ell\ell$, which are predicted to have branching fractions of $\mathcal{O}(10^{-6})$ [46, 47]. The effect of these resonances can be separated from short-distance contributions by studying the invariant mass of the $\ell^+\ell^-$ pair. In radiative charm decays, $c \to u\gamma$, uncertainties in calculating the long-distance terms make it impossible to study the underlying short-distance physics [48]. The decay $D^0 \to \gamma\gamma$ is dominated by long-distance contributions and the theoretical braching fraction is estimated to be $\mathcal{O}(10^{-11}$ to $10^{-8})$ [46]. However, in the context of the Minimal Supersymmetric Standard Model, gluino exchange can enhance the SM rate by up to a factor of 200 [52].

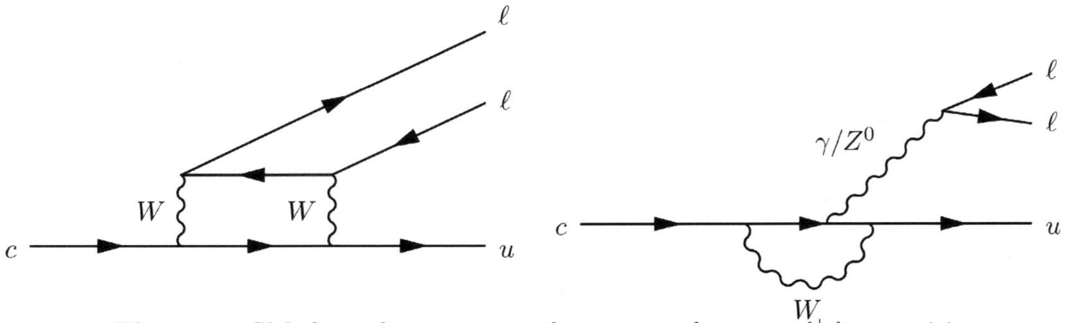

Figure 8. SM short-distance contributions to the $c \to u\ell^+\ell^-$ transition.

6.1. Search for Flavor Changing Neutral Current, Lepton-Flavor Violating, and Lepton-Number Violating Semileptonic Charm Decays [38]

BABAR searched for semileptonic charm hadron decays that are either forbidden or heavily suppressed in the SM of particle physics. The decays are of the form $X_c^+ \to h^\pm \ell^\mp \ell^{(\prime)+}$, where X_c^+ is a charm hadron (D^+, D_s^+, or Λ_c^+), h is a pion, kaon, or proton, and $\ell^{(\prime)}$ is an electron or muon. The decay modes with oppositely charged leptons but same lepton flavor are examples of FCNC processes, which are expected to be very rare because they cannot occur at tree level in the SM. The decay modes with two oppositely charged leptons of different flavor correspond to lepton-flavor violating (LFV) decays and are essentially forbidden in the SM because they can occur only through lepton mixing. The decay modes with two leptons of the same charge are lepton-number violating (LNV) decays and are forbidden in the SM. Hence, decays of the form $X_c^+ \to h^\pm \ell^\mp \ell^{(\prime)+}$ provide a sensitive tool to investigate physics beyond the SM. The most stringent existing upper limits [39, 40, 41, 42, 43] on the branching fractions for $X_c^+ \to h^\pm \ell^\mp \ell^{(\prime)+}$ decays range from 1 to 700×10^{-6} and do not exist for most of the Λ_c^+ decays.

Charm hadron candidates are formed from one track identified as either a pion, kaon, or proton (h) and two tracks, each of which is identified as an electron or a muon ($\ell\ell^{(\prime)}$). The final candidate selection is performed using a likelihood ratio test, combining probability distribution functions (PDFs) from the following three variables: the charm hadron candidate p^*, the total energy in the event, and the flight significance which is the ratio of the signed flight length to its uncertainty. The signed flight length is the scalar product between the direction of the charm-hadron candidate and the position vector that points from the beam spot to the charm-hadron decay vertex. The total energy in an event is calculated as the sum of the energies of all reconstructed tracks (assuming each track to be a charged pion) and neutral energy EMC clusters.

An extended, unbinned, ML fit is applied to the invariant-mass distributions for the $h^\pm \ell^\mp \ell^{(\prime)+}$ candidates. The PDF we use for signal events is the so-called Crystal Ball function [49], which has an asymmetric component to describe the radiative tail in the mass distribution. The shape parameters of the signal PDF are determined from fits to signal MC candidates and are fixed to these values in the fits to data, with only the overall normalization as a free parameter. The width of the Gaussian component (σ) is found to lie between 5 and 10 MeV/c^2, depending on the decay mode. The invariant mass distributions of the combinatoric background events for the signal modes are described by first-order polynomials. The background slope is a free parameter in all fits.

No signals are observed, yet for 32 of the 35 decay modes, this analysis is more sensitive than existing measurements. In most cases, the improvement is significant (factor of 2 to 60). The biggest improvements are seen for the lepton-flavor violating decays, which are all improved by at least a factor of 10. The only Λ_c^+ decay with a pre-existing limit is $\Lambda_c^+ \to p\mu^+\mu^-$, for which we improve the limit by roughly a factor of 8. For all other Λ_c^+ decays this study presents the first limits. The only modes that do not provide a more sensitive limit are $D^+ \to \pi^- e^+ e^+$, $D^+ \to \pi^+\mu^+\mu^-$, and $D_s^+ \to \pi^+\mu^+\mu^-$ where existing limits [39, 42, 43] are about a factor of two lower than the *BABAR* measurements (see table 5).

6.2. Search for the Decay $D^0 \to \gamma\gamma$ [50]

The only previous search of a neutral D meson into two photons was conducted by the CLEO collaboration in 2002 using 13.8 fb^{-1} [51]. The large number of charm decays contained in the *BABAR* dataset provide the opportunity to probe the upper limit of NP enhancements in this decay.

D^0 candidates are formed by combining pairs of photon candidates. A photon candidate is defined as energy deposited in the EMC, which is not associated with the trajectory of any charged track and exhibits the appropiate shower characteristics. The photon candidates are constrained to have a CM momentum between 0.74 and 4 GeV/c.

The signal yield is obtained using an unbinned ML fit to the D^0 mass distribution. The overall PDFs are sums of functions describing signal and background distributions obtained from MC studies. The relative normalization of these functions are free parameters while the shapes are fixed. The $D^0 \to \gamma\gamma$ signal PDF is the sum of a Crystal Ball function and a bifurcated Gaussian distribution. The background PDF is described by a second-order Chebychev polynomial.

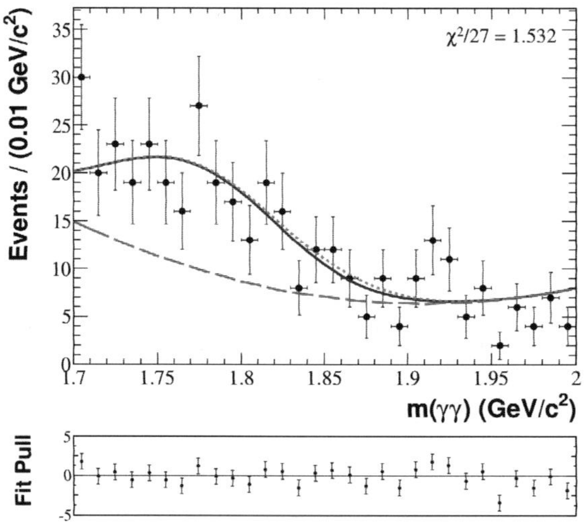

Figure 9. The full $D^0 \to \gamma\gamma$ PDF fit to the data showing the combinatoric background (long-dashed red), the combinatoric background plus $D^0 \to \pi^0\pi^0$ background shape (small-dash red), and both backgrounds and signal component (solid blue) which results in a negative signal. The fit is determined by unbinned maximum likelihood but the χ^2 value is determined from binned data and is provided as a goodness-of-fit measure. The pull distribution shows the differences between the data and the solid blue curve above with values and errors normalized to the Poisson error.

The invariant mass distribution of $D^0 \to \gamma\gamma$ is shown in figure 9 together with projections of the likelihood fit and the individual signal and background combinations. The signal yield is -6 ± 15 events. Including systematic uncertainties the upper limit for the branching ratio of the decay $D^0 \to \gamma\gamma$ is determined to be $B(D^0 \to \gamma\gamma) < 2.4 \times 10^{-6}$.

Table 5. Signal yields for the fits to the 35 $X_c^+ \to h^{\pm}\ell^{\mp}\ell^{(\prime)+}$ event samples. The first error is the statistical uncertainty and the second is the systematic uncertainty. The third column lists the estimated signal efficiency. The fourth column gives for each signal mode the 90% CL upper limit (UL) on the ratio of the branching fraction of the signal mode to that of the normalization mode (BR). The last column shows the 90% CL upper limit on the branching fraction for each signal mode (BF). The upper limits include all systematic uncertainties.

Decay Mode	Yield (events)	Eff.(%)	BR UL 90% CL (10^{-4})	BF UL 90% CL (10^{-6})
$D^+ \to \pi^+e^+e^-$	$-3.9 \pm 1.6 \pm 1.7$	1.56	3.9	1.1
$D^+ \to \pi^+\mu^+\mu^-$	$-0.2 \pm 2.8 \pm 0.9$	0.46	24	6.5
$D^+ \to \pi^+e^+\mu^-$	$-2.9 \pm 3.4 \pm 2.4$	1.21	11	2.9
$D^+ \to \pi^+\mu^+e^-$	$3.6 \pm 4.3 \pm 1.3$	1.54	13	3.6
$D_s^+ \to \pi^+e^+e^-$	$8 \pm 34 \pm 8$	6.36	5.4	13
$D_s^+ \to \pi^+\mu^+\mu^-$	$20 \pm 15 \pm 4$	1.21	18	43
$D_s^+ \to \pi^+e^+\mu^-$	$-3 \pm 11 \pm 3$	2.16	4.9	12
$D_s^+ \to \pi^+\mu^+e^-$	$9.3 \pm 7.3 \pm 2.8$	1.50	8.4	20
$D^+ \to K^+e^+e^-$	$-3.7 \pm 2.9 \pm 3.3$	2.88	3.7	1.0
$D^+ \to K^+\mu^+\mu^-$	$-1.3 \pm 2.8 \pm 1.1$	0.65	16	4.3
$D^+ \to K^+e^+\mu^-$	$-4.3 \pm 1.8 \pm 0.6$	1.44	4.3	1.2
$D^+ \to K^+\mu^+e^-$	$3.2 \pm 3.8 \pm 1.2$	1.74	9.9	2.8
$D_s^+ \to K^+e^+e^-$	$-5.7 \pm 5.8 \pm 2.0$	3.20	1.6	3.7
$D_s^+ \to K^+\mu^+\mu^-$	$4.8 \pm 5.9 \pm 1.2$	0.85	9.1	21
$D_s^+ \to K^+e^+\mu^-$	$9.1 \pm 6.0 \pm 2.8$	1.74	5.7	14
$D_s^+ \to K^+\mu^+e^-$	$3.4 \pm 6.4 \pm 3.5$	2.08	4.2	9.7
$\Lambda_c^+ \to pe^+e^-$	$4.0 \pm 6.5 \pm 2.8$	5.52	0.8	5.5
$\Lambda_c^+ \to p\mu^+\mu^-$	$11.1 \pm 5.0 \pm 2.5$	0.86	6.4	44
$\Lambda_c^+ \to pe^+\mu^-$	$-0.7 \pm 2.9 \pm 0.9$	1.10	1.6	9.9
$\Lambda_c^+ \to p\mu^+e^-$	$6.2 \pm 4.6 \pm 1.8$	1.37	2.9	19
$D^+ \to \pi^-e^+e^+$	$4.7 \pm 4.7 \pm 0.5$	3.16	6.8	1.9
$D^+ \to \pi^-\mu^+\mu^+$	$-3.1 \pm 1.2 \pm 0.5$	0.70	7.5	2.0
$D^+ \to \pi^-\mu^+e^+$	$-5.1 \pm 4.2 \pm 2.0$	1.72	7.4	2.0
$D_s^+ \to \pi^-e^+e^+$	$-5.7 \pm 14 \pm 3.4$	6.84	1.8	4.1
$D_s^+ \to \pi^-\mu^+\mu^+$	$0.6 \pm 5.1 \pm 2.7$	1.05	6.2	14
$D_s^+ \to \pi^-\mu^+e^+$	$-0.2 \pm 7.9 \pm 0.6$	2.23	3.6	8.4
$D^+ \to K^-e^+e^+$	$-2.8 \pm 2.4 \pm 0.2$	2.67	3.1	0.9
$D^+ \to K^-\mu^+\mu^+$	$7.2 \pm 5.4 \pm 1.6$	0.80	37	10
$D^+ \to K^-\mu^+e^+$	$-11.6 \pm 4.0 \pm 3.1$	1.52	6.8	1.9
$D_s^+ \to K^-e^+e^+$	$2.3 \pm 7.9 \pm 3.3$	4.10	2.1	5.2
$D_s^+ \to K^-\mu^+\mu^+$	$-2.3 \pm 5.0 \pm 2.8$	0.98	5.3	13
$D_s^+ \to K^-\mu^+e^+$	$-14.0 \pm 8.4 \pm 2.0$	2.26	2.4	6.1
$\Lambda_c^+ \to \bar{p}e^+e^+$	$-1.5 \pm 4.2 \pm 1.5$	5.14	0.4	2.7
$\Lambda_c^+ \to \bar{p}\mu^+\mu^+$	$0.0 \pm 2.1 \pm 0.6$	0.94	1.4	9.4
$\Lambda_c^+ \to \bar{p}\mu^+e^+$	$10.1 \pm 5.8 \pm 3.5$	2.50	2.3	16

7. Searches for CP Violation

In the SM, *CP* violation (CPV) arises from the complex phase of the CKM quark-mixing matrix [54]. Measurements of the CPV asymmetries in the K and B meson systems are consistent with expectations based on the SM and, together with theoretical inputs, lead to the determination of the parameters of the CKM matrix. CPV has not yet been observed in the charm sector, where the theoretical predictions based on the SM for CPV asymmetries are at the level of 10^{-3} or below [55].

7.1. Search for CP Violation in the decay $D^+ \to K^0_S \pi^+$ [53]

BABAR searched for CPV in the decay $D^\pm \to K^0_S \pi^\pm$ by measuring the CPV parameter A_{CP} defined as:

$$A_{CP} = \frac{\Gamma(D^+ \to K^0_S \pi^+) - \Gamma(D^- \to K^0_S \pi^-)}{\Gamma(D^+ \to K^0_S \pi^+) + \Gamma(D^- \to K^0_S \pi^-)}, \qquad (2)$$

where Γ is the partial decay width for this decay. This decay mode has been chosen because of its clean experimental signature. Although direct *CP* violation due to interference between Cabibbo-allowed and doubly Cabibbo-suppressed amplitudes is predicted to be negligible within the SM [56], $K^0 - \overline{K}^0$ mixing induces a time-integrated *CP* violating asymmetry of $(-0.332 \pm 0.006)\,\%$ [60]. Contributions from non-SM processes may reduce the value of the measured A_{CP} or enhance it up to the level of one percent [56, 57]. Therefore, a significant deviation of the A_{CP} measurement from pure $K^0 - \overline{K}^0$ mixing effects would be evidence for the presence of new physics beyond the SM. Due to the smallness of the expected value, this measurement requires a large data sample and precise control of the systematic uncertainties. Previous measurements of A_{CP} have been reported by the CLEO-c ($(-0.6 \pm 1.0(\mathrm{stat}) \pm 0.3(\mathrm{syst}))\%$ [58]) and Belle collaborations ($(-0.71 \pm 0.19(\mathrm{stat}) \pm 0.20(\mathrm{syst}))\%$ [59]).

We select $D^\pm \to K^0_S \pi^\pm$ decays by combining a K^0_S candidate reconstructed in the decay mode $K^0_S \to \pi^+\pi^-$ with a charged pion candidate. A K^0_S candidate is reconstructed from two oppositely charged tracks with an invariant mass within ± 10 MeV/c^2 of the nominal K^0_S mass [60]. To obtain the final candidate events, a Boosted Decision Tree (BDT) algorithm [61] is constructed from seven discriminating variables for each $D^\pm$ candidate: the measured proper decay time $\tau(D^\pm)$, the decay distance in the transverse plane $L_{xy}(D^\pm)$, the CM momentum magnitude $p^*(D^\pm)$, the momentum magnitudes and transverse components with respect to the beam axis for both the K^0_S and pion candidates.

A binned ML fit to the $m(K^0_S\pi^\pm)$ distribution for the retained $D^\pm$ candidates is used to extract the signal yield. The total PDF is the sum of signal and background components. The signal PDF is modeled as a sum of three Gaussian functions, the first two of them with common mean. The background PDF is taken as a sum of two components: a background from $D^\pm_s \to K^0_S K^\pm$, where the $K^\pm$ is misidentified as $\pi^\pm$, and a combinatorial background from other sources. The data and the fit are shown in figure 10. All of the fit parameters are extracted from the fit to the data sample apart from the normalization of the $D^\pm_s \to K^0_S K^\pm$ background, which is fixed to the value predicted by the MC simulation. We determine A_{CP} by measuring the signal yield asymmetry A defined as:

$$A = \frac{N_{D^+} - N_{D^-}}{N_{D^+} + N_{D^-}}, \qquad (3)$$

where $N_{D^+}(N_{D^-})$ is the number of fitted $D^+ \to K^0_S\pi^+ (D^- \to K^0_S\pi^-)$ decays. The quantity A is the result of two other contributions in addition to A_{CP}. There is a physics component due to the forward-backward (FB) asymmetry (A_{FB}) in $e^+e^- \to c\bar{c}$, arising from γ^*-Z^0 interference and high order QED processes in $e^+e^- \to c\bar{c}$. This asymmetry will create a difference in the number of reconstructed D^+ and D^- decays due to the FB detection asymmetries arising

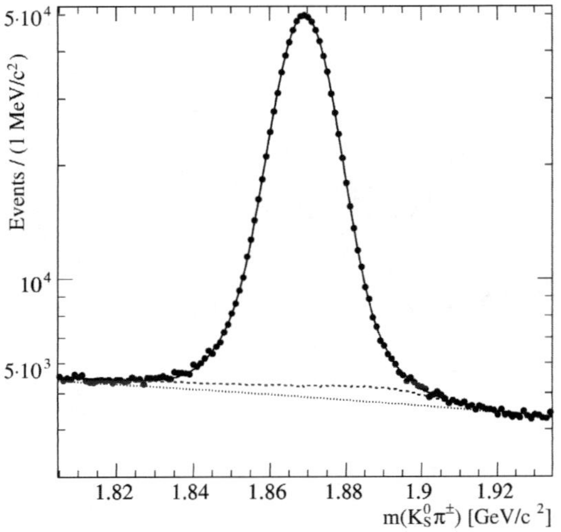

Figure 10. Invariant mass distribution for $K_S^0\pi^\pm$ candidates in the data (black points). The solid curve shows the fit to the data. The dashed line is the sum of all backgrounds, while the dotted line is combinatorial background only. The vertical scale of the plot is logarithmic.

from the boost of the CM system relative to the laboratory frame. There is also a detector-induced component due to the difference in the reconstruction efficiencies of $D^+ \to K_s^0\pi^+$ and $D^- \to K_s^0\pi^-$ generated by differences in the track reconstruction and identification efficiencies for π^+ and π^-. While A_{FB} is measured together with A_{CP} using the selected dataset, we correct the dataset itself for the reconstruction and identification effects using control data sets. *BABAR* developed a data-driven method to determine the charge asymmetry in track reconstruction as a function of the magnitude of the track momentum and its polar angle which is shown along with the associated errors in figure 11.

Neglecting the second-order terms that contain the product of A_{CP} and A_{FB}, the resulting asymmetry can be expressed simply as the sum of the two. The parameter A_{CP} is independent of kinematic variables, while A_{FB} is an odd function of $\cos\theta_D^*$, where θ_D^* is the polar angle of the $D^\pm$ candidate momentum in the e^+e^- CM frame. If we compute $A(+|\cos\theta_D^*|)$ for the $D^\pm$ candidates in a positive $\cos\theta_D^*$ bin and $A(-|\cos\theta_D^*|)$ for the candidates in its negative counterpart, the contribution to the two asymmetries from A_{CP} is the same, while the contribution from A_{FB} has the same magnitude but opposite sign. Therefore A_{CP} and A_{FB} can be written as a function of $|\cos\theta_D^*|$ as follows:

$$A_{FB}(|\cos\theta_D^*|) = \frac{A(+|\cos\theta_D^*|) - A(-|\cos\theta_D^*|)}{2} \tag{4}$$

and

$$A_{CP}(|\cos\theta_D^*|) = \frac{A(+|\cos\theta_D^*|) + A(-|\cos\theta_D^*|)}{2}. \tag{5}$$

The selected sample is divided into ten subsamples corresponding to ten $\cos\theta_D^*$ bins of equal width and a simultaneous binned ML fit is performed on the invariant mass distributions of D^+ and D^- candidates for each subsample to extract the signal yield asymmetries. Using the asymmetry measurements in five positive and in five negative $\cos\theta_D^*$ bins, we obtain five A_{FB} and five A_{CP} values. As A_{CP} does not depend upon $\cos\theta_D^*$, we compute a central value of this parameter using a χ^2 minimization to a constant. The A_{CP} and A_{FB} values are shown in

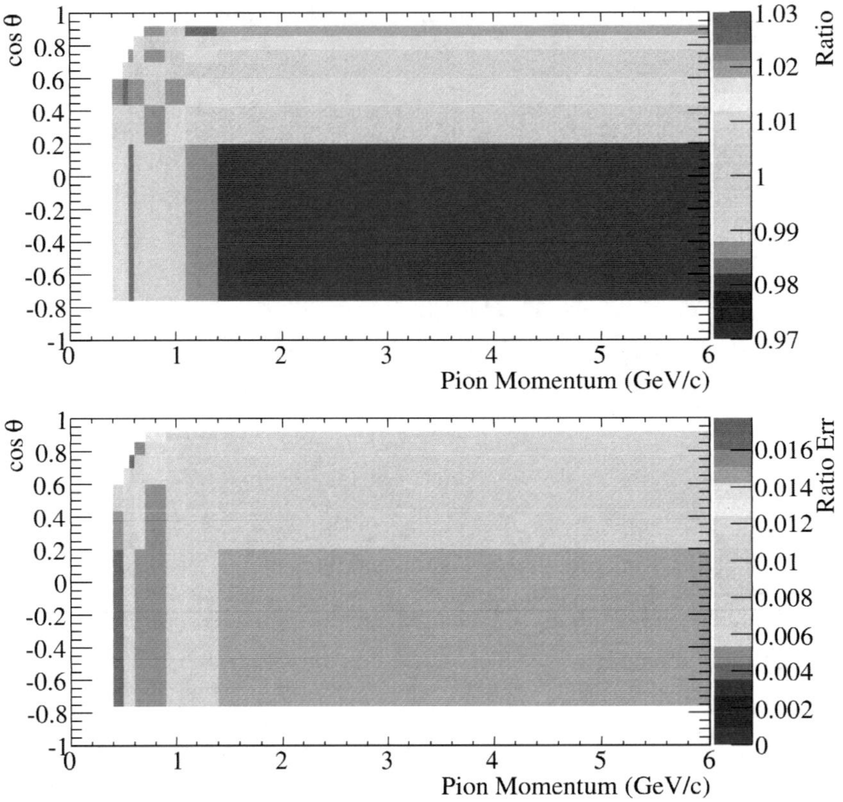

Figure 11. Map of the ratio between detection efficiency for π^+ and π^- (top) and the corresponding statistical errors (bottom). The map is produced using the numbers of π^- and π^+ tracks in a control sample in data.

figure 12, together with the central value and $\pm 1\,\sigma$ confidence interval for A_{CP}. We determine A_{CP} to be:

$$A_{CP} = (-0.39 \pm 0.13 \pm 0.10)\% \tag{6}$$

where the first error is statistical and the second systematic.

7.2. Search for T-Violation using T-Odd Correlation in $D^+_{(s)} \to K^0_S K^+ \pi^+ \pi^-$

A search for *CP* violation in the decays $D^+ \to K^+ K^0_S \pi^+ \pi^-$ and $D^+_s \to K^+ K^0_S \pi^+ \pi^-$ using *T*-odd correlations is described here. We define a kinematic triple product that is odd under time reversal using the vector momenta of the final state particles in the $D^+_{(s)}$ rest frame as

$$C_T \equiv \vec{p}_{K^+} \cdot (\vec{p}_{\pi^+} \times \vec{p}_{\pi^-})\,. \tag{7}$$

Under the assumption of *CPT* invariance, *T* violation is equivalent to *CP* violation.

We study the *T*-odd correlations by measuring the observable expressed in Eq. (7) and then evaluating the asymmetry

$$A_T \equiv \frac{\Gamma(C_T > 0) - \Gamma(C_T < 0)}{\Gamma(C_T > 0) + \Gamma(C_T < 0)}, \tag{8}$$

where Γ is the decay rate for the process under study. The observable defined in Eq. (8) can have a non-zero value due to final state interactions even if the weak phases are zero [62]. The

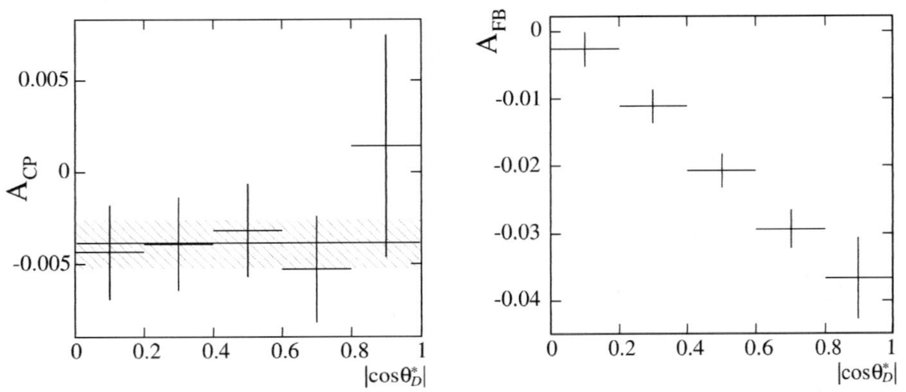

Figure 12. A_{CP} (top) and A_{FB} (bottom) asymmetries for $D^{\pm} \to K^0_S\pi^{\pm}$ candidates as a function of $|\cos\theta^*_D|$ in the data sample. The solid line represents the central value of A_{CP} and the hatched region is the $\pm 1\,\sigma$ interval, both obtained from a χ^2 minimization assuming no dependence on $|\cos\theta^*_D|$.

T-odd asymmetry measured in the CP-conjugate decay process, $\bar{A}_T$, is defined as:

$$\bar{A}_T \equiv \frac{\Gamma(-\bar{C}_T > 0) - \Gamma(-\bar{C}_T < 0)}{\Gamma(-\bar{C}_T > 0) + \Gamma(-\bar{C}_T < 0)}, \tag{9}$$

where $\bar{C}_T \equiv \vec{p}_{K^-} \cdot (\vec{p}_{\pi^-} \times \vec{p}_{\pi^+})$. We can then construct:

$$\mathcal{A}_T \equiv \frac{1}{2}\left(A_T - \bar{A}_T\right), \tag{10}$$

which is an asymmetry that characterizes T-violation in the weak decay process [63, 64, 65].

At least four different particles are required in the final state so that the triple product may be defined using momentum vectors only [66]. The D meson decays suitable for this analysis method are $D^+ \to K^+K^0_S\pi^+\pi^-$, $D^+_s \to K^+K^0_S\pi^+\pi^-$ and $D^0 \to K^+K^-\pi^+\pi^-$. The search for CP violation using T-odd correlations in $D^0 \to K^+K^-\pi^+\pi^-$ has recently been carried out by the *BABAR* Collaboration, and no evidence of CP violation has been observed [67].

The D^+ and D^+_s meson decay candidates are reconstructed in the production and decay sequence:

$$e^+e^- \to XD^+_{(s)};\, D^+_{(s)} \to K^+K^0_S\pi^+\pi^-;\, K^0_S \to \pi^+\pi^-, \tag{11}$$

using the events with at least five charged particles. To obtain the final set of signal candidates, the p^* $(D^+_{(s)})$, the difference in vertex probabilities that the parent meson originates from a common vertex and the primary vertex, and the signed transverse decay length are combined in a likelihood-ratio test. Figure 13 shows the resulting $K^+K^0_S\pi^+\pi^-$ mass spectra in the D^+ and D^+_s regions. For each region, the signal is described by the superposition of two Gaussian functions with a common mean value. The background is parametrized by a first-order polynomial in the D^+ region, and by a second-order polynomial in the D^+_s region. We extract the integrated yields $N(D^+) = 21210 \pm 392$ and $N(D^+_s) = 29791 \pm 337$ from the fits, where the uncertainties are statistical only.

We next divide the data sample into four sub-samples depending on $D_{(s)}$ charge and whether C_T $(\bar{C}_T)$ is greater or less than zero, and fit the corresponding mass spectra simultaneously to extract the yields and the values of the asymmetry parameters A_T and $\bar{A}_T$. The T-violating parameter $\mathcal{A}_T$ is then computed using Eq. (10). We obtain the T violation parameter values:

$$\mathcal{A}_T(D^+) = (-12.0 \pm 10.0_{(\mathrm{stat})} \pm 4.6_{(\mathrm{syst})}) \times 10^{-3} \tag{12}$$

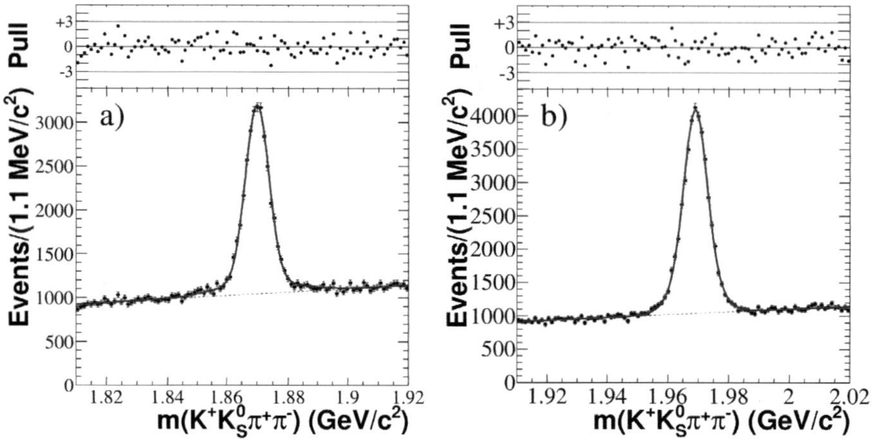

Figure 13. The $K^+K^0_S\pi^+\pi^-$ mass spectrum a) in the D^+, and b) in the D_s^+ mass region. The curves result from the fits described in the text. The distributions of the Pull values are also shown.

and

$$\mathcal{A}_T(D_s^+) = (-13.6 \pm 7.7_{(\mathrm{stat})} \pm 3.4_{(\mathrm{syst})}) \times 10^{-3}. \tag{13}$$

For comparison, the value obtained for D^0 decay was [67]

$$\mathcal{A}_T(D^0) = (+1.0 \pm 5.1_{(\mathrm{stat})} \pm 4.4_{(\mathrm{syst})}) \times 10^{-3}. \tag{14}$$

8. Acknowledgments

The *BABAR* collaboration would like to thank the conference organizers of the 2nd International Conference on Particle Physics in Istanbul, Turkey for the opportunity to present recent results of charm physics.

9. References

[1] B. Aubert *et al.* (*BABAR* Collaboration), Nucl. Instrum. Methods in Phys. Res. Sect. A **479**, 1 (2002).
[2] P. del Amo Sanchex *et al.* (*BABAR* Collaboration), Phys. Rev. D 83 052001 (2011).
[3] E. M. Aitala *et al.* (E791 Collaboration), Phys. Rev. Lett. **89**, 121801 (2002). M. Ablikim *et al.* (BES Collaboration), Phys. Lett. B **633**, 681 (2006).
[4] E. M. Aitala *et al.* (E791 Collaboration), Phys. Rev. Lett. **86**, 765 (2001). M. Ablikim *et al.* (BES Collaboration), Phys. Lett. B **598**, 149 (2004).
[5] See for example F. E. Close and N. A. Tornqvist
[6] S. Stone and L. Zhang, Phys. Rev. D **79**, 074024 (2009).
[7] Y. Xie, P. Clarke, G. Cowan and F. Muheim, JHEP **0909**, 074 (2009).
[8] P. del Amo Sanchex *et al.* (*BABAR* Collaboration), Phys. Rev. D 82 091003(R) (2010).
[9] E. Follana *et al.*, Phys. Rev. D **82**, 114504 (2010);
[10] E. Follana *et al.*, Phys. Rev. Lett. **100**, 062002 (2008);
[11] Heavy Flavor Averaging Group, `www.slac.stanford.edu/xorg/hfag/charm/index.html` (2010).
[12] A. G. Akeroyd and C. H. Chen, Phys. Rev. D **75**, 075004 (2007).
[13] B. A. Dobrescu and A. S. Kronfeld, Phys. Rev. Lett. **100**, 241802 (2008).
[14] L. Widhalm *et al.* (Belle Collaboration), Phys. Rev. Lett. **100**, 241801 (2008).
[15] L. Widhalm *et al.* (Belle Collaboration), Phys. Rev. Lett. **97**, 061804 (2006).
[16] C. Amsler *et al.* (Particle Data Group), Phys. Lett. B **667**, 1 (2008).).
[17] P. del Amo Sanchex *et al.* (*BABAR* Collaboration), Phys. Rev. D 83 072003 (2011).
[18] B. Aubert *et al.* (*BABAR* Collaboration), Phys. Rev. Lett. **90**, 242001 (2003).

[19] B. Aubert *et al.* (*BABAR* Collaboration), Phys. Rev. Lett. **93**, 181801 (2004).
[20] V. Mikani *et al.* (Belle Collaboration), Phys. Rev. Lett. **92**, 012002 (2004).
[21] B. Aubert *et al.* (*BABAR* Collaboration), Phys. Rev. D **74**, 032007 (2006).
[22] D. Benson *et al.* (CLEO Collaboration), Phys. Rev. D **68**, 032002 (2003).
[23] P. Krokovny *et al.* (Belle Collaboration), Phys. Rev. Lett. **91**, 262002 (2003).
[24] R.N. Cahn and J.D. Jackson, Phys. Rev. D **68**, 037502 (2003).
[25] T. Barnes, F.E. Close, and H.J. Lipkin, Phys. Rev. D **68**, 054006 (2003).
[26] W.A. Bardeen, E.J. Eichten, and C.T. Hill, Phys. Rev. D **68**, 054024 (2003).
[27] M.A. Nowak, M. Rho, and I. Zahed, Acta Phys. Polon. **B35**, 2377 (2004).
[28] E. van Beveren and G. Rupp, Phys. Rev. Lett. **91**, 012003 (2003).
[29] E.E. Kolomeitsev and M.F.M. Lutz, Phys. Lett. B **582**, 39 (2004).
[30] L. Maiani, F. Piccinini, A.D. Polosa, and V. Riquer, Phys. Rev. D **71**, 014028 (2005).
[31] K. Terasaki, Phys. Rev. D **68**, 011501 (2003).
[32] A. Dougall, R.D. Kenway, C.M. Maynard and C. McNeile (UKQCD Collaboration), Phys. Lett. B **569**, 41 (2003).
[33] G.S. Bali, Phys. Rev. D **68**, 071501 (2003).
[34] P. Colangelo, F. de Fazio, and R. Ferrandes, Mod. Phys. Lett. **A19**, 2083 (2004).
[35] E.S. Swanson, Phys. Rep. **429**, 243 (2006).,J. Phys. G **28**, R249 (2002).
[36] P. del Amo Sanchex *et al.* (*BABAR* Collaboration), Phys. Rev. D 82 111101 (2010).
[37] S. Godfrey and N. Isgur, Phys. Rev. D **32**, 189 (1985).
[38] J. P. Lees *et al.* (*BABAR* Collaboration) [arXiv:1107.4465 [hep-ex]].
[39] P. Rubin *et al.* (CLEO Collaboration), Phys. Rev. D **82**, 092007 (2010).
[40] E. M. Aitala *et al.* (E791 Collaboration), Phys. Lett. B **462**, 401 (1999).
[41] K. Kodama *et al.* (E653 Collaboration), Phys. Lett. B **345**, 85 (1995).
[42] D. Buchholz *el al.* (D0 Collaboration), Contribution to FPCP2006, hep-ex/0605090.
[43] J. M. Link *et al.* (FOCUS Collaboration), Phys. Lett. B **573**, 21 (2003).
[44] B. Aubert *et al.* (*BABAR* Collaboration), Phys. Rev. Lett. **98**, 211802 (2007).
[45] M. Staric *et al.* (Belle Collaboration), Phys. Rev. Lett. **98**, 211803 (2007).
[46] G. Burdman, E. Golowich, J. Hewett and S. Pakvasa, Phys. Rev. D **66**, 014009 (2002).
[47] S. Fajfer, S. Prelovsek and P. Singer, Phys. Rev. D **64**, 114009 (2001).
[48] G. Burdman, E. Golowich, J. Hewett and S. Pakvasa, Phys. Rev. D **52**, 6383 (1995).
[49] T. Skwarnicki (Crystal Ball Collaboration), Ph.D. thesis, Cracow Institute of Nuclear Physics (1986), DESY-F31-86-02.
[50] J. P. Lees *et al.* (*BABAR* Collaboration) [arXiv:1110.6480 [hep-ex]]
[51] T. E. Coan *et al.* (CLEO Collaboration), Phys. Rev. Lett. **90**, 101801 (2003).
[52] S. Prelovsek and D. Wyler, Phys. Lett. **B500**, 304 (2001).
[53] P. del Amo Sanchex *et al.* (*BABAR* Collaboration), Phys. Rev. D 83 071103 (2011).
[54] N. Cabibbo, Phys. Rev. Lett. **10**, 531 (1963); M. Kobayashi and T. Maskawa, Prog. Theor. Phys. **49**, 652 (1973).
[55] F. Buccella *et al.*, Phys. Rev. D **51**, 3478 (1995).
[56] H. J. Lipkin and Z. Xing, Phys. Lett. B **450**, 405 (1999).
[57] I. I. Bigi and H. Yamamoto, Phys. Lett. B **349**, 363 (1995).
[58] S. Dobbs *et al.* (CLEO Collaboration), Phys. Rev. D **76**, 112001 (2007) [arXiv:0709.3783 [hep-ex]].
[59] B. R. Ko *et al.* (Belle Collaboration), Phys. Rev. Lett. **104**, 181602 (2010).
[60] K. Nakamura *et al.* (Particle Data Group), J. Phys. G **37**, 075021 (2010).
[61] P. Speckmayer, A. Hocker, J. Stelzer and H. Voss, J. Phys. Conf. Ser. **219**, 032057 (2010).
[62] I. I. Bigi and H.-B. Li, Int. J. Mod. Phys. A **24S1**, 657 (2009).
[63] W. Bensalem, A. Datta, D. London, Phys. Rev. D **66**, 094004 (2002).
[64] W. Bensalem, A. Datta, D. London, Phys. Lett. B **538**, 309 (2002).
[65] W. Bensalem and D. London, Phys. Rev. D **64**, 116003 (2001).
[66] E. Golowich and G. Valencia, Phys. Rev. D **40**, 112 (1989).
[67] P. del Amo Sanchez *et al.* (*BABAR* Collaboration), Phys. Rev. D **81**, 111103 (2010).

Journal of Physics: Conference Series **347** (2012) 012027 doi:10.1088/1742-6596/347/1/012027

Properties of bound states containing fourth family quarks

V Bashiry[1], K Azizi[2], S Sultansoy[3,4]

[1] Engineering Faculty, Cyprus International University, Via Mersin 10, Turkey
[2] Department of Physics, Faculty of Arts and Sciences, Doğuş University, Acıbadem-Kadıköy, 34722 Istanbul, Turkey
[3] Physics Division, TOBB University of Economics and Technology, Ankara, Turkey
[4] Institute of Physics, National Academy of Sciences, Baku, Azerbaijan

E-mail: bashiry@ciu.edu.tr, kazizi@dogus.edu.tr, ssultansoy@etu.edu.tr

Abstract. The heavy fourth generation of quarks that have sufficiently small mixing with the three known SM families form hadrons. In the present work, we calculate the masses and decay constants of mesons containing either both quarks from the fourth generation or one from fourth family and the other from observed SM quarks, namely charm or bottom quark, in the framework of the QCD sum rules. In the calculations, the two gluon condensate diagrams as nonperturbative contributions are taken into account. The obtained numerical results are reduced to the known masses and decay constants of the $\bar{b}b$ and $\bar{c}c$ quarkonia, when the fourth family quark is replaced by the bottom or charm quark.

1. Introduction

In the standard model (SM), we have three generation of quarks experimentally observed. Among these quarks, the top (t) quark does not form bound states (hadrons) as a consequence of the high value of its mass. The top quark immediately decays to the bottom quark giving a W boson and this transition has full strength. The number of quark and lepton generations is one of the mysteries of nature and can not be addressed by the SM. There are flavor democracy arguments that predict the existence of the fourth generation of quarks [1, 2, 3]. It is expected that the masses of the fourth generation quarks be in the interval $(300 - 700)$ GeV [4]. The last value coincides with upper limit following from partial-wave unitarity at high energies [5]. Within the flavor democracy approach, the Dirac masses of the fourth family fermions are almost equal, whereas masses of the first three family fermions, as well as CKM and PMNS mixings are obtained via small violations of democracy [6, 7]. For the recent status of the SM with four generations (SM_4), see e.g. [8, 9, 10] and references therein.

Although the masses of fourth generation quarks are larger than the top quark mass (the last analysis of the Tevatron data implies $m_{d_4} > 372$ GeV [11] and $m_{u_4} > 358$ GeV [12]), they can form bound states as a result of the smallness of mixing between these quarks and ordinary SM quarks [13, 14, 15, 16, 17, 18, 19]. As the mass difference between these two quarks is small, we will refer to both members of the fourth family by u_4. The condition for formation of new hadrons containing ultra-heavy quarks (Q) is given by [20]:

$$|V_{Qq}| \leq \left(\frac{100\ GeV}{m_Q}\right)^{3/2}. \tag{1}$$

For t-quark with $m_t = 172\ GeV$, Eq. (1) leads to $V_{tq} < 0.44$, whereas single top production at the Tevatron gives $V_{tb} > 0.74$ [21]. When the fourth family quarks have sufficiently small mixing with the ordinary quarks, the hadrons made up from these quarks can be long enough lived and the bound state $\bar{u}_4 u_4$ decays through its annihilation and not via u_4 decays to a lower family quark plus a W boson [19]. Concerning flavor democracy approach, this situation is realized for parameterizations proposed in [7] and [22], whereas parameterization [6] predicts $V_{u_4 q} \sim 0.2$ which does not allow formation of the fourth family quarkonia for $m_{u_4} > 300GeV$.

Considering the above discussions, the production of such bound states if exist will be possible at LHC. The conditions for observation of the fourth SM family quarks at the LHC has been discussed in [13, 23, 24, 25, 26, 27, 28, 29, 30]. As there is a possibility to observe the bound states containing fourth family quarks at the LHC, it is reasonable to investigate their properties, theoretically and phenomenologically.

In the present work, we calculate the masses and decay constants of the bound state mesons containing two heavy quarks with either both quarks from the SM_4 or one from the heavy fourth family and the other from ordinary heavy b or c quark. Here, we consider ground state mesons with different quantum numbers, namely scalar ($\bar{u}_4 u_4$, $\bar{u}_4 b$ and $\bar{u}_4 c$), pseudoscalar ($\bar{u}_4 \gamma_5 u_4$, $\bar{u}_4 \gamma_5 b$ and $\bar{u}_4 \gamma_5 c$), vector ($\bar{u}_4 \gamma_\mu u_4$, $\bar{u}_4 \gamma_\mu b$ and $\bar{u}_4 \gamma_\mu c$) and axial vector ($\bar{u}_4 \gamma_\mu \gamma_5 u_4$, $\bar{u}_4 \gamma_\mu \gamma_5 b$ and $\bar{u}_4 \gamma_\mu \gamma_5 c$) mesons. These mesons, similar to the ordinary hadrons, are formed in a region of energy very far from the asymptotic region. Hence, perturbation theory can not be used in this region since the coupling constant between quarks and gluons is large. Therefore, to calculate the hadronic parameters such as the mass and leptonic decay constant, we need to consult to some nonperturbative approaches. Among the nonperturbative methods, the QCD sum rules [31], which is based on QCD Lagrangian and is free from the model dependent parameter, is one of the most applicable and predictive approaches to hadron physics. This method has been successfully used to calculate the masses and decay constants of mesons both in vacuum and at finite temperature (see for instance [32, 33, 34, 35, 36, 37, 38, 39, 40, 41]). Now, we extend the application of this method to calculate the masses and decay constants of the considered mesons containing fourth family quarkonia. For details see the original work [42].

2. QCD sum rules for the mass and decay constant

We start this section considering the sufficient correlation functions responsible for calculation of the masses and decay constants of the bound states containing heavy fourth generation quarks in the framework of the QCD sum rules. The two point correlation function corresponding to the scalar (S) and pseudoscalar (PS) cases can be written as:

$$\Pi^{S(PS)} = i \int d^4x e^{ip.x} \langle 0 \mid \mathcal{T}\left(J^{S(PS)}(x)\bar{J}^{S(PS)}(0)\right) \mid 0\rangle, \tag{2}$$

where $\mathcal{T}$ is the time ordering product and $J^S(x) = \overline{u}_4(x)q(x)$ and $J^{PS}(x) = \overline{u}_4(x)\gamma_5 q(x)$ are the interpolating currents of the heavy scalar and pseudoscalar bound states, respectively. Here, we consider the q to be either fourth family u_4 quark or ordinary heavy b or c quark. Similarly for the vector (V) and axial vector (AV), the correlation function can be written as:

$$\Pi^{V(AV)}_{\mu\nu} = i \int d^4x e^{ip.x} \langle 0 \mid \mathcal{T}\left(J^{V(AV)}_\mu(x)\bar{J}^{V(AV)}_\nu(0)\right) \mid 0\rangle, \tag{3}$$

where, the currents $J_\mu^V = \overline{u}_4(x)\gamma_\mu q(x)$ and $J_\mu^{AV} = \overline{u}_4(x)\gamma_\mu\gamma_5 q(x)$ are responsible for creating the vector and axial vector quarkonia, respectively from the vacuum with the same quantum numbers as the interpolating currents.

From the general philosophy of the QCD sum rules, we calculate the aforesaid correlation functions in two alternative ways. From the physical or phenomenological side, we calculate them in terms of hadronic parameters such as masses and decay constants. In QCD or theoretical side, they are calculated in terms of QCD degrees of freedom such as quark masses and gluon condensates by the help of operator product expansion (OPE) in deep Euclidean region. Equating these two representations of the correlation functions through dispersion relations, we acquire the QCD sum rules for the masses and decay constants. These sum rules relate the hadronic parameters to the fundamental QCD parameters. To suppress the contribution of the higher states and continuum, the Borel transformation with respect to the momentum squared is applied to both sides of the correlation functions.

First, to calculate the phenomenological part, we insert a complete set of intermediate states having the same quantum numbers as the interpolating currents. Performing the integral over x and isolating the ground state, we obtain

$$\Pi^{S(PS)} = \frac{\langle 0 \mid J^{S(PS)}(0) \mid S(PS)\rangle\langle S(PS) \mid J^{S(PS)}(0) \mid 0\rangle}{m^2_{S(PS)} - p^2} + \cdots, \tag{4}$$

where $\cdots$ represents the contributions of the higher states and continuum and $m_{S(PS)}$ is mass of the heavy scalar (pseudoscalar) meson. Similarly, for the vector (axial vector) case, we obtain

$$\Pi^{V(AV)}_{\mu\nu} = \frac{\langle 0 \mid J^{V(AV)}_\mu(0) \mid V(AV)\rangle\langle V(AV) \mid J^{V(AV)}_\nu(0) \mid 0\rangle}{m^2_{V(AV)} - p^2} + \cdots, \tag{5}$$

To proceed, we need to know the matrix elements of the interpolating currents between the vacuum and mesonic states. These matrix elements are parametrized in terms of the leptonic decay constants as:

$$\begin{aligned}
\langle 0 \mid J(0) \mid S\rangle &= f_S m_S, \\
\langle 0 \mid J(0) \mid PS\rangle &= f_{PS}\frac{m^2_{PS}}{m_{u4}+m_q}, \\
\langle 0 \mid J(0) \mid V(AV)\rangle &= f_{V(AV)} m_{V(AV)} \varepsilon_\mu,
\end{aligned} \tag{6}$$

where f_i are the leptonic decay constants of the considered bound state mesons. Using the summation over the polarization vectors in the $V(AV)$ case via

$$\epsilon_\mu \epsilon^*_\nu = -g_{\mu\nu} + \frac{p_\mu p_\nu}{m^2_{V(AV)}}, \tag{7}$$

we get, the final expressions of the physical sides of the correlation functions as:

$$\begin{aligned}
\Pi^S &= \frac{f_S^2 m_S^2}{m_S^2 - p^2} + \cdots \\
\Pi^{PS} &= \frac{f^2_{PS}(\frac{m^2_{PS}}{m_{u4}+m_q})^2}{m^2_{PS} - p^2} + \cdots \\
\Pi^{V(AV)}_{\mu\nu} &= \frac{f^2_{V(AV)} m^2_{V(AV)}}{m^2_{V(AV)} - p^2}\left[-g_{\mu\nu} + \frac{p_\mu p_\nu}{m^2_{V(AV)}}\right] + \cdots,
\end{aligned} \tag{8}$$

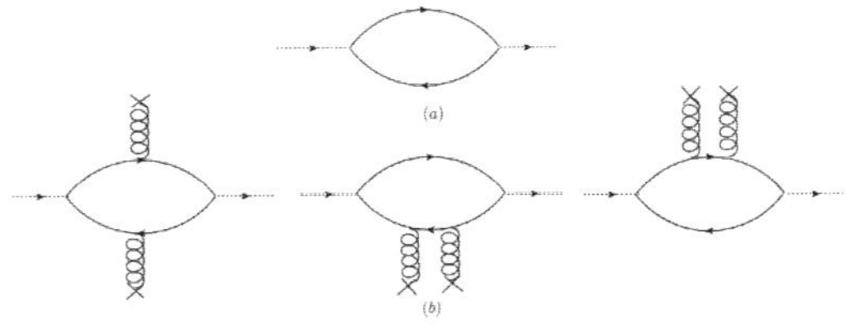

Figure 1. (a): Bare loop diagram (b): Diagrams corresponding to gluon condensates.

where to calculate the mass and decay constant in the $V(AV)$ case, we choose the structure $g_{\mu\nu}$.

In QCD side, the correlation functions are calculated in deep Euclidean region, $p^2 \ll -\Lambda^2_{QCD}$ via OPE where the short or perturbative and long distance or non-perturbative effects are separated. For each correlation function in $S(PS)$ case and coefficient of the selected structure in $V(AV)$ case, we write

$$\Pi^{QCD} = \Pi_{pert} + \Pi_{nonpert}. \tag{9}$$

The short distance contribution (bare loop diagram in figure (1) part (a)) in each case is calculated using the perturbation theory, whereas the long distance contributions (diagrams shown in figure (1) part (b)) are parameterized in terms of gluon condensates. To proceed, we write the perturbative part in terms of a dispersion integral,

$$\Pi^{QCD} = \int \frac{ds\rho(s)}{s-p^2} + \Pi_{nonpert}, \tag{10}$$

where, $\rho(s)$ is called the spectral density. To calculate the spectral density, we calculate the Feynman amplitude of the bare loop diagram by the help of the Cutkosky rules, where the quark propagators are replaced by Dirac delta function, i.e., $\frac{1}{p^2-m^2} \rightarrow (-2\pi i)\delta(p^2-m^2)$. As a result, the spectral density is obtained as follows:

$$\rho(s) = \frac{3s}{8\pi^2}\left(1 - \frac{(m_1 \pm m_2)^2}{s}\right)\sqrt{1 - 2\frac{m_1^2+m_2^2}{s} + \frac{(m_1^2-m_2^2)^2}{s^2}} \tag{11}$$

where $+$ sign in $(m_1 \pm m_2)$ is chosen for scalar and axial vector cases and $-$ sign is for pseudoscalar and vector channels. Here, $m_1 = m_{u_4}$ and m_2 is either m_{u_4} or $m_{c(b)}$.

To obtain the non-perturbative part, we calculate the gluon condensate diagrams represented in part (b) of figure (1). For this aim, we use Fock-Schwinger gauge, $x^\mu A^a_\mu(x) = 0$. In momentum space, the vacuum gluon field is expressed as:

$$A^a_\mu(k') = -\frac{i}{2}(2\pi)^4 G^a_{\rho\mu}(0)\frac{\partial}{\partial k'_\rho}\delta^{(4)}(k'), \tag{12}$$

where k' is the gluon momentum and in calculations, we use the quark-gluon-quark vertex as:

$$\Gamma^a_{ij\mu} = ig\gamma_\mu\left(\frac{\lambda^a}{2}\right)_{ij}, \tag{13}$$

After straightforward but lengthy calculations, the non-perturbative part for each case in momentum space is obtained as:

$$\Pi^i_{nonpert} = \int_0^1 \langle \alpha_s G^2 \rangle \frac{\Theta^i + \Theta^i(m_1 \leftrightarrow m_2)}{96\pi(m_2^2 + m_1^2 x - m_2^2 x - p^2 x + p^2 x^2)^4} dx \qquad (14)$$

where $\Theta^i(m_1 \leftrightarrow m_2)$ means that in Θ^i, we exchange m_1 and m_2. The explicit expressions for Θ^i are given in [42].

The next step is to match the phenomenological and QCD sides of the correlation functions to get sum rules for the masses and decay constants of the bound states. To suppress contribution of the higher states and continuum, Borel transformation over p^2 as well as continuum subtraction are performed. As a result of this procedure, we obtain the following sum rules:

$$\begin{aligned} m^2_{S(V)(AV)} f^2_{S(V)(AV)} e^{\frac{-m^2_{S(V)(AV)}}{M^2}} &= \int_{(m_1+m_2)^2}^{s_0} ds\ \rho^{S(V)(AV)}(s)\ e^{-\frac{s}{M^2}} + \hat{B}\Pi^{S(V)(AV)}_{nonpert}, \\ \frac{m^4_{PS} f^2_{PS}}{(m_{u_4} + m_q)^2} e^{\frac{-m^2_{PS}}{M^2}} &= \int_{(m_1+m_2)^2}^{s_0} ds\ \rho^{PS}(s)\ e^{-\frac{s}{M^2}} + \hat{B}\Pi^{PS}_{nonpert}, \end{aligned} \qquad (15)$$

where M^2 is the Borel mass parameter and s_0 is the continuum threshold. The sum rules for the masses are obtained applying derivative with respect to $-\frac{1}{M^2}$ to both sides of the above sum rules and dividing by themselves, i.e.,

$$m^2_{S(PS)(V)(AV)} = \frac{-\frac{d}{d(\frac{1}{M^2})}\left[\int_{(m_1+m_2)^2}^{s_0} ds\ \rho^{S(PS)(V)(AV)}(s)\ e^{-\frac{s}{M^2}} + \hat{B}\Pi^{S(PS)(V)(AV)}_{nonpert}\right]}{\int_{(m_1+m_2)^2}^{s_0} ds\ \rho^{S(PS)(V)(AV)}(s)\ e^{-\frac{s}{M^2}} + \hat{B}\Pi^{S(PS)(V)(AV)}_{nonpert}}, \qquad (16)$$

where

$$\hat{B}\Pi^i_{nonpert} = \int_0^1 e^{\frac{m_2^2 + x(m_1^2 - m_2^2)}{M^2 x(x-1)}} \frac{\Delta^i + \Delta^i(m_1 \leftrightarrow m_2)}{\pi 96 M^6 (x-1)^4 x^3} \langle \alpha_s G^2 \rangle dx, \qquad (17)$$

and explicit expressions for Δ^i are given in [42].

3. Numerical Results

To obtain numerical values for the decay constants and masses of the considered bound states containing heavy fourth family from the obtained QCD sum rules, we take the mass of the u_4 in the interval $m_{u_4} = (450 - 550)\ GeV$, $m_b = 4.8\ GeV$, $m_c = 1.3\ GeV$ and $\langle 0 \mid \frac{1}{\pi}\alpha_s G^2 \mid 0 \rangle = 0.012\ GeV^4$. The sum rules for the masses and decay constants also contain two auxiliary parameters, namely Borel mass parameter M^2 and continuum threshold s_0. The standard criteria in QCD sum rules is that the physical quantities should be independent of the auxiliary parameters. Therefore, we should look for working regions of these parameters such that our results be approximately insensitive to their variations. The working regions for the Borel mass parameter and the continuum threshold are found in [42].

As an example, let us consider the case of the bound state $\bar{u}_4 u_4$. The dependence of the masses of scalar $\bar{u}_4 u_4$, pseudoscalar $\bar{u}_4 \gamma_5 u_4$, vector $\bar{u}_4 \gamma_\mu u_4$ and axial vector $\bar{u}_4 \gamma_5 \gamma_\mu u_4$ are presented in figures (2-5) at three different fixed values from the considered working region for the continuum threshold. From these figures, we see a good stability of the masses with respect to the Borel mass parameter M^2. From these figures, it is also clear that the results do not depend on the continuum threshold in its working region. The dependence of the decay constants of the scalar $\bar{u}_4 u_4$, pseudoscalar $\bar{u}_4 \gamma_5 u_4$, vector $\bar{u}_4 \gamma_\mu u_4$ and axial vector $\bar{u}_4 \gamma_5 \gamma_\mu u_4$ are presented in figures (6-9) also at three different fixed values of the continuum threshold. These figures also

Journal of Physics: Conference Series **347** (2012) 012027 doi:10.1088/1742-6596/347/1/012027

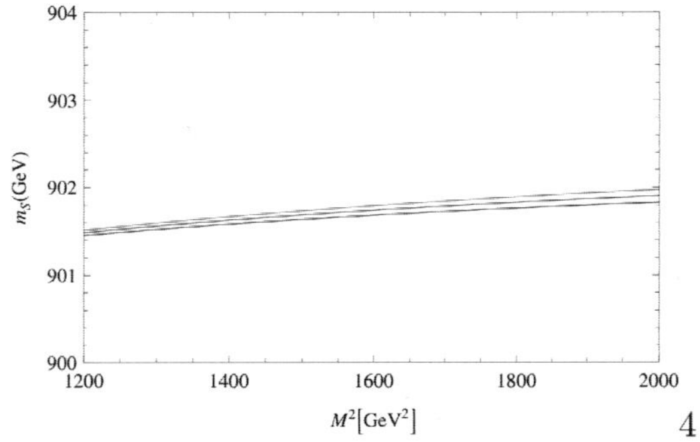

4

Figure 2. Dependence of mass of the scalar $\bar{u}_4 u_4$ on the Borel parameter, M^2 at three fixed values of the continuum threshold. The upper, middle and lower lines belong to the values $s_0 = (m_1 + m_2 + 3.7)^2\ GeV^2$, $s_0 = (m_1 + m_2 + 3.5)^2\ GeV^2$ and $s_0 = (m_1 + m_2 + 3.3)^2\ GeV^2$, respectively.

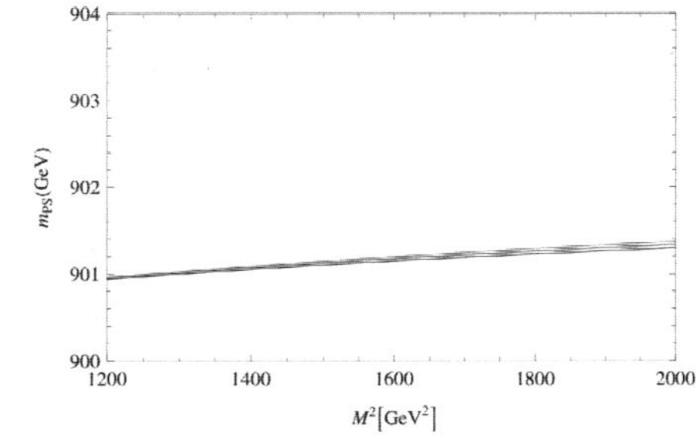

Figure 3. The same as Fig. 2 but for pseudoscalar $\bar{u}_4\gamma_5 u_4$.

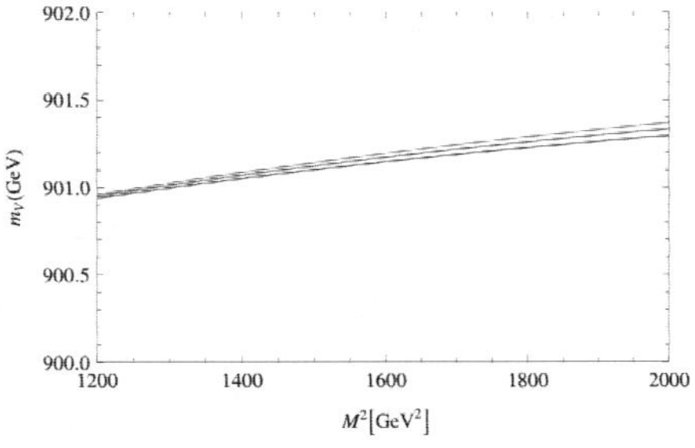

Figure 4. The same as Fig. 2 but for vector $\bar{u}_4\gamma_\mu u_4$.

depict approximately insensitivity of the results under variation of the Borel mass parameter in its working region. The results of decay constants also show very weak dependency on the continuum threshold in its working region. From the similar way we analyze the mass and decay constants of the cases when one of the quarks belong to the heavy fourth generation and the other is ordinary bottom or charm quark. The numerical results deduced from the figures are collected in Tables I-VI for three different values of the m_{u4}, namely $m_{u_4} = 450\ GeV$, $m_{u_4} = 500\ GeV$ and $m_{u_4} = 550\ GeV$. The errors presented in these tables are only due to the uncertainties coming from the determination of the working regions for the auxiliary parameters. Here, we should

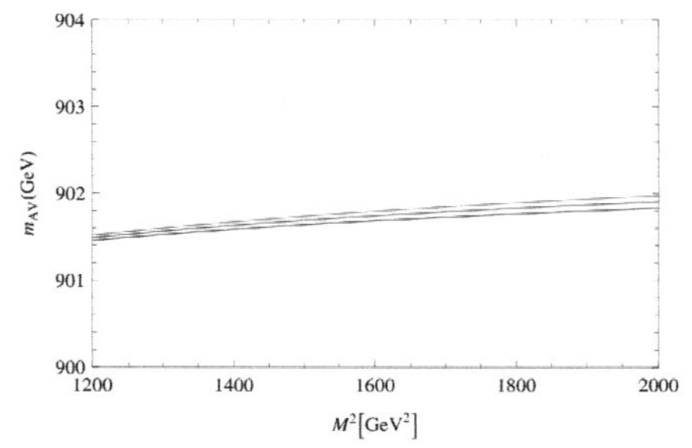

Figure 5. The same as Fig. 2 but for axial vector $\bar{u}_4\gamma_5\gamma_\mu u_4$.

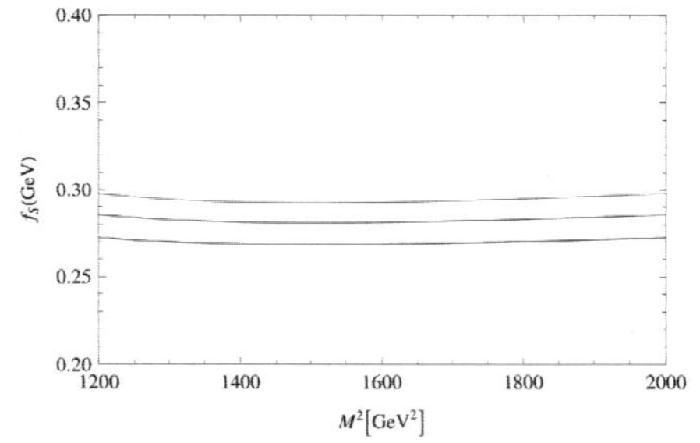

Figure 6. Dependence of the decay constant of the scalar $\bar{u}_4 u_4$ on the Borel parameter, M^2 at three fixed values of the continuum threshold. The upper, middle and lower lines belong to the values $s_0 = (m_1+m_2+3.7)^2\ GeV^2$, $s_0 = (m_1+m_2+3.5)^2\ GeV^2$ and $s_0 = (m_1+m_2+3.3)^2\ GeV^2$, respectively.

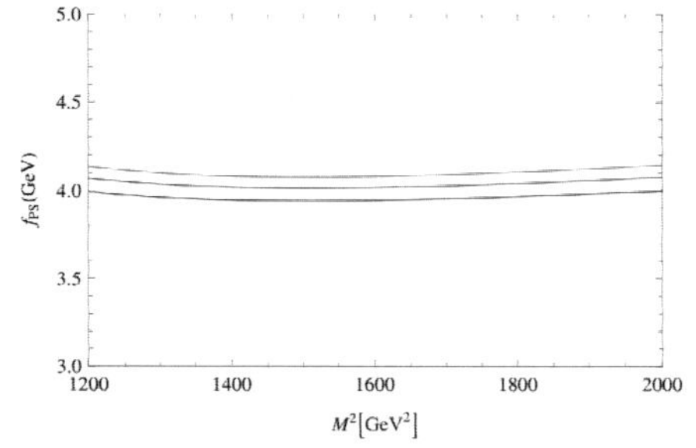

Figure 7. The same as Fig. 6 but for the decay constant of pseudoscalar $\bar{u}_4\gamma_5 u_4$.

stress that the obtained results in Tables I-VI are within QCD and do not include contributions coming from the Higgs couplings to the ultra heavy quarks. Such contributions to the binding energy have been calculated in [19], where it is shown that these contributions are more than several GeV in the case when both quarks belong to the fourth family. The Higgs contribution calculated in [19] is proportional to the product of two quark masses. When we replace one of the ultra heavy quarks by b or c quark, the binding energy obtained in [19] reduces to a value which is less than the QCD sum rules predictions in the present work. However, when both quarks belong to the fourth family, the binding energy obtained in the present work is very small comparing to the Higgs corrections in [19].

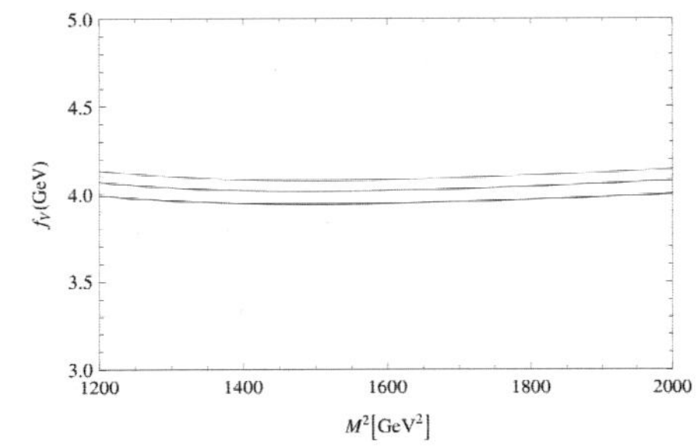

Figure 8. The same as Fig. 6 but for the decay constant of vector $\bar{u}_4\gamma_\mu u_4$.

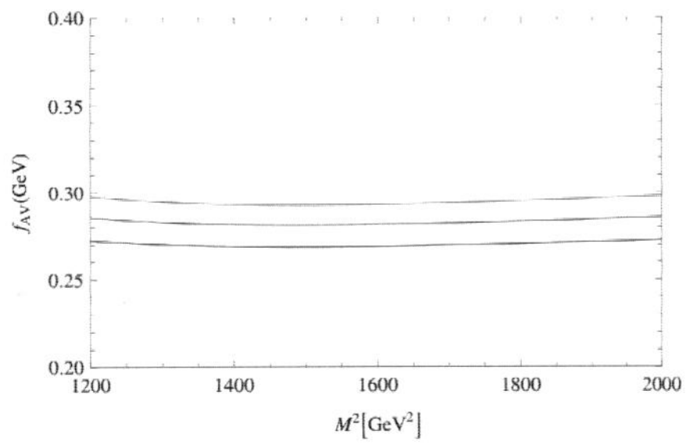

Figure 9. The same as Fig. 6 but for the decay constant of axial vector $\bar{u}_4\gamma_5\gamma_\mu u_4$.

Table 1. The values of masses of different bound states obtained using $m_{u_4} = 450\ GeV$.

mass (GeV)	$u_4\bar{c}$	u_4b	$u_4\bar{u}_4$
Scalar	453.01 ± 0.25	456.45 ± 0.25	901.68 ± 0.50
Pseudoscalar	452.62 ± 0.15	455.95 ± 0.15	901.12 ± 0.30
axial vector	453.00 ± 0.25	456.44 ± 0.25	901.70 ± 0.50
vector	452.62 ± 0.15	455.94 ± 0.15	901.13 ± 0.30

At the end of this part, we would like to mention that the obtained QCD sum rules in the present work reproduce the masses and decay constants of the ordinary $\bar{b}b(\bar{c}c)$ states when we set $u_4 \to b(c)$. The obtained numerical values in this limit are in a good consistency with the existing experimental data [43] and QCD sum rules predictions [40, 41].

To sum up, unlike the top quark, the heavy fourth generation of quarks that have sufficiently small mixing with the three known family SM quarks form hadrons. Considering the arguments mentioned in the text, the production of such bound states will be possible at LHC. Hoping this possibility, we calculated the masses and decay constants of the bound state objects containing two quarks with either both quarks from the SM_4 or one from heavy fourth generation and the other from observed SM bottom or charm quarks in the framework of the QCD sum rules. The obtained numerical results approach to the known masses and decay constants of the $\bar{b}b$ and $\bar{c}c$ heavy quarkonia, when the fourth family quark is replaced by the bottom or charm quark.

Table 2. The values of masses of different bound states obtained using $m_{u_4} = 500\ GeV$.

mass (GeV)	$u_4\bar{c}$	u_4b	$u_4\bar{u}_4$
Scalar	502.91 ± 0.28	506.36 ± 0.28	1001.61 ± 0.55
Pseudoscalar	502.52 ± 0.17	505.86 ± 0.17	1001.04 ± 0.33
Axial Vector	502.91 ± 0.28	506.35 ± 0.28	1001.60 ± 0.55
Vector	502.57 ± 0.17	505.85 ± 0.17	1001.04 ± 0.33

Table 3. The values of masses of different bound states obtained using $m_{u_4} = 550\ GeV$.

mass (GeV)	$u_4\bar{c}$	u_4b	$u_4\bar{u}_4$
Scalar	552.82 ± 0.31	556.27 ± 0.31	1101.67 ± 0.60
Pseudoscalar	552.43 ± 0.18	555.78 ± 0.18	1101.11 ± 0.36
Axial Vector	552.81 ± 0.31	556.25 ± 0.31	1101.68 ± 0.60
Vector	552.42 ± 0.18	555.77 ± 0.18	1101.12 ± 0.36

Table 4. The values of decay constants of different bound states obtained using $m_{u_4} = 450\ GeV$.

Leptonic decay constant f (GeV)	$u_4\bar{c}$	u_4b	$u_4\bar{u}_4$
Scalar	0.12 ± 0.01	0.15 ± 0.02	0.28 ± 0.03
Pseudoscalar	0.17 ± 0.01	0.34 ± 0.02	4.01 ± 0.20
Axial Vector	0.12 ± 0.01	0.15 ± 0.02	0.28 ± 0.03
Vector	0.17 ± 0.01	0.34 ± 0.02	4.01 ± 0.20

Table 5. The values of decay constants of different bound states obtained using $m_{u_4} = 500\ GeV$.

Leptonic decay constant f (GeV)	$u_4\bar{c}$	u_4b	$u_4\bar{u}_4$
Scalar	0.11 ± 0.01	0.13 ± 0.01	0.26 ± 0.03
Pseudoscalar	0.15 ± 0.01	0.30 ± 0.02	3.91 ± 0.19
Axial Vector	0.11 ± 0.01	0.13 ± 0.01	0.26 ± 0.03
Vector	0.15 ± 0.01	0.29 ± 0.02	3.91 ± 0.19

Table 6. The values of decay constants of different bound states obtained using $m_{u_4} = 550\ GeV$.

Leptonic decay constant f (GeV)	$u_4\bar{c}$	u_4b	$u_4\bar{u}_4$
Scalar	0.10 ± 0.01	0.12 ± 0.01	0.26 ± 0.03
Pseudoscalar	0.14 ± 0.01	0.27 ± 0.01	4.19 ± 0.20
Axial Vector	0.10 ± 0.01	0.12 ± 0.01	0.26 ± 0.03
Vector	0.14 ± 0.01	0.27 ± 0.01	4.18 ± 0.20

References

[1] Fritzsch H 1992 *Phys. Lett.* B **289** 92
[2] Datta A 1993 *Pramana* **40** L503
[3] Celikel A, Ciftci A K and Sultansoy S 1995 *Phys. Lett.* B **342** 257
[4] Sultansoy S 2000 Contributed paper to ICHEP 2000 *Preprint* hep-ph/0004271
[5] Chanowitz M S, Furman M A and Hinchliffe I 1978 *Phys. Lett.* B **78** 285; 1979 *Nucl. Phys.* B **153** 402
[6] Datta A and Raychaudhuri S 1994 *Phys. Rev.* D **49** 4762
[7] Atag S et al. 1996 *Phys. Rev.* D **54** 5745
[8] Holdom B et al. 2009 *PMC Phys.* A **3** 4
[9] Eberhardt O, Lenz A and Rohrwild J 2010 *Phys. Rev.* D **82** 095006
[10] Sahin M, Sultansoy S and Turkoz S 2011 *Phys. Rev.* D **83** 054022
[11] Altonen T et al., (CDF Collaboration) 2011 *Preprint* arXiv:1101.5782 [hep-ex]
[12] Convay J et al., CDF public conference note CDF/PUB/TOP/PUBLIC/10395
[13] ATLAS Detector and Physics Performance TDR 1999 CERN/LHCC/99-15; Jenni P et al. (ATLAS Collaboration) 1999 Report No CERN-LHCC-99-14/15, Sect. 18.2
[14] Ciftci A K, Ciftci R and Sultansoy S 2002 *Phys. Rev.* D **65** 055001
[15] Arik E et al. 2002 *Phys. Rev.* D **66** 116006
[16] Ciftci H and Sultansoy S 2003 *Modern Physics Letters* A **18** 859
[17] Ciftci R et al. 2003 *Turk. J. Phys.* **27** 179
[18] Accomando E et al. 2004 *Preprint* arXiv:hep-ph/0412251
[19] Ishiwata K and Wise M B 2011 *Preprint* arXiv:1103.0611 [hep-ph]
[20] Bigi I et al. 1986 *Phys. Lett.* B **181** 157
[21] Altonen T et al. (CDF Collaboration) 2009 *Phys. Rev. Lett.* **103** 092002
[22] Ciftci A K, Ciftci R and Sultansoy S 2005 *Phys. Rev.* D **72** 053006
[23] Arik E et al. 1998 *Phys. Rev.* D **58** 117701
[24] Holdom B 2007 *J. High Energy Phys.* **03** 063
[25] Holdom B 2007 *J. High Energy Phys.* **08** 069
[26] Cakir O et al. 2008 *Eur. Phys. J.* C **56** 537
[27] del Agulia Q F et al. 2008 *Eur. Phys. J.* C **57** 183
[28] Ozcan V E , Sultansoy S and Unel G 2008 *Eur. Phys. J.* C **57** 621
[29] Ciftci R 2008 *Phys. Rev.* D **78** 075018
[30] Cakir I T et al. 2009 *Phys. Rev.* D **80** 095009
[31] Shifman M A, Vainshtein A I and Zakharov V I 1979 *Nucl. Phys.* B **147** 385
[32] Vainshtein A I, Voloshin M B, Zakharov V I and Shifman M A 1978 *Sov. J. Nucl. Phys.* **28** 237 (1978 *Yad. Fiz.* **28** 465)
[33] Reinders L J, Rubinstein H and Yazaki S 1985 *Phys. Rep.* **127** 1
[34] Narison S 1989 *QCD Spectral Sum Rules* (World Scientific, Singapore)
[35] Jamin M and Lange B O 2002 *Phys. Rev.* D **65** 056005
[36] Penin A A and Steinhauser M 2002 *Phys. Rev.* D **65** 054006
[37] Du D-S, Li J-W and Yang M-Z 2005 *Phys. Lett.* B **619** 105–114
[38] Aliev T M, Azizi K and Bashiry V 2010 *J. Phys.* G **37** 025001
[39] Aliev T M, Azizi K and Savci M 2010 *Phys. Lett.* B **690** 164
[40] Veliev E V, Sundu H, Azizi K and Bayar M 2010 *Phys. Rev.* D **82** 056012
[41] Veliev E V, Azizi K, Sundu H and Aksit N 2010 *Preprint* arXiv:1010.3110 [hep-ph]
[42] Bashiry V, Azizi K and Sultansoy S 2011 *Preprint* arXiv:1104.2879 [hep-ph]
[43] Nakamura K et al. (Particle Data Group), 2010 *J. Phys.* G **37** 075021

Journal of Physics: Conference Series **347** (2012) 012028 doi:10.1088/1742-6596/347/1/012028

Search for the decay $\mu \to e\gamma$ in the MEG experiment

Yasuhiro Nishimura[1]

The University of Tokyo, 7-3-1 Hongo, Bunkyo-ku, Tokyo 113-0033, Japan

E-mail: nisimura@icepp.s.u-tokyo.ac.jp

Abstract. The MEG experiment is currently searching for the lepton-flavor-changing decay $\mu \to e\gamma$. The Standard Model implies the $\mu \to e\gamma$ by neutrino oscillations but it cannot be observed in experiments because of too small branching ratio. Many extensions of the Standard Model, however, predict the decay within a reach of the MEG experiment and it is a clear evidence of new physics. A liquid xenon gamma-ray detector and a positron spectrometer with a gradient magnetic field enables better sensitivity than the current limit by two orders of magnitude. The physics data taking started in 2008 and then upgrades of the detector brought a stable operation in 2009-year run. We report the preliminary result on the 2009 data together with a current status and a future of the experiment. The obtained upper limit of the branching ratio is $\mathrm{Br}(\mu^+ \to e^+\gamma) < 1.5 \times 10^{-11}$ at 90% confidence level (C.L.).

1. Search for the Decay, $\mu \to e\gamma$

In particle physics the Standard Model (SM) can clearly explain a large number of particle phenomena but an extended theory is desired to overcome the artificial features of the SM. One of the observations to imply a scenario of new physics is the $\mu \to e\gamma$.

Currently three generations of quarks and leptons are known. The SM describes mixing of flavors in the quark sector by the Cabibbo-Kobayashi-Maskawa (CKM or KM) matrix and in the lepton sector by the Pontecorvo-Maki-Nakagawa-Sakaga (PMNS or MNS) matrix. Currently there is no observation of flavor violation only in the charged lepton sector. The search for the rare muon decay, $\mu \to e\gamma$, is sensitive to many models of the SM extensions because they predict the branching ratio around the current upper limit, $\mathrm{Br}(\mu^+ \to e^+\gamma) < 1.2 \times 10^{-11}$ (90% C.L.) given by the MEGA experiment in 1999 [1]. Even without help of the extensions, the $\mu \to e\gamma$ decay occurs by the neutrino oscillation but the branching ratio is too small, $\mathrm{Br}(\mu \to e\gamma) \approx (\alpha/128\pi)\sin^2 2\theta_{12}\,(\Delta m^2_{21}/M^2_W)^2 < 10^{-54}$, where Δm_{21} is the mass difference of a muon neutrino and an electron neutrino and M_W is a W boson mass, and $\sin^2 2\theta_{12} = 0.86$ and $\Delta m^2_{21} = 8 \times 10^{-5}$ eV^2 are assumed. Therefore the $\mu \to e\gamma$ is a clear signal of the charged-lepton-flavor violation (cLFV) if it is observed in experiments.

The $\mu \to e\gamma$ signal is a two-body decay with 52.8 MeV energy corresponding to a half mass of a muon. Experiments that search for the $\mu^+ \to e^+\gamma$ are confronted with two types of backgrounds: an accidental background and a radiative decay background. The former is the most dominant background, and is due to the accidental pileup of positrons from Michel decays, $\mu^+ \to e^+\nu_e\bar{\nu}_\mu$, with gamma-rays that originate from radiative muon decays (RD), $\mu^+ \to e^+\nu_e\bar{\nu}_\mu\gamma$, or from the annihilation of positrons in flight. The radiative muon decays

[1] On behalf of the MEG Collaboration

Journal of Physics: Conference Series **347** (2012) 012028 doi:10.1088/1742-6596/347/1/012028

themselves make up the second type of background as well, when the energies and the opening angle of the positron and the gamma-ray occasionally end up around the signal region. The signal and the two backgrounds have different distributions of observables, which is taken into account in a likelihood function for the $\mu^+ \to e^+\gamma$ analysis.

A low muon-beam rate and good detector resolutions, especially for the gamma-ray energy, are essential to suppress the backgrounds, as well as a large number of muons and a sufficient detection efficiency. Intense muon beam at Paul Sherrer Institut (PSI) in Switzerland is suitable for the MEG experiment because provided DC surface muons are preferred to avoid pileups rather than AC beam. We developed a large liquid xenon (LXe) detector to detect gamma rays in a high resolution with an excellent performance of a rare-gas scintillator. The COBRA (COnstant-Bending-RAdius) magnet with a special gradient magnetic field enables to sweep positrons away quickly with hitting low-mass drift chambers, then finally timing counters measure the precise timing at the end of the track.

2. Detector

The overview of the detectors is shown in Figure 1. Detectors consist of the positron detector inside the COBRA magnet and the LXe gamma-ray detector outside the magnet. The positron detectors, which consist of a drift chamber (DCH) and a scintillation timing counter (TIC), are mounted around a lower part from a muon stopping target. Two sets of TIC are located at the upstream and downstream side, each of which consists of two layers of arrays along the beam axis (z) and the azimuthal axis (ϕ). The muon stopping target is a sheet of polyethylene and polyester with a 205 μm thickness corresponding to 18 mg/cm^2 on plane and gives 82% stopping efficiency thanks to a slanted angle with 20.5° relative to the beam axis.

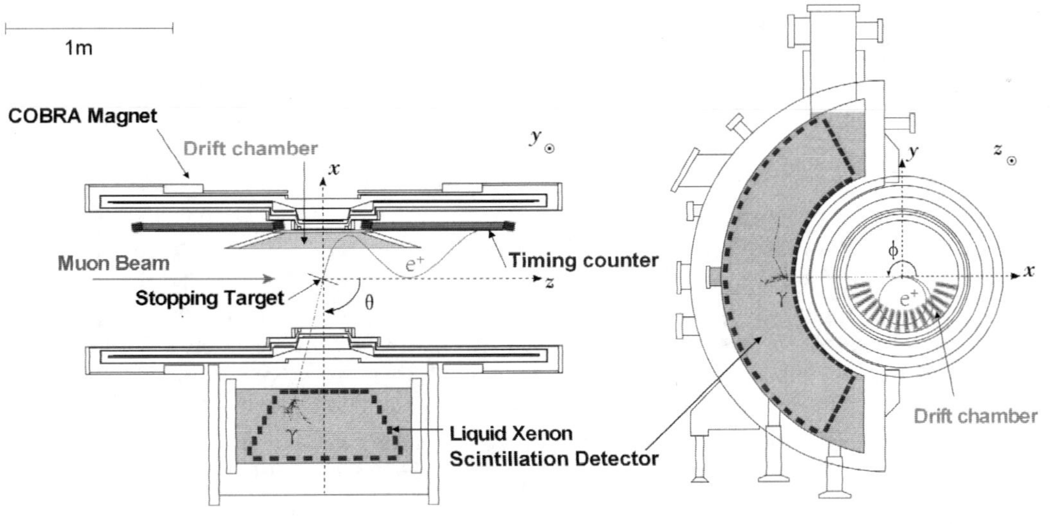

Figure 1. Layout of detectors in the MEG experiment.

2.1. Positron Spectrometer

The COBRA magnet has an excellent performance to select a momentum by a radius thanks to the gradient magnetic field, where tracks of the same momentum have the same radius irrelevant to emission angles. Furthermore, a non-zero radial component of the magnetic field forces to sweep positrons away quickly. As a result pileups of positrons reduce compared to those in a normal solenoidal magnet. Total thickness of the magnet including its cryostat is 0.197 X_0

and its light and thin structure enables gamma rays to pass through the magnet toward to the LXe detector with a good transmission efficiency of 85%. The thin structure can hold a strong magnetic field such that the central field is 1.27 T at $z = 0$ and slowly decreasing to 0.49 T at edge as $|z|$ increases.

The drift chambers consist of 16 modules that are radially aligned with 10.5° intervals in the azimuthal angle. The radius is from 19.3 cm to 27.9 cm in order to select only positrons around the signal region. Timing counter array detects a positron in the angular range of $0.08 < |\cos\theta| < 0.35$. The array consists of two layers of different plastic scintillators along z direction and ϕ direction. A timing ϕ-counter is an array of 15 plastic scintillation bars with a $4 \times 4 \times 80$ cm^3 dimension. It covers 160° ϕ-acceptance between −150° and 10°. A timing z-counter can tag hits of positrons although it is under preparation.

2.2. The Liquid Xenon γ-Ray Detector

The 900-liter LXe detector is designed to detect a gamma ray around 52.8 MeV signal after penetrating the COBRA magnet and the detector vessel. Liquid xenon has an excellent performance as a scintillator such as a high density of 2.95 g/cm^3 corresponding to $X_0 = 2.77$ (cm), a high light yield (80% of NaI(Tl)), a fast response of a 45-ns decay time for gamma rays, no absorption of the scintillation light, an uniform response of the scintillation light because of liquid and a particle discrimination by injecting particle thanks to different scintillation process. Although there are some difficulties to treat the liquid xenon, we developed a photo-multiplier tube (PMT) for the liquid xenon to detect high-rate vacuum-ultraviolet (VUV) photons ($\lambda \sim$ 178 nm), a pulse-tube refrigerator to keep xenon liquid stably and two purification systems for liquid and gaseous xenon. The LXe detector equips 846 PMTs immersed in xenon and attached on six faces with recording waveforms from all PMTs. The spatial distribution of the scintillation light allows energy, timing and position measurement.

To calibrate and monitor the LXe detector, many apparatus are equipped. In order to calibrate gains and quantum efficiencies of PMTs blue-light LEDs on the wall and alpha sources of ^{241}Am stretched on wires are immersed in the liquid xenon. For a constant light-yield monitor a Cockcroft-Walton (CW) proton accelerator provides gamma rays of 4.4 MeV, 12.0 MeV and 17.7 MeV energies after a nuclear reaction of $^{9}_{4}$Be(α, n)$^{12}_{6}$C and $^{7}_{3}$Li(p,γ)$^{8}_{4}$Be at a proton target. In addition, even without a special setup of a beam line, Am/Be source for a 4.8 MeV gamma ray enables the monitor. Meanwhile, the π^- beam from the same beam line as muons brings a charge exchange (CEX) reaction at a liquid hydrogen target, $\pi^- \mathrm{p} \rightarrow \pi^0 \mathrm{n}$. It produces two gamma rays in a π^0 decay at the energy range from 54.9 MeV to 82.9 MeV and gamma rays get monochromatic after selecting a back-to-back decay by a tagging counter opposite to the LXe detector. The 3×3 NaI(Tl) array with avalanche photo diodes allows the selection of two-gamma-rays opening angle and is mounted on a mover system in order to scan the whole acceptance of the LXe detector. The CEX calibration is the most important method because the 54.9 MeV gamma ray is near to the signal 52.8 MeV energy. It is suitable for a determination of the energy scale and a performance evaluation of the energy, timing and position at arbitrary position on the detector.

2.3. Data Acquisition

Trigger, waveform sampling and data acquisition (DAQ) systems have been developed for the MEG. Trigger logic is programmable on FPGA (Field-Programmable Gate Array) chips and signals from the detectors are arranged in triple layers, which enables a complicated physical selection and custom triggers for each event type. Domino ring sampler (DRS) chips allow a fast waveform sampling up to 6 GHz with 1024 cells and 11-bit resolution. It reads out at 800 MHz for the drift chamber and 1.6 GHz for the LXe detector and the positron timing counter. The DAQ system, MIDAS (Maximum Integration Data Acquisition System), collects and records data from all electronics and ROME (Root based Object oriented Midas Extension) framework offers an event display. ROME also performs an analysis for detector calibrations, an event reconstruction and a calculation for the final result.

3. Reconstruction and Performance

3.1. Reconstruction

Reconstructions of the positron and the gamma ray are separately performed in each detector after analyzing waveforms, then a vertex information can be acquired by extracting the positron track.

The reconstruction has different performances for the z and r positions because wires are stretched only along beam direction (z) in the drift chamber. Due to this difference, the relative angle is separately treated as two components of $\theta_{e\gamma} = \theta_e - (\pi - \theta_\gamma)$ and $\phi_{e\gamma} = \phi_e - (\pi - \phi_\gamma)$ on the analysis. In the radially aligned timing counter, z position is determined by a charge ratio of two PMTs at both ends of the scintillator. After hit clusters and track candidates are found in drift chambers, the positron track is reconstructed by using the Kalman filter that can include the multiple Coulomb scattering, the energy losses and the magnetic field effectively [2].

Time and position of gamma rays are obtained by least squares fitting. Defined χ^2 for the timing considers a time of flight and charge statistics. The reconstruction of the position uses local PMTs only on the detection face with taking solid angles into account. The energy is converted from the total number of scintillation photons in all PMTs weighted with the reciprocal of the coverage of the photo-cathode. Intrinsically non-uniform response of the energy scale arises from the reconstruction, though it is corrected by a measurement.

3.2. Run

In the previous run in 2008, some parts of the drift chamber had suffered from a discharge problem and the light yield of the LXe detector had been increasing by the purification. Throughout the 2009 run, the detector was quite stable for both the drift chamber and the LXe detector. The run consists of the physics run and dedicated periodic or long-term calibration runs. Twelve types of the trigger are mixed in the physics run. The signal candidate is taken at 5 Hz rate and totally 6.5 Hz with including triggers for a pedestal, positrons from Michel decays, LED flashing, cosmic rays and so on. Before the physics run started we had the CEX run to calibrate the LXe detector for twelve days. During physics data taking the gains of the PMTs were calibrated for times a week by dedicated runs using LEDs and alpha sources. The light yield was monitored by the 17.7 MeV gamma rays from the CW accelerator as well.

3.3. Event Selection

The event candidates of $\mu^+ \to e^+\gamma$ is finally analyzed in an analysis region. The region is blinded until the calibration and the analysis are fixed. As soon as physics data is taken, a first analysis is performed and discards unnecessary events by using only timing information. Simultaneously the blind region is selected by the gamma-ray energy ($|E_\gamma - m_\mu/2| < 4.8$ MeV) and the relative timing ($|t_{e\gamma}| < 1$ ns) as shown in Figure 2. In addition, we defined the analysis region as 50 MeV $< E_e <$ 56 MeV, 48 MeV $< E_\gamma <$ 58 MeV, $|\phi_{e\gamma}| < 50$ mrad, $|\theta_{e\gamma}| < 50$ mrad and $|t_{e\gamma}| <$ 0.7 ns. The time window corresponds to ten-sigma window in detector resolutions. Sideband data beside the blind region is useful to calibrate the detectors and to estimate backgrounds and sensitivities. The E_γ sideband of the lower gamma-ray energy region contains radiative muon decays, therefore the gamma ray and the positron coincidently make a peak on the $t_{e\gamma}$ distribution. On the other hand the signal and the radiative muon decay are absent in two $t_{e\gamma}$ sidebands, where only accidental background forms a flat $t_{e\gamma}$ distribution.

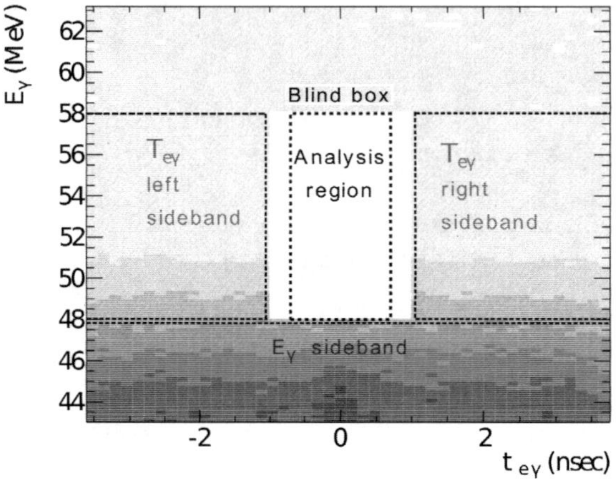

Figure 2. Event distribution on E_γ - $t_{e\gamma}$ plane. The white blank box is the blind region.

3.4. Performance Evaluation

The $\mu^+ \to e^+\gamma$ analysis needs the likelihood function, which is based on probability density functions (PDFs) of the signal, the RD background and the accidental background. It is important to measure resolutions and distributions in order to construct the PDFs and determine the likelihood on the events.

The timing resolution is evaluated by the peak from the radiative muon decay in E_γ sideband outside the blind region at the range of 40 MeV to 47 MeV. Figure 3 shows the peak fitted with double Gaussian function and the timing resolution resulted in 149±10 ps in 70% fraction as core sigma and 250 ps sigma in tail. The timing resolution at the signal energy is estimated to be 142 ps in sigma by correcting for the small energy dependence.

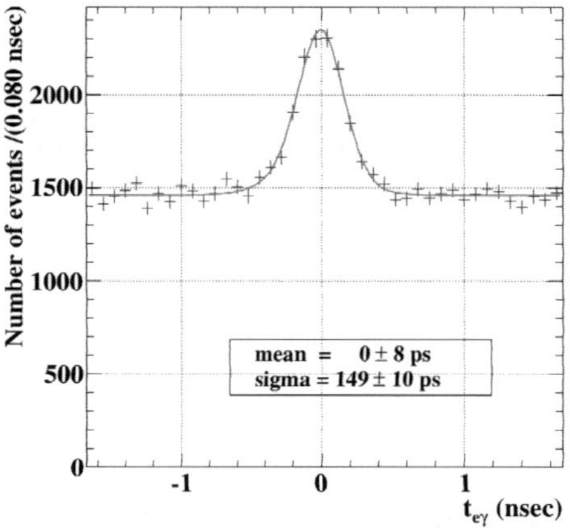

Figure 3. Time difference between positron and gamma ray in radiative muon decays.

Figure 4 shows the 54.9 MeV gamma-ray peak obtained from the π^0 decay in CEX run. The peak of the energy has a low-end tail because of interactions before reaching liquid xenon and leakage around the face, therefore a combined function of Gaussian and exponential function is used for the fitting. The high-end energy resolution is essential to identify the signal because the background gamma-ray energy of the radiative decay has an upper edge and is lower than the signal energy. We construct the position-dependent resolution map by the scanning whole detection face, then the typical energy resolutions are estimated by depths (w) to be 2.1% (2 cm $\leq w <$ 38 cm), 2.8% (1 cm $\leq w <$ 2 cm) and 3.3% (0 cm $\leq w <$1 cm). To evaluate the position resolution, collimators made of lead with 1 cm slits are put on the vessel of the LXe detector to project gamma rays from the π^0 decay. By means of a Monte Carlo (MC) simulation, the depth resolution and a position dependence of the resolution is considered by taking the difference between true and reconstruction. The resolutions of the position reconstruction are evaluated to be 5 mm and 6 mm along the direction orthogonal to the LXe front face and for the depth, respectively.

In Figure 5 the background energy distribution of positrons from the Michel decays is fitted with a theoretical function, that is due to the kinematics of the decay, but smeared by a double Gaussian to account for the detector resolution. The resolution component of the function is separately shown in dashed line. The energy and position resolution for the signal PDF is estimated by tracks with two full turns in the drift chambers such that two turns are separately reconstructed as independent one turn and then a difference of each reconstruction indicates a resolution. The energy resolution is described with double Gaussian function, estimated to be 0.39 MeV as the core resolution in 79% fraction with 1.71 MeV sigma as a tail in 21%. The positron angular resolutions are also evaluated along two directions to be σ_{θ_e} = 11.2 mrad and σ_{ϕ_e} = 7.1 mrad, where σ_{ϕ_e} is an 85% component of double Gaussian.

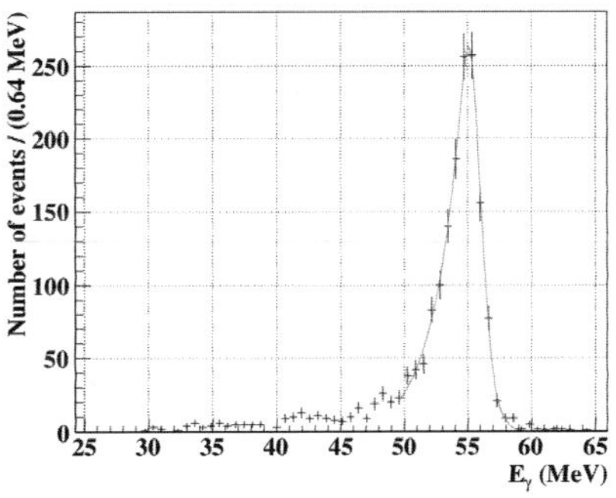

Figure 4. Gamma-ray energy peak of 54.9 MeV from π^0 decays, obtained by selecting two back-to-back gamma rays.

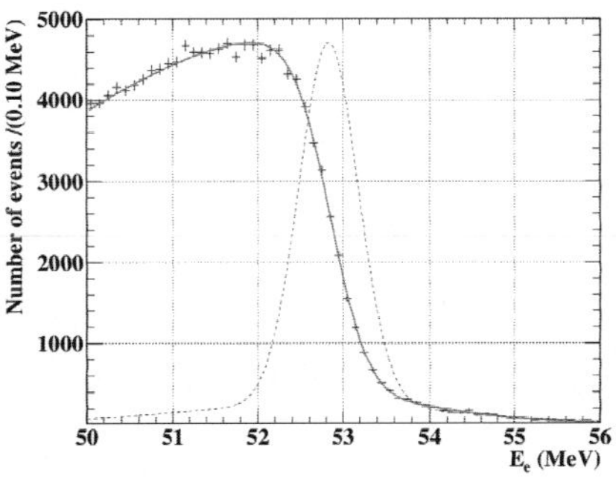

Figure 5. Energy distribution of positrons from the Michel decays, $\mu^+ \rightarrow e^+\nu_e\overline{\nu}_\mu$.

4. Analysis for the $\mu \rightarrow e\gamma$

4.1. Likelihood Analysis

The goal of the maximum likelihood analysis is to obtain the best fit value of the number of the $\mu^+ \rightarrow e^+\gamma$ signal (N_{sig}), as well as the RD background (N_{RD}) and the accidental background (N_{BG}). Based on the measured performance or background distributions, three types of PDFs are defined for the signal (S), the RD background (R) and the accidental background (B) as a function of observables. Main observables are the following set, $\vec{x}_i = (E_e, E_\gamma, \theta_{e\gamma}, \phi_{e\gamma}, t_{e\gamma}, \delta E_e, \delta E_\gamma, \delta\theta_{e\gamma}, \delta\phi_{e\gamma}, \delta t_{e\gamma})$, accordingly PDFs are separated for each independent observable. The variables with δ indicate resolutions of the detectors used in the construction of the PDFs. The likelihood function is then defined in Equation 1,

$$L(N_{sig}, N_{RD}, N_{BG}) = \frac{N^{N_{obs}} \exp(-N)}{N_{obs}!} \prod_{i=1}^{N_{obs}} \left(\frac{N_{sig}}{N} \cdot S(\vec{x}_i) + \frac{N_{RD}}{N} \cdot R(\vec{x}_i) + \frac{N_{BG}}{N} \cdot B(\vec{x}_i) \right), \quad (1)$$

where N_{obs} is the observed number of events in the analysis region and N is the sum, $N = N_{sig} + N_{RD} + N_{BG}$.

The best way to convert the number of signals to the branching ratio is to count the Michel decays instead of the decayed muons because instability of the beam and uncertainty of the muon stopping efficiency can be ignored and the detection efficiency of positrons is cancelled out. Relatively the signal is normalized as described in Equation 2,

$$\frac{\text{Br}(\mu^+ \to e^+\gamma)}{\text{Br}(\mu^+ \to e^+\nu_e\overline{\nu}_\mu(\gamma))} = \frac{N_{sig}}{N_{e\nu\overline{\nu}}} \times \frac{f^{E_e}_{e\nu\overline{\nu}}}{P_{e\nu\overline{\nu}}} \times \frac{\epsilon^{trig}_{e\nu\overline{\nu}}}{\epsilon^{trig}_{e\gamma}} \times \frac{A^{TIC}_{e\nu\overline{\nu}}}{A^{TIC}_{e\gamma}} \times \frac{\epsilon^{DCH}_{e\nu\overline{\nu}}}{\epsilon^{DCH}_{e\gamma}} \times \frac{1}{\epsilon^{LXe}_{e\gamma}} \times \frac{1}{A^{LXe}_{e\gamma}}, \quad (2)$$

where $N_{e\nu\overline{\nu}} = 18,096$ is the number of observed Michel decays, $f^{E_e}_{e\nu\overline{\nu}} = 0.114 \pm 0.002$ is the fraction of Michel spectrum after the analysis cuts, $P_{e\nu\overline{\nu}}$ is a 10^7 pre-scale factor for the trigger, $\epsilon^{trig}_{e\gamma}/\epsilon^{trig}_{e\nu\overline{\nu}} = 0.84 \pm 0.02$ is the trigger efficiency of a direction match for positrons, $A^{TIC}_{e\nu\overline{\nu}}/A^{TIC}_{e\gamma} \times \epsilon^{DCH}_{e\nu\overline{\nu}}/\epsilon^{DCH}_{e\gamma} = 1.12 \pm 0.06$ is the signal-to-Michel ratio of an acceptance at the timing counter and the reconstruction efficiency, $\epsilon^{LXe}_{e\gamma} = 0.58 \pm 0.02$ is the efficiency for the detection of gamma rays and $A^{LXe}_{e\gamma} = 0.99 \pm 0.01$ is the acceptance of the LXe detector after detecting the positron. As a result the branching ratio is obtained from the number of signal events by the formula: $\text{Br}(\mu^+ \to e^+\gamma) = N_{sig} \times (1.01 \pm 0.08) \times 10^{-12}$, which corresponds to the decays of 6×10^{13} muons yields.

4.2. Result

The sensitivity in the MEG 2009 run is evaluated by means of the averaged results in the toy-MC simulation such that many toy experiments are generated based on the PDFs and data set in 2009, as well as the fit in the $t_{e\gamma}$ sideband for a consistency check. With using the likelihood ratio ordering principle [3], the expected 90% C.L. branching-ratio upper limit is calculated to be

$$S_{2009} = 6.1 \times 10^{-12}. \quad (3)$$

The obtained sensitivity improves from the previous 2008 result, $S_{2008} = 1.3 \times 10^{-11}$ [4].

The analysis region was opened and no strange reconstruction was found by checking each event in detail. The event distribution is plotted on the E_γ-E_e plane in Figure 6 (*a*) and on the $t_{e\gamma}$-$\cos\Theta_{e\gamma}$ plane in Figure 6 (*b*).

In total, 370 events are observed in the analysis region. The best fit value of the signal is calculated to be 3.0 events, and that of the RD background is $N_{RD} = 35^{+24}_{-22}$. The 90% C.L. upper limit of the N_{sig} is estimated to be 14.5, while $N_{sig} = 0$ is inside the confidence interval of 90%. In this calculation the systematic uncertainties are included, but the result is still preliminary. The upper limit of the branching ratio is finally obtained as

$$\text{Br}(\mu^+ \to e^+\gamma) < 1.5 \times 10^{-11} \text{ at 90\% C.L.} \quad (4)$$

5. Update and Plan

After the analysis for the 2009 data, we have improved the reconstruction and understood several uncertainties, which is involved in the next result. Detector alignment had a large uncertainty for opening angles, therefore it was investigated in detail. Gamma rays from Am/Be source scanning

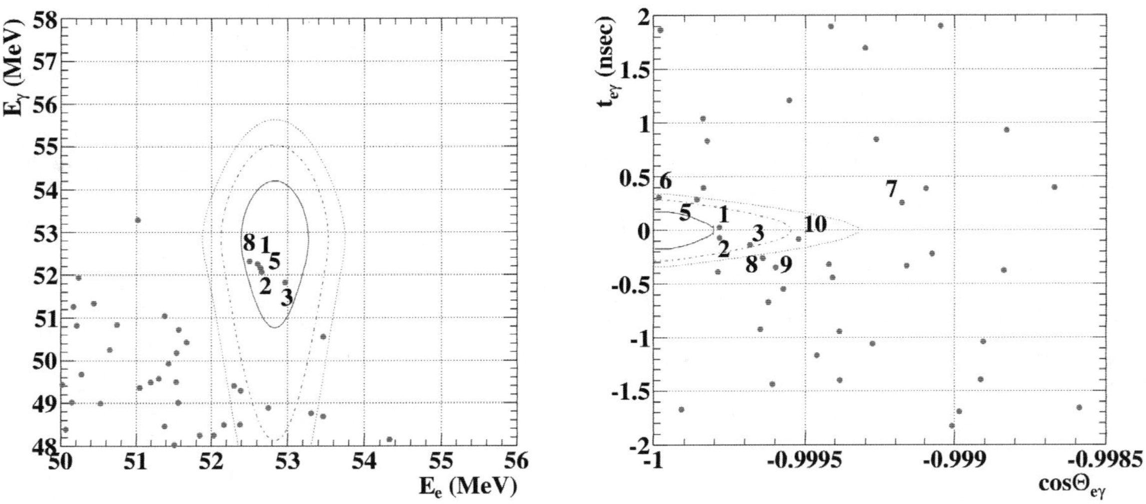

(a) E_γ and E_e (b) $t_{e\gamma}$ and opening angle $\cos\Theta_{e\gamma}$

Figure 6. Distribution in analysis region. The signal PDFs are superimposed as contours at 1, 1.64 and 2 sigma as blue solid, dot-dashed, and dashed lines respectively. The number shows the rank given by $S/(0.1R + 0.9B)$.

on the detector or collimated gamma rays from the nuclear reactions of protons by the CW suggest the actual position of PMTs of the LXe detector. Globally cosmic-ray muons through both the LXe detector and the drift chamber allow a relative alignment. These alignments reduced systematic uncertainties of opening angles by a factor of 0.4-0.5.

In addition, the magnetic field mapping, that is used for the positron tracking, was reconsidered to identify and resolve uncertainties previously unknown. Intrinsic bias in the Kalman filter was also studied in the MC simulation and is now well understood. In the reconstruction, a magnetic-field component along ϕ direction had previously been neglected because it is tiny and much less than 1% of that of z direction, however, it is now taken into account. Accordingly the positron resolution of the energy and the angles improved as well as the uncertainties.

After the unblinding, we observe that the backgrounds estimated from the sidebands are in good agreement with the number of background events extracted by the fit in the analysis region. Given the larger background statistics in the sidebands, we decide to use the sideband data to determine the amount of background in the analysis region. Therefore, in the calculation of the branching ratio, we apply the constraints from the side-band background information to the fitting and the profile likelihood method is adopted for the toy-MC generation.

After the 2009 run new calibration systems were installed such as a neutron generator for the 9-MeV gamma ray, a target dedicated to Mott scattering for monochromatic positrons. For three months in 2010 we have accumulated twice as many muon decays as the events in 2009 with stable conditions. Performance of triggers and DAQ improved especially on $\mu^+ \to e^+\gamma$ direction matching. Thus our next result will include 2010 data and will appear with the combined statistics of the years.

In 2011 a long run for four months is planned with an upgrade of high-voltage module with less noise and a more efficient live time by a buffer readout. Additionally BGO crystals instead of NaI arrays for the CEX run and the timing z-counter will be installed to make data taking more efficient. For a few years a long and stable run will be expected toward to the $\mu^+ \to e^+\gamma$ observation.

6. Summary

The MEG experiment reported the upper limit of the $\mu^+ \to e^+\gamma$ branching ratio on the 2009 run. With the improved analysis, next result will be presented by combining the 2009 run with the 2010 run, corresponding to three times larger statistics than 2009 data. Currently the statistics limit the sensitivity and the physics data taking will be going on for the next few years. The performance of the detector and analysis are still improving toward to the branching ratio of 10^{-13}.

References

[1] Brooks M L *et al* 1999 *Phys. Rev. Lett.* **83** 1521-1524
[2] Kalman R E 1985 *Transaction of the AMSE-J. Basic Engineering* D **82** 35-45
[3] Feldman G J and Cousins R D 1998 *Phys. Rev.* D **57** 3873-3889
[4] Adam J *et al* 2010 *Nucl. Phys.* B **834** 1-12

Journal of Physics: Conference Series **347** (2012) 012029 doi:10.1088/1742-6596/347/1/012029

Performance of micromegas detectors in the CAST Experiment

S C Yildiz[1,2], S Aune[3], T Dafni[4], M Davenport[5], G Fanourakis[6], E Ferrer-Ribas[3], J Galán[3], J A García[4], A Gardikiotis[7], J G Garza[4], T Geralis[6], I Giomataris[3], H Gómez[4], F J Iguaz[3], I G Irastorza[4], T Papaevangelou[3], A Rodríguez[4], J Ruz[4,5], A Tomás[4] and T Vafeiadis[5,7,8]

[1] Doğuş University, Istanbul, Turkey
[2] Boğaziçi University, Istanbul, Turkey
[3] IRFU, Centre d Études de Saclay, CEA, France
[4] Laboratorio de Física Nuclear y Astropartículas, Universidad de Zaragoza, Spain
[5] CERN, European Organization for Particle Physics and Nuclear Research, Switzerland
[6] Institute of Nuclear Physics, NCSR Demokritos, Greece
[7] University of Patras, Greece
[8] Aristotle University of Thessaloniki, Greece

E-mail: cenk.yildiz@cern.ch

Abstract. CERN Axion Solar Telescope (CAST) experiment is searching for axions coming from the Sun. Inside transverse magnetic fields, axions can be converted into X-rays, which can be detected by X-ray detectors. The expected energy of the signal in CAST is in the 1-10 keV range. Low noise and low background detectors are necessary to increase the sensitivity of the experiment. Micro Mesh Gaseous Structure (micromegas) detectors have been used in CAST since the beginning, providing good energy and spatial resolution for CAST's needs. CAST has been intensely studying micromegas detectors to develop new technologies. Initially, CAST detectors consisted of a micromegas, a Time Projection Chamber (TPC) and a Charged Couple Device (CCD), however the improvements in micromegas technologies encouraged CAST to replace the TPC with 2 new micromegas detectors. In some periods during CAST run, ultra low background has been observed in one of the micromegas detectors and it is being investigated through simulations and laboratory tests carried out at Canfranc Underground Laboratory. If this low background is indeed not a systematic effect, it can open new possibilities on rare event searches.

1. CAST Experiment

CERN Axion Solar Telescope (CAST) has been running at CERN since 2003. CAST is searching for particles called axions. To solve the strong CP problem in the Standard Model, Peccei-Quinn mechanism is suggested, adding a global U(1) symmetry to the Standard Model. Axions arise from the breaking of this symmetry. Through Primakoff effect, axions can be converted into photons and vice versa in transverse magnetic and electric fields. The Sun is a good source for axions. The photons created in the core of the Sun can be converted into axions in the high electric fields of the solar plasma and travel to the Earth without interacting with ordinary matter. If one can create a high transverse magnetic field at the Earth, axions can be converted back into photons and detected. The spectrum of photons coming from the hot core of the Sun is in the x-ray region, thus x-ray detectors should be used [1, 2].

CAST is using an LHC prototype dipole magnet, which is mounted on a pivot that can move ±40° horizontally, ±8° vertically to track the Sun for 1.5 hours both during sunset and sunrise. CAST has not yet discovered axions, but has been putting the best limits in the parameter space of axion mass and coupling constant for axion - photon interaction. Both for the discovery potential and better upper limits in coupling constant, sensitivity of CAST depends on several factors: strength and length of the magnetic field, exposure time, detector efficiency and detector background levels. Magnet parameters and exposure time are limited by physical properties and constraints of the magnet. Detector hardware efficiency is limited by the cross section of the gas and windows that are between the detector and vacuum system. Among these, only the detector background levels are improvable. Hence low background detectors are crucial for the results of the experiment.

CAST magnet has 2 sides: sunset(taking tracking data during sunsets), and sunrise(taking tracking data during sunrise). 4 detectors are mounted on ends of two magnet bores. Currently in the sunset side there are 2 micromegas, while sunrise side is operational with a micromegas and a CCD detector, which is coupled to an x-ray telescope. A picture of CAST magnet and the exclusion plot can be seen in Figure 1.

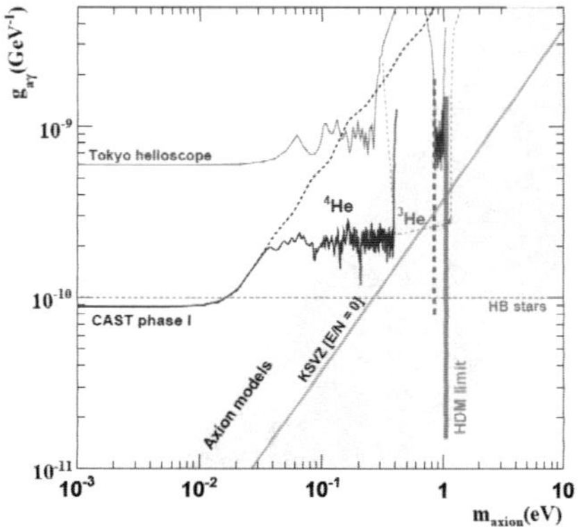

Figure 1. Left: CAST magnet, installed on a pivot. It can move vertically and horizontally to track the Sun every day. Right: Axion coupling constant exclusion plot. CAST has been putting strictest limits on the axion - photon coupling.

2. Micromegas Detector

Micromegas detector was invented in 1996 by Charpak and Giomataris [3]. It is a gaseous detector that has two regions, 3 different voltage levels and 2 readout levels (Figure 2). On the top, there is the drift window, which is the first high voltage level (around −600V). Below the drift window there is the drift gap filled with gas. When charged particles or photons pass through the chamber, primary electrons are created by excitation of this gas. After the drift gap there is micromesh, a transparent mesh, which is both the second high voltage level (around −300V) and first readout level. The primary electrons are accelerated with the electric field directed from mesh to drift and excite more gas atoms. Most of the electrons pass through the mesh. The region below the mesh is called the amplification gap. Electrons are multiplied in this region with higher electric fields and create a signal in the anode pads/strips which is the third voltage level (at ground level) and second readout level. They are placed a short distance from the mesh (50-150 μm) and provide a 2-dimensional readout. The ions that are created in amplification gap drift back, and create a signal in the mesh.

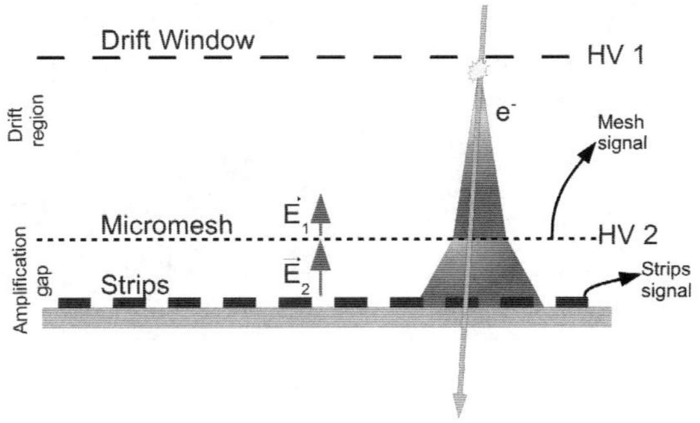

Figure 2. Illustration of micromegas detector's working principle. Incoming particles excite gas atoms, and the free electrons move to the readout layers with the help of the electric field, stripping more electrons from the gas.

2.1. Micromegas Technologies
In CAST, different micromegas technologies have been used since the beginning of data taking.

2.1.1. Conventional Micromegas In conventional micromegas, the strips and mesh are attached to each other mechanically. It was the first model to be used at CAST. The limit in the achievable resolution and rejection power directed CAST to develop better technologies, in which mesh and strips can be attached more uniformly by advanced methods.

2.1.2. Bulk Micromegas Bulk micromegas was developed in 2005 and has been used in CAST for many years [4]. The main difference from conventional micromegas is that the 2 readout levels are being manufactured together. The 30μm thick stainless steel mesh and the strips are attached together with pillars in between, using PCB technology. Bulk detectors can provide energy resolutions as good as 18% (FWHM at 5.96 keV).

2.1.3. Microbulk Micromegas Similar to bulk micromegas, mesh and strips are produced together [4]. In microbulk detectors, 5μm copper mesh is used and it is produced together with strips by photo-lithography technique. Mesh holes are aligned with the shape of the readout pads/strips, and due to the advanced production technique, amplification gap is smaller. Microbulk detectors have achieved better energy resolution than bulk detectors, as good as 12% (FWHM at 5.96 keV). The differences between bulk and microbulk micromegas can be seen in Figure 3.

3. Micromegas Detectors in CAST
CAST experiment is one of the first experiments to use micromegas detectors and has many collaborating institutes, contributing to the development of micromegas technologies. Micromegas detectors are preferred for their ease of production, good background levels and stable behaviour.

3.1. Sunrise and Sunset Lines
CAST uses 3 micromegas detectors, 2 in the sunset line and 1 in the sunrise line. Each of the current micromegas has a readout plane consisting of 106 strips in both x and y directions, covering an area of 36 cm^2.

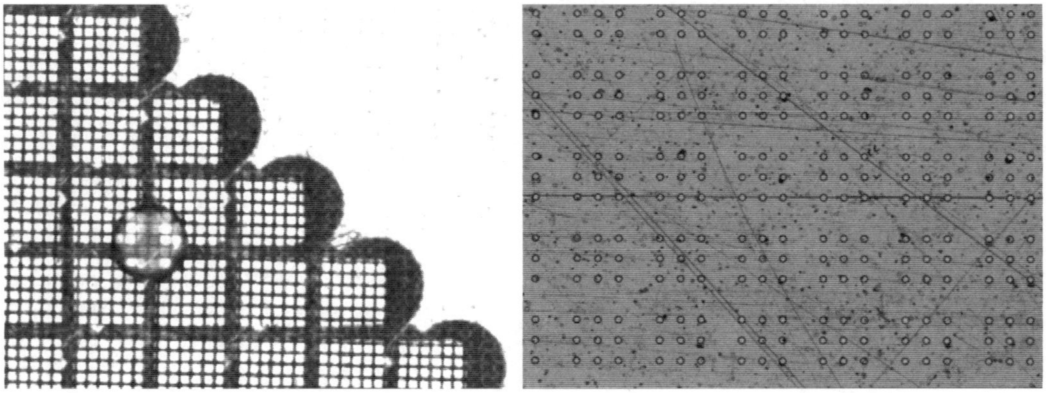

Figure 3. Left: Bulk micromegas. The cylindrical pillars are placed between the mesh and readout pads. Right: Microbulk micromegas. Mesh holes are grouped in 9 and positioned exactly where the readout pads are.

In the sunrise line, there has always been a micromegas installed. All micromegas technologies were used, and currently a microbulk micromegas is installed. Sunrise line is operating with 97.7% Argon, 2.3% isobutan mixture at a pressure of 1.4 bar. It is shielded with lead, copper and polyethylene. In 2011, a special extra shielding was installed to cover the pipe near the detector (Figure 4).

Figure 4. From left to right, sunrise micromegas in different shielding setups. Left: The detector with the lead, cadmium and copper shielding. Middle: The detector with the addition of polyethylene shielding. Right: New shielding implemented in 2011 to block an important contribution of the background events that come from the pipes.

In the sunset line, a TPC covering both bores was operational until 2007. The performance that was shown by the micromegas detector encouraged the replacement of the TPC with 2 micromegas detectors. A compact design was implemented to have common shielding and vacuum pipeline for both detectors (Figure 5). Initially, a bulk and a microbulk detector was used and currently 2 microbulk detectors are installed. The shielding and grounding improvements lowered the electronic noise levels well under the threshold, and enabled stable detector operation. Sunset detectors are operating with 98% Argon, 2% isobutan mixture at a pressure of 1.4 bar. The shielding is similar to sunrise detectors.

3.2. Data Acquisition System

CAST Micromegas has been using a data acquisition chain made of NIM and VME modules and software written in Labview [11]. For every event that passes a certain threshold, time

Figure 5. Left: Drawing of the sunset micromegas detectors, showing their common Faraday cage. Right: Sunset side of the magnet with the full shielding.

information of the mesh pulse, and integrated charge on each strip are recorded, giving a 2 dimensional charge distribution in the strips.

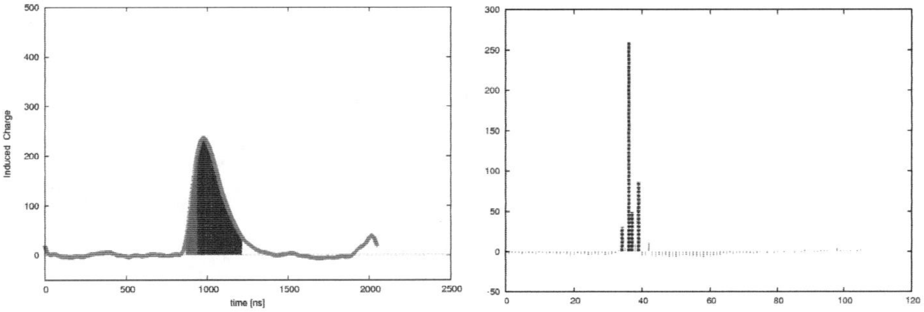

Figure 6. Left: Time evolution of the pulse. Right: Integrated charge on each strip for one direction of the readout.

3.3. Background Discrimination

There are two types of observables in micromegas detectors. Pulse observables like risetime, width and integral are extracted from the time evolution of the pulse. The strip observables like multiplicity, cluster size and sigma of the cluster charge are extracted from charge distribution in the x and y strips.

Characteristics of x-rays signatures are obtained from Fe^{55} calibration, which is done at least once a day. Calibrations prove the stability of the detectors (Figure 7) and show that x-rays are localized events (creating only one cluster in the strips) with narrow distributions. Using pulse and strip observables, a decision on recognizing an event to be an x-ray signal is made. There are two different approaches to event selection at CAST: multivariate analysis and sequential cut analysis.

In the multivariate analysis, few observables are used to create an n-dimensional hyper ellipsoid. Information of the ellipsoid is used to create a single variable, and each event is evaluated using this variable. This method takes the correlations between the variables into account. In sequential cut analysis, various cuts can be defined and applied one after another. Cuts can be intervals in one dimensional distributions, or regions in 2 dimensional parameter space of observables. Multivariate method is known to be better when the observables are correlated, while sequential cuts give greater control in the selection process. Despite the differences, both methods have similar rejection capability and 99.9% of the background events are rejected and a stable background evolution is obtained (Figure 8).

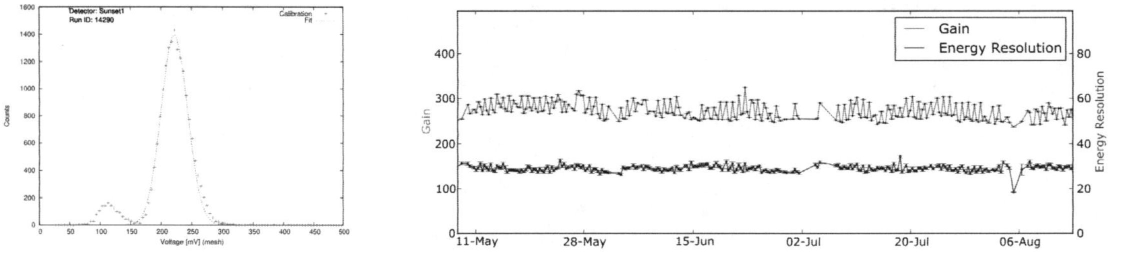

Figure 7. Left: Calibration spectrum. Right: Stability of the gain and of the energy resolution in 2008.

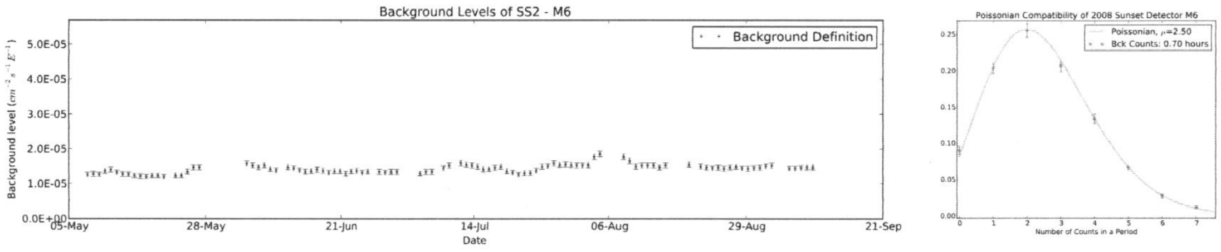

Figure 8. Left: Stability of the estimated background level in year 2008 in Sunset detector 2. Right: Compatibility of observed counts in periods of 0.7 hours with Poissonian distribution.

3.4. Evolution of Micromegas Detectors in CAST

Since the beginning of CAST, several different micromegas detectors have been used. The first detectors were conventional micromegas and had a less complete shielding compared to current detectors. The improvements in shielding together with the invention of bulk and microbulk technologies enabled better stability and lower background levels throughout the years.

3.5. Ultra Low Background Periods

Nominal background levels are around $8 \cdot 10^{-6}\text{keV}^{-1}\text{cm}^{-2}\text{s}^{-1}$ in CAST micromegas. However, in sunrise micromegas, there were some periods when this level decreased to $1 \cdot 10^{-7}\text{keV}^{-1}\text{cm}^{-2}\text{s}^{-1}$ for a few days. In 2011, with the implementation of a new shielding, a longer ultra low background period was registered. However, the phenomenon is not yet fully understood. It can either be a real drop in the background caused by radon abundance, or a systematic error that we have yet not been able to detect.

The simulations and laboratory controlled background tests that are underway will improve the understanding of the real potential of micromegas detectors and reasons behind ultra low background levels. CAST micromegas detectors are currently being tested at CEA Saclay in Paris and Canfranc Underground Laboratory in Zaragoza. Furthermore a CAST detector lab is being established at CERN, and will have advanced instrumentation to test micromegas detectors soon. If ultra low background is a real effect, its comprehension may help us develop detectors that will stably have ultra low levels.

4. Background Simulations

There are several possible sources of background events in CAST. Cosmic rays, external gamma radiation, radon contamination and internal radioactivity of the detectors are the most likely ones. CAST micromegas team has initiated background simulations to understand the nature of the background events. The complete geometry of the sunrise micromegas, the pipeline around it and the shielding is implemented in Geant4 [12] (Figure 9). Using the environmental flux values measured at Canfranc with a Germanium detector, the response of the detectors are being simulated. Simulated and experimental data show a good agreement. The simulations so far show that most of the background of micromegas detectors are coming from the interaction of external gammas with material close to the detector (windows and pipeline). This information led to the implementation of the new shielding in sunrise micromegas in 2011.

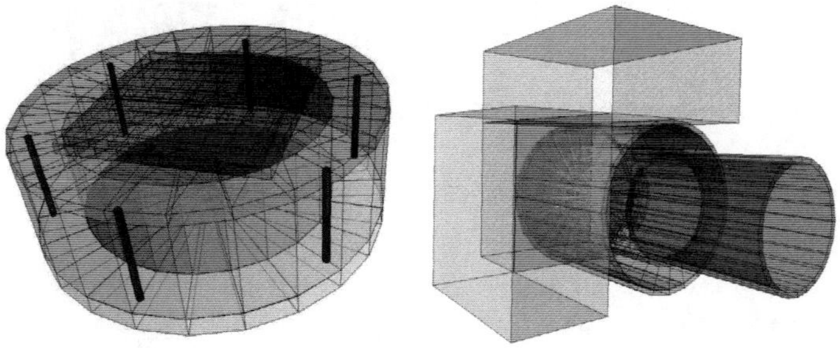

Figure 9. Left: Geometry of the gaseous micromegas chamber. Right: Implementation of the shielding in Geant4.

5. Background Tests

A CAST microbulk micromegas is installed in Canfranc Underground Laboratory, Huesca, Spain. The laboratory is at a depth of 2500 m.w.e (meters water equivalent), where the muon flux is 4 orders of magnitude lower than that at the surface, the environment is stable and the environmental gamma flux is well known. The tests at Canfranc aim to understand the limits of internal radioactivity of CAST detectors, and explore the effects of different shielding setups on them. The detector at Canfranc was tested in different setups, primarily a CAST-like shielding, after a half closed lead shielding, and finally a fully closed lead shielding. Later, another thin layer of copper was installed inside the shielding, to prevent an internal contamination due to Fe^{55} source. With the CAST-like shielding although trigger rate was lower, background levels were similar to CAST, showing that cosmic rays are well discriminated by the offline analysis. As more shielding was added, background levels dropped more. The studies show that, by blocking the radiation from outside of the detector totally, the background levels can be brought down to be as low as $1.5 \cdot 10^{-7}\mathrm{keV}^{-1}\mathrm{cm}^{-2}\mathrm{s}^{-1}$ (Figure 10). This can be considered as a first upper limit for the internal radioactivity of the detectors, and confirm that detectors are built from quite pure materials [10].

Evolution of the background levels at CAST, and comparison with the levels obtained in Canfranc can be seen in Figure 11.

6. Conclusions

CAST is continuing its search for axions with 3 micromegas and 1 CCD detectors. Micromegas has become popular in rare event search experiments for their stability in the long run and capacity for low background levels. The improvements in the micromegas technologies, especially the development of microbulk micromegas, made them even more valuable. The ultra low background phenomenon that CAST has observed few times during the recent years is under investigation by simulations and laboratory tests. The goal of the CAST micromegas team is to use the experience gained from the CAST and other controlled tests to develop the next generation micromegas technology with lower background levels, and use it at the next generation axion helioscopes.

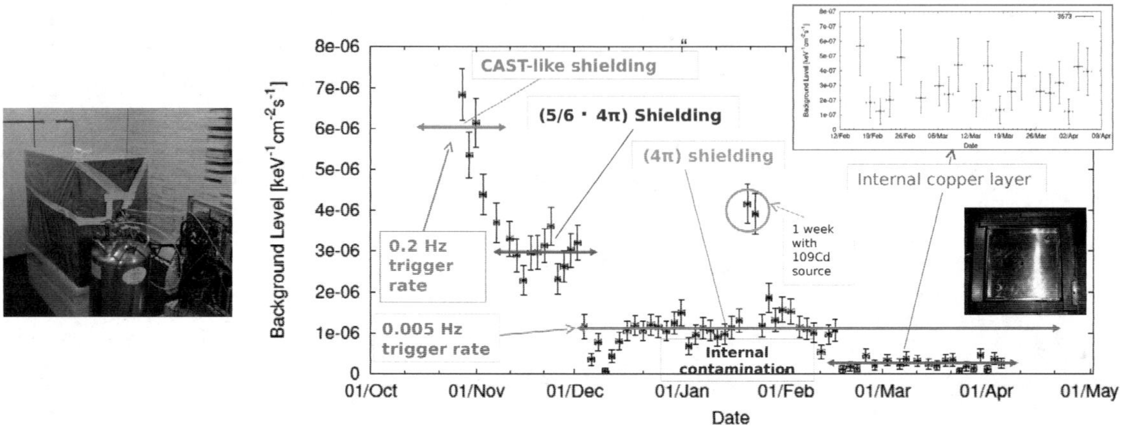

Figure 10. Left: 4π shielding with 20cm of lead. It shields the detector from most of the cosmic particles and environmental gammas. Right: Evolution of the background levels at the Canfranc Underground Laboratory. After the installation of the full shielding, levels dropped to $1.5 \cdot 10^{-7}\text{keV}^{-1}\text{cm}^{-2}\text{s}^{-1}$. By adding a thin copper layer between the calibration source and the detector window, the level dropped to its minimum. The effect of the copper layer reveals that, an important contribution to the background was coming from contamination due the Fe^{55} source.

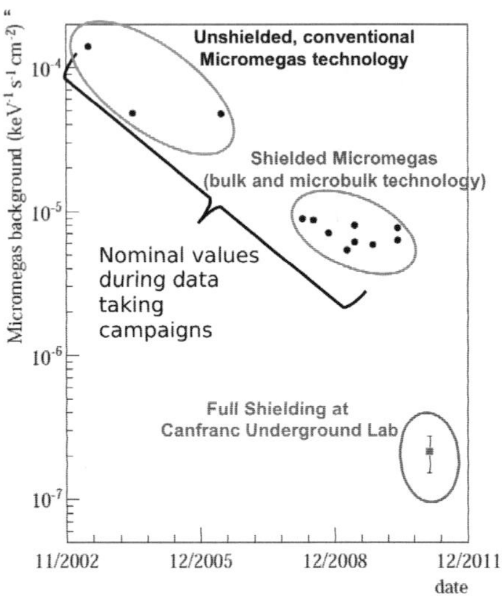

Figure 11. Evolution of CAST micromegas background levels throughout the years. Each point represent nominal values during different periods by different detectors. The rightmost point shows the final result of Canfranc tests.

7. Acknowledgements

We acknowledge support from the Spanish Ministry of Science and Innovation (MICINN) under contract Ref. FPA2008-03456, as well the CPAN project Ref. CSD2007-00042 from the Consolider-Ingenio2010 program of the MICINN. We also acknowledge support from the European Commission under the European Research Council T-REX Starting Grant Ref. ERC-2009-StG-240054 of the IDEAS program of the 7th EU Framework Program, and Turkish Atomic Energy Authority (TAEK).

Finally we would like to acknowledge members of CAST experiment for their hard work, making CAST a pioneer in the search for axions.

References

[1] Zioutas K *et al.* 1999 A decomissioned LHC model magnet as an axion telescope *Nucl. Instrum. Meth. Phys. Res.* A **425**, 480-9

[2] Arik E *et al.* 2009 Probing eV-scale axions with CAST *J. Cosmol. Astropar. Phys.* JCAP02(2009)008

[3] Giomataris Y, Rebourgeard P, Robert J P and Charpak G 1996 A high-granularity position-sensitive gaseous detector for high particle-flux environments *Nucl. Instrum. Methods Phys. Res.* A **376**, 29-35

[4] Giomataris Y *et al.* 2006 Micromegas in a bulk *Nucl. Instrum. Methods Phys. Res.* A **560** 405-408

[5] Aune S *et al.* 2009 New Micromegas detectors in the CAST experiment *Nucl. Instrum. Methods Phys. Res.* A **604** 15-19

[6] Tomás A *et al.* 2011 The new micromegas X-ray detectors in CAST *X-Ray Spectrometry* **40** 240-246

[7] Aune S *et al.* 2009 An ultra-low-background detector for axion searches *J. Phys.: Conf. Ser.* **179** 012015

[8] Andriamonje S *et al.* 2010 Development and performance of Microbulk Micromegas detectors *JINST* **5** P02001

[9] Galán J *et al.* 2010 MICROMEGAS detectors in the CAST experiment *JINST* **5** P01009

[10] Cebrian S *et al.* 2011 Radiopurity of Micromegas readout planes *Astropart. Phys.* **34** 345-359

[11] Geralis T, Fanourakis G, Giomataris Y and Zachariadou K 2003 The data acquisition of the Micromegas detector for the CAST experiment *Nuclear Science Symp. Conf. Record (Portland)* vol 5, 3455-3459

[12] Agostinelli S *et al.* 2003 Geant4 - a simulation toolkit *Nucl. Instrum. Methods Phys. Res.* A **506** 250-303

Journal of Physics: Conference Series **347** (2012) 012030 doi:10.1088/1742-6596/347/1/012030

The T-REX project: Micromegas for Rare Event Searches

T Dafni[1], S Aune[2], J Castel[1], S Cebrián[1], G Fanourakis[3], E Ferrer-Ribas[2], D Fortuño[1], J Galán[2], J A García[1], A Gardikiotis[4], J G Garza[1], T Geralis[3], I Giomataris[2], H Gómez[1], D C Herrera[1], F J Iguaz[2], I G Irastorza[1], G Luzón[1], J P Mols[2], A Ortiz[1], T Papaevangelou[2], A Rodríguez[1], J Ruz[5], L Segui[1], A Tomás[1], T Vafeiadis[6,5,4] and S C Yildiz[7,8]

[1] Grupo de Física Nuclear y Astropartículas, University of Zaragoza, Zaragoza, Spain
[2] IRFU, Centre d' Études de Saclay, CEA, Gif-sur-Yvette, France
[3] Institute of Nuclear Physics, NCSR Demokritos, Athens, Greece
[4] University of Patras, Patras, Greece
[5] CERN, European Organization for Particle Physics and Nuclear Research, Switzerland
[6] Aristotle University of Thessaloniki, Thessaloniki, Greece
[7] Boğaziçi University, Istanbul, Turkey
[8] Doğuş University, Istanbul, Turkey

E-mail: tdafni@unizar.es

Abstract. The T-REX project aims at developing novel readout techniques for Time Projection Chambers for experiments searching for Rare Events. The Micromegas detectors are a good option, because of their good performance regarding low background levels, energy and time resolution, gain and stability of operation. In the present we will shortly refer to two particular cases, on one hand their performance in the CAST experiment and on the other the studies carried out within NEXT, a neutrinoless double-beta decay experiment.

1. The T-REX project

The field of Rare Event Searches is composed of searches involving processes with very low expected rates like the axion, dark matter or the neutrinoless double-beta decay ($0\nu\beta\beta$). It is exactly because of the low expected rates that these searches dictate challenging experimental requirements, the main one being to bring down the experimental backgrounds to the lowest levels possible. The ultra-low background techniques to achieve this include the use of active and passive shielding, operation in underground sites, event discrimination and a careful selection of the detector materials from the radiopurity point of view.

It is only lately that the gaseous time projection chambers (TPC) are being seriously considered as options for the Rare Event Searches, despite the fact that they could offer unique features to their detection. One of them is the topological information, and therefore the directionality of an event, that gas TPC can provide allowing a powerful tool to discriminate signal events from background ones. The complexity rose mainly because of the large volumes and long exposures necessary and all the complications that might arise from that in terms of mechanics, electronics and operation of the chamber itself mainly of the detector readout used.

Journal of Physics: Conference Series **347** (2012) 012030 doi:10.1088/1742-6596/347/1/012030

This change has been produced mainly because of the latest developments on TPC readouts, the micropattern gas detectors (MPGD), that are overcoming the limitations of the previous generations. The technological boost provided by these readouts has been the use of metallic strips or pads, precisely printed on plastic supports with photolithography techniques, substituting the traditional wires of the previous generation of detectors. The simplicity, robustness and mechanical precision are much higher than those of conventional planes of wires. One of the most attractive MPGD for application in Rare Events, is the Micromegas readout plane[1]. The Micromegas detectors have attracted a lot of attention since their invention in 1996, for their potential use in this field of particle physics, among others. Since then it has been constantly and substantially improved, especially regarding the manufacturing techniques.

The T-REX project proposes to merge the latest gas TPC readout concepts and the ultra-low background techniques know-how in order to develop practical ideas that could be used in the field of Rare Event Searches. The main component of these studies is the use of the latest generation Micromegas microbulk readout planes and their further development in order to meet the rather strict requirements of these searches. We give examples in the frame of two different applications, within the CAST experiment [2, 3, 4, 5] for axion searches and within the NEXT project [6, 7] for the neutrinoless double beta decay.

Micromegas technologies The Micromegas readouts are composed of a metallic micromesh suspended over an anode plane by means of isolator pillars, thus defining an amplification gap of the order of 50–150 μm. Electrons drifting towards the readout, go through the micromesh holes and trigger an avalanche inside the gap, inducing detectable signals both in the anode pixels and in the mesh. This amplification geometry has advantages that are of special interest in the context of Rare Events, like good energy resolution and a very low dependance of gain on geometrical or environmental factors, which have been important drawbacks of the previous generation of gas detectors.

The classic Micromegas design [1] has been used in several applications, including nuclear physics experiments at CERN, medical applications, etc. In the last years a lot of effort has been invested to improve the detectors' performance starting with its construction technique. These efforts have resulted in two new types of detectors, build with the bulk [8] and microbulk [9] technologies. The improvement they presented, is that the mesh is no longer suspended above the anode mechanically, instead it is united with the active area by gluing the two parts, as is the case of the bulk, or built together with the active area during the construction process, as is the case of the microbulk. In this way, they provide better robustness, uniformity and therefore stability of operation.

Detectors of both types were constructed and tested in the lab, before they were installed in the CAST experiment, achieving very good background rejection factors.

2. Micromegas for Solar axions

The CAST (CERN Axion Solar Telescope) experiment is searching for axions. The energy range of the expected signal is between 1 and 10 keV and therefore low background detectors are necessary for the experiments sensitivity. The experiment is counting with a powerful dipole magnet and and x-ray focusing device, both of which make CAST the most sensitive helioscope so far.

However, the possibility to improve the key characteristics of a helioscope, with respect to CAST, has lead to a study of the New Generation Axion Helioscope (NGAH) [10]. These characteristics, include the magnetic field volume, the use of x-ray focusing optics and, of most interest for the present work, detector backgrounds. The new helioscope will improve the sensitivity of the technique by 1-1.5 orders of magnitude, enabling it to probe a large range of axion models. Within the scope of the NGAH, the developments described in the present work are further motivated.

Micromegas detectors have been taking data since the beginning of the experiment in 2003. CAST has been a test ground for these detectors, where they have been combined with low background techniques like the use of shielding, radiopurity screening of detector components or advanced event discrimination techniques.

All three types of Micromegas have been used in the experiment through the years. The CAST Micromegas have typically an amplification gap of 50 μm, hole diameters of 30 μm and a pitch of 550 μm and a total active area of 36 cm^2. They were the first to be constructed with a 2d readout, where the anode consists of square pads interconnected in the diagonal direction through vias. Since the upgrade in 2007, the number of Micromegas installed in the experiment was increased from one to three, being now all three of them of the microbulk type. During the upgrade there were improvements in the shielding of all detectors, which in combination with the better characteristics of the microbulks, has lead to an improvement of the detector performance. The effect on the background levels can be appreciated in the plot of Figure 1: they have dropped by at least a factor of 20, with the current detectors showing levels of $6\text{–}9\times10^{-6}s^{-1}cm^{-2}keV^{-1}$.

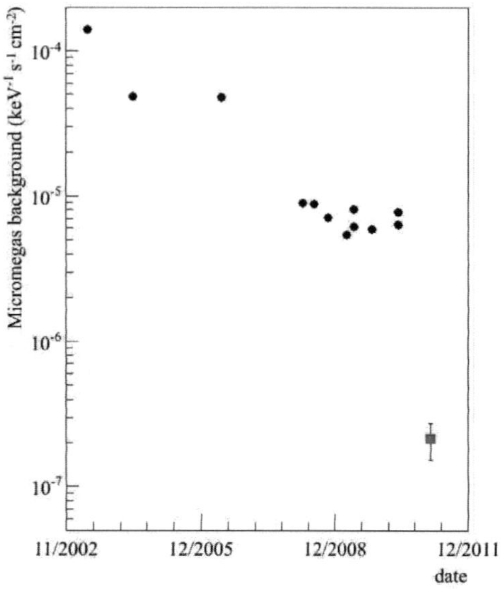

Figure 1. A history of the background levels of the Micromegas detectors registered in different periods of data taking in the CAST experiment. The effect of the different technologies and shielding can be appreciated, as the background levels have improved by at least a factor of 20 with respect to the first detectors. The last point on the right represents the background level recorded in the LSC when the detector was covered with 20 cm of lead. This level is approx. 20 times lower than the standard levels registered in the CAST experiment.

2.1. Studies underground

The experience gathered within CAST and the good performance of the detector, encouraged more dedicated studies for the full understanding of the background levels observed with the aim to push these at the minimum.

On one hand, part of the efforts are focused on experimental background measurements in an environment with well controlled parameters, in order to systematically explore the diverse parameters and configurations in search for a correlation with backgrounds levels. To begin with, a copy of one of the systems in CAST showing very low background levels was prepared and installed to take data in the lab in Zaragoza. The setup is housed inside a Faraday cage, with an inner shielding of copper and lead, an outer shielding of polyethylene and it has been completed with a nitrogen flux in the vicinity of the chamber, imitating the setup of the CAST detectors (Figures 2 and 3).

Figure 2. A photo of the setup in the Zaragoza lab, where the detector can be seen mounted inside the copper and lead shielding and the electronics housed in a Faraday cage.

Figure 3. The same setup in the Zaragoza lab shown in Figure 2, with the Faraday cage surrounded by the polyethylene shielding.

After a short period of data-taking, the system was moved underground, in the Canfranc Underground Laboratory (LSC) in the Spanish Pyrenees. There are several advantages in working underground; among them, the fact that the flux of the cosmic muons is greatly reduced (a factor 10^4), that the environmental conditions like humidity, temperature and pressure are stable, and most importantly, that the flux of gammas in the laboratory is well known. An additional advantage is that the amount of external shielding (lead, primarily) can be largely varied and its effects studied, something that could not take place in the experiment itself because of space and weight restrictions. A continuous Nitrogen flux is provided to purge the radon that could stay close to the detector (Figure 4). With these parameters under control, important conclusions can be drawn for the contribution of the individual background sources (cosmics, external gammas, internal background from contamination) as well as possible systematic effects that could be related with the environmental conditions.

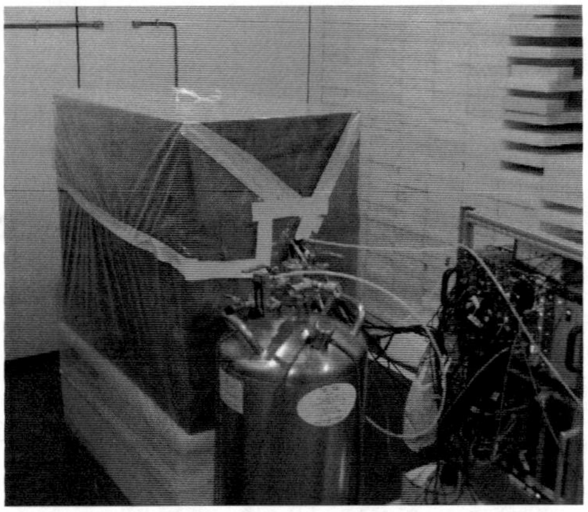

Figure 4. A photo of the setup in the Canfranc Underground laboratory: The Faraday cage that houses the detector is placed inside a lead shielding, here covered in 4π. The dewar placed in front of the shielding provides the Nitrogen flux.

The studies are still ongoing, but levels as low as $2\times10^{-7}\mathrm{s}^{-1}\mathrm{cm}^{-2}\mathrm{keV}^{-1}$ are reached by adding more lead all around the detector (Figure 1). This level corresponds to the intrinsic level of the radiopurity of the detector components. The results obtained, indicate that the background at CAST is most probably driven by the external gamma radiation. These studies provide the needed information in order to design improvements on the shielding of some of the detectors in CAST. Another conclusion, which will be commented in a later section, is that the levels of radioactivity in these detectors (and the surrounding materials) are very low. Furthermore, all the information obtained by these studies, can be used to devise the design of the future helioscope.

The second branch of the work was focused on a general and complete simulation of the detector. Based on Geant4, simulations have been prepared with a detailed description of the geometry of the detector and its shielding. Special care has been taken to include information about radiopurity measurements that have been performed at the LSC, using the materials that compose the detector. A big effort was invested in implementing all the ingredients from signal generation to the response of the electronic chain as is currently used in CAST. The first results show that the simulation seems to describe successfully the system. As an example, Figure 5 shows the comparison of two parameters of the detector signal as given by the experiment (left column) and the simulation (right column) for a calibration run where an ^{55}Fe source is used.

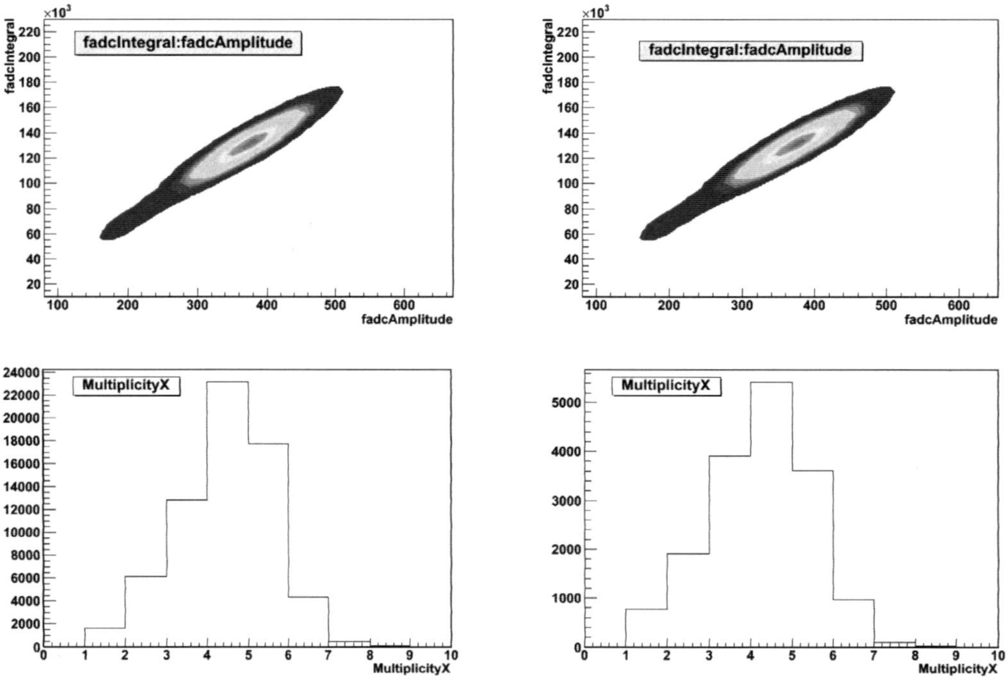

Figure 5. Comparison of two parameters of the detector performance as observed in the experiment (left column) and by the simulation (right column). At the top is seen the integral versus the amplitude of the pulse signal collected in the mesh, while at the bottom the number of consecutive strips or pads activated in an X-ray event, also know as *multiplicity* of the event.

More details on this work can be found in [11, 12, 13, 14].

3. Studies within the NEXT project

The neutrinoless double-beta decay experiments are another part of the Rare Event Searches where Micromegas readouts could be used. NEXT (Neutrino Experiment with a Xenon TPC) is one of these experiments. It is a project approved by the LSC in the Spanish Pyrenees. It has been funded by the Consolider Program of the Spanish Ministry of Science to carry out the construction of a 100 kg enriched high-pressure xenon gas (HPGXe) TPC. The proposal has good prospects regarding the requirements of such an experiment: the excellent energy resolution which can be achieved, the background rejection power provided by the topological information of the electron tracks, the ability to define a t0 (thus a fiducial region away from any surface), and the scaling-up to large masses. The final detector, NEXT-100, will hold 100 kg of Xe gas enriched in ^{136}Xe, its $\beta\beta$-decaying isotope. The project baseline is to build the TPC based on the electroluminescence technique using a Photomultiplier tubes (PMT) plane for the energy measurement and SiPMs for the topological information [15]. Another option that was studied was using APDs[16], while here we will discuss the option of the charge collection with Micromegas [17], studied within the T-REX project.

As mentioned before, the requirements that the readout planes should meet regard operation at high pressure, good energy resolution, low radioactivity. The primary goal of the first tests were to establish that the Micromegas planes can operate satisfactorily in high pressure pure Xenon. As a second step, to explore their energy resolution performance, whether it can reach the levels required for a neutrinoless double beta decay search with Xe.

Two prototypes were built to test the performance of Micromegas readouts at high pressure in pure Xe. The first one, known as NEXT-0-MM, is a Stainless Steel vessel of 2 litres, with a diameter of approx. 14 cm and a drift region of 6 cm. In the interior a series of 6 copper rings interconnected with resistors form a field cage. The cathode consists of a copper disk with a hole to accommodate the sources used in the tests (Figures 6, 7).

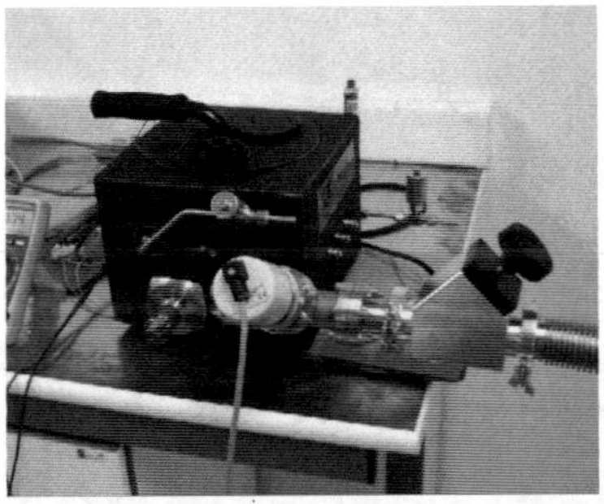

Figure 6. A photo of NEXT-0-MM. The vessel is covered with an insulator layer (black).

Figure 7. The field cage of NEXT-0-MM: six copper rings interconnected via resistors, held in place by three PEEK bars and a copper disk as a cathode on top.

While this was testing Micromegas detectors at high pressures, a second prototype was designed and built with the aim to study electron tracks, which would be the expected signal. These would reach a length of the order of 30 cm in a 10 bar Xe environment. Therefore, NEXT-1-MM, has an inner volume of approximately 75 litres, measuring 600 mm of height and almost 400 mm in diameter (Figure 8). The vessel should hold 1 kg of Xenon at a pressure of 10 bar. This vessel, like the smaller prototype, is also equipped with a field cage which defines a drift length of 35 cm. The structures is composed of 34 copper rings, interconnected with 10 MΩ resistors, which can be seen in Figure 9.

The Micromegas readouts used in the tests were of the microbulk type and they were mainly circular readouts with an amplification gap of 50 μm and 35 mm diameter whose anode was not segmented (Figure 10). When the manufacturing technology permitted bigger areas, tests were made with a readout with an active area of 10$\times$10 cm^2 (Figure 11). This has a pixelized anode

Figure 8. A photo of the NEXT-1-MM prototype and its supporting structure.

Figure 9. The field cage of NEXT-1-MM: 34 copper rings, connected through resistors and held by PEEK bars.

with a total of 144 pixels. However, bigger readouts were needed to equip NEXT-1-MM, which has an active area corresponding to a diameter of 30 cm. The first readout prepared for that vessel was a circular bulk detector, seen in Figure 12, with a segmented anode of 1252 pixels. Lately a microbulk detector with a side of 15 cm was manufactured (Figure 13). It has a shape of a quarter of a circle, such as that we need four of these to cover the active area of NEXT-1-MM. Each of the sectors counts 288 pixels. During the tests both signals of the anode and of the mesh are recorded.

Figure 10. The first generation of microbulk Micromegas used. They had a diameter of 35 mm and an amplification gap of 50 μm.

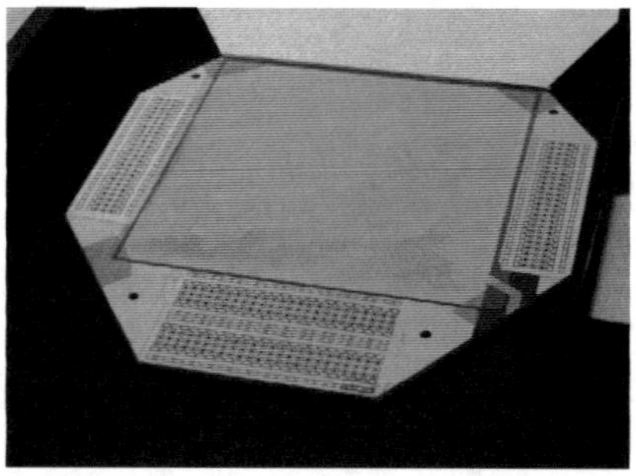

Figure 11. The first large microbulk built, with a segmented anode. It has an active area of $10 \times 10\,\mathrm{cm}^2$ divided into 144 pixels.

Examples of tracks In the cases where the Micromegas readouts have a segmented anode, the pixels are individually read. This is done employing the electronics used in the T2K experiment. To read the first ever pixelized microbulk Micromegas, a reduced version of the electonics was connected to 72 out of the 144 pixels. An example of a track can be seen in Figure 14. In NEXT-1-MM, the bulk installed had 1252 pixels and therefore the full chain of the T2K electronics was necessary to record tracks like the one in Figure 15.

Journal of Physics: Conference Series **347** (2012) 012030 doi:10.1088/1742-6596/347/1/012030

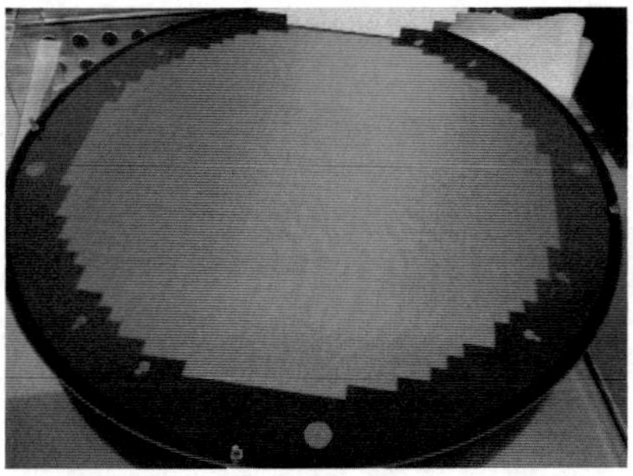

Figure 12. A photo of the bulk Micromegas detector installed in the NEXT-1-MM. It has a diameter of 30 cm and its anode is segmented to ~1200 pixels.

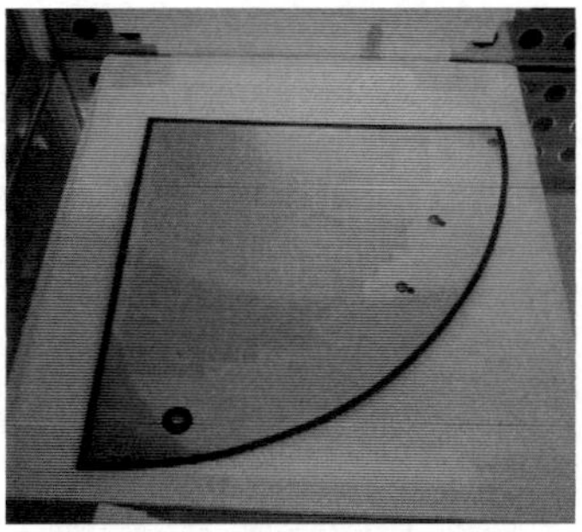

Figure 13. A photo of the new microbulk Micromegas recently constructed. It is one of the four sectors that will be installed in order to cover the sensitive surface of NEXT-1-MM.

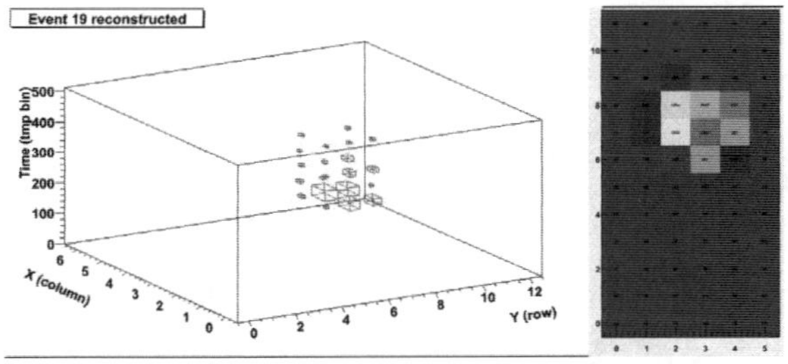

Figure 14. Projection of an alpha track obtained with the ^{241}Am source in NEXT-0-MM. A reduced version of the T2K electronics was used to read the pixels. On the right is the 2D mapping of the event while on the left, a 3D reconstruction of the same event.

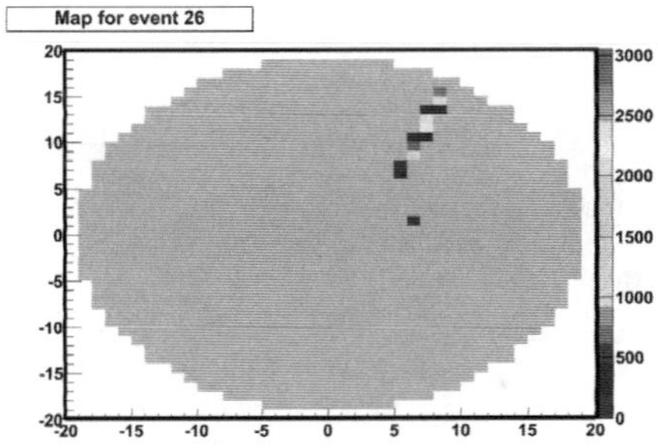

Figure 15. An alpha track of approximately 4 cm obtained with the ^{222}Rn source in NEXT-1-MM. The shaded area corresponds to the active surface of the bulk detector.

Studies at high pressure with various gases In order to determine the absolute position of an event along the drift direction, a measurements of the absolute time of interaction, or t_0, is needed. This can be obtained by the primary scintillation of Xe, that is why operation in pure Xenon would be required. The first tests performed were oriented towards this goal.

Different types of tests were performed with microbulk Micromegas readouts in NEXT-0-MM. The starting point of all microbulk measurements is testing them in Argon-iC_4H_{10} mixtures. Later, data in pure Argon and in pure Xenon were taken. All these measurements were repeated in various pressure settings, reaching 10 bar. In most of the cases the ^{241}Am was used as a source, which emits alphas at an energy of 5.5 MeV. In others, gamma sources of low energy were employed. Here we will summarize the best results obtained.

In Ar-2%iC_4H_{10} gas mixture, the best energy resolution achieved was 0.7% (FWHM) for the 5.5 MeV of the alphas coming from an source, at 4.75 bar [18]. In the measurements performed

Journal of Physics: Conference Series **347** (2012) 012030 doi:10.1088/1742-6596/347/1/012030

with pure Ar and Xe, a protocol of purifying the system with long bake-out periods and warm Nitrogen circulation was followed. As seen in Figures 16 and 17, the best results achieved respectively, for the 5.5 MeV α of the ^{241}Am, were 2.03% (FWHM) at 8 bar of Ar and 1.94% (FWHM) at 4 bar of Xe. Data with the pure Xe were taken as well at pressures of up to 8 bar, although the results are not so good due to attachment in the gas.

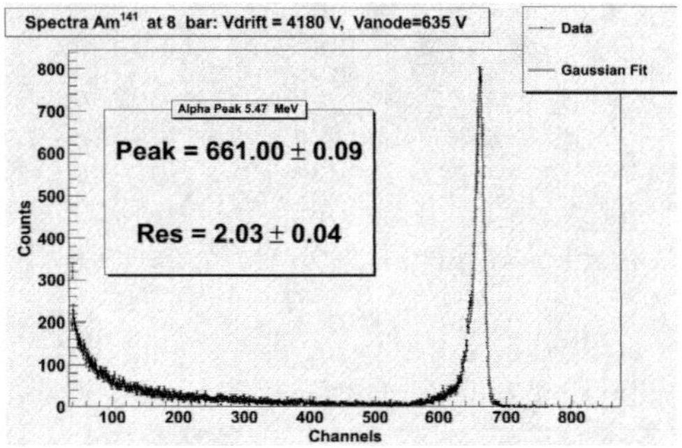

Figure 16. Energy spectrum of the ^{241}Am 5.5 MeV alpha peak measured at 8 bar of pure Argon, showing an energy resolution of 2.03% FWHM.

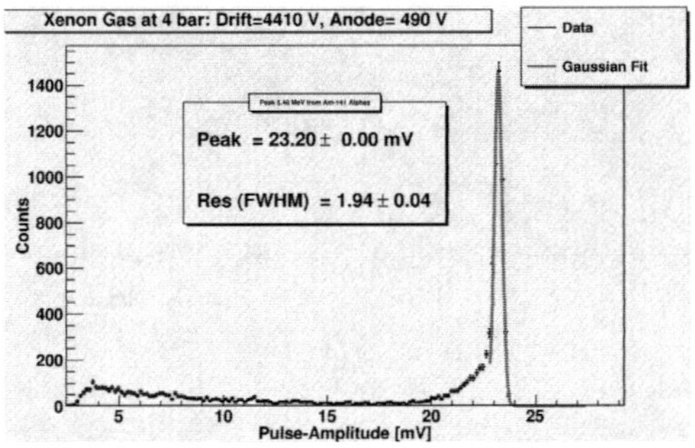

Figure 17. Energy spectrum of the ^{241}Am 5.5 MeV alpha peak measured at 4 bar of pure Xenon, showing an energy resolution of 1.94% FWHM.

Interesting results were also obtained when using the ^{241}Am source but blocking the alphas, thus probing lower energy region with the 59.54 keV photons that are also emitted. The few measurements at high pressure in these conditions, have yielded an energy resolution of 7.8% at 2 bar and 9.3% (FWHM) for the 60 keV γ of the ^{241}Am at 3.5 bar (Figure 18). Other colleagues in a smaller setup obtained more results in high pressures [19].

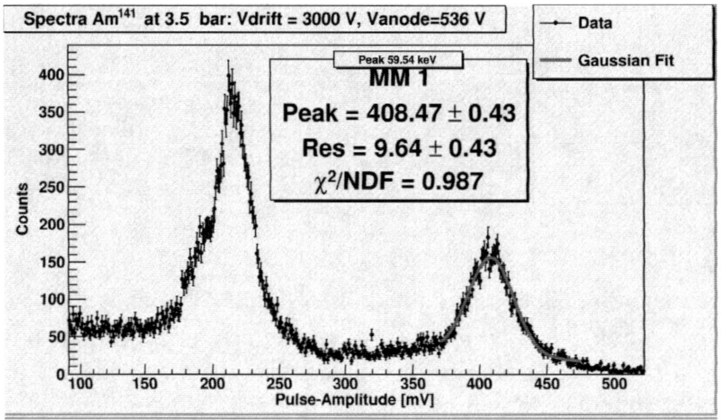

Figure 18. Energy spectrum of 59.54 keV γ in pure Xenon at 3.5 bar with an energy resolution of 9.6% (FWHM). The peak on the left is composed mainly of escape peaks from the 60 keV α, due to the fluorescence of Xe at 30 and 33 keV and partly of the 26 keV γ line of ^{241}Am.

Although the first inclination was to work in pure Xe, the possibility of using a quencher gas is still very interesting. An extensive campaign of tests with other gas mixtures is ongoing. As an example, studies are being done with Xe-Ne mixtures, and Xe-TMA, with which higher gains than in pure Xe have been achieved. In the case of Xe-TMA, very good energy resolutions seem to be at hand. Figure 19 shows an example of a measurement with a ^{109}Cd source in Xe-2.4% TMA at 4 bar with an energy resolution better than 13% (FWHM) for the 22 keV photons. More results on these studies will soon be published.

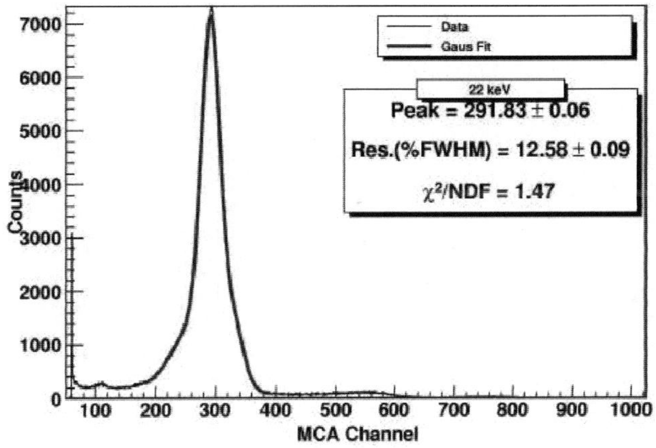

Figure 19. Energy spectrum of the ^{109}Cd 22 keV gamma peak measured at 4 bar of a mixture of Xenon with 2.4%TMA. The energy resolution achieved is 12.6% FWHM.

Commissioning NEXT-1-MM Work in the bigger prototype is progressing slowly. There are currently two microbulks installed and being tested with alpha sources. The next steps include installing the four readouts in order to cover the complete active area and start the tests towards the goal of seeing electron tracks in the system.

4. Radiopurity of Micromegas planes

As mentioned before, in the field of Rare-Event searches the detectors, apart from other performance characteristics, are required to follow a strict budget on radiopurity; looking for the lowest background possible, a limiting factor could be the readout plane introduced. We performed the first measurements of the radiopurity of microbulk Micromegas planes in the low background facilities of the LSC with a high purity germanium detector [20]. The results obtained show radioactivity levels below 30 μBq/cm^2 for Th and U chains and $\sim$60μBq/cm^2 for ^{40}K. These results are comparable to the cleanest detector components of the most stringent low background experiments at present, despite the fact that the readouts measured were manufactured without any special care from the radiopurity point of view. Therefore it is reasonable to believe that after a dedicated study and development, these results will be improved.

Currently there is an effort to identify the most radioactive raw materials or the fabrication processes which may introduce radioactive traces during the construction of the planes and try to use alternative ones, aiming at even better results.

5. Conclusion

The aim of the T-REX project is to develop microbulk Micromegas readouts suitable for Rare Event Searches, taking advantage of ultra-low background techniques. A good part of this work has been done within the frame of two Rare Event Searches, Solar axions and NEXT.

Journal of Physics: Conference Series **347** (2012) 012030 doi:10.1088/1742-6596/347/1/012030

The results of the studies show that the microbulk Micromegas are of most interest summing performance characteristics such as: stability of operation, very low background levels (6–$9\times10^{-5}s^{-1}cm^{-2}keV^{-1}$), good operation at high pressures and with very encouraging results regarding the energy resolution (1.96% (FWHM) at 4 bar of pure Xe for 5.5 MeV alphas, 9.5% (FWHM) at 3.5 bar of pure Xe for 60 keV gammas, 2.4%(FWHM) at 3.4 bar of Xe-2.4%TMA for 22 kev gammas). Moreover, the microbulk readouts show rather low radioactivity levels, a must for any readout plane to be used in such experiments.

Acknowledgments

We want to thank our colleagues of the groups of the University of Zaragoza, CEA/Saclay and our colleagues from the CAST, NEXT and RD-51 collaborations for helpful discussions and encouragement. We thank R. de Oliveira and his team at CERN for the manufacturing of the microbulk readouts. We acknowledge support from the European Commission under the European Research Council T-REX Starting Grant ref. ERC-2009-StG-240054 of the IDEAS program of the 7th EU Framework Program. We also acknowledge support from the Spanish Ministry of Science and Innovation (MICINN) under contract ref. FPA2008-03456, as well as under the CUP project ref. CSD2008-00037 and the CPAN project ref. CSD2007-00042 from the Consolider-Ingenio 2010 program of the MICINN. Part of these grants are funded by the European Regional Development Fund (ERDF/FEDER). Finally, we acknowledge support from the Regional Government of Aragón under contract PI001/08.

References

[1] Giomataris Y, Rebougeard P, Robert JP and Charpak G, 1996 *Nucl. Instr. and Meth.* A **376** 29

[2] Zioutas K et al. 1999 *Nucl. Instrum. Meth.* A **425**, 480.

[3] Zioutas K et al. (CAST Collaboration),2005 *Phys. Rev. Lett.* **94** 121301

[4] Adriamonje S et al (CAST Collaboration), 2007 *J. Cosmol. Astropart. Phys.* JCAP04(2007)01

[5] See talk on "CAST physics achievements and perspectives" by K. Zioutas et al. of this conference.

[6] Granena F et al. (The NEXT Collaboration) 2009 NEXT, a HPGXe TPC for neutrinoless double beta decay searches *Preprint hep-ex/0907.4054*

[7] Alvarez A et al. (The NEXT Collaboration) 2011 The NEXT-100 experiment for neutrinoless double beta decay searches (Conceptual Design Report) *Preprint arXiv:1106.3630v1 [physics.ins-det]*

[8] Giomataris Y et al., 2006 *Nucl. Instrum. Meth.* A **560** 405–8

[9] Andriamonje S et al. 2010 *J. Instrum.* JINST 5 P02001

[10] Irastorza I G et al. 2011 *J. Cosmol. Astropart. Phys.* JCAP06(2011)013

[11] See talk on "CAST Micromegas detector performance" by S.C. Yildiz et al. of this conference.

[12] See talk on "CAST Micromegas detectors in the Canfranc Underground Laboratory" by A. Tomás, TIPP 2011 http://indico.cern.ch/getFile.py/access?contribId=139&sessionId=16&resId=0&materialId=slides&confId=102998

[13] Galán J et al. 2010 *J. Instrum.* JINST 5 P01009

[14] Dafni T et al. 2011 *Nucl. Instr. and Meth.* A **628** 172–6

[15] Liubarsky I for the NEXT Collaboration 2010 2011 *J. Phys.: Conf Series* **309** 012007

[16] Lux T et al. 2010 2011 *J. Phys.: Conf Series* **309** 012008

[17] Cebrián S et al. 2010 *J. Cosmol. Astropart. Phys.* JCAP10(2010)010

[18] Dafni T, Ferrer-Ribas E, Giomataris I, Gorodetzky P, Iguaz F, Irastorza IG, Salin P and Tomás A, 2009 *Nucl. Instr. and Meth.* A **608** 259–66

[19] Balan C, Freitas EDC, Papaevangelou T, Giomataris I, Natal da Luz H, Monteiro CMB and dos Santos JMF, 2011 *J. Instrum.* JINST 6 P02006

[20] Cebrián S et al. 2011 *Astropart. Phys.* **34** 354–9 (Preprint arXiv:1005.2022v1)

2nd International Conference on Particle Physics IOP Publishing
Journal of Physics: Conference Series **347** (2012) 012031 doi:10.1088/1742-6596/347/1/012031

The CUORE experiment: a search for neutrinoless double beta decay

Marco Carrettoni[1]

Dipartimento di Fisica dell'Università degli studi di Milano "Bicocca" and Sezione INFN di Milano-Bicocca, Italy

E-mail: marco.carrettoni@mib.infn.it

Abstract. Neutrinoless Double Beta Decay is searched by physicists since tens of years. Fighting to reach higher and higher sensitivities, novel techniques have been suggested and large mass experiments have been proposed or are under construction. Within this picture, cryogenic particle detectors play a primary role. CUORICINO, a 40.7 kg array of low temperature TeO_2 calorimeters, provided one of the best limits on the effective neutrino Majorana mass. The final result of this - now closed - experiment is here presented. The heir of CUORICINO is CUORE: an array of 988 TeO_2 calorimeters with a total mass of 750 kg. The current status of CUORE - presently under construction at Laboratori Nazionali del Gran Sasso - is illustrated, the perspectives for the Neutrinoless Double Beta Decay search are discussed.

1. Introduction

The *Double Beta Decay* is a transition in which an (A, Z) nucleus decays into its (A, Z+2) isobar. This is the only decay channel for a group of isotopes whose single beta decay is forbidden. The interest in its detection stands in the possibility that the transition goes through a channel in which no neutrinos are emitted. The *Neutrinoless Double Beta Decay* ($\beta\beta(0\nu)$) is possible only if the neutrino is a massive Majorana particle [1] and it happens to be (at least today) the only way to prove this property. If the $\beta\beta(2\nu)$ exists, its half-life is inversely proportional to the square of the Majorana mass of neutrino, defined as $m_{ee} = |\sum m_i U_{ei}^2|$ (where U_{ei} are elements of the PNMS matrix and m_i are the three neutrino mass eigenstates). The $\beta\beta(2\nu)$ half-life ($T_{1/2}^{0\nu}$) can be written as:

$$\frac{1}{T_{1/2}^{0\nu}} = m_{ee}^2 F_N = m_{ee}^2 G^{0\nu} |M^{0\nu}|^2 \quad (1)$$

Here $G^{0\nu}$ is the two-body phase-space factor and $M^{0\nu}$ is the *Nuclear Matrix Element* (NME), their product F_N is called the *Nuclear Factor of Merit.* Present experimental upper bounds on $\langle m_{ee}\rangle$ are of the order of ~0.5 eV [2, 3, 4, 5], with only one positive value reported by part of the Heidelberg-Moscow collaboration [6]

The decay is detected on the basis of the two electron signal: given the negligible energy of the recoiling nucleus the sum kinetic energy of the two electrons is equal to the Q-value of the $\beta\beta(0\nu)$transition. This almost monochromatic signal is the main signature used by

[1] On behalf of the CUORE Collaboration

all the experiments. Depending on the detector type and on the set-up features, other characteristic information of the process can be used (as it is in tracking experiments) to identify $\beta\beta(2\nu)$ events rejecting background signals, thus improving the experimental sensitivity. We define sensitivity as the decay time corresponding to the minimum number of detectable events above a background at a given C.L., which can be expressed in terms of the detection efficiency (ϵ), the live-time (T), the energy resolution (ΔE) and the background counting rate (B) as:

$$S^{0\nu} = \ln 2 \frac{N_A \cdot a \cdot \eta \cdot \epsilon}{W} \sqrt{\frac{M \times T}{\Delta E \times B}} \tag{2}$$

where N_A is the Avogadro number, η is the stoichiometric coefficient of the candidate and a the isotopic abundance. W is the molecular weight of the absorber and M is the mass of the detector. In order to convert this expression to quantify the $\langle m_{ee} \rangle$ sensitivity we can use equation 1 by taking into account the spread between the different nuclear matrix elements .

The Q-value, the predicted F_N values, the natural isotopic abundance and the available detection techniques bias the choice of the $\beta\beta$ emitters used in experiments. High abundances (or isotopes with a viable enrichment) imply high $N_{\beta\beta}$. High Q-values imply low background (B) coming from environmental radioactive emissions. Finally the detection technique defines ϵ and ΔE and, in most cases, also restricts the number of isotopes that can be investigated.

$\beta\beta(2\nu)$ experiments can be divided into two major classes, conventionally referred to as *source=detector* experiments (the particle detector used to search the decay contains - in its active material - the $\beta\beta(2\nu)$ candidate isotopes) and *source≠detector* (the source is a passive material inserted into a suitable particle detector). The former technique bases its power on the use of high resolution detectors (low ΔE) with a rather simple set-up, having as a disadvantage the generally poor background rejection capability. In the latter technique only poor resolution detectors have been employed up to now, but the high value of ΔE is compensated by the presence of some supplementary information - like particle tracking - that allows to establish a powerful background rejection system.

2. CUORICINO

Experiments using low temperature calorimeters (bolometers) belong to the *source=detector* class. Here the isotope is contained in the molecule of the bolometer absorber which is a single crystal, generally with dielectric and diamagnetic properties. The working principle of a bolometer is quite simple: the energy deposited by a particle traversing or stopping in a crystal absorber produces a thermal signal that - when the temperature is low enough - can be detected with a proper sensor [7]. When compared with traditional solid state devices based on ionization (e.g. semiconductor devices or scintillators), bolometers offer a quite larger choice of materials. Size and energy resolution can be similar to those achieved with Ge diodes. The application of this technique to $\beta\beta(2\nu)$ searches was pioneered by the Milano group [8, 9, 10] with the study of ^{130}Te $\beta\beta(2\nu)$ by means of TeO_2 bolometers. Indeed, ^{130}Te is considered a *golden candidate* for $\beta\beta(2\nu)$ investigation because of the favorable F_N, the high transition energy 2527.5± 0.01 keV [11] and the high natural isotopic abundance (33.8%). This was a good reason to start with ^{130}Te, when the idea of applying the bolometric technique to $\beta\beta(2\nu)$ searches was proposed, about 20 years ago. The good performances of TeO_2 bolometers, their extremely good radiopurity, the reasonable cost of the crystals lead to a story of successes ending up with the result of the CUORICINO experiment that yields the most stringent bound on m_{ee} based on ^{130}Te studies, and one of the best in general [12].

CUORICINO was an array of 62 detectors (40.7 kg of total mass) of TeO_2 natural crystals operated as bolometers in a low temperature refrigerator. Two types of detectors were used for the construction of the array: 44 TeO_2 crystals of size 5x5x5 cm^3 and mass 790 g (*big crystals*)

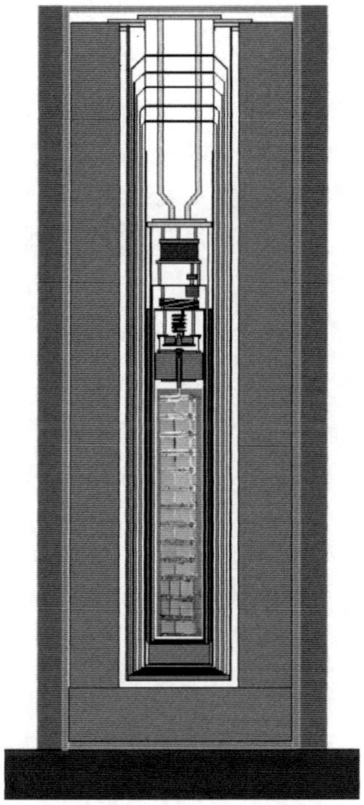

Figure 1. Left: CUORICINO experiment. The cryostat structure is visible with the bolometer array mounted inside the internal vacuum chamber, surrounding the cryostat there are the external lead and the neutron shields. Right: the bolometer array in its Copper structure.

and 18 TeO_2 crystals of size 3x3x6 cm^3 and mass 330 g (*small crystals*). All the crystals were made of natural tellurium but 4 of the small size ones. These were made with enriched materials: two of them were enriched to 75% in ^{130}Te (^{130}*Te-enriched crystals*) and two were enriched to 82.3% in ^{128}Te. The experiment was located underground, in Laboratori Nazionali del Gran Sasso (INFN - Italy), that provides an average coverage of 1400 m of rock (i.e. a depth of 3650 m.w.e.). The array and the experimental set-up are shown in figure 1.

CUORICINO operated between 2003 and 2008 in two slightly different configurations (run I and run II). The total ^{130}Te exposure was 19.75 kg·y. For the $\beta\beta(0\nu)$analysis the two ^{128}Te enriched crystals were not used and their data will not be discussed here. The statistics collected in the two runs is detailed in table 1: big crystals, small-natural crystals and ^{130}Te-enriched crystals are kept separate in this table as they have been treated separately for the $\beta\beta(2\nu)$ analysis. This is because they are characterized by different masses and consequently different efficiencies with respect both to the background sources and $\beta\beta(0\nu)$. Similarly run I and run II differ in the number of active detectors and in their energy resolutions, therefore they have been treated as two independent experiments whose likelihoods - for $\beta\beta(2\nu)$ analysis - have been combined.

In CUORICINO each detector was provided with its own read-out and DAQ channel and trigger, so that each was a completely independent device. Coincidences among the detectors could be invoked to reject background events (as discussed later) or for dedicated studies such as those aiming at the indentification of the background sources. The energy calibration, i.e. the conversion of the thermal pulse amplitude into energy, was provided by a periodic exposure of the entire array to a ^{232}Th source. The main full energy gamma lines emitted by the beta decaying isotopes of the ^{232}Th chain (ranging from 511 keV to 2615 keV) were used for this

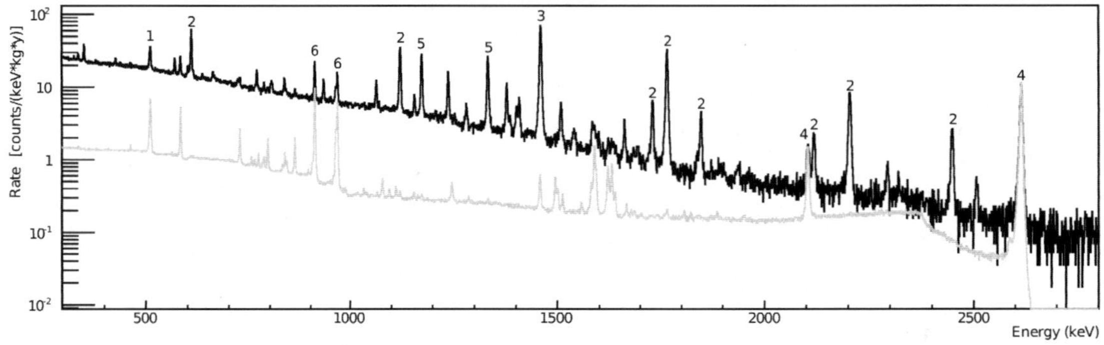

Figure 2. Anticoincidence total energy spectrum of all CUORICINO detectors (black). The most prominent peaks are labeled and come from known radioactive sources such as: e^+e^- annihilation (1), ^{214}Bi (2), ^{40}K (3), ^{208}Tl (4), ^{60}Co (5) and ^{228}Ac (6). The total energy spectrum of all CUORICINO detectors during calibration measurements is also shown (green). For convenience it is normalized to have the same intensity of the 2615 keV line of ^{208}Tl as measured in the background spectrum.

purpose. The periodic ^{232}Th measurements have been used also for a check of the variation of the energy resolution of the detectors in time.

In CUORICINO the detector responses were quite uniform within the 3 groups of crystals (big, small and ^{130}Te-enriched) with respect both to energy resolutions and to background counting rates. In run I we measured a FWHM resolution on the 2615 keV ^{208}Tl line of 7.8$\pm$ 2.8 keV for the big crystals, 9.1$\pm$ 3.1 keV for small natural crystals and about 20 keV for the two ^{130}Te enriched oness. In run II FWHM values were 6.3$\pm$ 2.5 keV for the big crystals, 9.9$\pm$ 4.2 keV for small crystals and about 13 keV on the two ^{130}Te enriched ones. Here the error is the spread over the values measured for the different detectors in the 5 calibration measurements done in run I and in the 32 done during run II. The 2615 keV line is used as a reference because it is the high intensity gamma line nearest to the $\beta\beta(2\nu)$ region.

The spectrum obtained by summing all the CUORICINO collected data is shown - below 2.7 MeV - in figure 2. The background recorded by the detectors is clearly dominated, in this region, by the gamma emissions due to radioactive contaminants of the detector and of the surrounding apparatus. The most intense gamma lines are listed in reference [12]. In figure 2, the spectrum corresponding to the sum of all calibration data is also shown. For convenience the calibration spectrum is normalized to have the same intensity of the 2615 keV line of ^{208}Tl as measured in the background spectrum. The $\beta\beta(2\nu)$ signal was searched for looking to the monochromatic peak that would be produced by the $\beta\beta(0\nu)$decays fully contained within the source crystal. These are a quite large fraction of the total number of decays ($\sim$87% for the big crystals and $\sim$84% for the smaller ones) and would produce a line at the transition energy. When looking to this particular signature, an efficient background suppression technique is that obtained operating the array in anticoincidence. This provides in CUORICINO a suppression of the background in the Region Of Interest (ROI) of about 20%.

In the CUORICINO spectrum no peak was observed in correspondence of the ^{130}Te $\beta\beta$ transition energy. The $\beta\beta(2\nu)$ half-life limit was evaluated as follows. The peaks and continuum in the region of the spectrum centered on the $\beta\beta(2\nu)$ energy were fit using a maximum likelihood analysis where free parameters were the $\beta\beta(2\nu)$ rate, the ^{60}Co rate and line position[2]

[2] Here we refer to the 2505 keV line visible in the ROI, which is due to the coincident detection - in a crystal - of the two gamma rays emitted by ^{60}Co.

Table 1. Cuoricino Run I and Run II crystal masses and exposures.

Crystal Type	Crystal Mass [g]	^{130}Te Mass [g]	Exposure Run II [kg(^{130}Te)·y]	Exposure Run I [kg(^{130}Te)·y]
big	790	217	15.80	0.94
small	330	91	2.02	0.094
^{130}Te-enriched	330	199	0.75	0.145

and the flat background rate. A Bayesian approach has been then used to evaluate the $T^{0\nu}_{1/2}$ lower bound for ^{130}Te decay that results to be $2.8{\cdot}10^{24}$ y at 90% C.L.[13]. In this value the systematic errors have been included. This limit was used to extract an upper limit on m_{ee} using the theoretical NME evaluations for the ^{130}Te nucleus that - depending on the nuclear model- range from 300 meV to 710 meV. Indeed:

- 300-570 meV is the range obtained using the Quasiparticle Random Phase Approximation (QRPA) evaluations of reference [14].
- 360-580 meV is the range obtained using the QRPA evaluations of reference [15].
- 570-710 meV is the range obtained using the Shell Model (SM) evaluations of reference [16].
- 350-370 meV is the range obtained using the Interacting Boson Model (IBM) evaluations of reference [17].

Note that, for each reference, a range (and not a single value) for m_{ee} is presented because of the different results on the NME obtained by the authors when e.g. varying the treatment of the short range correlations or the value of g_A (the axial-vector coupling).

3. CUORE

CUORE is the natural extension of the CUORICINO experiment toward higher $\beta\beta(2\nu)$ sensitivities. The CUORE Project [18] foresees the realization of a $\beta\beta(2\nu)$ experiment with an active mass of the order of 1 ton. CUORE will employ 988 natural TeO_2 bolometers each made of a cubic 5×5×5 cm^3 TeO_2 crystal with a mass of about 750 g. The goal of the CUORE collaboration is to reach, in the energy region of interest, a background level lower than 10^{-2} counts/keV/kg/y obtaining hence a sensitivity on the effective Majorana mass of the order of 50 meV. The background counting rate in the ROI has to be reduced by a factor ∼16 with respect to CUORICINO (in the ROI the 5×5×5 cm^3 crystals in CUORICINO had a counting rate of 0.161±0.006 counts/keV/kg/y). The CUORE array is designed in order to have the most compact structure, reducing to a minimum the distance between the crystals and the amount of inert material interposed. The 988 bolometers of the array are arranged in a cylindrical matrix organized into 19 *towers*, each made of 13 *planes*. Every plane contains four crystals supported inside a Copper frame. The entire array, surrounded by a 6 cm thick lead shield, will be operated at about 10 mK in a He^3/He^4 dilution refrigerator (see figure 3). A further thickness of 30 cm of low activity lead will be used to shield the array from the dilution unit of the refrigerator and from the environmental activity. A borated polyethylene shield and an air-tight cage will surround externally the cryostat. The experiment will be installed underground in the same experimental hall of LNGS where CUORICINO was operated.

While operating a 1000 bolometer array at 10 mK could look rather challenging, the technical feasibility of CUORE has been extensively proved by the good performances of the CUORICINO

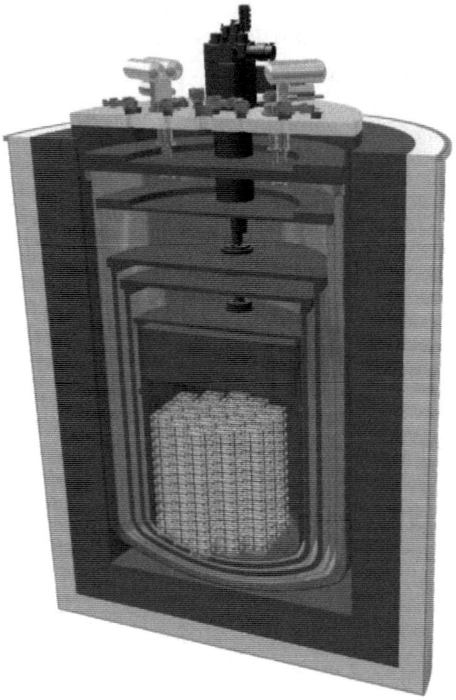

Figure 3. CUORE cryostat and the 988 crystals array.

experiment, and the possibility of cooling large masses in dilution refrigerators have been proved, for example, by the gravitational antenna experiments. With respect to CUORICINO, the CUORE array will benefit from an improved mechanical decoupling of the array from the cryogenic set-up, from the elimination of the LHe bath and from a holder for the crystal that has been specifically studied to reduce vibrational noise contribution to the energy resolution. All this should lead to a better reproducibility of the detectors from the point of view of their energy resolution and threshold, as well as to a better duty cycle in the data taking.

The true challenge in CUORE - as in all the next generation $\beta\beta(2\nu)$ experiments - will be in achieving the background target. The set-up of the experiment is designed in order to ensure a negligible contribution coming from external background sources such as cosmic rays (i.e. the residual underground muon flux), neutrons (muon-induced or radioactivity-induced) and gammas (from radioactive contamination or from neutron/muon-induced reactions) [19]. The severe selection of all the construction materials used in CUORE will allow to control the gamma contribution to the $\beta\beta(0\nu)$-decay background. This contribution is expected to be mainly due to the ^{208}Tl gamma line at 2615 keV. Indeed, this is the only gamma line due to environmental long-living contaminations (^{208}Tl belongs to the ^{232}Th chain) having an energy high enough to contribute to the ROI with a sizable branching ratio. In CUORICINO the contribution of this line is evaluated to be of about 30% of the total counting rate in the ROI, while the residual 70% of counts (i.e. ~0.11 counts/keV/kg/y) is ascribed mainly to degraded alpha particles. Indeed, ^{238}U and ^{232}Th contaminations in the TeO_2 crystal surfaces or in the materials directely facing the crystals [3] produce - in their decay - alpha particles with energies larger than 4 MeV. When the alpha deposits only a fraction of its energy in the crystal - either because the decay

[3] Note that the array is held in vacuum, fixed in a Copper skeleton by means of PTFE stands, and bolometers are fully active i.e. they do not have any dead layer.

originated in a facing inert material or in a nearby crystal - it contributes to the background below 4 MeV [20]. Such contributions can extend down to the lowest energies and certainly in the ROI. This kind of sources are responsible of the rather flat continuum - extending from 4 MeV down to the 2.615 keV line - which is measured by the CUORICINO experiment. The alpha background due to crystal surface contaminations is reduced by operating the detectors in anticoincidence, in CUORICINO a fraction of the mentioned 20% reduction obtained by the anticoincindence cut has to be ascribed to the rejection of this kind of events. In CUORE the same technique will be much more efficient, since a larger fraction of the crystal surface will be facing a nearby crystal. However anticoincidences have no effect on alpha events where a crystal and an inert material (as the Copper holder) are involved. For this reason huge efforts have been devoted by the CUORE collaboration in the study of procedures capable of control and reduce ^{238}U and ^{232}Th surface contaminations in all the parts used for the array construction, and mainly in TeO_2 crystals and Copper. For both of these materials important achievements have been obtained [21, 22] proving that a background of 10^{-2} counts/keV/kg/y is feasible. These techniques will be used to assemble the first CUORE tower that will be tested, as a stand alone experiment, in the CUORICINO cryostat within year 2011. Besides being a very important step in CUORE construction this first tower (CUORE-0) will be able to produce a meaningful improvement in the ^{130}Te $\beta\beta(2\nu)$ results of CUORICINO.

4. Conclusions

The result of the CUORICINO experiment has been presented: with an average background of 0.161 ± 0.006 in the ROI, it was possible to set an upper limit for the effective neutrino Majorana mass from 300 to 710 meV, depending on the nuclear model used for NME calculations.
The next generation bolometric experiment CUORE, which has CUORICINO as a fundamental prototype, has also been presented. CUORE, with a forseen aim of 10^{-2} counts/keV/kg/yr of background contributions in the region of interest and with an active mass of nearly 206 kg of ^{130}Te, will be able to probe the inverted hierarchy of neutrino mass in 5 years of live time.

Journal of Physics: Conference Series **347** (2012) 012031 doi:10.1088/1742-6596/347/1/012031

References

[1] Examples of reviews are:
V. I. Tretyak and Yu. G. Zdesenko, Atomic Data and Nuclear Data Tables, **80** (2002) 83.
S. Elliott and P. Vogel, Ann. Rev. Nucl. Part. Sci. **52** (2002) 115.
A. Morales and J. Morales, Nucl. Phys. B (Proc. Suppl.) **114** (2003) 141.
F. T. Avignone III, S. R. Elliott and J. Engel (2006), Rev. Mod. Phys. **80** (2008) 481.
[2] H. V. Klapdor-Kleingrothaus *et al.*, Eur. Phys. J. A **12** (2001) 147.
[3] C. E. Aalseth *et al.*, Phys. Rev. D **65** (2002) 092007.
C. E. Aalseth *et al.*, Phys. Rev. D **70** (2004) 078302.
[4] R. Arnold *et al.*, Phys. Rev. Let. **95** (2005) 182302. A. S. Barabash (NEMO Collaboration), Presentation at XXXIII International Conference on High Energy Physics (Moscow, July 26 - August 02, 2006), [arXiv:hepex/0610025].
[5] R. Bernabei *et al.*, Phys. Lett. B **546** (2002) 23.
[6] H. V. Klapdor-Kleingrothaus *et al.*, Mod. Phys. Lett. A, **16** (2001) 2409.
H. V. Klapdor-Kleingrothaus *et al.*, Nucl. Instr. Meth. A, **522** (2004) 371.
[7] N. Booth, B. Cabrera and E. Fiorini, Ann. Rev. Nucl. Part. Sci. **46**(1996) 471.
A. Alessandello *et al.*, Czech. J. Phys. **48** (1998) 133.
C. Enss and D. McCammon, J. of Low Temp. Phys. **151**(2008) 5;
[8] A. Alessandrello *et al.*, Phys. Lett. B **285** (1992) 176;
A. Alessandrello *et al.*, Phys. Lett. B **335** (1994) 519-525.
[9] A. Alessandrello *et al.*, Nucl. Instr. Meth. A **360** (1995) 363.
A. Alessandrello *et al.*, Nucl. Instr. Meth. A **370** (1996) 241.
[10] A. Alessandrello *et al.*, Phys. Lett. B **433** (1998) 156.
[11] M. Redshaw, B. J. Mount, E. G. Myers, and F. T. Avignone III, Phys. Rev. Lett. **102** (2009) 212502.
[12] C. Arnaboldi *et al.*, Phys. Rev. Lett. **95** (2005) 14501.
C. Arnaboldi *et al.*, Phys. Rev. C **78** (2008) 035502.
C. Arnaboldi *et al.*, submitted to AstroparticlePhysics (2010).
[13] Andreotti *et al.*, Astropart. Phys. 34 (2011) 822831.
[14] F. Simkovic *et al.*, Phys.Rev. C **77** (2008) 045503.
[15] O. Civitarese *et al.*, JoP Conference series **173** (2009) 012012.
[16] J. Menendez *et al.*, Nucl. Phys. A **818** (2009) 139.
[17] J. Barea and F. Iachello, Phys.Rev. C **79** (2009) 044301.
[18] C. Arnaboldi *et al.*, Astropart. Phys. **20** (2003) 91.
C. Arnaboldi *et al.*, Nucl. Instr. Meth. A **518**(2004) 775.
[19] F. Bellini *et al.*, Astropart. Phys. **33** (2010) 169.
[20] M. Pavan *et al.*, Eur. Phys. J A **36** (2008) 159.
C. Bucci *et al.*, Eur. Phys. J A **41** (2009) 155.
[21] C. Arnaboldi *et al.*, J. Cryst. Growth **312** (2010) 2999.
[22] C. Arnaboldi *et al.*, LNGS Annual Report year 2009.

2nd International Conference on Particle Physics IOP Publishing
Journal of Physics: Conference Series **347** (2012) 012032 doi:10.1088/1742-6596/347/1/012032

QCD and Top physics studies in proton-proton collisions at 7 TeV centre-of-mass energy with the ATLAS detector

M Barisonzi
on behalf of the ATLAS Collaboration

Bergische Universität Wuppertal, Fachbereich C – Physik, Gaußstraße 20, D-42119 Wuppertal, Germany

E-mail: marcello.barisonzi@physik.uni-wuppertal.de

Abstract. Beginning from 2010, ATLAS collected proton-proton collision data at a centre-of-mass energy of 7 TeV, accumulating by the Spring of 2011 an integrated luminosity of $\sim$200 pb^{-1}. The high energy gives the possibility to produce a sizeable amount of top quarks, allowing for measurements of top quark properties and searches for heavy exotic resonances decaying to top-antitop pairs. Measurements of jet cross-sections confirm the predictions of perturbative QCD at the high energy of the Large Hadron Collider.

1. Introduction

The Large Hadron Collider (LHC) reached in 2010 the record centre-of-mass energy of $\sqrt{s} =$ 7 TeV, inaugurating the phase of physics studies at the TeV scale, with the possibility of solving outstanding issues of the Standard Model — such as the mechanism of Electroweak Symmetry Breaking or the origin of Dark Matter and Dark Energy — but also of finding new physics phenomena beyond the Standard Model itself. Before such studies can be accomplished, however, it is important to test the validity of the theoretical predictions and the simulation tools at the TeV scale. Section 2 describes the studies on the predictions of Next-to-Leading-Order perturbative QCD (NLO pQCD) models applied to the measurement of inclusive jet and dijet cross-sections.

Another important sector of physics studies at the LHC is top physics. The high beam energy and luminosity make the LHC a top factory, with $\sim$13000 $t\bar{t}$ pairs created daily at a luminosity of 10^{33}cm^{-2}s^{-1}. The large amount of statistics will facilitate precision measurements of the top quark that may help to explain the role that a particle of such a large mass plays in Electroweak Symmetry Breaking mechanisms — if any of these mechanisms is found. The measurements of the top quark cross-sections for the QCD and Electroweak production processes at ATLAS are presented in Section 3, which also contains a search for heavy reasonances decaying to a $t\bar{t}$ pair.

2. Jet cross-section measurements

The aim of these studies is to compare the NLO pQCD predictions for inclusive jet and dijet production. Inclusive single-jet double-differential cross sections are measured as a function of jet transverse momentum p_{T} and rapidity y, while the dijet double-differential cross section is

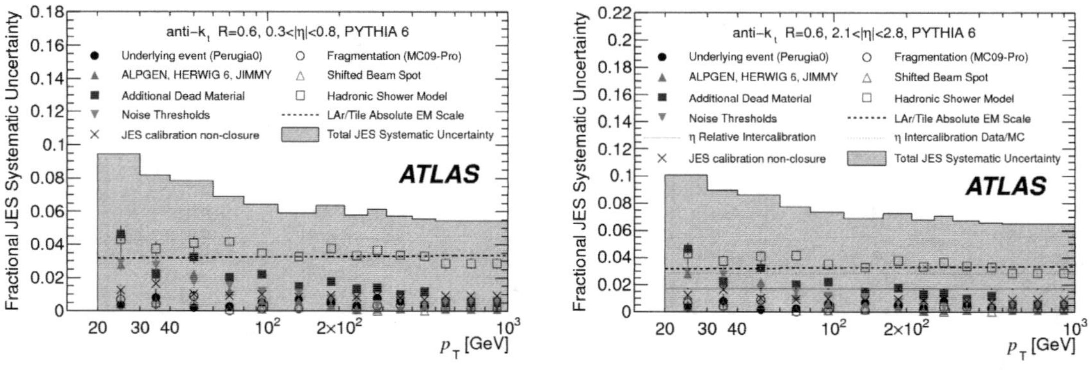

Figure 1. Fractional jet energy scale systematic uncertainty as a function of p_T for jets in the pseudorapidity region $0.3 < |\eta| < 0.8$ (left) and $2.1 < |\eta| < 2.8$ (right) [1].

measured as a function of the invariant mass of the dijet system, m_{12}, binned in the maximum rapidity of the two leading jets, $|y|_{max} = \max(|y_1|, |y_2|)$ [1].

The cross section measurements are performed using jets reconstructed with the anti–k_T algorithm [2]. For the inclusive jet measurements, jets are required to have $p_T > 20$ GeV and be within $|y| < 4.4$. For the dijet mass measurements, the selection requires a leading jet with $p_T > 30$ GeV and $|y| < 4.4$, and at least one sub-leading jet with $p_T > 20$ GeV and $|y| < 4.4$ [3]. The energy of the reconstructed jets is calibrated to obtain the energy of the Monte Carlo particle jets; the calibration of the Jet Energy Scale (JES) is in fact the largest factor contributing to systematic uncertainties, and can be affected by [1]:

- material and geometrical modelling of the detector;
- noise thresholds in the calorimeter;
- electromagnetic response of the calorimeter;
- beamspot position.

The jets unfolded to particle level are compared to the NLO pQCD prediction, calculated using the NLOJet++ 4.1.2 program [4] along with the CTEQ 6.6 NLO parton distribution functions [5]. The fixed-order NLO matrix element predictions are combined with a leading-log showering and fragmentation model — Pythia with LHC-specific tuning — to better describe the data. This introduces additional uncertainties in the JES estimation, due to [1]:

- hadronic shower modelling;
- event generator model;
- modelling of pile-up effects;
- effect of decorrelated JES uncertainty on dijet observables.

All JES uncertainties are added in quadrature; any deviation from unity of the closure test for JES corrections is treated as a fully correlated uncertainty and added linearly. The total uncertainty is estimated to be below 10% in the jet p_T range 20-1000 GeV (see Figure 1).

The double-differential inclusive jet cross-section, binned as a function of jet p_T and $|y|$, is shown in Figure 2. The ratio between the measured data and the theoretical predictions is compatible with unity, within the systematic and theoretical uncertainties.

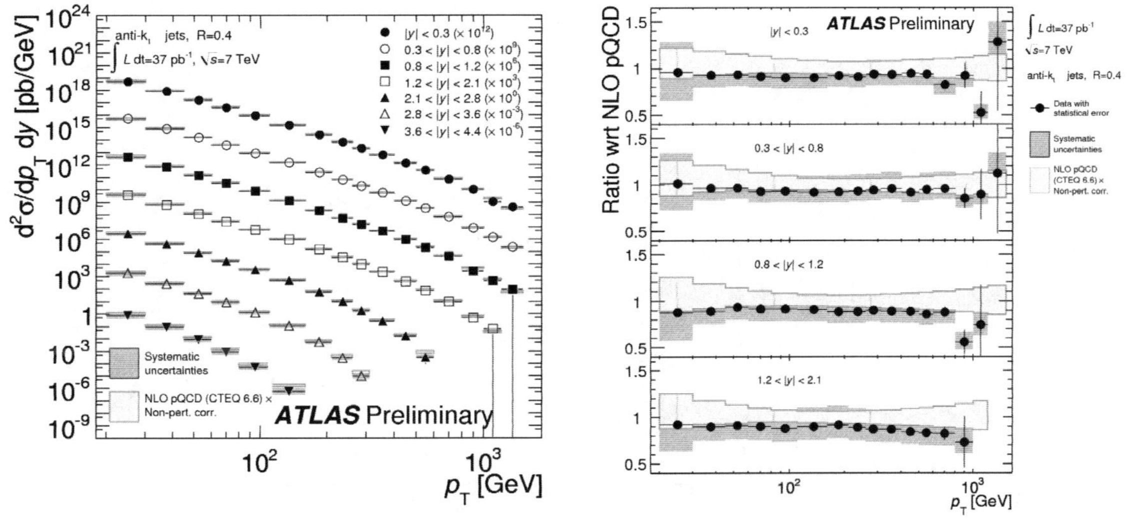

Figure 2. Left: inclusive jet double-differential cross-section as a function of jet p_T in different regions of $|y|$ for jets identified using the anti-k_T algorithm with $R = 0.4$. Right: ratio of the data to the theoretical prediction, with total systematic uncertainties on the theory and measurement [3].

The double-differential dijet cross-section, binned as a function of dijet mass and maximum jet rapidity $|y|_{max}$, is shown in Figure 3. Again, the ratio between the measured data and the theoretical predictions does not present any significant deviation from unity.

The good agreement between the NLO pQCD predictions and the data allows to search for the presence of resonances decaying into quark pairs. With 2010 data, the following resonant processes are searched for [6]:

- excited quarks;
- axigluons;
- Quantum Black Holes (QBH);
- Randall-Sundrum gravitons;
- quark contact interactions.

The study does not find any significant localized excess in the dijet mass spectrum (see Figure 4). Therefore, the following 95% CL limits can be set [6; 7]:

- excited quarks with masses in the interval $0.80 < m_q < 2.49$ TeV are excluded;
- axigluons with masses in the interval $0.80 < m < 2.67$ TeV are excluded;
- assuming six extra dimensions, Quantum Black Hole models with quantum gravity mass scale $0.75 < M_D < 3.67$ TeV are excluded;
- quark contact interactions with a scale $\Lambda < 9.5$ TeV are excluded.

No limits can be placed on Randall-Sundrum gravitons yet.

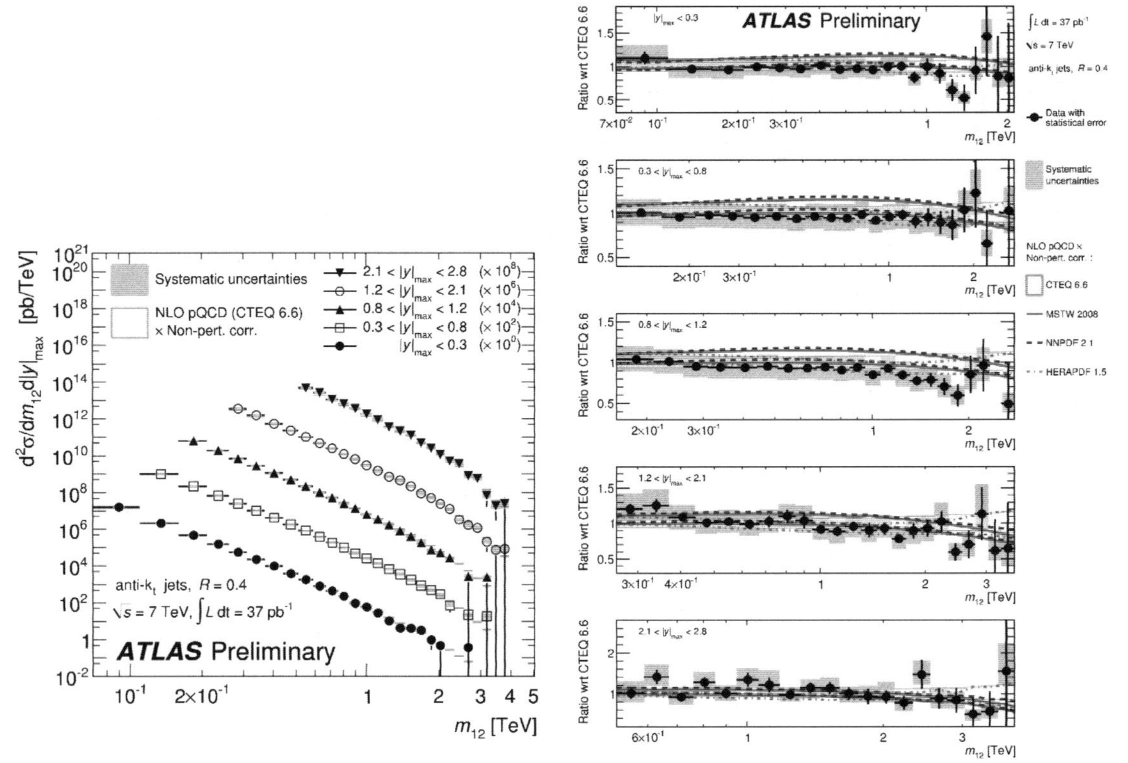

Figure 3. Left: dijet double-differential cross-section as a function of dijet mass, binned in the maximum rapidity of the two leading jets $|y|_{max}$. The results are shown for jets identified using the anti-k_T algorithm with $R = 0.4$. Right: the data points and the error bands are normalized to the theoretical predictions obtained by using the CTEQ 6.6 PDF set. The theoretical error bands obtained by using different PDF sets (MSTW 2008, NNPDF 2.1, HERAPDF 1.5) are also shown [3].

3. Top Physics

Top quark production at the LHC occurs mainly via two processes: top-antitop pair production from gluon-gluon fusion and single top electroweak production. Electroweak production can be broken down into three different subprocesses. Of these subprocesses, only the gluon-boson fusion (also known as t-channel) guarantees sufficient statistics and signal-to-background ratio to be studied with $O(100\ \mathrm{pb}^{-1})$ of data[1].

Top quark events are characterised by a complex signature: the top quark decays almost exclusively to Wb, and the W boson can in turn decay to either quarks or leptons. The vast majority of top studies at ATLAS require the decay of at least one W boson into either (e, ν) or (μ, ν); in the case of top pair production, events are classified as *semi-leptonic* or *di-leptonic* if, respectively, one or both W bosons decay leptonically. The selection of di-lepton events guarantees better purity at the cost of statistics, while semi-leptonic events offer a reasonable trade-off between event yield and background rejection.

All the top physics studies presented in this article implement a common selection:

[1] At Tevatron, the dominant production process was the s-channel quark-antiquark annihilation, which is negligible at LHC energies in 2011.

Journal of Physics: Conference Series **347** (2012) 012032 doi:10.1088/1742-6596/347/1/012032

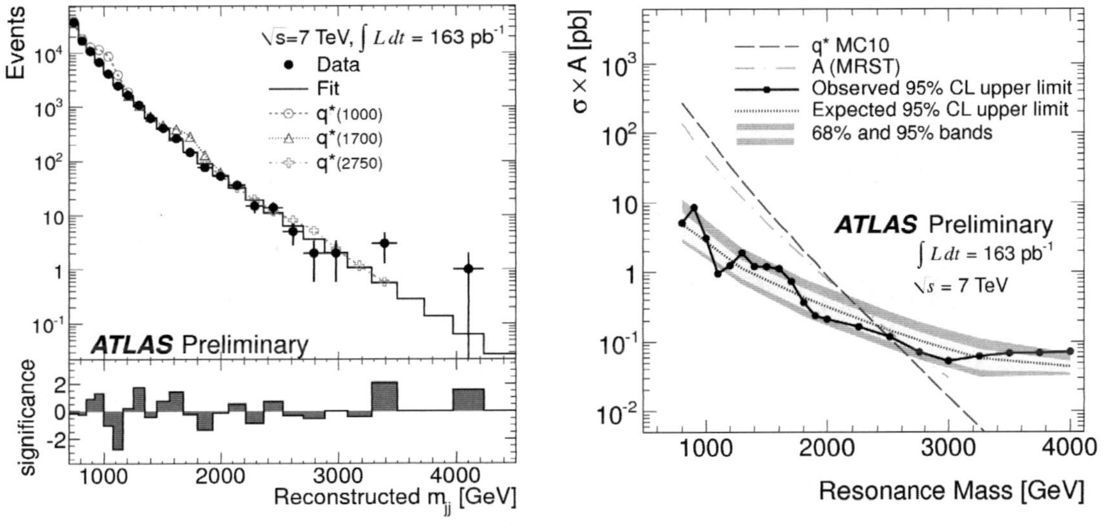

Figure 4. Left: the data dijet mass distribution with the predicted q^* signals for excited-quark masses of 1000, 1700, and 2750 GeV; for bins where the discrepancy has a Poisson p-value of less than 0.5, the bin-by-bin significance of the data-background difference is shown, in units of σ. Right: the 95% CL upper limit on $\sigma \cdot \mathcal{A}$ (the product of the production cross-section and acceptance) as a function of dijet resonance mass [7].

- single lepton (electron or muon) trigger fired;
- at least one lepton (electron or muon) with $p_{\mathrm{T}} > 25$ GeV, lepton matched with the object firing the trigger[2];
- jets with $p_{\mathrm{T}} > 25$ GeV; pair production studies require four or more central in rapidity ($|y| < 2.5$) jets, while single top studies require exactly two jets: one central and one forward ($|y| > 2.5$);
- at least one jet must be tagged as coming from a b-quark decay; the working point of the b-tagging algorithm is chosen to yield 50% tagging efficiency;
- missing transverse energy $E_{\mathrm{T}}^{\mathrm{miss}} > 25$ GeV.

To further reduce the QCD multijet backgrounds, the reconstruction of a leptonically-decaying W boson is used in some studies, with the additional cut: $E_{\mathrm{T}}^{\mathrm{miss}} + M_T(W) > 60$ GeV; where $M_T(W)$ is the transverse mass of the reconstructed W boson [8; 9]. QCD background manifests itself as jets misidentified by the detector as leptons. Templates that model the behaviour of lepton-faking jets can be obtained from data; these templates require the presence of either a lepton failing some of the quality cuts or a jet with a large fraction of energy deposited in the electromagnetic calorimeter. A fit of the shape of the $E_{\mathrm{T}}^{\mathrm{miss}}$ distribution from these events is then used to obtain the magnitude of the QCD background (see Figure 5).

3.1. Measurement of the $t\bar{t}$ cross-section

The measurement of the top quark pair production cross-section relies on a multivariate analysis. Additional methods, including cut-and-count methods and methods using fits to the kinematic

[2] The trigger threshold for 2010 data was set to 20 GeV.

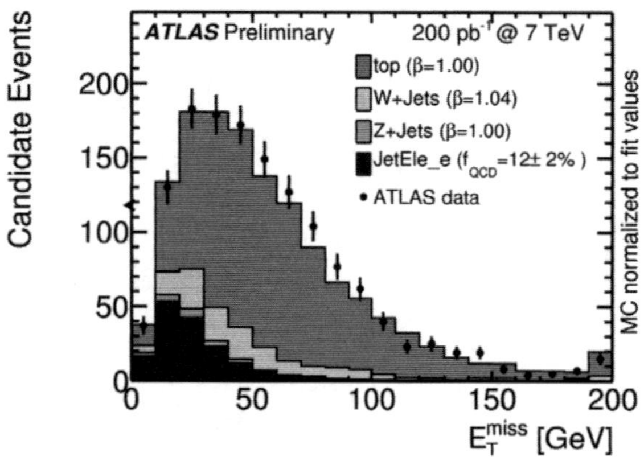

Figure 5. Fit of the E_T^{miss} distribution. The QCD template (blue) is obtained by selecting from data jets with an electromagnetic energy fraction between 0.8 and 0.95 [10], while the other templates are simulated using Monte Carlo generators. The parameter β is the scaling factor to be applied to the individual channels to fit the data [9].

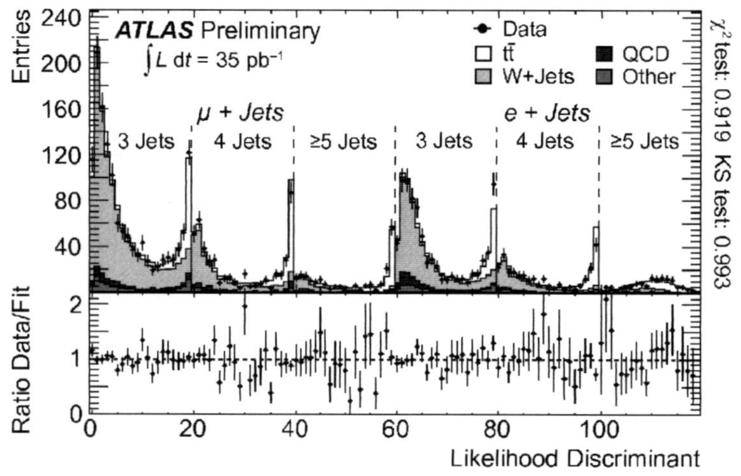

Figure 6. Fit of the likelihood discriminant D for the 3,4,5 jet bins in the semileptonic channels μ+jets, e+jets [11].

distributions of the reconstructed top quark pairs were used as cross-checks, but are not included in this article.

The multivariate analysis constructs a likelihood discriminant D from four variables [11]:

- the pseudorapidity of the lepton;
- the aplanarity, defined as 1.5 times the smallest eigenvalue of the momentum tensor $M_{ij} = \sum_{k=1}^{N_{objects}} p_{ik}p_{jk}/\sum_{k=1}^{N_{objects}} p_k^2$, where p_{ik} is the i-th momentum component and p_k is the modulus of the momentum of object k;
- the sum of the square of the transverse energy of all jets except the two leading ones, normalized to the sum of absolute values of all longitudinal momenta in the event;
- the average of the two lowest light-jet probabilities in the event, as computed by the b-tagging algorithm.

The discriminant D is fitted in the 3, 4 and 5 jet bins for both semileptonic channels e+jets and μ+jets (see Figure 6). The fit yields a cross-section for $t\bar{t}$ production of [11]:

$$\sigma_{t\bar{t}} = 186 \pm 10\,(\text{stat.})\,^{+21}_{-20}\,(\text{syst.})\,\pm 6\,(\text{lumi.})\text{ pb}.$$

The results of the semileptonic and dilepton channels are combined by maximizing the likelihood as a function of $\sigma_{t\bar{t}}$, giving as a result [12]:

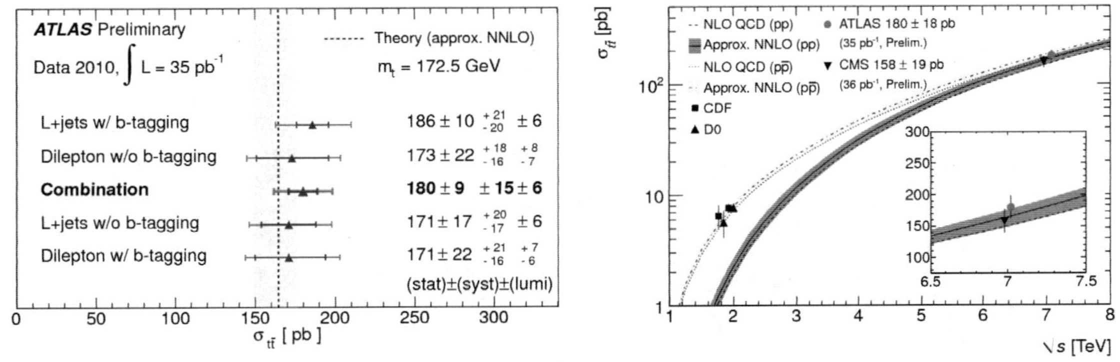

Figure 7. Left: the measured value of $\sigma_{t\bar{t}}$ in the combined single-lepton and dilepton channels, including error bars for both statistical uncertainties only (blue) and with full systematics (red). The approximate NNLO prediction is shown as a vertical dotted line with its uncertainty. Right: measurements of $\sigma_{t\bar{t}}$ from ATLAS and CMS in proton-proton collisions, and CDF and DØ in proton-antiproton collisions, compared to theoretical predictions assuming a top mass of 172.5 GeV as a function of $\sqrt{s}$. The presented result is indicated by the red circle [12].

$$\sigma_{t\bar{t}} = 180 \pm 9\,(\text{stat.}) \pm 15\,(\text{syst.}) \pm 6\,(\text{lumi.})\ \text{pb}.$$

The result shows excellent agreement with a similar study from CMS [13] and with the extrapolation from the Tevatron measurements, and is consistent with the Standard Model predictions (see Figure 7).

3.2. Measurement of the single top cross-section

The measurement of the electroweak t-channel single top cross section is performed in the lepton+jets final state using two different techniques: a cut-and-count method and a neural network approach.

Both methods use a combination of selection variables that give good discrimination power against the underlying background processes. The most important variables are [8]:

1. the invariant mass of the b-tagged jet, the lepton and the neutrino, $M_{l\nu b}$;
2. the pseudorapidity of the untagged jet, η(l-jet);
3. the pseudorapidity of the b-tagged jet, η(b-jet);
4. the scalar sum of the transverse energies of all jets H_T;
5. the $\Delta\eta$ distance between the b-tagged jet and the lepton, $\Delta\eta(b\text{-jet, lepton})$;
6. the invariant mass of the two selected jets, $M(\text{jet1, jet2})$.

Variables 1-5 are used in the cut-and-count method. The neural network uses a combination of many more variables, where variables 1, 2 and 6 are the most important ones.

For the cut-and-count method, the measured cross-section and combined statistical and systematic uncertainty is [8]:

$$\sigma_t = 97^{+54}_{-30}\ \text{pb}.$$

The measurement has a significance of 6.1σ and the expected sensitivity is 4.4σ.

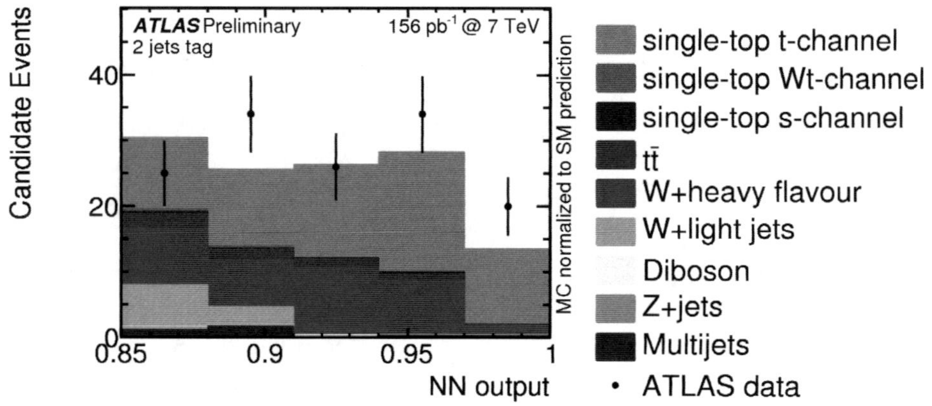

Figure 8. Neural network output distribution after the final selection cut normalized to number of expected events [8].

The neural network method gives a better sensitivity than the cut-and-count analysis. The output of the neural network for data and Monte Carlo is shown on Figure 8. The observed cross-section measurement and combined statistical and systematic uncertainty is [8]:

$$\sigma_t = 76^{+41}_{-21} \text{ pb}.$$

The measurement has a significance of 6.2σ and the expected sensitivity is 5.7σ.

3.3. Search for heavy reasonances

The search for heavy resonances focuses on two benchmark processes that allow to estimate the experimental sensitivity in both the narrow- and wide-resonance regimes. As a narrow resonance, the top-colour Z' boson is chosen. In the leptophobic scenario, this resonance couples preferably to top quark pairs, with a resonance width of 1.2% of the Z' mass [9]. The model used for wide resonances is a Kaluza-Klein gluon g_{KK}, which appears in Randall-Sundrum (RS) models with a warped extra dimension; in this case, the resonance is predicted to be significantly wider than the detector and reconstruction algorithm's resolution [9].

As resonances are identified by excess data in the spectrum of the invariant mass of the $t\bar{t}$ pair, mass resolution effects in the tail of the spectrum must be kept under control. The dominant source of these effects is the use of a jet from initial- or final-state radiation in the reconstruction of the invariant mass. To reduce this contribution, the *dRmin* algorithm selects the jets with $p_{\mathrm{T}} > 20$ GeV and $|\eta| < 2.5$, which also satisfy the condition $\Delta R_{min} > 2.5 - 0.015 \times m_j$, where m_j is the jet's mass and ΔR_{min} is computed with respect to the closest lepton or surviving jet. The combination of three or four jets surviving the cuts enters in the top pair mass spectrum, allowing for merged jets or jets outside the acceptance [9]. The method is tested with Monte Carlo samples of Z' bosons, shown in Figure 9 (left).

The study searches for an excess of data in the reconstructed $t\bar{t}$ mass spectrum, with respect to Standard Model predictions. As no statistically significant excess can be found in data (see Figure 9, right), the study sets the following 95% CL exclusion limits (see Figure 10):

- for narrow leptophobic Z' resonances, the maximum allowed cross-section ranges from 38 pb at $m = 500$ GeV to 3.2 pb at $m = 1300$ GeV;
- in Randall-Sundrum models, Kaluza-Klein gluons with masses below 650 GeV are excluded.

Journal of Physics: Conference Series **347** (2012) 012032 doi:10.1088/1742-6596/347/1/012032

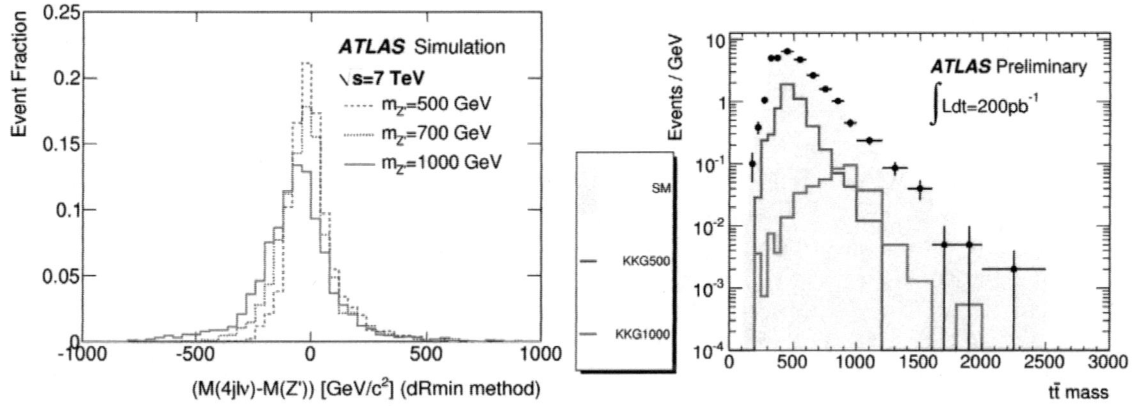

Figure 9. Left: reconstructed $t\bar{t}$ pair invariant mass resolution for three Z' boson masses: $m_{Z'}$ = 500 GeV, $m_{Z'}$ = 700 GeV and $m_{Z'}$ = 1000 GeV for the *dRmin* algorithm. Right: $t\bar{t}$ mass spectrum reconstructed from data using the *dRmin* algorithm after selection cuts with expected Kaluza-Klein gluon spectrum for KK-gluon masses of 500 and 1000 GeV [9].

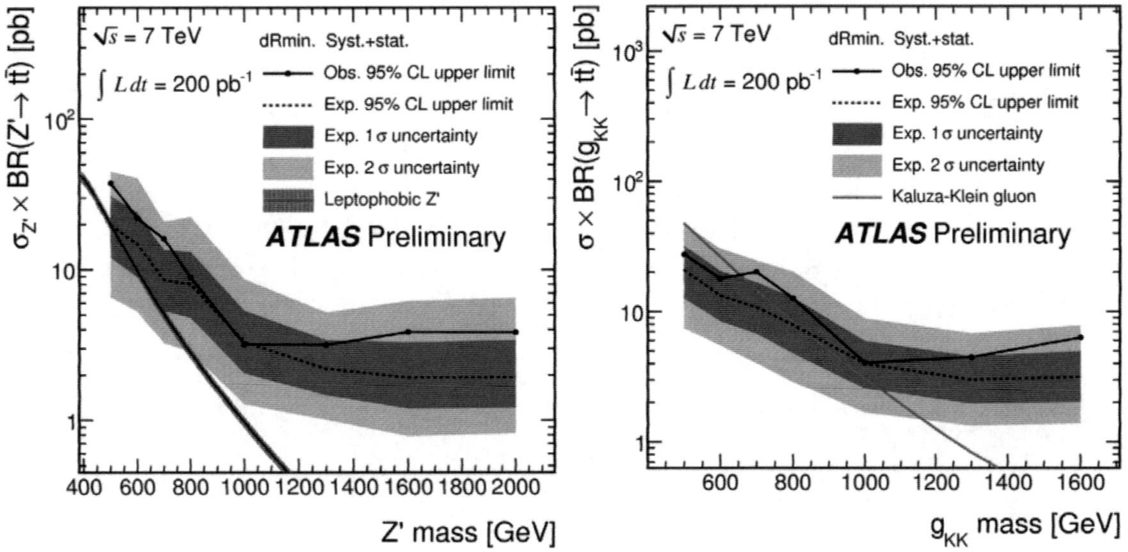

Figure 10. Expected (dashed line) and observed (black dots) upper limits on $\sigma \times BR(Z' \to t\bar{t})$ (left) and $\sigma \times BR(g_{KK} \to t\bar{t})$ (right) using the *dRmin* algorithm [9].

Journal of Physics: Conference Series **347** (2012) 012032 doi:10.1088/1742-6596/347/1/012032

4. Conclusions

The first year and a half of data taking at LHC allowed to study the Standard Model predictions at the unprecedented center of mass energy of 7 TeV. Next-to-Leading-Order Perturbative QCD predictions have been tested in the inclusive jet and dijet channels and show good agreement with data. No resonances were found in the dijet invariant mass spectrum.

In top physics, ATLAS made a first measurement of the $t\bar{t}$ production cross-section, to compare against pQCD predictions at the new unprecented LHC *pp* collision energy. The measurement confirms at this energy the good agreement — already observed at the lower energy of Tevatron — between data and theory on $t\bar{t}$ production cross-section. Moreover, ATLAS measured with a level of significance of 6.2σ the cross-section for the electroweak production of single top quarks in the t-channel, complementing the Tevatron measurement in the s-channel. The invariant mass spectrum of top quark pairs in $t\bar{t}$ production was searched for the presence of both narrow and wide resonances; no resonance has been found yet.

References

[1] The ATLAS Collaboration 2011 *The European Physical Journal C - Particles and Fields* **71**(2) 1–59

[2] Cacciari M, Salam G P and Soyez G 2008 *JHEP* **0804** 063 (*Preprint* 0802.1189)

[3] The ATLAS Collaboration 2011 Measurement of inclusive jet and dijet cross sections in proton-proton collision data at 7 TeV centre-of-mass energy using the ATLAS detector Tech. Rep. ATLAS-CONF-2011-047 CERN Geneva

[4] Nagy Z 2003 *Phys. Rev.* **D68** 094002 (*Preprint* hep-ph/0307268)

[5] Nadolsky P M *et al.* 2008 *Phys. Rev.* **D78** 013004 (*Preprint* 0802.0007)

[6] The ATLAS Collaboration 2011 *New J. Phys.* **13** 053044. 51 p

[7] The ATLAS Collaboration 2011 Update of the search for new physics in the dijet mass distribution in 163 pb^{-1} of *pp* collisions at $\sqrt(s) = 7$ TeV measured with the ATLAS detector Tech. Rep. ATLAS-CONF-2011-081 CERN Geneva

[8] The ATLAS Collaboration 2011 Observation of t channel single top-quark production in *pp* collisions at $\sqrt{s} = 7$ TeV with the ATLAS detector Tech. Rep. ATLAS-CONF-2011-088 CERN Geneva

[9] The ATLAS Collaboration 2011 A search for $t\bar{t}$ resonances in the lepton plus jets channel in 200 pb^{-1} of *pp* collisions at $\sqrt{s} = 7$ TeV Tech. Rep. ATLAS-CONF-2011-087 CERN Geneva

[10] The ATLAS Collaboration 2011 Searches for single top-quark production with the ATLAS detector in *pp* collisions at $\sqrt{s} = 7$ TeV Tech. Rep. ATLAS-CONF-2011-027 CERN Geneva

[11] The ATLAS Collaboration 2011 Measurement of the top quark-pair cross-section with ATLAS in *pp* collisions at $\sqrt(s) = 7$ TeV in the single-lepton channel using b-tagging Tech. Rep. ATLAS-CONF-2011-035 CERN Geneva

[12] The ATLAS Collaboration 2011 A combined measurement of the top quark pair production cross-section using dilepton and single-lepton final states Tech. Rep. ATLAS-CONF-2011-040 CERN Geneva

[13] The CMS Collaboration 2011 Combination of top pair production cross sections in *pp* collisions at 7 TeV and comparisons with theory Tech. Rep. CMS-PAS-TOP-11-001 CERN Geneva

2nd International Conference on Particle Physics IOP Publishing
Journal of Physics: Conference Series **347** (2012) 012033 doi:10.1088/1742-6596/347/1/012033

Measurements of forward energy flow and forward jet production with CMS

Deniz Sunar Cerci[1]

Adiyaman University, Science and Letters Faculty, Physics Department 02040 Adiyaman, TURKEY

E-mail: deniz.sunar.cerci@cern.ch

Abstract. We present measurements of the forward ($3 < |\eta| < 5$) energy flow in minimum bias events and in events with either hard jets or W and Z bosons produced at central rapidities, as well as measurements of the inclusive forward jet cross section and of associated production of forward and central jets. Results are compared to MC models with different parameter tunes for the description of the underlying event and to perturbative QCD calculations, the PYTHIA and HERWIG parton shower event generators, as well as to the CASCADE Monte Carlo.

1. Introduction

At high energies, the $pp \rightarrow XY$ reaction can be described in terms of a perturbative parton-parton interaction and of proton remnants carrying a large fraction of the total energy. This picture leads to a final state X and an underlying event Y consisting of mostly soft hadrons, originating from parton showers and non-perturbative multi-parton interactions. The latter are largely independent of the hard interaction, increase the particle multiplicity and are leading to correlations between the energy flow in the central and forward regions.

The measurement of energy flow in the forward region is directly sensitive to the parton radiation at large η and to multiparton interactions (MPI) [1]. Such a measurement in pp collisions not only discriminates between different models of MPI but also improves the understanding of the basic process responsible for multiparton radiation and thus can provide additional input to the determination of the parameters in MPI models. The currently available models of MPI have mainly been tuned to minimum bias data and to final states including jets with large transverse momentum (p_T), using the observed central track multiplicities and the transverse momentum spectra of hadrons which are not associated to the hard jets. Models for a detailed simulation of MPI are under a rapid development and new features, including diffractive components in the energy flow, are extensively studied. The analysis of the underlying event structure and energy flow correlation in processes with colorless final states, such as $pp \rightarrow W(Z)X \rightarrow l\nu(ll)X$, can provide new insights into so far unexplored aspects of multi-parton interactions. In particular, the study of such processes with colorless final states allows for a straightforward separation of the hard interaction and the underlying event. $W(Z)$ events with high track multiplicities in the central detector indicate a possibly large contribution from multi-parton interactions. Studies of correlations between the track multiplicity in the central

[1] For the CMS Collaboration

rapidity range and energy deposits at large rapidities, in the forward regions of the experiment, should give additional information. Fig. 1 shows a prediction for the energy flow as a function of pseudorapidity for minimum bias events and for events which have a dijet with $E_{T,jet} > 20$ GeV in the central pseudorapidity region $|\eta| < 2.5$ at $\sqrt{s} = 7$ TeV. The energy flow is defined as: $(1/N_{event})(d\sum E_i/d\eta)$ where N_{event} is the number of events. The energy E_i of charged and neutral stable particles i is summed in bins of η. The predictions from the PYTHIA [2] Monte Carlo (MC) event generator using different sets of parameters which were tuned to describe the Tevatron measurements are shown. The solid line uses the Perugia tune [3] and the dashed line the D6T tune [4]. The prediction without multiparton interaction (dotted line) is also shown. Significant differences of the energy flow in the forward region ($|\eta| > 3$) can be observed.

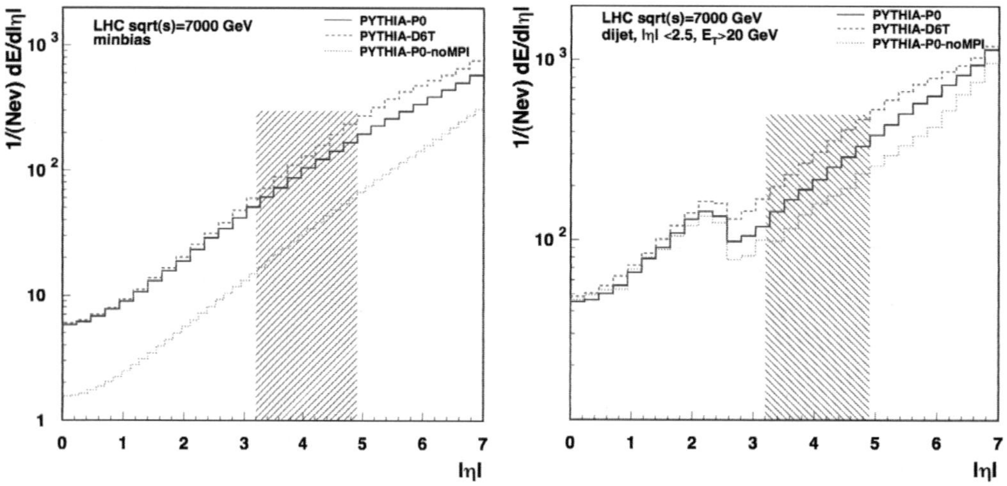

Figure 1. The predicted energy flow by the PYTHIA MC event generator at $\sqrt{s} = 7$ TeV for particles with $E > 0.1$ GeV as a function of the pseudo-rapidity for minimum bias events (left) and for events containing a dijet with $E_{T,jet} > 20$ GeV in the central rapidity region of $|\eta| < 2.5$ (right). The shaded area indicates the region of measurement.

Jet production in hadron-hadron collisions is sensitive to the underlying partonic QCD processes, to the details of parton radiation and to the parton density functions (PDF) of the colliding hadrons. At previous colliders, the measured jet cross sections at large transverse momenta (p_T) are successfully described by perturbative QCD (pQCD) calculations over several orders of magnitude [5] . The measurements, however, are limited to central pseudorapidities, $|\eta| < 2.5$, where the momentum fractions x_1, x_2 of the incoming partons are of the same order. Jets produced in the forward/backward hemispheres result from scatterings between colliding partons with increasingly different momentum fractions $x_2 \ll x_1$, and thus allows one to study QCD in the small-x region where the PDFs are well constrained by deep-inelastic electron-proton data, and where deviations of the parton dynamics beyond the standard Dokshitzer-Gribov-Lipatov-Altarelli-Parisi (DGLAP) evolution [6] e.g. as modelled in the Balitski-Fadin-Kuraev-Lipatov (BFKL) [7], Ciafaloni-Catani-Fiorani-Marchesini (CCFM) [8], or gluon saturation [9] approaches are expected. However, we do not see any deviations until now. Understanding the dynamics of forward jet production accompanied with or without central jets is also essential for the control of the QCD backgrounds in searches of the Higgs boson produced via the vector-boson-fusion (VBF) mechanism as well as for general VBF scattering cross sections which are fundamental to understand the Electroweak Symmetry Breaking mechanism (see for example [10]).

The large calorimetric coverage of CMS allows to study jet production over a region of pseudorapidity extending to $|\eta| < 5.2$ never investigated before. The production of forward jets [11] alone or in conjunction with a central jet [12] is measured using the early data collected by the CMS detector at LHC during 2010 in proton-proton ($p-p$) collisions at a center of mass energy of 7 TeV. The differential cross section $d^2\sigma/dp_T d\eta$ for single inclusive forward jets, as well as the differential cross sections $d^2\sigma/dp_T^f d\eta^f$ and $d^2\sigma/dp_T^c d\eta^c$ for the simultaneous production of at least one forward jet (f) in conjunction with at least one central jet (c) are presented. The *axis* of the forward jet is required to be in the fiducial Hadronic-Forward (HF) calorimeters acceptance ($3.2 < |\eta| < 4.7$), and that of the central jet is measured within $|\eta| < 2.8$. In all cases, jets are reconstructed with the calorimeter energy deposits for transverse momenta above $p_T = 35$ GeV/c.

2. Experimental Aspect

A complete description of the CMS detector can be found elsewhere [13]. A brief outline of the components which are the most relevant for the present measurement is given here. Two hadron forward calorimeters (HF), one on each side of the CMS interaction point (IP), at ~ 11 m, cover the pseudo-rapidity range $3 < |\eta| < 5$. Two elements of the CMS monitoring system, the Beam Scintillator Counters (BSC) and the Beam Pick-up Timing eXperiment (BPTX) devices, are used to trigger the CMS readout. The two BSC are located at a distance of ± 10.86 m from the nominal interaction point (IP) and are sensitive in the $|\eta|$ range from 3.23 to 4.65. The two BPTX, located around the beam pipe at a distance of ± 175 m from the IP on either side, are designed to provide precise information on the bunch structure and timing of the incoming beam, with better than 0.2 ns time resolution.

3. Analysis Strategy

The average energy flow at forward rapidities is determined separately in two different event classes: in minimum bias events and in events with a hard scale provided by a dijet system at central rapidities [14]. The analysis is carried out at two different centre-of-mass energies, 900 GeV and 7 TeV. The amount of activity in the underlying event is expected to vary strongly with both the QCD scale of the reaction and the centre-of-mass energy. The forward energy flow is measured by the HF calorimeters, by summing up all energy deposits of the HF towers above a noise threshold. In the minimum bias sample a search for dijets in the central region of the CMS detector is performed. Jets are reconstructed by means of the anti-k_T jet algorithm (with $R = 0.5$). The dijet sample consists of events in which the two hardest jets have $|\eta| < 2.5$ and $|\Delta\phi(jet_1, jet_2) - \pi| < 1.0$. The η cut ensures that the jets are contained in a region of CMS outside of the HF calorimeters. At $\sqrt{s} = 900$ GeV, the leading and the sub-leading jets are required to have $p_T > 8$ GeV, while at $\sqrt{s} = 7$ TeV, the jets are required to have $p_T > 20$ GeV.

Next we look at W and Z production where the identification of W and Z bosons is based on the presence of isolated electrons and muons with high transverse momentum [15]. The transverse momentum of the electron candidate is required to be larger than 25 GeV and $|\eta| < 2.5$. Two algorithms are used to identify muons. One is based on the matching of a reconstructed high transverse momentum track candidate with a track candidate found in the CMS muons system. The second applies a global fit of tracker and muon system hits. Muons in this analysis have to pass two different selection algorithms. The reconstructed muon candidate transverse momentum has to be larger than 25 GeV and $|\eta| < 2.5$.

The observables used to study the underlying event structure in W and Z events are the track multiplicity in the central detector, the energy deposits in both HF calorimeters and the correlations between them. In the following only the distributions for $W \to l\nu$ events are discussed. No significant differences are observed between W events selected with decays to electrons or muons. The same analysis has been performed with the $pp \to ZX$ data samples and, taking the

Journal of Physics: Conference Series **347** (2012) 012033 doi:10.1088/1742-6596/347/1/012033

smaller number of Z events into account, essentially identical distributions and conclusions are obtained.

As a further investigation of the forward energy distribution, the jets in the forward region are taken into account. The anti-k_T jet clustering algorithm [16] is used to reconstruct all jets with a distance parameter of R = 0.5, with their reconstructed axis within $3.2 < |\eta| < 4.7$, and with $p_T > 20$ GeV.

4. Results

The energy flow is measured with the CMS HF calorimeters in the region $3.15 < |\eta_{obs}| < 4.9 <$, and corrected to hadron level. The hadron level is defined by Monte Carlo generated events in which unstable particles ($\tau < 10^{-12}$ s) decay. The particles on hadron level are required to have the same rapidity range as in the measurement ($3.15 < |\eta| < 4.9$), and do not have any selection on their energy. Moreover, at least one particle each is required on both sides of the BSC acceptance in order to suppress the diffractive events. The result is shown in Fig. 2 for minimum bias events and in Fig. 3 for dijet events, at $\sqrt{s}$ = 900 GeV and $\sqrt{s}$ = 7 TeV. The systematic uncertainties are indicated as error bars, while the statistical errors are not shown since they are comparably small. The data are compared to various Monte Carlo predictions on hadron level.

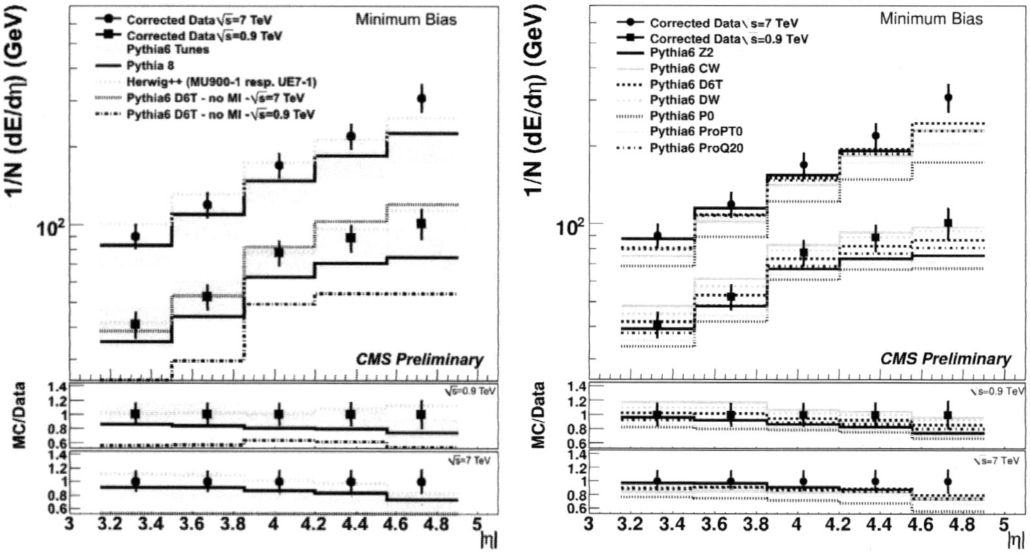

(a) (b)

Figure 2. Energy flow as a function of η for the Minimum Bias analyses at $\sqrt{s}$ = 900 GeV and at $\sqrt{s}$ = 7 TeV, corresponding to integrated luminosities of 239 μb^{-1} and 206 μb^{-1}, respectively. The corrected data are shown as points. a) The histograms are the predictions from various MC generators. The yellow bands illustrates predictions from the different Pythia 6 tunes. The error bars on the data represent the systematic uncertainties, which are strongly correlated between the bins. The statistical errors of the data are less than 0.1%. b) The histograms are the predictions from Pythia 6 with various tunes.

Fig. 4 shows the HF energy sum for $W \to \mu\nu$ events and for several different MC tunes of the underlying event. As can be seen from Fig. 4, besides the ProQ20 tune [17], none of the studied MC models provides a good description of the HF energy distribution as observed in

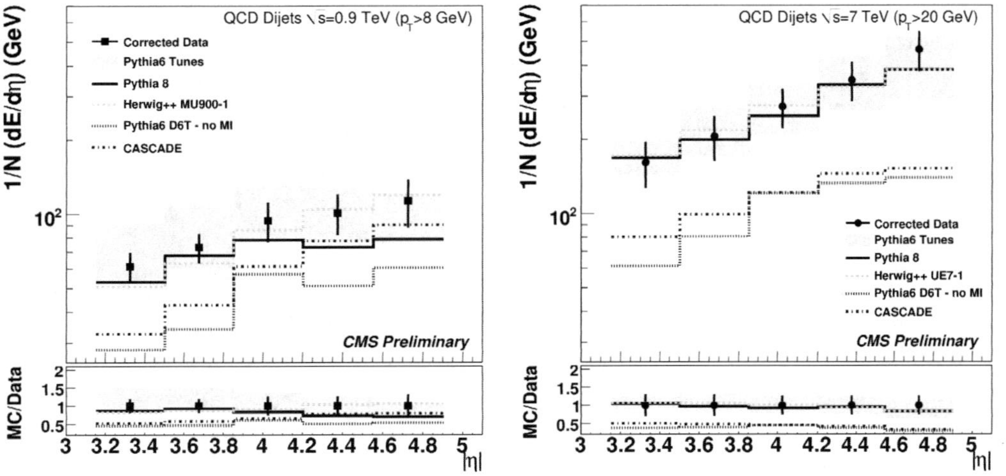

(a) (b)

Figure 3. Energy flow as a function of η for the dijet analyses at $\sqrt{s} = 900$ GeV and at $\sqrt{s} = 7$ TeV, corresponding to integrated luminosities of 239 μb^{-1} and 206 μb^{-1}, respectively. The corrected data are shown as points, the histograms are the predictions from various MC generators. The yellow bands illustrate predictions from the different Pythia 6 tunes.

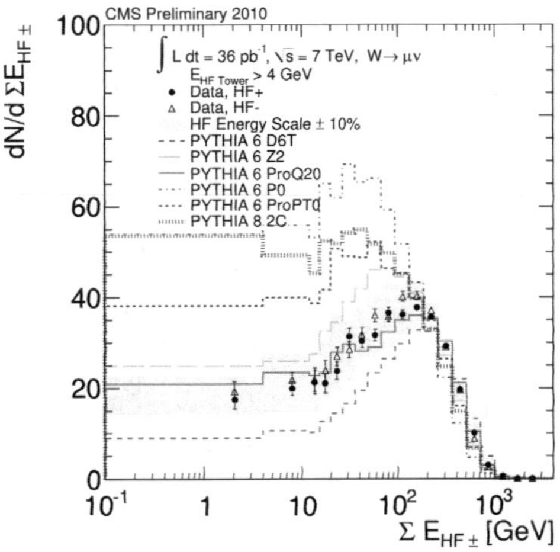

Figure 4. The summed HF energy distribution for $W \rightarrow \mu\nu X$ candidate events are shown for data and MC simulations with different tunes for the underlying event.

the data. For energy deposits between $10 - 150$ GeV, large differences between the data and the different tunes are observed. In particular, the number of events in data is about 30% to 50% higher than predicted by the D6T tune, and 50% lower than predicted by the Z2 tune. In order to investigate the correlations of the forward energy flow and the central track multiplicity the energy on one side of HF is summed and checked how the energy on the other side of HF behaves. For this study events with energy deposits in the HF$-$ calorimeter between $20 - 100$ GeV (low), $200 - 400$ GeV (medium) and above 500 GeV (high) are selected. Analyzing the HF+ energy distributions, the distributions corresponding to the medium HF$-$ energy interval

(Fig. 5b) shows smaller differences between the data and the different tunes. On the other hand, requiring a low HF$-$ energy deposit results in a measured HF+ energy distribution, which is badly modeled by all MC tunes (Fig. 5a). Finally, when selecting events with high energy deposits in HF$-$ we find that PYTHIA 8 [18], which in the inclusive case underestimates the number of events with such large HF$-$ energy deposits, provides a rather good description of the energy in HF+. In contrast to that, all other tunes predict more events with large HF+ energy than observed in data (Fig. 5c).

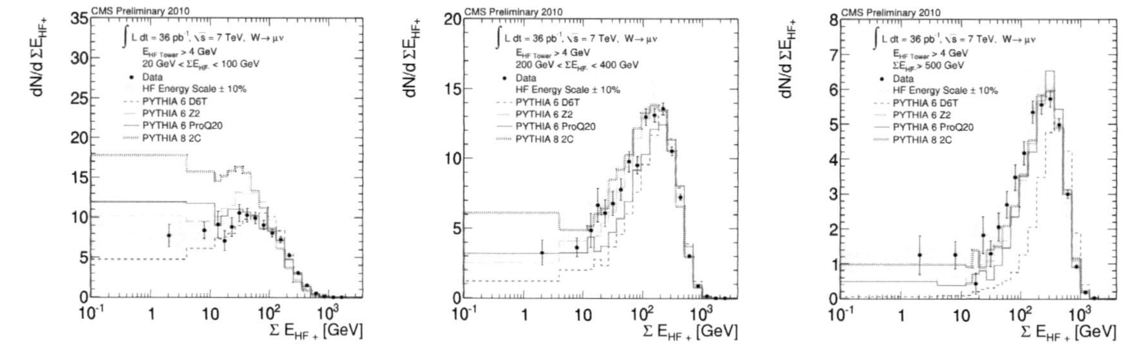

(a) (b) (c)

Figure 5. The observed energy distribution in the HF+ calorimeter, shown for different HF$-$ energy intervals: a) 20 $-$ 100 GeV, b) 200 $-$ 400 GeV, c) $>$ 500 GeV, for data and different MC tunes. The plots are shown for $pp \rightarrow W^{\pm}X \rightarrow \mu^{\pm}\nu X$.

The differential inclusive jet cross-section is measured as a function of transverse momentum p_T with the total integrated luminosity of $\mathcal{L} = 3.14$ pb^{-1}. The statistical uncertainties are small compared to the systematic uncertainties at all p_Ts. The primary sources of systematic uncertainties in the cross-section measurement are the jet energy scale (JES) and resolution, as well as the estimated integrated luminosity. The total systematic uncertainty, obtained adding in quadrature these three uncertainties, is shown in Fig. 6a. Fig. 6b shows all the sources of theoretical uncertainty: the non-perturbative (NP) corrections dominate for $p_T < 50$ GeV whereas PDF+α_S uncertainties dominate above that p_T; scale uncertainties are subleading at all transverse momenta. We have added in quadrature these three sources of uncertainty into a single theoretical uncertainty band. Fig. 6c shows the same NP and scale uncertainties with the PDF envelope obtained using the HERAPDF1.0 parton densities [19] which provides a different uncertainty estimate.

The jet cross section as a function of p_T is shown in Fig. 7a compared to all the theoretical models considered. The total experimental systematic uncertainty is shown as a yellow band. Fig. 7b shows the fractional difference between the experimental jet cross section and the theoretical predictions following the PDF4LHC prescription (see Fig. 6b). The predictions agree reasonably with the measurements within the theoretical and experimental uncertainties. However, CASCADE [20] appears to have a more concave spectral shape than the data in the intermediate p_T region.

The fully corrected cross section for simultaneous production of at least one central and at least one forward jet is presented in Fig. 8 as a function of the forward and central jet p_T. The uncertainty bands take into account both the statistical and the systematic uncertainties, summed in quadrature. In all p_T bins of the measured cross section, the statistical uncertainty (of the order of 1 $-$ 2% in the low p_T bin and 5 $-$ 10% in the highest) is subdominant with respect to the systematic one, which amounts to $\approx$ 30%, dominated by the uncertainty on the

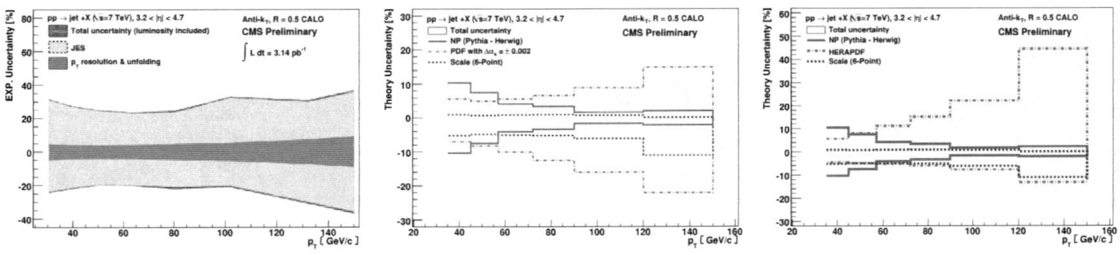

(a) (b) (c)

Figure 6. a) Systematic uncertainties in the forward jet measurement, b) theoretical uncertainties on the forward jet spectrum including non-perturbative (NP), PDF4LHC and scale uncertainties, c) including NP, HERAPDF and scale uncertainties.

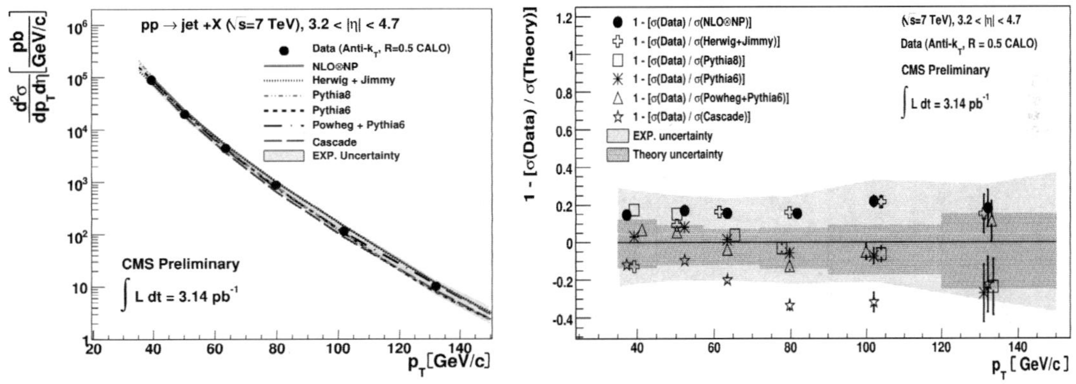

(a) (b)

Figure 7. a) Inclusive jet cross section (anti-k_T, R= 0.5) measured at forward pseudorapidities ($3.2 < |\eta| < 4.7$), fully corrected and unfolded, b) fractional differences between the fully-corrected forward jet spectra.

jet enegy scale (JES).

To better evaluate the compatibility of the Monte Carlo predictions with the measured cross section, the ratio of the data over various Monte Carlo simulations are plotted on top of the band corresponding to the total uncertainty in Fig. 9. The ratios are shown as a function of the jet p_T. Discrepancies between data and MC models are observed. The prediction from Z2 model, which is tuned to reproduce the underlying event data at the LHC and reproducing more satisfactorily the central jet spectrum than D6T or PYTHIA 8, still lies above the data (the same holds true for the lower p_T range of the forward jet spectrum). The NLO MC POWHEG matched with HERWIG parton shower describes the shape well, but not the normalization. CASCADE predicts a different shape of the p_T cross sections, indicating a non trivial correlation between the forward and central jet transverse momenta. The HEJ event generator, which provides all-order description of the dominant radiative corrections for hard, wide angle emissions, is used here only at parton level. It describes the data reasonably well.

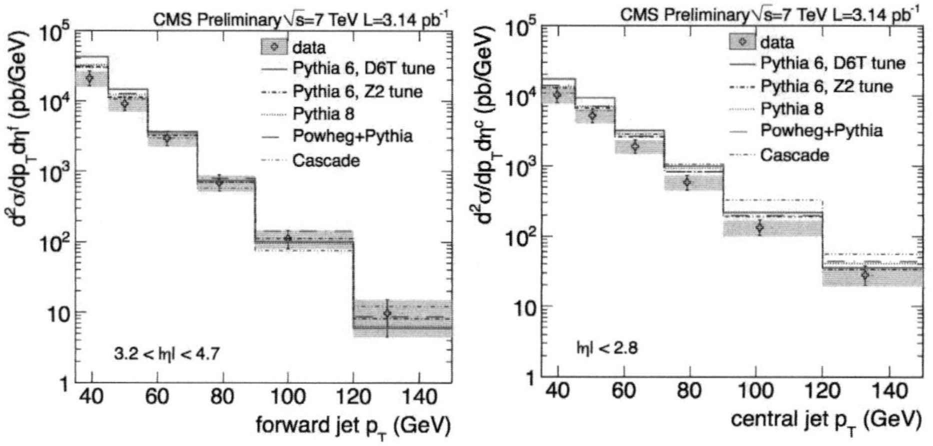

Figure 8. The cross sections for the hadronic final state, as a function of p_T, for the forward jets and the central jets.

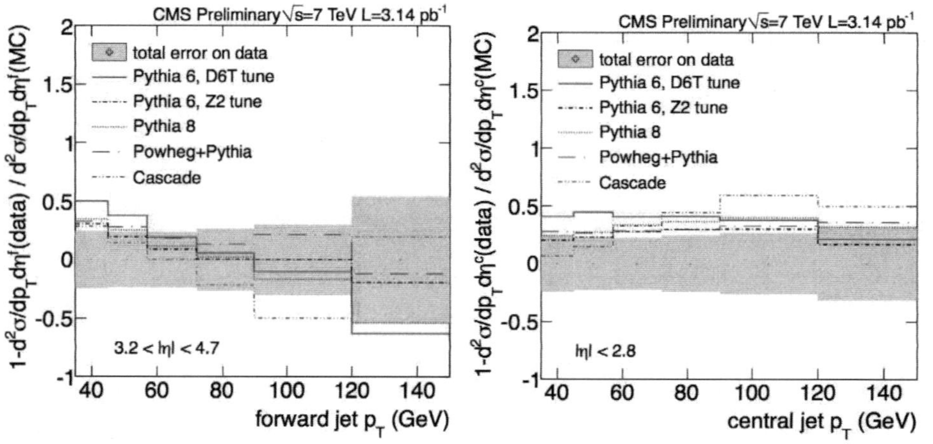

Figure 9. Ratio of the fully corrected p_T differential jet cross section in data over the various Monte Carlo models considered. The yellow band corresponds to the total uncertainty for the forward region (left) and the central one (right).

5. Conclusions

The energy flow in the forward region, $3.15 < |\eta_{obs}| < 4.9$, is measured in hadron-hadron collisions for minimum bias events and events having a hard scale defined by a dijet system in the central region, $|\eta_{jet}| < 2.5$, with $E_{T,jet} > 8$ GeV ($E_{T,jet} > 20$ GeV) for $\sqrt{s} = 900$ GeV ($\sqrt{s} = 7$ TeV). Requiring a hard sub-system in the central region works as a lever arm for the forward energy flow. Thus, the forward energy flow is significantly higher in the dijet measurement compared to the minimum bias measurement. The increase in energy flow in the forward region with increasing centre-of-mass energy is significant and is reproduced by the Monte Carlo for minimum bias and dijet events. Besides, it is shown that including multiple interactions helps to describe the data better. The MC predictions without multiple interactions significantly underestimate the data.

Central track multiplicities, forward energy flow and correlations among them have been studied in W and Z events, identified by the vector boson decays to electrons and muons. None of the studied MC tunes provides simultaneously a satisfactory description of the track multiplicity in the central rapidity region ($|\eta| < 2.5$) and of the forward energy flow ($3 < |\eta| < 4.9$). Strong positive correlations between the energy measured on one side HF, the track multiplicity and on the other side energy deposits are observed in data and in Monte Carlo models with different tunes. However, the correlations in the various MC tunes are different than those seen in the data.

The forward jet production in the pseudorapidity range $3.2 < |\eta| < 4.7$ is measured. The jet transverse momentum spectrum is obtained with the anti-k_T algorithm ($R = 0.5$) in the Hadronic Forward calorimeters in the p_T range of $35 - 150$ GeV. The total systematic uncertainties are of the order of $\pm 25\%$ and dominated by the absolute jet energy scale. Within the current experimental and theoretical uncertainties, perturbative calculations – as implemented in parton shower Monte Carlos PYTHIA and HERWIG, and at NLO accuracy both in fixed-order (NLOJET++) and parton-shower-matched (POWHEG) frameworks, as well as accounting for combined DGLAP+BFKL resummations in the CASCADE model – reproduce globally well the measured forward jet cross section. The cross section for the production of one central and one forward jet, covering the region in $|\eta| < 5.2$ at $\sqrt{s} = 7$ TeV is presented and compared to various Monte Carlo predictions. The data-model comparison shows that, in general, all PYTHIA tunes overestimate the jet spectra when requiring the simultaneous production of jets in both regions. The disagreement is larger for the central jet spectrum and for the lowest p_T bins at forward rapidity. Calculations including resummation of low-x logarithms, as in the CASCADE Monte Carlo, do not reproduce well the central jet spectrum, whereas the HEJ model is in good agreement with the measurement.

Acknowledgments

The author would like to acknowledge the organizing committee for providing the partial support to present this talk during the conference.

References

[1] T. Sjöstrand and M. van Zijl, *Phys. Rev.* D **36** (1987) 2019
[2] T. Sjostrand, S. Mrenna and P. Z. Skands, *JHEP* **0605** (2006) 026
[3] P. Z. Skands, *Phys. Rev.* D**82** (2010) 074018
[4] R. Field [CDF Collaboration], In *Perugia 2008, Multiple Partonic Interactions at the LHC (MPI08)* 12-31
[5] A. Bhatti and D. Lincoln, "Jet Physics at the Tevatron", (2010) (arXiv-hep-ex:1002.1708)
CMS Collaboration, "Measurement of the inclusive jet cross section in pp collisions at 7 TeV", *CMS PAS* QCD-10-011 (2010)
[6] V. N. Gribov and L. N. Lipatov, *Sov. J. Nucl. Phys.* **15** (1972) 438-450
L. N. Lipatov, *Sov. J. Nucl. Phys.* **20** (1975) 94-102
G. Altarelli and G. Parisi, *Nucl. Phys.* B **126** (1977) 298
Y. L. Dokshitzer, *Sov. Phys. JETP* **46** (1977) 641-653
[7] E. A. Kuraev, L. N. Lipatov, and V. S. Fadin, *Sov. Phys. JETP* **44** (1976) 443-450
E. A. Kuraev, L. N. Lipatov, and V. S. Fadin, *Sov. Phys. JETP* **45** (1977) 199-204
I. I. Balitsky and L. N. Lipatov, *Sov. J. Nucl. Phys.* **28** (1978) 822-829
[8] M. Ciafaloni, *Nucl. Phys.* B **296** (1988) 49
S. Catani, F. Fiorani, and G. Marchesini, *Phys. Lett.* B **234** (1990) 339
S. Catani, F. Fiorani, and G. Marchesini, *Nucl. Phys.* B **336** (1990) 18
G. Marchesini, *Nucl. Phys.* B **445** (1995) 49-80
[9] F. Gelis, E. Iancu, J. Jalilian-Marian et al., "The Color Glass Condensate", (2010) (arXiv:1002.0333)
[10] F. Ambroglini et al., "Proceedings of the Workshop on Monte Carlo's, Physics and Simulations at the LHC", (2009) (arXiv:0902.0293)
[11] CMS Collaboration, "Inclusive forward jet production cross sections in proton-proton collisions at $\sqrt{s} = 7$ TeV", *CMS PAS* FWD-10-003 (2010)
[12] CMS Collaboration, "Cross section measurement for simultaneous production of a central and a forward jet in proton-proton collisions at $\sqrt{s} = 7$ TeV", *CMS PAS* FWD-10-006 (2010)
[13] CMS Collaboration, *JINST* **0803** (2008) S08004

[14] CMS Collaboration, "Forward energy flow in the CMS detector", *CMS PAS* FWD-10-011 (2010)
[15] CMS Collaboration, "Forward energy flow, central track multiplicities and large rapidity gaps in W and Z boson events at 7 TeV pp collisions", *CMS PAS* FWD-10-008 (2010)
[16] M. Cacciari, G. P. Salam, and G. Soyez, *JHEP* **04** (2008) 063
[17] A. Buckley et al., *Eur. Phys. J.* **C65** (2010) 331-357
[18] T. Sjostrand, S. Mrenna, and P. Z. Skands, *Comput. Phys. Commun.* **178** (2008) 852-867
[19] H1 and ZEUS Collaborations, *JHEP* 01 (2010) 109
[20] H. Jung, S. Baranov, M. Deak et al., *The European Physical Journal C - Particles and Fields* **79** (2010) 1237-1249

2nd International Conference on Particle Physics IOP Publishing
Journal of Physics: Conference Series **347** (2012) 012034 doi:10.1088/1742-6596/347/1/012034

Decay constants of heavy vector mesons at finite temperature

E Veli Veliev*[1], K Azizi[2], H Sundu[1], G Kaya[1] and A Türkan[1]

[1] Department of Physics, Kocaeli University, 41380 Izmit, Turkey
[2] Department of Physics, Faculty of Arts and Sciences, Doğuş University, Acıbadem-Kadıköy, 34722 Istanbul, Turkey

E-mail: *[1]elsen@kocaeli.edu.tr

Abstract. This study deals with determination of the decay constants of heavy vector mesons in the framework of the thermal QCD sum rules. We calculate both thermal spectral density and non-perturbative contributions taking into account the traditionally existing operators at $T = 0$ and also additional operators appearing at finite temperature. Analysis of the obtained thermal sum rules shows that the decay constants almost remain unchanged with respect to the variation of temperature up to $T \cong 100MeV$, however after this point, they start to decrease sharply with increasing temperature.

1. Introduction

It is believed that investigation of thermal behavior of hadronic properties in medium provides testing ground for the Standard Model and helps understanding the results of the heavy ion collision experiments. In particular, investigation of hadronic properties of J/ψ and Υ as heavy vector mesons in medium plays important role in the study of quark gluon plasma [1]. One of the most powerful and applicable approaches to calculate the hadronic parameters is the QCD sum rules [2]. This method was later extended to finite temperature and density [3]. In formulation of thermal sum rules, the complications are breakdown of Lorentz invariance in medium by the choice of the thermal rest frame and appearance of additional operators in operator product expansion (OPE) compared to the QCD sum rules in vacuum [4]-[6]. This method has been extensively used to predict the behavior of hadronic parameters such as masses, widths, decay constants at zero [7, 8] and finite temperature [9]-[13].

In the present work, we calculate the decay constants f_V of the heavy vector quarkonia $J/\psi(\bar{c}c)$ and $\Upsilon(\bar{b}b)$, which are defined by the matrix element of the vector current J_μ between the vacuum and the vector-meson state,

$$\langle 0|J_\mu|V(q,\lambda)\rangle = f_V m_V \varepsilon_\mu^{(\lambda)}. \tag{1}$$

In our calculations, we use the values of the energy density and gluon condensates obtained via Chiral perturbation theory [14] and lattice QCD [15]-[17]. We observe that the values of the decay constants decrease considerably near to the critical or deconfinement temperature comparing to their values in vacuum.

Journal of Physics: Conference Series **347** (2012) 012034 doi:10.1088/1742-6596/347/1/012034

2. Thermal QCD sum rules for vector mesons

We consider the thermal correlation function,

$$\Pi_{\mu\nu}\Big(q,T\Big) = i\int d^4x\, e^{iq\cdot x}Tr\Big(\rho\, \mathcal{T}\Big(J_\mu(x)J^\dagger_\nu(0)\Big)\Big), \tag{2}$$

with $J_\mu(x) =: \overline{Q}(x)\gamma_\mu Q(x) :$. Here, $Q(x)$ is heavy (charm or bottom) quark field, $\mathcal{T}$ indicates the time ordered product and $\rho = e^{-\beta H}/Tre^{-\beta H}$ is the thermal density matrix of QCD at temperature $T = 1/\beta$. The correlation function in thermal field theory is given by Lorentz invariant functions, $\Pi_l\Big(q^2,\omega\Big) = \frac{1}{\overline{q}^2}\Pi_2$ and $\Pi_t\Big(q^2,\omega\Big) = -\frac{1}{2}(\Pi_1 + \frac{q^2}{\overline{q}^2}\Pi_2)$. Here $\Pi_1 = g^{\mu\nu}\Pi_{\mu\nu}$, $\Pi_2 = u^\mu\Pi_{\mu\nu}u^\nu$, $\overline{q}^2 = \omega^2 - q^2$, $\omega = u\cdot q$ and u_μ is four-velocity.

It can be shown that, in the fixed value of $|\mathbf{q}|$, the spectral representation of the thermal correlation function is given by [6]:

$$\Pi_{l,t}\Big(q_0^2,T\Big) = \int_0^\infty dq_0'^2\, \frac{\rho_{l,t}\Big(q_0'^2,T\Big)}{q_0'^2 + Q_0^2}, \tag{3}$$

where $Q_0^2 = -q_0^2$, and

$$\rho_{l,t}\Big(q_0^2,T\Big) = \frac{1}{\pi}Im\Pi_{l,t}\Big(q_0^2,T\Big)\tanh\frac{\beta q_0}{2}. \tag{4}$$

In order to obtain thermal sum rules, we equate the spectral representation and results of operator product expansion for amplitudes $\Pi_l(q^2,\omega)$ or $\Pi_t(q^2,\omega)$ at sufficiently high Q_0^2 to the hadronic representation of the correlation function. When performing numerical results, we should exchange our reference to one at which the particle is at rest, i.e., we shall set $|\mathbf{q}| \to 0$. In this limit, the functions Π_l and Π_t are related to each other so it is enough to use one of them to acquire thermal sum rules. Here, we use the function Π_t. The fundamental assumption of the thermal QCD sum rule approach is the principle of duality, i.e., it is assumed that there is an interval over which this thermal correlation function may be equivalently described at the quark representation and at the hadronic representation. Therefore the hadronic spectral density is expressed by the ground state vector meson pole plus the contribution of the higher states and continuum,

$$\rho_t^{had}(s) = f_V^2(T)m_V^2(T)\delta(s - m_V^2) + \theta(s - s_0)\rho_t^{pert}(s). \tag{5}$$

Firstly, we consider $\Pi_2(q,T) = u^\mu\Pi_{\mu\nu}(q,T)u^\nu$ function. After some simplifications, we obtain the imaginary part of the $\Pi_2(q,T)$ in the form

$$\Pi_2(q,T) = -4iN_c\int\frac{d^4k}{(2\pi)^4}(k^2 - q\cdot k - m^2 + 2q_0k_0 - 2k_0^2)D(k)D(k-q), \tag{6}$$

where $D(k) = 1/(k^2 - m^2 + i\varepsilon) + 2\pi in(|k_0|)\delta(k^2 - m^2)$. Carrying out the integral over k_0 and angles we obtain annihilation and scattering parts of $Im\Pi_2(q,T)$ as follows:

$$Im\Pi_{2,a} = N_c\int_{\omega_-}^{\omega_+}\frac{d\omega_1}{8\pi|\mathbf{q}|}(4q_0\omega_1 - q^2 - 4\omega_1^2)F(\omega_1), \tag{7}$$

$$Im\Pi_{2,s} = N_c\int_{\omega_+}^{\infty}\frac{d\omega_1}{8\pi|\mathbf{q}|}(4q_0\omega_1 - q^2 - 4\omega_1^2)G(\omega_1). \tag{8}$$

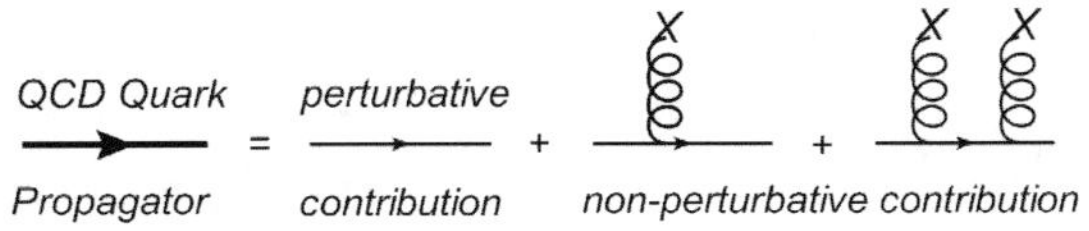

Figure 1. The quark propagator in the gluon background fields.

Here $F(\omega_1) = 1 - n(\omega_1) - n(q_0 - \omega_1) + 2n(\omega_1)n(q_0 - \omega_1)$, $G(\omega_1) = 2n(\omega_1)n(q_0 - \omega_1) - n(\omega_1) - n(q_0 - \omega_1)$ and $\omega_\pm = \frac{1}{2}(q_0 \pm |\mathbf{q}|v)$. From a similar way, the annihilation and scattering parts of $\Pi_1(q, T)$ are also calculated.

As we also previously mentioned, when doing numerical analysis, we will set $|\mathbf{q}| \to 0$ representing the rest frame of the particle. In this case, the annihilation part of spectral density is expressed as follows:

$$\rho_{t,a}(s) = \frac{1}{8\pi^2} s v(s)(3 - v^2(s)) \left[1 - 2n\left(\frac{\sqrt{s}}{2}\right)\right], \tag{9}$$

for $4m^2 \leq s < \infty$ and $v(s) = \sqrt{1 - 4m^2/s}$. Note that the scattering cut shrinks to a point in the considered limit and this part of the spectral density does not contribute to the thermal QCD sum rule.

In our calculations, we also take into account the perturbative two-loop order α_s correction to the spectral density. This correction at zero temperature can be written as [7]:

$$\rho_{\alpha_s}(s) = \alpha_s \frac{s}{6\pi^2} v(s)\Big(3 - v^2(s)\Big)\Big[\frac{\pi}{2v(s)} - \frac{1}{4}\Big(3 + v(s)\Big)\Big(\frac{\pi}{2} - \frac{3}{4\pi}\Big)\Big], \tag{10}$$

where we replace the strong coupling α_s in Eq. (10) with its temperature dependent lattice improved expression [12, 17]. Now, we proceed to calculate the non-perturbative part in QCD side. Taking into account one and two gluon lines attached to the quark line as shown in Fig. 1, up to terms required for our calculations, the non-perturbative part of the massive quark propagator at finite temperature is obtained as:

$$\begin{aligned} S^{aa'nonpert}(k) &= -\frac{i}{4} g(t^c)^{aa'} G^c_{\kappa\lambda} \frac{1}{(k^2 - m^2)^2}\Big[\sigma_{\kappa\lambda}(\not{k} + m) + (\not{k} + m)\sigma_{\kappa\lambda}\Big] \\ &+ \frac{i\, g^2\, \delta^{aa'}}{9\,(k^2 - m^2)^4}\Big\{\frac{3m(k^2 + m\,\not{k})}{4}\langle G^c_{\alpha\beta}G^{c\alpha\beta}\rangle + \Big[m\Big(k^2 - 4(k\cdot u)^2\Big) \\ &+ \Big(m^2 - 4(k\cdot u)^2\Big)\not{k} + 4(k\cdot u)(k^2 - m^2)\,\not{u}\Big]\langle u^\alpha \Theta^g_{\alpha\beta} u^\beta\rangle\Big\}. \end{aligned} \tag{11}$$

where $\Theta^g_{\alpha\beta}$ is the traceless gluonic part of the energy-momentum tensor of the QCD.

Applying Borel transformation with respect to Q_0^2 in Eq. (3) and taking into account hadronic representation in Eq. (5), we obtain (for details see [18])

$$f_V^2 m_V^2 \exp\Big(-\frac{m_V^2}{M^2}\Big) = \int_{4m^2}^{s_0} ds\, [\rho_{t,a}(s) + \rho_{\alpha_s}(s)] \exp\Big(-\frac{s}{M^2}\Big) + \widehat{B}\Pi_t^{nonpert}. \tag{12}$$

Here $\widehat{B}\Pi_t^{nonpert}$ shows the nonperturbative part of QCD side in Borel transformed scheme, which is given by:

$$\hat{B}\Pi_t^{nonpert} = \int_0^1 dx \frac{1}{144\ \pi\ M^6\ x^4\ (-1+x)^4} \exp\Big[\frac{m^2}{M^2\ x\ (-1+x)}\Big]\Big\{\langle\alpha_s G^2\rangle\Big[12\ M^6\ x^4\ (-1+x)^4$$
$$-\ m^6\ (1-2x)^2(-1-x+x^2) - 12\ m^2\ M^4\ x^2\ (-1+x)^2\ (1-3x+3x^2) + m^4\ M^2\ x\ (-2+19x$$
$$-\ 32x^2 + 11x^3 + 6x^4 - 2x^5)\Big] + 4\ \alpha_s\langle\Theta^g\rangle\Big[-8\ M^6\ x^3\ (1-2x)^2(-1+x)^3 + m^6\ (1-2x)^2(-1-x$$
$$+\ x^2) - 2\ m^2\ M^4\ x^2\ (-1+x)^2(-1-6x+8x^2-4x^3+2x^4) + m^4\ M^2\ x\ (-2+3x-12x^2+31x^3$$
$$-\ 30x^4 + 10x^5)\Big]\Big\}, \tag{13}$$

where $\Theta^g = \Theta^g_{00}$.

In further analysis, we use the values, $m_c = (1.3 \pm 0.05)\ GeV$, $m_b = (4.7 \pm 0.1)\ GeV$ and $\langle 0 \mid \frac{1}{\pi}\alpha_s G^2 \mid 0\rangle = (0.012 \pm 0.004)\ GeV^4$ for quarks masses and gluon condensate at zero temperature. We choose the values $s_0 = (11-12)\ GeV^2$ and $s_0 = (98-100)\ GeV^2$ for continuum threshold at J/ψ and Υ channels, respectively.

Our final task is to discuss the temperature dependence of the leptonic decay constant of the considered particles. For this aim, we plot this quantity in terms of temperature in Fig. 2 and Fig. 3 using the total energy density from both chiral perturbation theory [14] and lattice QCD (valid only for $T \geq 100\ MeV$) [15]-[17] at different fixed values of s_0. As shown in these graphs, at $T = 0$, the values of the decay constants of the J/ψ and Υ are obtained as $f_{J/\psi} = (0.460 \pm 0.022)\ GeV$ and $f_\Upsilon = (0.715 \pm 0.032)\ GeV$. These results are in good consistency with the existing experimental data and predictions of other nonperturbative models[19]-[21]. Also, we observe that the decay constants remain insensitive to the variation of the temperature up to $T \cong 100\ MeV$, however after this point, they start to diminish with increasing temperature. At deconfinement or critical temperature, the decay constants approach roughly to 45% of their values at zero temperature.

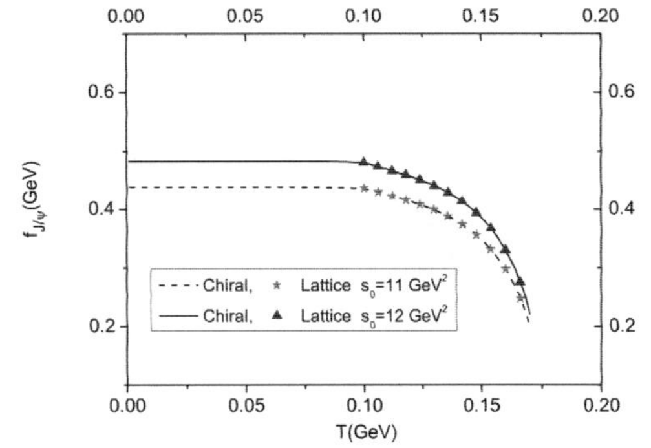

Figure 2. The dependence of the leptonic decay constant of J/ψ in GeV on temperature at $M^2 = 10\ GeV^2$

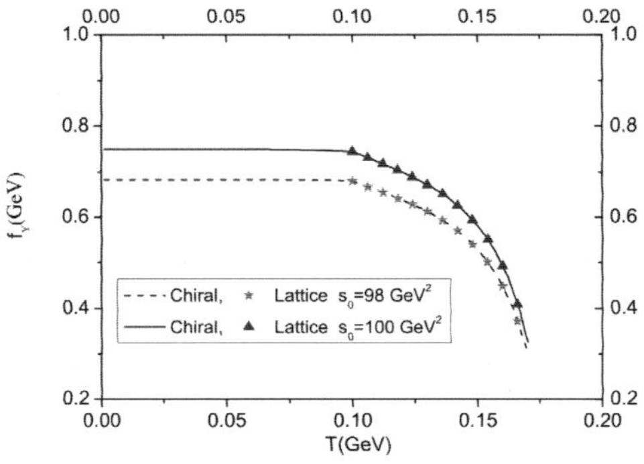

Figure 3. The dependence of the leptonic decay constant of Υ in GeV on temperature at $M^2 = 20\ GeV^2$.

Our results at zero temperature as well as the behavior of the mass and decay constant with respect to the temperature can be checked in future experiments. Also the temperature dependence of the considered quantities can be used in analysis of heavy ion collision experiments.

Journal of Physics: Conference Series **347** (2012) 012034 doi:10.1088/1742-6596/347/1/012034

Acknowledgments
The authors are grateful to T. M. Aliev for useful discussions. This work is supported in part by the Scientific and Technological Research Council of Turkey (TUBITAK) under the research project No. 110T284 and Research Fund of Kocaeli University under grant no. 2011/029.

References

[1] Matsui T and Satz H 1986 *Phys. Lett.* B **178** 416
[2] Shifman M A, Vainstein A I and Zakharov V I 1979 *Nucl. Phys.* B **147** 385 ; Shifman M A, Vainstein A I and Zakharov V I 1979 *Nucl. Phys.* B **147** 448
[3] Bochkarev A I and Shaposhnikov M E 1986 *Nucl. Phys.* B **268** 220
[4] Hatsuda T, Koike Y and Lee S H 1993 *Nucl. Phys.* B **394** 221
[5] Mallik S 1998 *Phys. Lett.* B **416** 373
[6] Mallik S and Mukherjee K 1998 *Phys. Rev.* D **58**, 096011
[7] Reinders L J, Rubinstein H and Yazaki S 1985 *Phys. Rep.* **127**, 1
[8] Colangelo P and Khodjamirian A 2001 *in At the Frontier of Particle Physics/Handbook of QCD, edited by M. Shifman* vol 3 (Singapore: World Scientific) p 1495
[9] Veliev E V 2008 *J. Phys.* G, **35** 035004, Veliev E V and Aliev T M 2008 *J. Phys.* G **35** 125002
[10] Dominguez C A, Loewe M, Rojas J C and Zhang Y 2010 *Phys. Rev.* D **81** 014007
[11] Veliev E V and Kaya G 2009 *Eur. Phys. J.* C **63** 87; Veliev E V and Kaya G 2010 *Acta Phys. Polon.* B **41** 1905
[12] Morita K and Lee S H 2008 *Phys. Rev.* C **77** 064904
[13] Veliev E V, Sundu H, Azizi K and Bayar M 2010 *Phys. Rev.* D **82** 056012
[14] Gerber P and Leutwyler H 1989 *Nucl. Phys.* B **321** 387
[15] Cheng M et.al 2008 *Phys. Rev.* D **77** 014511
[16] Miller D E 2007 *Phys. Rept.* **443** 55
[17] Kaczmarek O, Karsch F, Zantow F and Petreczky P 2004 *Phys. Rev.* D **70** 074505
[18] Veliev E V, Azizi K, Sundu H, Kaya G and Turkan A 2011 *Preprint* arXiv:1103.4330 [hep-ph].
[19] Kiselev V V, Likhoded A K, Pakhomova O N and Saleev V A 2002 *Phys. Rev.* D **65** 034013
[20] Lakhina O and Swanson E S 2006 *Phys. Rev.* D **74** 014012
[21] Hwang C W and Wei Z T 2007 *J. Phys.* G **34** 687

2nd International Conference on Particle Physics IOP Publishing
Journal of Physics: Conference Series **347** (2012) 012035 doi:10.1088/1742-6596/347/1/012035

Analyses of $g_{D_s^* DK}$ and $g_{B_s^* BK}$ vertices in QCD sum rules

J Y Süngü*[1], H Sundu[1] and K Azizi[2]

[1] Department of Physics, Kocaeli University, 41380 Izmit, Turkey

[2] Department of Physics, Faculty of Arts and Sciences, Doğuş University, Acıbadem-Kadıköy, 34722 Istanbul, Turkey

E-mail: *[1]jyilmazkaya@kocaeli.edu.tr

Abstract. The coupling constants $g_{D_s^* DK}$ and $g_{B_s^* BK}$ are calculated in the framework of QCD sum rules. We evaluate the correlation functions of these vertices considering both D(B) and K mesons off-shell and obtain the results as $g_{D_s^* DK} = (2.89 \pm 0.25)$ and $g_{B_s^* BK} = (3.01 \pm 0.28)$.

1. Introduction

A precise determination of the coupling constants is a vital task to get knowledge about cross sections and the nature and structure of the encountered particles. Heavy-light pseudoscalar mesons strong coupling constants are of great importance especially in evaluating charmonium cross sections. Experimentally, it is believed that in the production of the charmonium states like J/ψ and ψ' from the B_c or newly discovered charmonium X, Y and Z states by the BaBar and BELLE collaborations, there are intermediate two body states containing D, D_s, D^* and D_s^* mesons (for example, the kaon can annihilate the charmonium in a nuclear medium to give D and D_s mesons), which decay to the final J/ψ and ψ' states exchanging one or more virtual mesons. A similar story would happen in decays of heavy bottomonium. To exactly follow and analyze the procedure in the experiment, we need to have knowledge about the coupling constants among the particles involved.

In the literature, there have been series of works on coupling constants such as $D^*D\pi$ [1, 2], $DD\rho$ [3], DDJ/ψ [4], D^*DJ/ψ [5], $D^*D^*\pi$ [6, 7], D^*D^*J/ψ [8], D_sD^*K, D_s^*DK [9, 10], $D_0D_sK, D_{s0}DK$[10], $DD\omega$ [11], $D^*D^*\rho$ [12], $D^*D\rho$ [13], $B_{s0}BK, B_{s1}B^*K$ [14], $D_s^*DK^*(892)$, and $B_s^*BK^*(892)$ [15] in the framework of QCD sum rules (QCDSR) technique [16].

In this work, we calculate the D_s^*DK and B_s^*BK vertices using QCDSR (for details see [17]). These coupling constants belong to the low energy sector of QCD, which is far from perturbative regime. Therefore, for calculation of these coupling constants some nonperturbative methods are needed. QCDSR is one of the most promising and predictive one among all existing non-perturbing methods in studying the properties of hadrons.

The paper is organized as follows. In section II, the model is shortly described. In this section, we calculate the correlation function when both the $D(B)$ and K mesons are off-shell. Then we obtain QCD sum rules for the strong coupling constants of the $D_s^* - D - K$ and $B_s^* - B - K$ vertices. Finally in section III, we numerically analyze the obtained strong coupling constant sum rules for the considered vertices. We will obtain the numerical values for each coupling constant when both the $D(B)$ and K states are off-shell. Then taking the average of the two off-shell cases, we will obtain final numerical values for each coupling constant. In this section, we also compare our result on $g_{D_s^* DK}$ with existing predictions in the literature [9].

2. QCD sum rules for the coupling constants

The aim of this section is to calculate the coupling constants $g_{D_s^* DK}$ and $g_{B_s^* BK}$ which characterize the $D_s^* DK$ and $B_s^* BK$ decays, respectively. We start by considering the three-point correlation function. The three-point function associated to $D_s^* DK(B_s^* BK)$ vertex for both D meson off-shell and K meson off-shell states is given respectively by

$$\Pi_\mu^{D(B)}(p',q) = i^2 \int d^4x \; d^4y \; e^{ip'\cdot x} \; e^{iq\cdot y} \langle 0|\mathcal{T}\left(\eta^K(x) \; \eta^{D(B)}(y) \; \eta_\mu^{D_s^*(B_s^*)}(0)\right)|0\rangle, \tag{1}$$

$$\Pi_\mu^{K}(p',q) = i^2 \int d^4x \; d^4y \; e^{ip'\cdot x} \; e^{iq\cdot y} \langle 0|\mathcal{T}\left(\eta^{D(B)}(x) \; \eta^{K}(y) \; \eta_\mu^{D_s^*(B_s^*)}(0)\right)|0\rangle. \tag{2}$$

Here $\mathcal{T}$ is the time ordering product. Each meson interpolating field can be written in terms of the quark field operators as following:

$$\begin{aligned} \eta^K(x) &= \overline{s}(x)\gamma_5 u(x), \\ \eta^{D(B)}(x) &= \overline{u}(x)\gamma_5 c(b)(x), \\ \eta_\mu^{D_s^*(B_s^*)}(x) &= \overline{s}(x)\gamma_\mu c(b)(x). \end{aligned} \tag{3}$$

where u, s, c and b are the up, strange, charm and bottom quark field, respectively. Each current has the same quantum numbers of the associated meson.

According to the idea of the QCDSR, we should calculate this correlator both in terms of hadrons and in quark-gluon language, and then equate these representations. The first side, called phenomenological or physical side, is obtained using hadronic degrees of freedom. The second, so called QCD or theoretical side is calculated using quark and gluon degrees of freedom by the help of the operator product expansion (OPE) in deep Euclidean region.

Firstly, let us focus on the calculation of the physical side of the correlation function Eq.(1) for an off-shell D(B) meson. The physical part can be obtained by saturating Eq.(1) with the appropriate $D(B)$, $D_s^*(B_s^*)$ and K states. After some straightforward calculation, we obtain:

$$\begin{aligned} \Pi_\mu^{D(B)}(p',p) &= \frac{\langle 0|\eta^K|K(p')\rangle\langle 0|\eta^{D(B)}|D(B)(q)\rangle\langle K(p')D(B)(q)|D_s^*(B_s^*)(p,\epsilon)\rangle\langle D_s^*(B_s^*)(p,\epsilon)|\eta_\mu^{D_s^*(B_s^*)}|0\rangle}{(q^2-m_{D(B)}^2)(p^2-m_{D_s^*(B_s^*)}^2)(p'^2-m_K^2)} \\ &+ \; ... \end{aligned} \tag{4}$$

where .... represents the contribution of the higher states and the continuum.

The phenomenological side of the sum rule is defined in terms of meson masses, meson decay constants and coupling constants. We introduce the meson decay constants f_K, $f_{D_s^*(B_s^*)}$ and $f_{D(B)}$ defined by the following matrix elements.

$$\begin{aligned} \langle 0|\eta^K|K(p')\rangle &= i\frac{m_K^2 f_K}{m_u+m_s}, \\ \langle 0|\eta^{D(B)}|D(B)(q)\rangle &= i\frac{m_{D(B)}^2 \, f_{D(B)}}{m_{c(b)}+m_u}, \\ \langle D_s^*(B_s^*)(p,\epsilon)|\eta_\mu^{D_s^*(B_s^*)}|0\rangle &= m_{D_s^*(B_s^*)} f_{D_s^*(B_s^*)} {\epsilon^*}_\mu, \\ \langle K(p')D(B)(q)|D_s^*(B_s^*)(p,\epsilon')\rangle &= g_{D_s^* DK(B_s^* BK)}(p'-q)\cdot\epsilon, \end{aligned} \tag{5}$$

where ϵ are the polarization vectors associated with the $D_s^*(B_s^*)$.

Finally, using Eqs.(4)-(5), the physical side of the correlation function for an off-shell D(B) meson can be written as:

$$\Pi_\mu^{D(B)}(p',p) = g^{D(B)}_{D_s^*DK(B_s^*BK)}(q^2)\frac{f_{D_s^*(B_s^*)}f_{D(B)}f_K m_K^2 m_D^2 m_{D_s^*(B_s^*)}}{(q^2-m_{D(B)}^2)(p^2-m_{D_s^*(B_s^*)}^2)(p'^2-m_K^2)(m_{c(b)}+m_u)(m_s+m_u)} \times \Big[\Big(1+\frac{m_K^2-q^2}{m_{D_s^*}^2}\Big)p_\mu - 2p'_\mu\Big]. \quad (6)$$

Similarly, we get the final expression of the physical side of the correlation function for an off-shell K meson as:

$$\Pi_\mu^{K}(p,p') = g^{K}_{D_s^*DK(B_s^*BK)}(q^2)\frac{f_{D_s^*(B_s^*)}f_{D(B)}f_K m_K^2 m_D^2 m_{D_s^*(B_s^*)}}{(q^2-m_{D(B)}^2)(p^2-m_{D_s^*(B_s^*)}^2)(p'^2-m_K^2)(m_{c(b)}+m_u)(m_s+m_u)} \times \Big[\Big(1+\frac{m_{D(B)}^2-q^2}{m_{D_s^*}^2}\Big)p_\mu - 2p'_\mu\Big]. \quad (7)$$

To calculate the coupling constant, we will choose the structure, p_μ from both sides of the correlation functions. Now, we concentrate on the QCD side, the correlation function is calculated at deep Euclidean space, where $p^2 \to -\infty$ and $p'^2 \to -\infty$ in terms of the operator product expansion. For this aim, each correlation function, $\Pi_\mu^i(p,p')$, where i stands for $D(B)$ or K, can be written in terms of perturbative and non-perturbative parts as:

$$\Pi^{QCD} = \Pi_{per} + \Pi_{nonper}, \quad (8)$$

where the perturbative part is defined in terms of double dispersion integral as:

$$\Pi_{per} = -\frac{1}{4\pi^2}\int ds'\int ds\frac{\rho(s,s',q^2)}{(s-p^2)(s'-p'^2)} + subtraction\ \ terms, \quad (9)$$

where $\rho(s,s',q^2)$ is called spectral density. In order to obtain the spectral density, we need to calculate the bare loop diagram (a) and (d) in Fig.(1) for $D(B)$ and K off-shell, respectively. We calculate these diagrams in terms of the usual Feynman integration technique by the help of the Cutkosky rules, i.e., by replacing the quark propagators with Dirac delta function: $\frac{1}{q^2-m^2} \to (-2\pi i)\delta(q^2-m^2)$. After some straightforward calculations, we obtain the spectral densities as following:

$$\begin{aligned}\rho^{D(B)}(s,s',q^2) &= \frac{N_c}{\lambda^{3/2}(s,s',q^2)}\Big[(m_u-m_s)(q^2-s)\Big(m_{c(b)}m_s^2+m_u\Big(s-m_s^2-q^2\Big)\Big)\\ &- s'\Big(-m_s^3m_u+2m_{c(b)}^3(m_u-m_s)-2m_s^2q^2+m_{c(b)}^2(2m_sm_u+q^2-s)\\ &+ q^2(s-q^2)+m_sm_u(s+q^2)+m_{c(b)}(m_s-m_u)(m_s^2+q^2+s)\Big)\\ &- s'^2(m_{c(b)}^2-m_{c(b)}m_s+m_{c(b)}m_u+q^2)\Big],\end{aligned} \quad (10)$$

$$\begin{aligned}\rho^{K}(s,s',q^2) &= \frac{N_c}{\lambda^{3/2}(s,s',q^2)}\Big[(m_{c(b)}-m_u)(q^2-s)\Big(m_{c(b)}^2(m_{c(b)}-m_u)+m_u(-m_sm_u-q^2\\ &+ s)\Big)+\Big(m_{c(b)}^3(m_s-m_u)+2m_s^3m_u+m_{c(b)}^2(-m_sm_u-2q^2)+m_s^2(q^2-s)\\ &+ q^2(s-q^2)-m_sm_u(q^2+s)+m_{c(b)}(-2m_s^3+2m_s^2m_u+m_u(q^2+s)\\ &+ m_s(q^2+s))\Big)s'+(-m_{c(b)}m_s+m_s^2+m_sm_u+q^2)s'^2\Big],\end{aligned} \quad (11)$$

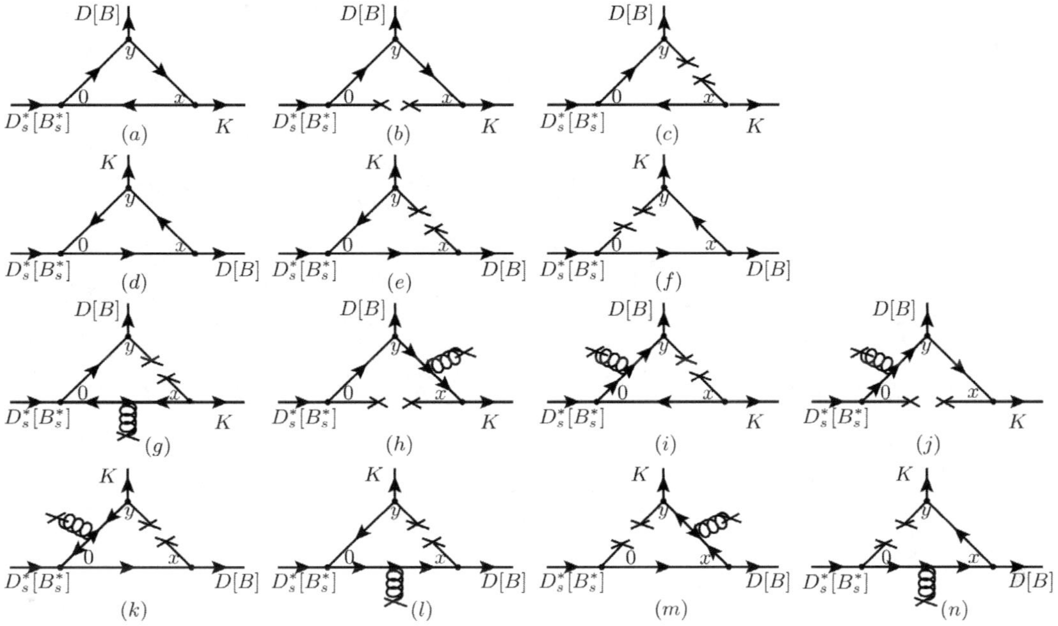

Figure 1. (a) and (d): Bare loop diagram for the $D(B)$ and K off-shell, respectively; (b) and (c): Diagrams corresponding to quark condensate for the $D(B)$ off-shell; (e) and (f): Diagrams corresponding to quark condensate for the K off-shell; (g), (h), (i), (j): Diagrams corresponding to gluon-quark condensate for the $D(B)$ off-shell; (k), (l), (m), (n): Diagrams corresponding to gluon-quark condensate for the K off-shell.

for the $D_s^* DK$ and $B_s^* BK$ vertex associated with the off-shell D and K meson, respectively. Here $\lambda(a,b,c) = a^2 + b^2 + c^2 - 2ac - 2bc - 2ab$ and $N_c = 3$ is the color number.

To calculate the nonperturbative contributions in QCD side, we consider the quark condensate diagrams presented in (b), (c), (e), (f), (g), (h), (i), (j), (k), (l), (m) and (n) parts of Fig. (1). There is also a numerically negligible contribution from the heavy quark condensates, which we will not take into account in this calculation. Therefore, for the nonperturbative part, we only encounter contributions coming from light quark condensates. Contributions of the diagrams (c), (e), (f), (g), (i), (k), (l), (m) and (n) in Fig. (1) are zero since applying double Borel transformation with respect to both of the variables p^2 and p'^2 will kill them because only one variable appears in the denominator in these cases. Hence, we calculate the diagrams (b), (h) and (j) in Fig. (1) for the off-shell D(B) meson. As a result, we obtain:

$$\Pi^{D(B)}_{nonper} = -\langle \bar{s}s \rangle \Big\{ \frac{m_u}{r r'} + \frac{m_0^2\, m_u}{4\, r^2\, r'} + \frac{m_0^2\, m_u}{2\, r\, r'^2} \Big\}, \qquad (12)$$

for the off-shell D and B meson and

$$\Pi^{K}_{nonper} = 0, \qquad (13)$$

for the off-shell K meson. Here $r = p^2 - m^2_{c(b)}$ and $r' = p'^2 - m_u^2$.

Now, it is time to apply the double Borel transformations with respect to $p^2 (p^2 \longrightarrow M^2)$ and $p'^2 (p'^2 \longrightarrow M'^2)$ to the physical as well as the QCD sides and equate the coefficient of the

Journal of Physics: Conference Series **347** (2012) 012035 doi:10.1088/1742-6596/347/1/012035

selected structure p_μ from two representations. Finally, we get the following sum rules for the corresponding coupling constants:

$$g^{D(B)}_{D_s^*DK(B_s^*BK)}(q^2) = \frac{(q^2 - m^2_{D(B)})(m_{c(b)} + m_u)(m_s + m_u)}{f_{D_s^*(B_s^*)} f_{D(B)} f_K m_{D_s^*(B_s^*)} m_K^2 m^2_{D(B)} (1 + \frac{m_K^2 - q^2}{m^2_{D_s^*(B_s^*)}})} e^{\frac{m^2_{D_s^*(B_s^*)}}{M^2}} e^{\frac{m_K^2}{M'^2}}$$
$$\times \left[-\frac{1}{4\,\pi^2} \int_{(m_{c(b)}+m_s)^2}^{s_0} ds \int_{(m_s+m_u)^2}^{s'_0} ds' \rho^{D(B)}(s,s',q^2) \theta[1 - (f^{D(B)}(s,s'))^2] e^{\frac{-s}{M^2}} e^{\frac{-s'}{M'^2}} \right.$$
$$\left. + \quad \widehat{B}\Pi^{D(B)}_{nonper} \right] \tag{14}$$

$$g^{K}_{D_s^*DK(B_s^*BK)}(q^2) = \frac{(q^2 - m^2_K)(m_{c(b)} + m_u)(m_s + m_u)}{f_{D_s^*(B_s^*)} f_{D(B)} f_K m_{D_s^*(B_s^*)} m_K^2 m^2_{D(B)} (1 + \frac{m^2_{D(B)} - q^2}{m^2_{D_s^*(B_s^*)}})} e^{\frac{m^2_{D_s^*(B_s^*)}}{M^2}} e^{\frac{m^2_{D(B)}}{M'^2}}$$
$$\times \left[-\frac{1}{4\,\pi^2} \int_{(m_{c(b)}+m_s)^2}^{s_0} ds \int_{(m_{c(b)}+m_u)^2}^{s'_0} ds' \rho^{K}(s,s',q^2)\ \theta[1 - (f^{K}(s,s'))^2] e^{\frac{-s}{M^2}} e^{\frac{-s'}{M'^2}} \right] \tag{15}$$

for the off-shell D(B) and K meson associated with the D_s^*DK (B_s^*BK) vertex, respectively.

The integration regions for the perturbative part in Eqs.(14)-(15) are determined requiring that the arguments of the three δ functions coming from Cutkosky rule vanish simultaneously. So, the physical region of the s and s' planes are described by the following non-equalities:

$$-1 \le f^{D(B)}(s,s') = \frac{2\ s\ (m_s^2 - m_u^2 + s') + (m^2_{c(b)} - m_s^2 - s)(-q^2 + s + s')}{\lambda^{1/2}(m^2_{c(b)}, m_s^2, s)\lambda^{1/2}(s, s', q^2)} \le 1 \tag{16}$$

$$-1 \le f^{K}_{D(B)}(s,s') = \frac{2\ s\ (-m^2_{c(b)} + m_u^2 - s') + (m^2_{c(b)} - m_s^2 + s)(-q^2 + s + s')}{\lambda^{1/2}(m^2_{c(b)}, m_s^2, s)\lambda^{1/2}(s, s', q^2)} \le 1 \tag{17}$$

for the D(B) and K off-shell meson associated with the $D_s^*DK(B_s^*BK)$ vertex, respectively. These physical regions are imposed by the limits on the integrals and step functions in the integrands of the sum rules. In order to subtract the contributions of the higher states and continuum, the quark-hadron duality assumption is used, i.e., it is assumed that,

$$\rho^{higherstates}(s,s') = \rho^{OPE}(s,s')\theta(s - s_0)\theta(s' - s'_0) \tag{18}$$

where s_0 and s'_0 are the continuum thresholds.

Note that, the double Borel transformation used in the calculations is written as:

$$\hat{B}\frac{1}{(p^2 - m_1^2)^m}\frac{1}{(p'^2 - m_2^2)^n} \to (-1)^{m+n}\frac{1}{\Gamma(m)}\frac{1}{\Gamma(n)}e^{-m_1^2/M_1^2}e^{-m_2^2/M_2^2}\frac{1}{(M_1^2)^{m-1}(M_2^2)^{n-1}}. \tag{19}$$

3. Numerical analysis

This section is devoted to the numerical analysis of the sum rules for the coupling constant. To obtain numerical values of the considered coupling constants, the following input parameters are used in calculations: $m_K = (0.493677 \pm 0.000016)\ GeV$, $m_D = (1.86480 \pm 0.00014)\ GeV$, $m_{D_s^*} = (2.1123 \pm 0.0005)\ GeV$, $m_B = (5.2792 \pm 0.0003)\ GeV$, $m_{B_s^*} = (5.4154 \pm 0.0014)\ GeV$

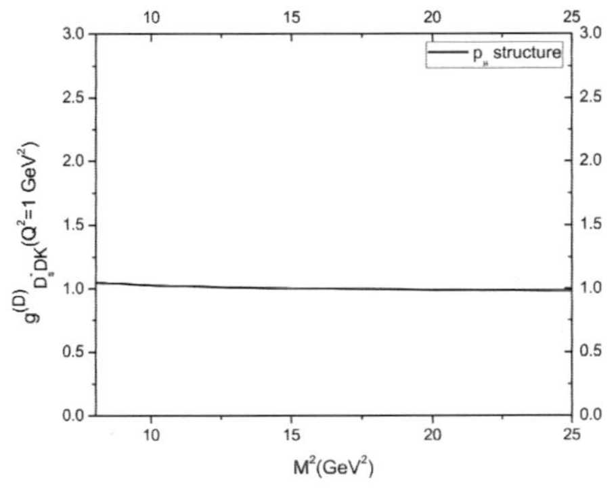

Figure 2. $g^D_{D_s^*DK}(Q^2 = 1\ GeV^2)$ as a function of the Borel mass M^2. The continuum thresholds, $s_0 = 6.82\ GeV^2$, $s'_0 = 0.99\ GeV^2$ and $M^2 = 8\ GeV^2$ have been used.

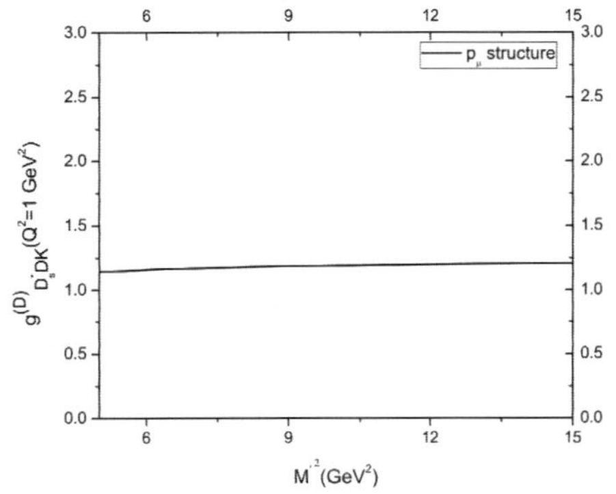

Figure 3. $g^D_{D_s^*DK^*}(Q^2 = 1\ GeV^2)$ as a function of the Borel mass M'^2. The continuum thresholds, $s_0 = 6.82\ GeV^2$, $s'_0 = 0.99\ GeV^2$ and $M'^2 = 5\ GeV^2$ have been used.

[18], $m_c = 1.3\ GeV$, $m_b = 4.7\ GeV$, $m_s = 0.14\ GeV$[19, 20], $f_K = 160\ MeV$[21], $f_{D_s^*} = (272 \pm 16^{0}_{-20})\ MeV$, $f_{B_s^*} = (229 \pm 20^{31}_{-16})\ MeV$ [22], $f_B = (190 \pm 13)\ MeV$ [23], $f_D = (202 \pm 41 \pm 17)\ MeV$ [24] , $\langle \bar{s}s \rangle = -0.8(0.24 \pm 0.01)^3\ GeV^3$ [25], $m_0^2 = (0.8 \pm 0.2)\ GeV^2$ [26]. The sum rule contains the four auxiliary parameters, namely the continuum thresholds, s_0 and s'_0 and Borel mass parameters, M^2 and M'^2. Since these parameters are not physical quantities, our results should be independent of them. Therefore, the working regions for the Borel mass parameters M^2 and M'^2 are determined requiring that both the contributions of the higher states and continuum are sufficiently suppressed and the contributions coming from higher dimensions are small. As a result, we can show that D off-shell stabilizes for $8GeV^2 \leq M^2 \leq 25GeV^2$ and $5GeV^2 \leq M'^2 \leq 15GeV^2$ and K off-shell for $6GeV^2 \leq M^2 \leq 15GeV^2$ and $4GeV^2 \leq M'^2 \leq 12GeV^2$ associated with the D_s^*DK vertex. Similarly, the regions, $14GeV^2 \leq M^2 \leq 30GeV^2$ and $5GeV^2 \leq M'^2 \leq 20GeV^2$ for B off-shell, and $6GeV^2 \leq M^2 \leq 20GeV^2$ and $5GeV^2 \leq M'^2 \leq 15GeV^2$ for K off-shell are obtained for the B_s^*BK vertex. The dependence of considered coupling constants on Borel parameters for different cases are shown in Figs.(2)-(9). From these figures, we see a good stability of the results with respect to the Borel mass parameters in the working regions. The continuum thresholds, s_0 and s'_0 are not completely arbitrary but they are correlated to the energy of the first excited states with the same quantum numbers. Our numerical calculations show that in the regions $(m_i+0.3)^2 \leq s_0 \leq (m_i+0.7)^2$ and $(m_f+0.3)^2 \leq s'_0 \leq (m_f+0.7)^2$, respectively for the continuum thresholds s and s', our results have weak dependence on these parameters. Here, m_i is the mass of initial particle and the m_f stands for the mass of the final on-shell state.

Now, using the working region for auxiliary parameters and other input parameters, we would like to discuss the behavior of the strong coupling constants in terms of q^2. In the case of off-shell D meson related to the D_s^*DK vertex, our numerical result is described well by the following mono-polar fit parametrization

$$g^{(D)}_{D_s^*DK}(Q^2) = \frac{8.76(GeV^2)}{Q^2 + 7.12(GeV^2)} \tag{20}$$

where $Q^2 = -q^2$. The coupling constants are defined as the values of the form factors at $Q^2 = -m^2_{meson}$ (see also [13]), where m_{meson} is the mass of the off-shell meson. Using $Q^2 = -m_D^2$

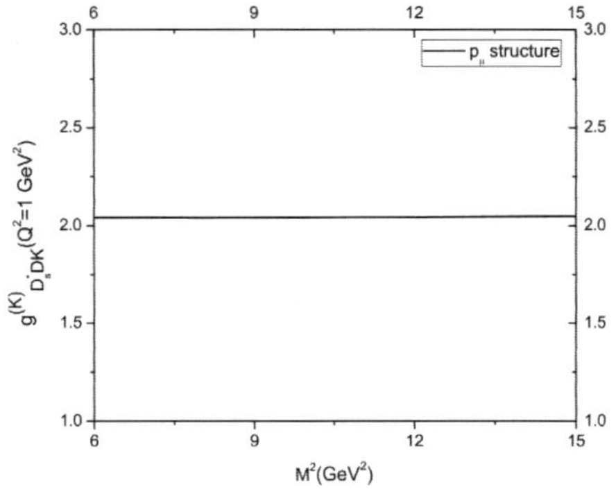

Figure 4. $g^K_{D_s^* DK}(Q^2 = 1\ GeV^2)$ as a function of the Borel mass M^2. The continuum thresholds, $s_0 = 6.83\ GeV^2$, $s_0' = 5.97\ GeV^2$ and $M^2 = 7\ GeV^2$ have been used.

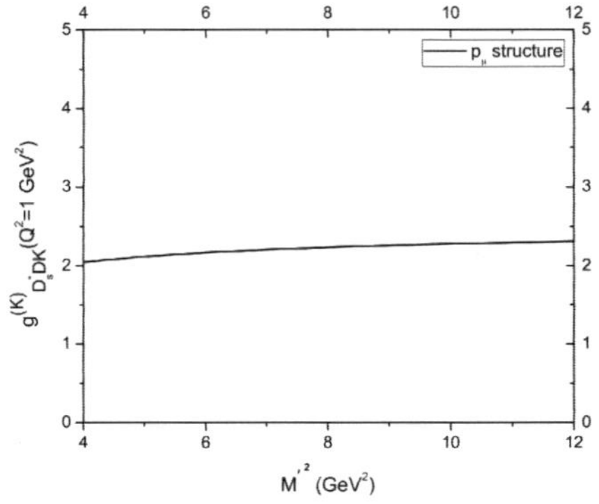

Figure 5. $g^K_{D_s^* DK}(Q^2 = 1\ GeV^2)$ as a function of the Borel mass M'^2. The continuum thresholds, $s_0 = 6.83\ GeV^2$, $s_0' = 5.59\ GeV^2$ and $M'^2 = 5\ GeV^2$ have been used.

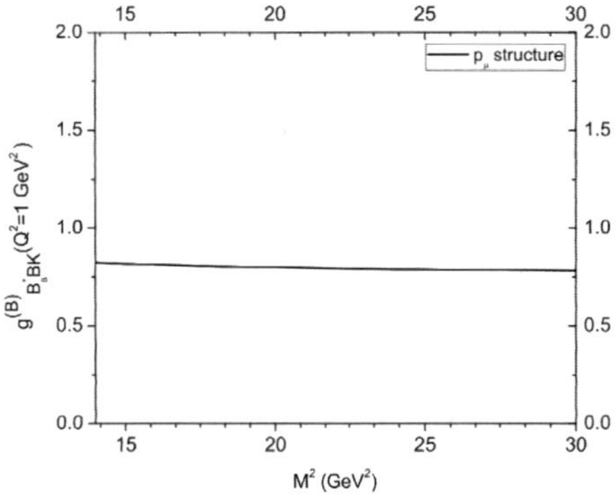

Figure 6. $g^B_{B_s^* BK}(Q^2 = 1\ GeV^2)$ as a function of the Borel mass M^2. The continuum thresholds, $s_0 = 34.99\ GeV^2$, $s_0' = 0.99\ GeV^2$ and $M^2 = 15\ GeV^2$ have been used.

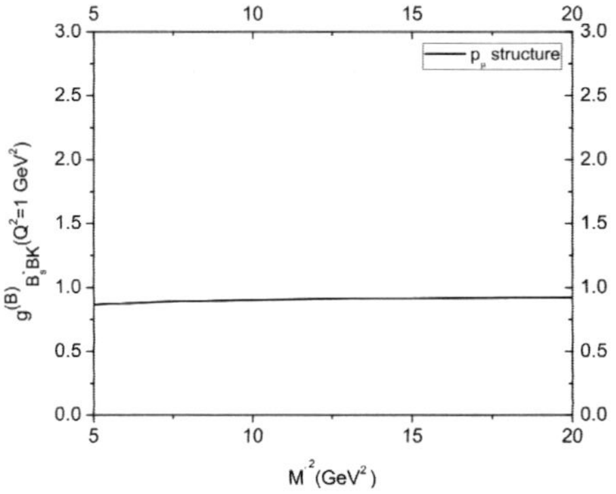

Figure 7. $g^B_{B_s^* BK}(Q^2 = 1\ GeV^2)$ as a function of the Borel mass M'^2. The continuum thresholds, $s_0 = 34.99\ GeV^2$, $s_0' = 0.99\ GeV^2$ and $M'^2 = 5\ GeV^2$ have been used.

in Eq.(20), $g^D_{D_s^* DK} = 2.79 \pm 0.24$ is obtained. The result for an off-shell K meson can be well fitted by the exponential parametrization:

$$g^{(K)}_{D_s^* DK}(Q^2) = 3.55\ e^{\frac{-Q^2}{7.25(GeV^2)}} - 0.88. \tag{21}$$

Using $Q^2 = -m_K^2$ in Eq.(21), $g^K_{D_s^* DK} = 2.99 \pm 0.26$ is obtained. Taking the average of two above obtained values, finally we get the value of the $g_{D_s^* DK}$ coupling constant as:

$$g_{D_s^* DK}(Q^2) = (2.89 \pm 0.25). \tag{22}$$

This result is consistent with the result obtained in [9]as $g_{D_s^* DK} = 2.84 \pm 0.31$.

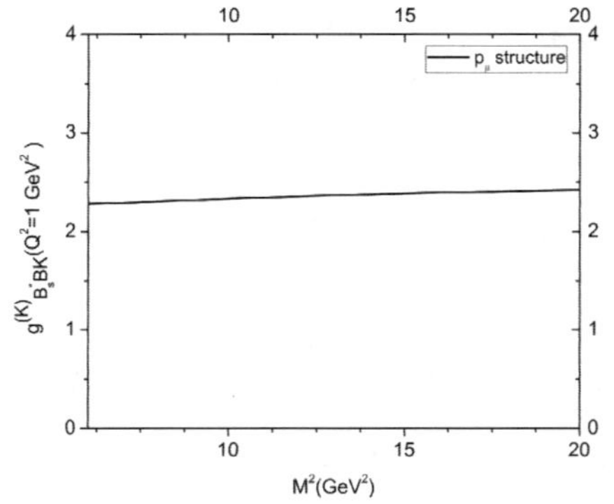

Figure 8. $g^K_{B_s^*BK}(Q^2 = 1\ GeV^2)$ as a function of the Borel mass M^2. The continuum thresholds, $s_0 = 34.99\ GeV^2$, $s_0' = 33.39\ GeV^2$ and $M^2 = 10\ GeV^2$ have been used.

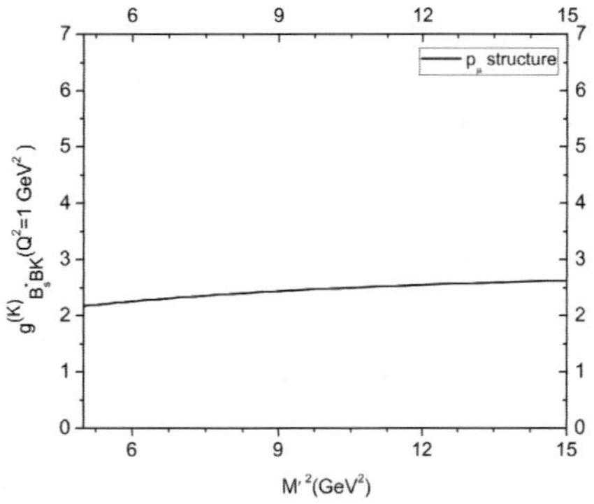

Figure 9. $g^K_{B_s^*BK}(Q^2 = 1\ GeV^2)$ as a function of the Borel mass M'^2. The continuum thresholds, $s_0 = 34.99\ GeV^2$, $s_0' = 33.39\ GeV^2$ and $M'^2 = 7\ GeV^2$ have been used.

Similarly, for B_s^*BK vertex, our result for B off-shell is better extrapolated by the exponential fit parametrization,

$$g^{(B)}_{B_s^*BK}(Q^2) = 0.66\ e^{\frac{-Q^2}{23.34(GeV^2)}} + 0.23 \tag{23}$$

and for K off-shell case, the parametrization is

$$g^{(K)}_{B_s^*BK}(Q^2) = 4.39\ e^{\frac{-Q^2}{4.02(GeV^2)}} - 1.03. \tag{24}$$

Using $Q^2 = -m_B^2$ in Eq.(23), the coupling constant is obtained as $g^B_{B_s^*BK} = 2.40 \pm 0.22$. Also $g^K_{B_s^*BK} = 3.62 \pm 0.34$ is obtained at $Q^2 = -m_K^2$ in Eq.(24). Taking the average of these results, we get the following result

$$g_{B_s^*BK}(Q^2) = (3.01 \pm 0.28). \tag{25}$$

The errors in the results are due to the uncertainties in determination of the working regions for the auxiliary parameters as well as the errors in the input parameters.

4. Acknowledgments

The authors thank E. Veli Veliev for his useful discussions. This work has been supported partly by the Scientific and Technological Research Council of Turkey (TUBITAK) under the research project 110T284.

References

[1] Navarra F S, Nielsen M, Bracco M E, Chiapparini M E and Schat C L 2000 *Phys. Lett.* B **489** 319
[2] Navarra F S, Nielsen M, Bracco M E 2002 *Phys. Rev.* D **65** 037502
[3] Bracco M E, Chiapparini M, Lozea A, Navarra F S and Nielsen M 2001 *Phys. Lett.* B **521** 1
[4] Matheus R D, Navarra F S, Nielsen M and da Silva R R 2002 *Phys. Lett.* B **541** 265
[5] Matheus R D, Navarra F S, Nielsen M and da Silva R R 2005 *Int. J. Mod. Phys.* E **14** 555
[6] Wang Z G 2007 *Nucl. Phys.* A **796** 61; 2007 *Eur. Phys. J.* C **52** 553; 2007 *Phys. Rev.* D **74** 014017

[7] Carvalho F, Duraes F O, Navarra F S and Nielsen M 2005 *Phys. Rev.* C **72** 024902.
[8] Bracco M E, Chiapparini M, Navarra F S and Nielsen M 2005 *Phys. Lett.* B **605** 326
[9] Bracco M E, Cerqueira Jr A, Chiapparini M, Lozea A, Nielsen M 2006 *Phys. Lett.* B **641** 286–293
[10] Wang Z G and Wan S L 2006 *Phys. Rev.* D **74** 014017
[11] Holanda L B, Marques de Carvalho R S and Mihara A 2007 *Phys. Lett.* B **644** 232.
[12] Bracco M E, Chiapparini M, Navarra F S and Nielsen M 2008 *Phys. Lett.* B **659** 559
[13] Rodrigues B O, Bracco M E, Nielsen M and Navarra F S 2011 *Nucl.Phys.* A **852** 127–140
[14] Wang Z G 2008 *Phys. Rev.* D **77** 054024
[15] Azizi K and Sundu H 2011 *J.Phys.* G **38** 045005
[16] Shifman M A, Vainshtein A I and Zakharov V I 1979 *Nucl. Phys.* B **147** 385; Reinders L J, Rubinstein L J and Yazaki S 1985 *Phys. Rept.* **127** 1; Narison S 1989 *QCD spectral sum rules*, World Sci. Lect. Notes Phys. **26** 1
[17] Sundu H, Sungu J Y, Sahin S, Yinelek N and Azizi K 2011 *Phys. Rev.* D **83** 114009
[18] Nakamura K and et al (Particle Data Group) 2010 *J. Phys.* G **37** 075021
[19] Montanet L and et al 1994 *Phys. Rev.* D **50** 1173
[20] Eidelman S and et al (Particle Data Group) 2004 *Phys. Lett.* B **592** 1
[21] Eidelman S and et al (Fermilab E653 Collaboration) 2004 *Phys. Lett.* B **592** 1
[22] Becirevic D and et al 1999 *Phys. Rev.* D **60** 074501
[23] Gamiz V and et al (HPQCD Collab.) 2009 *Phys. Rev.* D **80** 014503
[24] Danko I et al (CLEO Collaboration) 2005 *J. Phys. Conf. Ser.* **9** 91
[25] Ioffe B L 2006 *Prog. Part. Nucl. Phys.* **56** 232
[26] Dosch H G, Jamin M and Narison S 1989 *Phys. Lett.* B **220** 251; Belyaev V M and Ioffe B L 1982 *Sov. Phys. JETP* **57** 716

Journal of Physics: Conference Series **347** (2012) 012036 doi:10.1088/1742-6596/347/1/012036

PHIN photo-injector as the CLIC drive beam source

Eric Chevallay[1], Marta Csatari[1], Anne Dabrowski[1], Steffen Doebert[1], Daniel Egger[1,2], Valentine Fedosseev[1], Oznur Mete* [1], Maja Olvegaard[1], Massimo Petrarca[1,3]

[1]CERN, 1211 Geneva, Switzerland, [2]EPFL, 1015 Lausanne, Switzerland, [3] University of Geneva

E-mail: oznur.mete@cern.ch

Abstract. The Compact Linear Collider (CLIC) study proposes a multi-TeV, high luminosity, electron-positron linear collider in order to fulfill the current need for a lepton collider. The study has been started in the late 80s at CERN and currently is a joint effort with a collaboration of 40 institutes. An innovative scheme of high peak RF power production for the high accelerating gradient has been proposed for CLIC. The so called "two-beam scheme" consists of two beams that are running parallel to each other. One of the beams is to be accelerated for the collision experiments and called "the main beam". The second beam of the CLIC scheme is "the drive beam" and will be employed for the power production. The quality of the main beam acceleration depends on the stability of the power that is generated by the drive beam. Therefore, the optimization of the drive beam production with the proper time structure and within the required beam dynamics tolerances is one of the most important accelerator physics aspects of the project. Currently in the conceptual level, the baseline design of the drive beam injector consists of a thermionic gun. This electron source has to be combined with a sub-harmonic bunching system in order to provide the required time structure of the drive beam. However, a big disadvantage of this scheme is the parasitic satellite bunches that are produced due to the sub-harmonic bunching system. PHIN photoinjector has been raised as another option in order to replace the existing thermionic gun of CLIC test facility (CTF3) and to form the bases of a source for the CLIC drive beam. The PHIN project is in the framework of the European CARE (Coordinated Accelerator Research in Europe) program.

1. Introduction

Within the CLIC study [1, 2], an RF system has been proposed by using normal accelerator cavities operating at 12 GHz, in pulses 239 ns long, with the nominal accelerating gradient of 100 MV/m. The implementation of such cavities would result a total machine length of 48 km for a 3 TeV centre of mass energy of colliding beams.

The choice of an optimum acceleration frequency is an essential item of optimization for the linear collider studies. This optimization is done concerning the RF power consumption and the achievable accelerating gradient. In the last fifty years, a significant number of research activities have been carried out under the assumption that the achievable accelerating gradient extends with the increasing RF frequency [3, 4]. The limitation comes from the phenomenon called RF breakdown and the early investigations has been done by Kilpatrick [5] in 1957. The RF breakdown is a localized dissipation of energy on the accelerating structure surface due to high electric or magnetic fields. The process causes distortions such as crater forming and melting on the structure surface. The experimental results confirm that the gradient limit increases

Journal of Physics: Conference Series **347** (2012) 012036 doi:10.1088/1742-6596/347/1/012036

with the square root of the frequency in a range of 300 MHz to 3 GHz in agreement with the Kilpatrick criteria in Eq. (1),

$$f[MHz] = 1.64E^2[MV/m]e^{-\frac{8.5}{E[MV/m]}}, \tag{1}$$

where f is the RF frequency in MHz and **E** is the maximum electric field in MV/m.

The accelerating structure development is a crucial research activity within the CLIC study. The aspects of this research are studying the RF breakdown phenomenology and determining the limitations on the achievable accelerating gradient. The experiments focus on the RF frequency and pulse length dependence of the phenomenon as well as the impact of the material type and the surface treatment of the structures. The target structure should operate at low RF breakdown rates and remain undamaged against the inevitable RF breakdowns and RF heating. Among several designs, the candidate structure for CLIC has demonstrated an extremely low RF breakdown probability of less than one in 10^7 pulses. The tests have been done for 100 MV/m at the nominal pulse length. The mentioned accelerating structure can be seen in Figure 1.

Figure 1. An accelerating structure for CLIC that has been tested to 100 MV/m. Designed at CERN, the components were manufactured by KEK and the assembly and bonding as well as testing with high-power RF was done by SLAC (Courtesy SLAC).

2. Description of the CLIC General Layout

The general CLIC layout is shown in Figure 2. An innovative scheme of high peak RF power production for the high accelerating gradient has been proposed for CLIC. The so called "ttwo-beam scheme" consists of two beams that are running parallel to each other. The overall CLIC parameters as indicated in the conceptual design studies in 2010 is given in Table 1.

One of the beams is to be accelerated for the collision experiments and called "the main beam". For a linear collider providing the colliding beams of high charge and ultra-low emittance is essential to obtain high collision luminosity. This implies the requirement for an injection system that has been designed, efficiently. The lower part of the layout illustrates the injector system for the CLIC main beams. The pre-damping and the damping rings are shown as a part of the system. These rings have been designed to reduce the beam emittance of the high charged electron and positron beams. The circumference, therefore the design of the optical lattice of the damping rings can be accomplished by considering the beam parameters such as injection emittance, number of bunches in the bunch trains, bunch spacing and bunch charge. The necessary extraction emittance is obtained depending on the damping time, which determines the energy of the ring. More details on the damping ring design and issues can be found in reference [6]. The pre-damping rings of CLIC has been designed to receive a 2.86 GeV beam with a large emittance. They reduces emittance of electron and positron beams before the

Table 1. Overall CLIC parameters from 2010 conceptual design study.

Centre of Mass Energy (GeV)	3000
Main Linac RF Frequency (GHz)	11.994
Luminosity $(10^4 cm^{-2} s^{-1})$	5.9
Linac Repetition Rate (Hz)	50
Number of Particles / Bunch	3.72×10^9
Number of Bunches / Train	312
Bunch Separation (ns)	0.5
Bunch Train Length (ns)	156
Beam Power / Beam (MW)	14
Unloaded / Loaded Gradient (MW/m)	120 / 100
Overall Two Linac Length (km)	42.16
Total Beam Delivery System Length (km)	2×2.75
Proposed Site Length (km)	48.4
Total Site AC Power (MW)	415
Wall Plug to Main Beam Power Efficiency (%)	7

injection to the damping rings. Damping rings produce the ultra-low emittance needed for the luminosity performance of CLIC [7, 8]. Some of the beam parameters before and after the pre-damping rings, and damping rings are given in the Table 2.

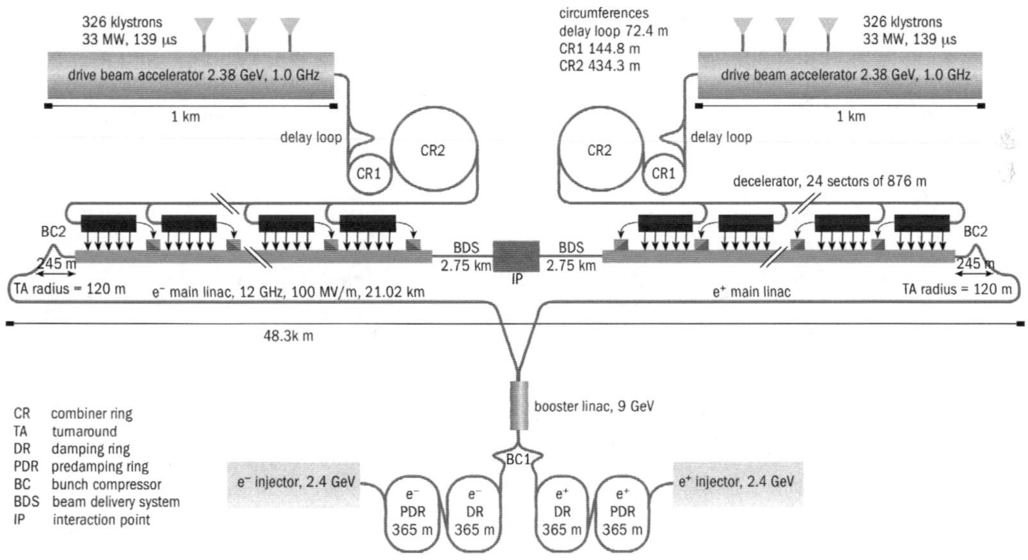

Figure 2. CLIC general layout.

Additionally, further emittance reduction is possible by minimizing the emittance growth due to wake fields that are generated by the beam itself. The accelerating cavities of the main linac have been optimized in order to minimize this effect. Eventually, after the beam delivery system, the beam has the emittances of 1 nm rms and 40 nm rms, in the vertical and horizontal planes, respectively.

The second beam of the CLIC scheme is "the drive beam" and will be employed for the power production. The two-beam scheme should deliver a power of 275 MW per active accelerating

Table 2. The parameters for the CLIC main beams before and after the pre-damping rings.

	Pre-Damping Rings			Damping Rings
	Injected		Extracted	
	e^-	e^+		
Bunch Population (10^9)	4.7	6.4	4.5	4.1
Bunch Length (mm)	1	9	10	1.4
Bunch Spacing (ns)				0.5
Number of Bunches / Train				312
Energy Spread (%)	0.07	1	0.5	0.13
Longitudinal Emittance (keV.m)	2	257	121	6
Horizontal Normalized Emittance (μm)	100	7.10^3	63	6.10^3
Vertical Normalized Emittance (μm)	100	7.10^3	1.5	5.10^3

metre, which is necessary to produce the electric fields of 100 MV/m. The drive beam generation system can be seen in the upper part of the layout. The stability of the main beam acceleration depends on the stability of the power that is generated by the drive beam. Therefore, the optimization of the drive beam production with the proper time structure and within the required beam dynamics tolerances is one of the most important accelerator physics aspects of the project.

3. Power Generation for CLIC: Drive Beam and Its Injector Specifications

The two-beam acceleration scheme of CLIC Study proposes a unique power production technique. The scheme eliminates the utilization of the conventional klystrons which are not capable of providing high peak power required by the CLIC accelerating structures (275 MW/m), efficiently. This also prevents the additional underground building of a scheme with klystrons.

The drive beam consists of a high current (101 A), low energy (2.37 GeV) beam with a repetition rate of 12 GHz. The design specifications of drive beam is shown in Table 3. The high current is needed for the high peak power production in the so called "Power Extraction and Transfer Structures (PETS)" [9].

According to the CLIC scheme, the drive beam is generated as a long train (139 μs) of electron bunches. The separation between the individual bunches is 60 cm. This beam with a large bunch separation is accelerated up to 2.37 GeV in a linac with high efficiency, using the fully-beam-loaded acceleration mode [10]. In this mode, the most of the RF power is transferred to the beam so that the accelerating gradient in the end of the structure reduces nearly down to zero. Short RF structures are used in order to minimize the ohmic losses.

Table 3. The baseline parameters of the CLIC drive beam from the conceptual design studies (2010).

Energy (decelerator injection)(GeV)	2.37
Energy (final, minimum) (MeV)	237
Average Current in the Pulse (A)	101
Train Duration (ns)	243.7
Number of Bunches / Train	2922
Bunch Charge (nC)	8.4
Bunch Separation (ns)	0.083
Bunch Length (rms) (mm)	1
Normalized Emittance (rms) (mm mrad)	150

The drive beam linac is built by using normal conducting accelerating structures that operate at 1 GHz. Those structures are powered by the conventional klystrons. A series of rings are located in the drive beam generation system. These rings can be seen in Figure 2 as "delay loop" and two successive "combiner rings (CR1 and CR2)". This system of three rings is designed to achieve the frequency multiplication of the drive beam in order to increase its peak current up to 101 A. This innovative technique is used to increase the beam current and it is called "the beam recombination". In the delay loop, the 139 μs long bunch train is divided into 24 sub-trains with the length of 243.7 ns. The 500 MHz RF deflectors are used to separate the "even" and "odd" buckets to be transferred to the delay loop or toward CR1 and CR2 that are equipped with RF deflectors of 1 GHz and 3 GHz, respectively. The "even" buckets are delayed while they are circulating the delay loop. Then they are interleaved between the "odd" buckets by means of the combiner rings. After the whole procedure, the 24 sub-trains form a burst with the average beam current of 101 A. The time structure of the drive beam before and after the "beam recombination process" is shown in Figure 3.

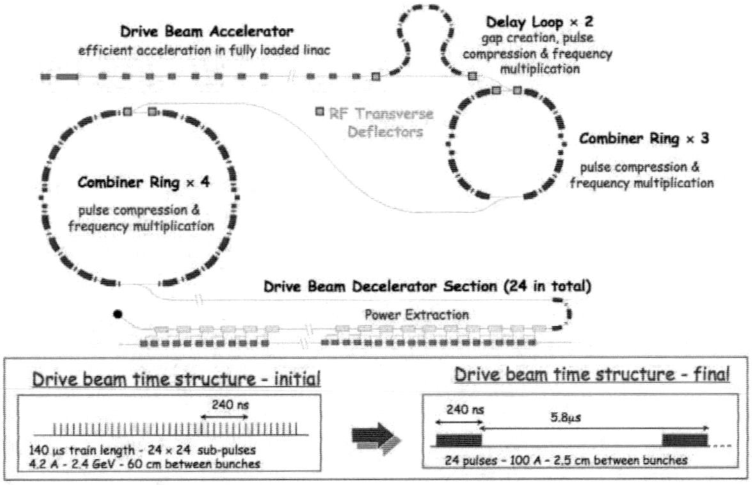

Figure 3. The schematic of the beam recombination and the time structure of the drive beam before and the beam recombination process.

The drive beam system generates 24 of such bursts with the separation of 5.8 μs and sends them to the decelerators for the power production. A decelerator module houses PETS, quadrupoles and high power RF networks. The layout of the CLIC two-beam module can be seen in Figure 4-a. When the bursts of drive beam enters one of the PETS, it excites preferentially the synchronous mode by interacting with the impedance of the periodically loaded waveguide. In this way, a single PETS generates 12 GHz RF power and drives two accelerating structures of the main beam, as shown in Figure 4-b.

4. Photo-injector Option for the Drive Beam

A photoinjector is an electron source that uses laser pulses in order to extract electrons from the surface of a metallic or a semiconductor cathode. The electrons can escape the cathode surface provided that the laser pulses satisfies the energy to decrease the potential barrier of the surface. The cathode plug is placed in one end of an RF cavity. This RF cavity is used for the rapid acceleration of the electrons after the production. The photoinjectors are high brightness, low emittance electron beam sources. The theory and implementation of them date back to 80s.

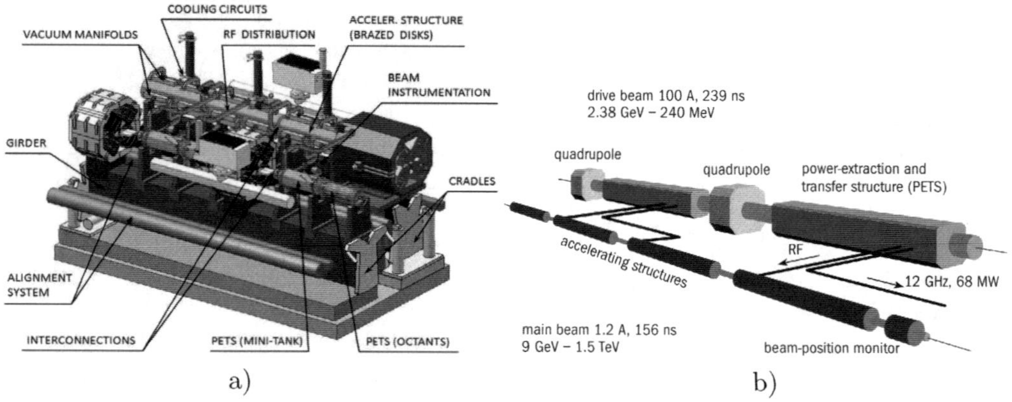

Figure 4. a) Schematic of the CLIC two-beam acceleration module. b) The illustration of two-beam acceleration.

A photo-injector option for CLIC drive beam has been also discussed during the early stages of the test facility studies [11]. The previous studies for the CLIC photo-injector can be found in the reference [12]. Indeed, there would be certain advantages of implementing an RF gun as the CLIC drive beam injector.

First of all, as a laser-driven system, a photo-injector can provide flexibility in manipulating the time structure of the beam. This manipulation is provided by a phase coding technique of a laser system. Moreover, in drive beam linac every second bucket is populated and the interleaving buckets are supposed to be empty. However, the parasitic charge in the supposedly-empty buckets is inevitable due to the bunching system. Those bunches are called "the satellite bunches". The charge in the satellite bunches can cause radiation problems and compromises the power production efficiency. The necessary time structure can be provided by a photo-injector without producing this parasitic charge.

A photo-injector can provide a beam with small transverse and longitudinal emittances compared to a thermionic gun. Therefore, the beam transportation and bunch lenght manipulation become easier throughout the machine. A photo-injector is capable of producing long electron pulse trains by ensuring the stability of the beam parameters within the defined tolerances.

The feasibility research on the implementation of a photoinjector as the current CTF3 and future drive beam source is the main objective of the commissioning of the PHIN photoinjector. During the PHIN commissioning, the produced electron beam has been characterized in terms of the longitudinal and transverse properties.

5. Instrumentation and Experimental Methods for PHIN

The PHIN photoinjector consists of a semiconductor photocathode, a laser system and the RF gun and a set of solenoids for focusing. The test-stand at CERN also houses a section with various instruments to measure the transverse and the longitudinal beam parameters with different techniques.

The measurement section has been equipped with the beam current monitoring instruments such as a fast current transformer, a beam position monitor, a wall current monitor and a Faraday cup. The OTR profiling has been widely used in the system in order to measure the size, the transverse emittance and the energy of the beam. A magnetic spectrometer has been installed in the beam line. In addition, the spectrometer set-up includes a segmented beam dump.

The demonstration of the stability along the $1.3\,\mu s$ bunch train is one of the main focuses of the PHIN project. Therefore, the time-resolved measurements give the substantial results of the PHIN commissioning. In the beamline, the time resolved measurements were done by using gated-intensified CCD (Charge Coupled Device) cameras in order to be used for the OTR monitoring. The segmented dump monitors the energy and the energy spread as a function of time.

This section introduces the specifications of the building blocks for the PHIN photoinjector.

5.1. Cs_2Te Cathode Preparation for PHIN at CERN Photo-Emission Laboratory

The electrons of the PHIN photoinjector are extracted from a semiconductor Cs_2Te cathode, which is mounted on one end of a $2 + 1/2$ cell RF gun. The cathode has been developed by the CERN photoemission laboratory and they have demonstrated a lifetime to allow> $100h$ run at a 3% quantum efficiency for a $262\,nm$ laser wavelength [13]. Excellent vacuum conditions, below $10^{-11}\,mbar$ can be obtained by applying a bake-out at $300°C$ to achieve this lifetime.

The cathodes are prepared by co-evaporation of Cs and Te on a copper plug. After the preparation, they have to be transferred from the photoemission laboratory to the experimental area. The cathode transfer chamber, the so-called "Transport Carrier" (or T.C.), is a mechanical device designed to receive and deliver up to four photocathodes, under Ultra High Vacuum conditions ($U.H.V.$), from the photo-emission laboratory to the PHIN photo-injector.

Before the measurements at the PHIN set-up, a bake-out at 130°C has been done. The vacuum value of $\sim 3 \text{ x } 10^{-10}$ has been measured in the set-up without the RF and the beam. During the measurements a vacuum value of $\sim 10^{-9}$ mbar have been maintained in the existence of the RF and the beam.

5.2. Laser System

In order to provide the laser pulses for PHIN, initially a Nd:YLF oscillator produces laser pulses at a repetition rate of $1.5\,GHz$ with an average power of $P = 10\,W$ in a continues wave (CW) train.

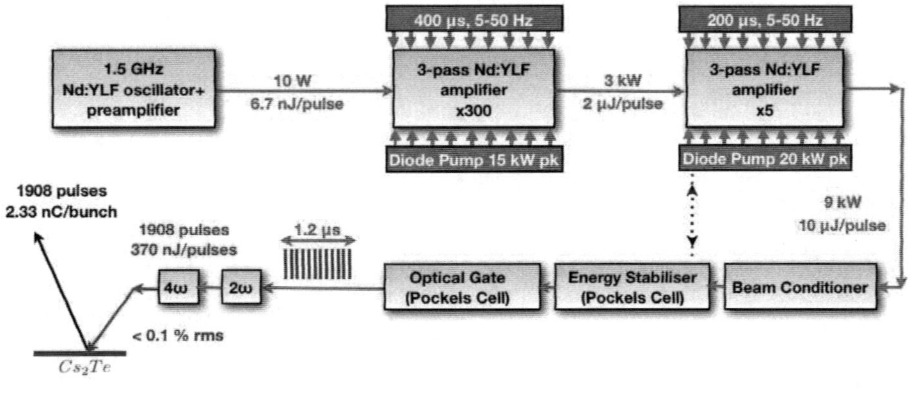

Figure 5. Layout of the laser system.

The layout of the laser is shown in Figure 5. In the layout, after the pre-amplification, two additional amplification stages follow in order to increase the laser power from $10\,W$ CW power to $9\,kW$ of quasi continues wave (QCW) mean power.

The length of the pulse train can be adjusted between $50\,ns$ and $170\,\mu s$ by using the Pockels cells [14, 15, 16]. The nominal pulse train length for the PHIN photoinjector is $1.27\,\mu s$. The

PHIN laser has been designed to produce pulse trains at a repetition rate of up to $50\,Hz$. A temporal structure that consists of a macropulse repetition rate of $1-5\,Hz$ has been used during the measurements.

After the UV conversion the cathode is illuminated by the laser train of 1908 pulses ($1.27\,\mu s$) at the wavelength of $262\,nm$ and with the energy of $370\,nJ/Pulse$. This is the nominal energy in order to extract the bunch charge of $2.33\,nC$ assuming the quantum efficiency of 3%. The laser spot size on the photocathode can be changed within a range of $0.1 - 5mm\,(4\sigma)$. The micropulse width of $\sim 6.5\,(\pm 0.41)$ ps (FWHM) has been measured for the UV laser.

During the commissioning runs a laser with the pulse train length of $200-1300\,ns$ was used, alternatively, obtaining 300 – 1950 bunches. The laser spot sizes of 2, 3, $4\,mm$ have been used for different measurements.

5.3. Design and Simulations for the PHIN RF Gun

The laser-driven PHIN RF gun consists of a 2 + 1/2 cell normal conducting S-band standing wave cavity [17]. The gun has been designed by LAL and it has been installed on the PHIN beamline at CERN. The design has been based on a previous proto-type CERN RF gun. The previous design has been modified for the PHIN gun in order to provide the specifications that are needed by a possible CTF3 photoinjector. The specifications for the PHIN RF gun are summarized in the Table 4. The design of the RF gun aims to maximize the vacuum pumping speed which increases the dynamic vacuum. Therefore, a Non Evaporable Getter (NEG) thin film has been implemented on the wall of the anti-chamber that is placed around the cavity. A photograph of the PHIN RF gun is shown in Figure 6.

Table 4. The design specifications for the PHIN laser-driven RF gun.

Parameter	Value
RF Frequency (GHz)	2.99855
RF Power (MW)	30
Gradient (MV/m)	85
Vacuum Pressure ($mbar$)	2.10^{-10}
Beam Energy (MeV)	5.5
Beam Current (A)	3.5
Charge / Bunch (nC)	2.33
Bunch Length (ps)	10
Energy Spread (%)	¡1
Normalized Transverse Emittance ($mm\,mrad$)	¡25
Pulse Length Duration (μs)	1.27
Vacuum Pressure ($mbar$)	2×10^{-10}

The π resonance mode is used for the acceleration in the PHIN RF gun which corresponds to a frequency of $2.99855\,GHz$. This mode has been presented in Figure 7-a from the result of SUPERFISH simulation program. The on-axis electric field that is excited on the cavity is presented in Figure 7-b.

6. Experimental Results of the PHIN Commissioning

The PHIN photoinjector has been commissioned with the intermittent runs between 2008 and 2010 with a total beam time of 60 workdays. The studies led to an eventual success by fulfilling the design specification which are given in Table 5. The earlier results of the PHIN commissioning can be found in reference [18]. The latest commissioning measurements have been focused on

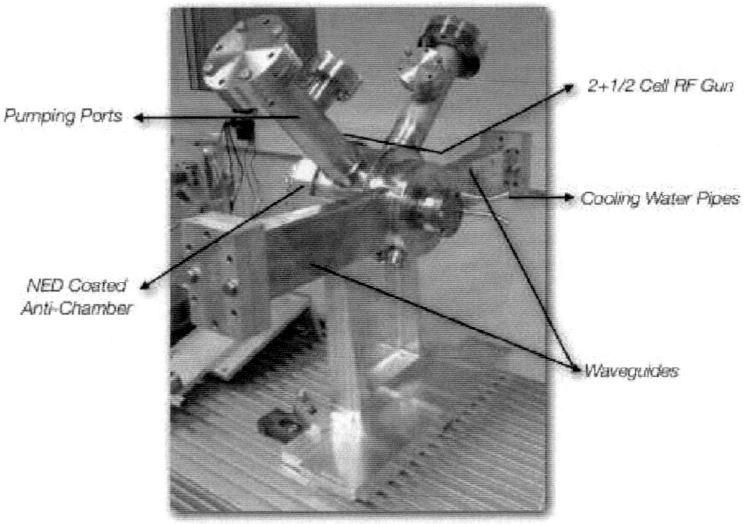

Figure 6. A photograph from the laser-driven PHIN RF gun that consists of a 2 + 1/2 cell normal conducting S-band standing wave cavity.

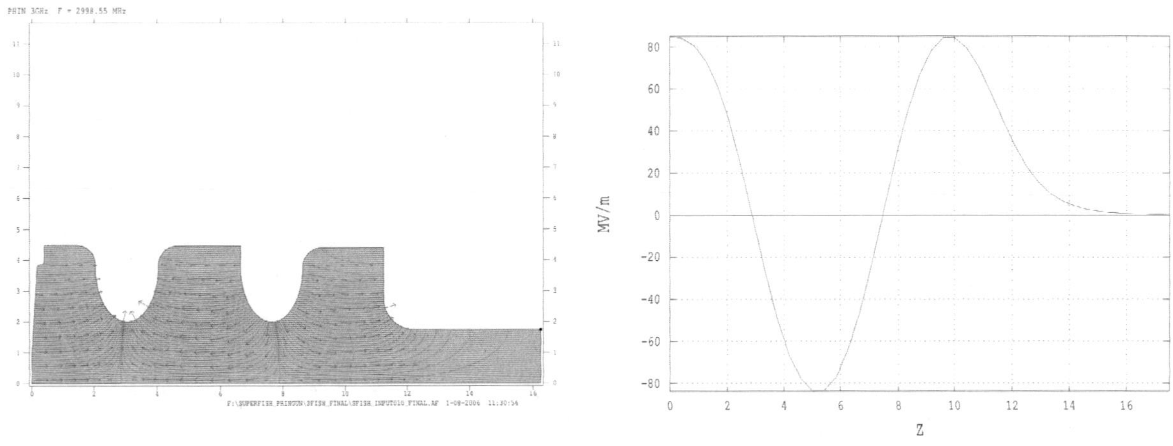

a) b)

Figure 7. a) Illustration of the electric field in the cavity that is excited in the π-mode. The figure obtained as the result of the SUPERFISH simulation for the 2+1/2 cell normal conducting S-band standing wave PHIN cavity. b) The on-axis electric field that is excited in the PHIN RF gun for the π mode.

the charaterization of the PHIN beam along the pulse train. Those results have been presented in [19].

In oder to provide the drive beam requirements in terms of the time structure of the beam, a phase coding system has been developed and installed in the laser system. The details of the phase coding stage and the measurement on the phase coded electron beam will be soon presented in the International Particle Accelerators Conference (IPAC 2011) [20, 21].

Table 5. The specifications for the PHIN photoinjector.

Parameter	**Specification**
Laser	
UV Laser Pulse Energy (nJ)	370
Micropulse Repetition Rate (GHz)	1.5
Macropulse Repetition Rate	1-5
Train Length (ns)	1273
Electron Beam	
Charge per Bunch (nC)	2.33
Charge per Train (nC)	4446
Current (A)	3.5
Transverse Normalized Emittance (mm mrad)	25
Energy Spread (%)	1
Energy (MeV)	5.5
Charge Stability (%, rms)	0.25
RF Gun	
RF Gradient (MV/m)	85
RF Frequency (GHz)	2.99855
Cathode	Cs_2Te
Quantum Efficiency (%)	3

7. Conclusions

Throughout this paper the photoinjector concept has been introduced. The PHIN experimental set-up and a variety of accompanying beam diagnostics instrumentation have been described. The PHIN commissioning has been continued with the intermittent runs between 2008 and 2010. The experimental program has achieved the full characterization of the PHIN photoinjector for the short and the long bunch trains. The results table of the beam characterization measurements have been presented in comparison with the design specifications.

References

[1] H. H. Braun, J. P. Delahaye, A. De Roeck, and G. Geschonke. CLIC Here for the Future. CERN COURIER, Geneva, September 19, 2008.

[2] Marco Battaglia, A de Roeck, Jonathan Richard Ellis, and Daniel Schulte. Physics at the CLIC Multi-TeV Linear Collider : Report of the CLIC Physics Working Group. CERN, Geneva, 2004.

[3] S. Doebert. Status of Very High-Gradient Cavity Tests. Proceedings of LINAC 2002, Gyeongju, Korea, 2002.

[4] S. Doebert. RF-Breakdown in High-Frequency Accelerators. SLAC-PUB-10463, 2004.

[5] W. D. Kilpatrick. Criterion for Vacuum Sparking Designed to Include Both RF and DC. Rev. Sci. Instrum. 28, 824 (1957); doi:10.1063/1.1715731, 1957.

[6] A. Wolski and W. Decking. Damping Ring Designs and Issues. Lawrence Berkeley National Laboratory. LBNL Paper LBNL-52580. Retrieved from: http://escholarship.org/uc/item/2766w0b8, 2003.

[7] Y. Papaphilippou and et al. Design Optimization of the CLIC Damping Rings. Proceedings of IPAC 2010 Kyoto, Japan, 2006.

[8] F. Antoniou, Y. Papaphilippou, and F. Zimmermann. Design considerations for the clic pre-damping rings. oai:cds.cern.ch:1212964. Technical Report CERN-ATS-2009-081. CLIC-Note-784, CERN, Geneva, Oct 2009.

[9] I. Syratchev, D. Schulte, E. Adli, and M. Taborelli. High rf power production for clic. oai:cds.cern.ch:1055564. 2007.

[10] H. H. Braun, R. Corsini, S. Doebert, E. Jensen, F. Tecker, and P. Urschutz. Efcient long-pulse fully- loaded ctf3 linac operation. oai:cds.cern.ch:999488. Technical Report CERN-OPEN-2006-070. CLIC- Note-697, CERN, Geneva, Nov 2006.

[11] G. Geschonke and A Ghigo. Ctf3 design report, chapter 8. Technical Report CERN-PS-2002-008-RF. CTF-3-NOTE-2002-047. LNF-2002-008-IR, CERN, Geneva, May 2002.

[12] G. Bienvenu, H. Braun, E. Chevallay, G. Cheymol, M. Divall, S. Doebert, K. Elsener, V. Fedosseev, A. Ghigo,

G. Hirst, M. Jore, N. Lebas, R. Losito, B. Mercier, O. Mete, M. Petrarca, C. Prevost, L. Ri- nol(Editor), R. Roux, A. Variola, and C. Vicario. CTF3 3 GHz RF Gun Test at CERN. CTF3 Note 093.

[13] A. Barbiero, E. Chevallay, K. Elsener, and R. Losito. Cesium-Telluride Photocathode No. 166. AB-Note-2007-023 ATB, CARE-Report-07-018-PHIN, CTF3-Note-089, 2007.

[14] M. Petrarca, V. Fedosseev, and N. Lebas. CTF3 Photo-Injector Laser Amplier Construction. CARE- Report-2008-012-PHIN, CARE-Report-2008-012-PHIN, 2008.

[15] M. Petrarca and et al. CTF3 Photo-Injector Laser. CLEO/IQEC Proceedings, 2009, Baltimore, Maryland.

[16] M. Petrarca, K. Elsener, V. Fedosseev, and N. Lebas. CTF3 Photo-Injector: RF Synchronization of the Laser System. CARE-Report-2008-030-PHIN.

[17] R. Roux, G. Bienvenu, and B. Mercier. Design of a RF Photo-Gun. CARE Note-2004-034-PHIN.

[18] O. Mete, E Chevallay, A Dabrowski, M Divall, S Doebert, D Egger, K Elsener, V Fedosseev, T Lefevre, and M Petrarca. The transverse and longitudinal beam characteristics of the phin photo-injector at cern. oai:cds.cern.ch:1262856. Technical Report CERN-OPEN-2010-013. CLIC-Note-809, CERN, Geneva, Mar 2010.

[19] M. Csatari Divall, E., E. Chevallay, A. Dabrowski, S. Doebert, D. Egger, V. Fedosseev, O. Mete, M. Olvegaard, and M. Petrarca. Production of long bunch trains with 4.5 C total charge using the PHIN photo-injector (submitted). 2011.

[20] M. Csatari Divall, A. Andersson, E. Bolzon B., Bravin, E. Chevallay, A. Dabrowski, S. Doebert, A. Drozdy, D. Egger, V. C. Fedosseev, Hessler, T. Lefevre, S. Livesley, R. Losito, O. Mete, M. Olvegaard, M. Petrarca, and A. N. Rabiller. High charge PHIN photo injector at CERN with fast phase switching within the bunch train for beam combination, IPAC2011 Proceedings (in progress). 2011.

[21] M. Csatari Divall, A. Andersson, E. Bolzon B., Bravin, E. Chevallay, A. Dabrowski, S. Doebert, A. Drozdy, D. Egger, V. C. Fedosseev, Hessler, T. Lefevre, S. Livesley, R. Losito, O. Mete, M. Olvegaard, M. Petrarca, and A. N. Rabiller. Fast phase switching within the bunch train of the phin photo injector at cern using ber-optic modulators on the drive laser (in progress). Nuclear Instruments and Methods in Physics Research Section A: Accelerators, Spectrometers, Detectors and Associated Equipment.

Journal of Physics: Conference Series **347** (2012) 012037 doi:10.1088/1742-6596/347/1/012037

Start-to-end simulation of TAC SASE FEL facility

H Duran Yıldız
Dumlupınar University, Faculty of Arts and Sciences, Physics Department, Merkez Kampüs, 43100, Kütahya, TURKEY

E-mail: Hatice.Yildiz@cern.ch

Abstract. We study start-to-end simulation of the proposed Turkish Accelerator Center (TAC) Self Amplified Spontaneous Emission (SASE) Free Electron Laser (FEL) Facility. Astra Code is used to simulate the electron gun, including space charge effects and Elegant Code is used to track particle distribution from the accelerator modules through the entrance of the undulator including weak fields but not including space charge effects. Undulator optimization is performed using Genesis 1.3. In this study, tentative electron beam and laser parameters of the SASE FEL Facility are given.

1. Introduction

The goal of the Turkish Self Amplified Spontaneous Emission Free Electron Laser (TAC SASE-FEL) Project is to provide a source of bright, coherent, and tunable free electron laser light that enables diverse range of application areas from science to industry. The project aims to cover VUV to soft X-Ray region with an electron beam energy 1 GeV. Throughout the paper, start-to-end simulation studies of the project is explained in a detailed way. From the electron gun until the end of the undulator, three main programs Astra [1], Elegant [2], and Genesis 1.3 [3] are used, where in between, two interface programs and converters (astra2elegant and elegant2genesis) take care of the outputs from previous program and prepare it for the next one respectively. SDDS (Self Describing Data Set) [4] and IDL (Interactive Data Language) [5] are used in order to display the graphical outputs of the simulations.

2. Beam Dynamics and Start-To-End Simulations

For the start-to-end simulation of the Proposed Facility, 1 GeV electron beam energy with photo-cathode RF gun and superconducting linac is considered. TAC SASE-FEL Facility layout is displayed in Figure 1.

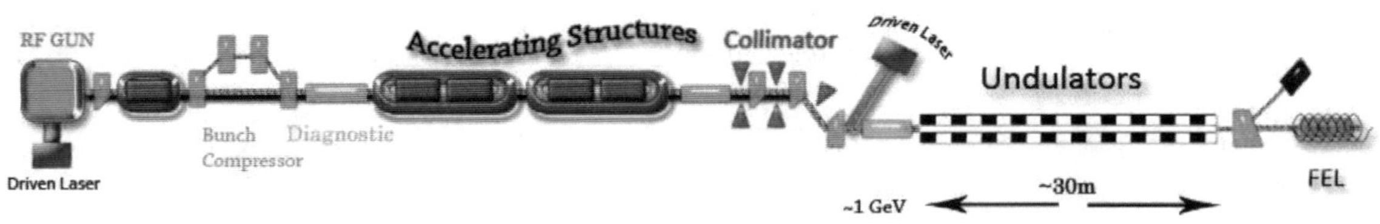

Figure 1. Proposed SASE-FEL Facility Layout

Beam parameters for the proposed SASE-FEL Facility are shown in Table 1. Astra is used to simulate the beam dynamics from the cathode up to the end of the first linac module. In the SASE-FEL system, accelerating RF field with 1.3 GHz gun is considered for modelling. Some important Astra features are given below:

- Astra is nominally cylindrically symmetric code,
- Astra tracks the particles by taking into account space charge, thus charge in time, current behaves like the Gaussian shape statistically,
- Astra does not consider weak-field effects.

Table 1. Proposed Beam Parameters for SASE FEL Facility.

Parameter	**Unit**	**Value**
Electron beam energy	GeV	1
Bunch charge	nC	1
Normalized emittance	πmm.mrad	< 2
Transverse emittances	Nm	3.2
FWHM bunch length	μm	150
Peak current	kA	2
Energy spread	-	< 0.02
Macropulse Repetition Rate	Hz	10

By using Astra - Ps_Viewer [6] interface codes, particles are simulated after being released from the gun. Bunch current is shown in Figure 2 for 2 nC charge and 100000 particles.

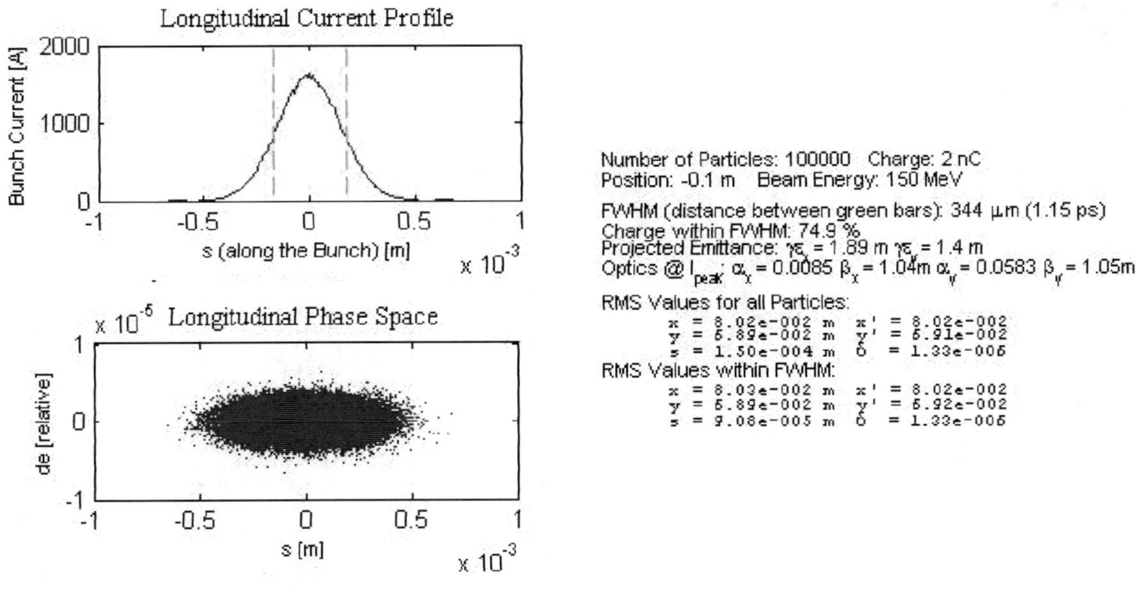

Figure 2. Longitudinal current profile along the bunch

Behaviour of the bunch current is Gaussian shape as expected starting from the cathode. An initial bunch was produced with both the gun and linac operating on-crest with gun's solenoid peak field of 0.226 T. Figure 3 shows the evolution of rms parameters throughout the beamline. Beam emittance, beam size, bunch length, energy spread, particle velocity etc versus the longitudinal beam position as meter are displayed. In the first plot, horizontal and vertical emittances are shown along the z-axis. Beam starts to emit after cathode warm up. In the second graph, a constant emittance is not achieved because of the high solenoid field. There is one solenoid in the RF gun around 0.1 m, but emittance seems diverging around 0.4 m. In order to prevent the divergence, one can reduce solenoid magnetic field.

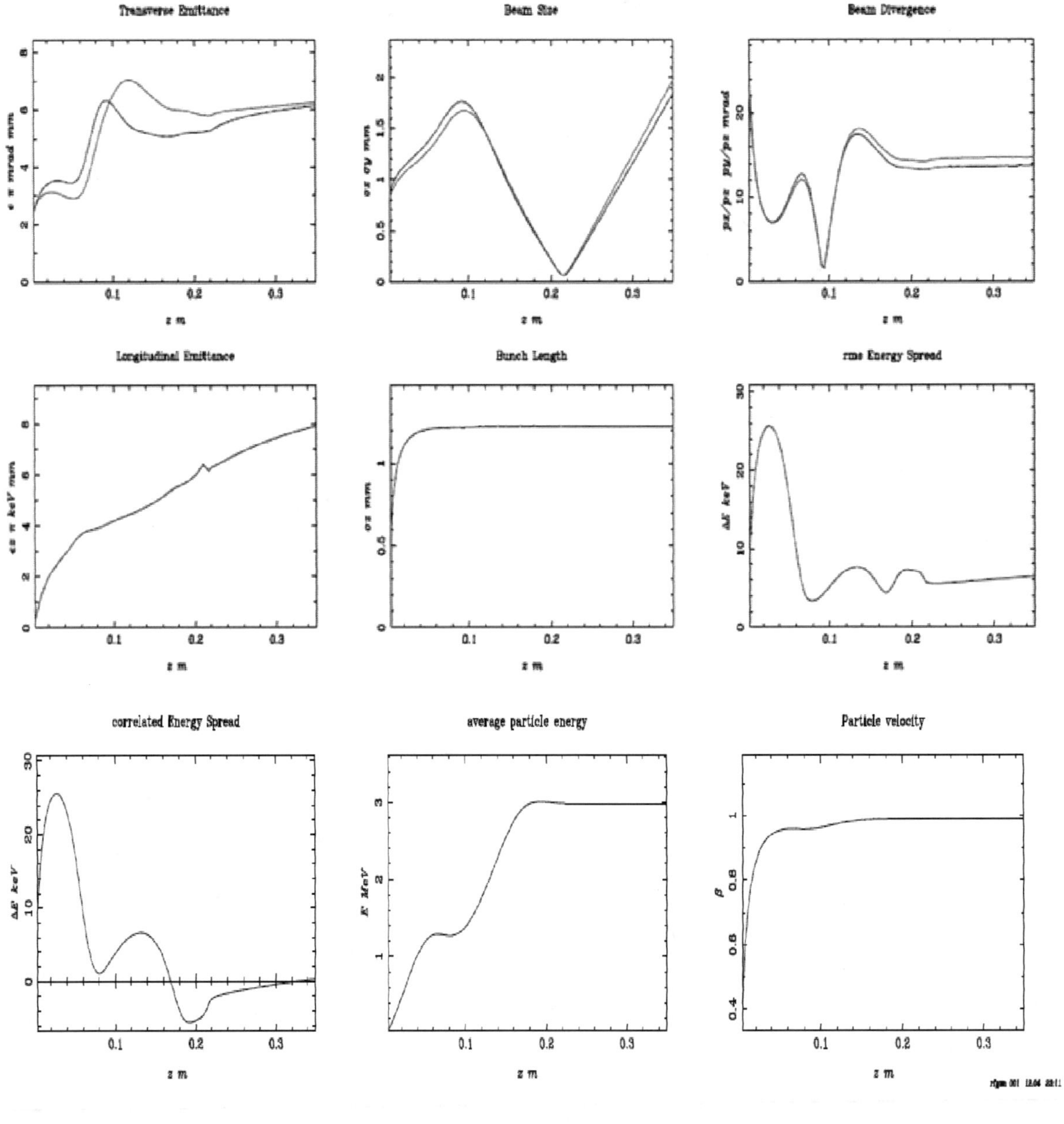

Figure 3. Astra results - evolution of rms beam parameters along the beamline.

3. Linac Beam Dynamics Simulations and Laser Optimizations

SASE-FEL start-to-end simulations are performed using Elegant and Genesis 1.3 for linear accelerations and undulator sections respectively.

3.1. Particle Tracking with Elegant

Once the Astra output is obtained, the bunch is converted and fed into Elegant for tracking through the linac and the bunch compressors. Elegant uses six dimensional phase space. These are transverse coordinates, slopes, total-equivalent distance travelled, fractional momentum deviation. Elegant code includes RF magnet, wakefield effects, all strength parameters, power supplies, electrical connectors, accelerators, quadrupoles, dipoles and drift properties. In Elegant simulation, graphical outputs are obtained using SDDS interface. Figure 4 can be obtained directly from Elegant by using SDDS. This figure shows the twiss parameters to explain beam ellipses (horizontal and vertical beta functions) without matching.

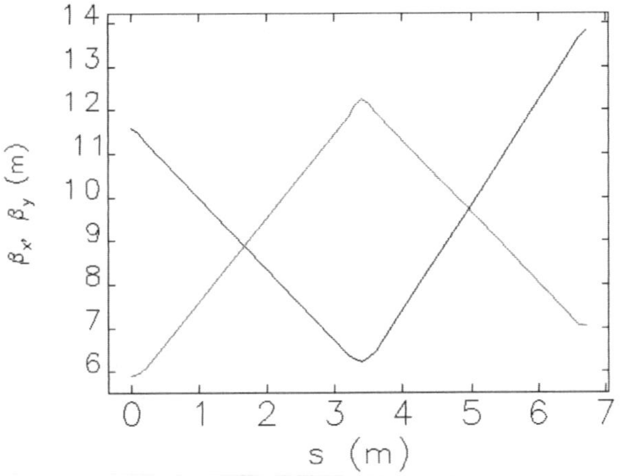

Figure 4. Beta function along the beamline

Between two lattice locations, transfer matrix for twiss parameters is given by beta, alpha, and gamma. Red shows β_y while black shows β_x. To get the size of the particle bunch, we determine the beta function. One can get tranverse beam size by using the $\sigma = \sqrt{}$ emittance $\times\ \beta\ /\ \gamma$. Although superconducting type of accelerator is foreseen for the proposed facility simulations, both superconducting and normal conducting type of accelerator parameter sets are given in the following table because normal conducting accelerator case is yet not out of consideration.

Table 2. Proposed accelerator parameters for SASE FEL Facility.

Parameter	**Tesla Structure**	**S Band Linac**
Frequency (GHz)	1.3	3
Gradient (MV/m)	30	30
Input Power (MW)	10	35
Pulse Length (μs)	~ 10	~ 10
Number of Structure	35	12
Linac Length (m)	~ 100	45

3.2. Undulator and Laser Optimization

Hybrid with Iron [7] is chosen as undulator material because it produces the magnetic field that enables coverage from VUV to soft X-Ray. Undulator gap is very important for lasing; smaller gap, lower current or smaller volume of magnetic material is needed to reach a given magnetic field. In vacuum undulator studies and for seeding HHG, HGHG possibilities are still understudy. One and three dimensional gain length, saturation power and beam power are both analytically calculated and simulated using Genesis 1.3 Code. Genesis can be used for steady-state and time dependent FEL simulations. Lattice, electron beam, radiation field steps are needed to figure out propagation of the beam. The external magnetic field input file has four sections: as ID, Strength, length, and offset to the previous element. Radiation power, ponderomotive phase, radiation growth rate, energy, growth of bunching, radiaton size, vertical and horizontal sigma are obtained by using IDL (see Figures 5 and 6). Proposed SASE-FEL Facility laser parameters are shown in the Table 3.

Table 3. Laser paramaters for SASE-FEL Facility.

Laser Parameters	
Wavelength Range (nm)	3 – 60
Maximum Peak Power (GW)	1.3
Peak Brilliance (photons/s/mrad2/mm2/0.1%bw)	~10^{29}
Peak Energy (μJ)	130

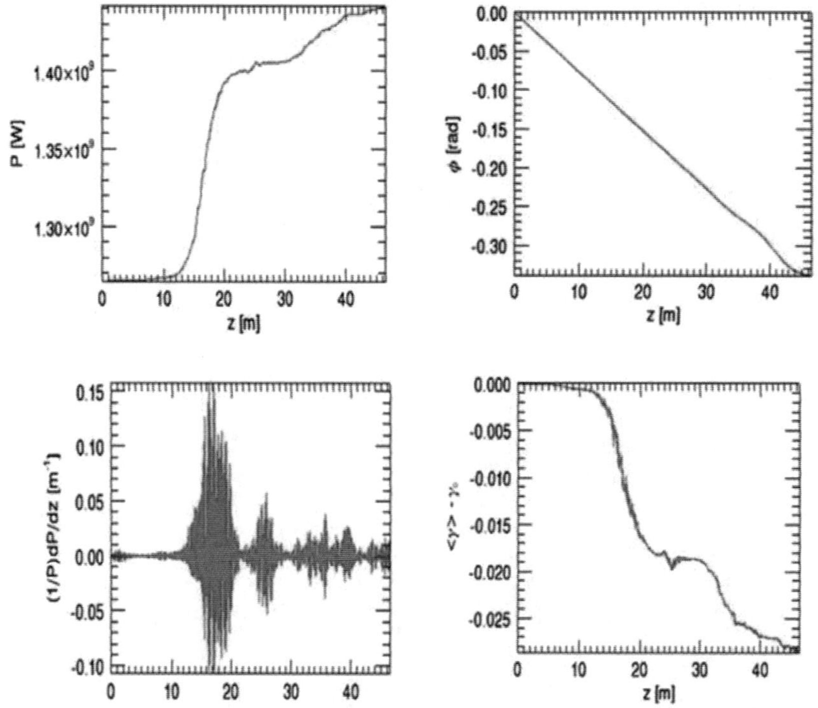

Figure 5. End of the undulator, laser properties (power, growth, energy, and bunching) as a result of Genesis 1.3.

Journal of Physics: Conference Series **347** (2012) 012037 doi:10.1088/1742-6596/347/1/012037

The growth rate tends to decrease after 20 m, this can be explained by the electron motion and with its period length. In Genesis 1.3, we consider 1 GeV electron beam energy with Rayleigh range 1.29196×10^2 m and current peak of 2×10^3 A with a slice of 8192. As a result of Genesis 1.3 Code, we have obtained saturation power as 1.3977×10^9 W with a Saturation length 21.48 m.

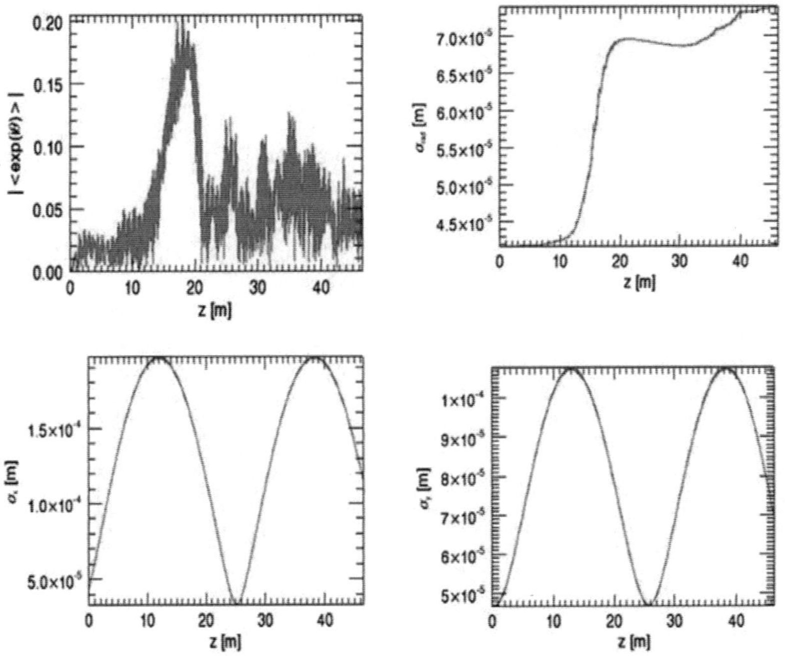

Figure 6. Laser properties (photon beam size and sigma) as a result of Genesis 1.3.

As a result, simulation studies for the Proposed SASE-FEL Facility continue. In the future, seeding properties and some other new features that can give shorter wavelengths to cover soft x-ray region will be included.

Acknowledgments
This work is supported by Turkish State Planning Organization under grant no. DPT2006K-120470. The author thanks İ. Yıldız (METU) and B. Faatz (DESY) for valuable discussions.

References

[1] http://www.desy.de/~mpyflo/

[2] http://www.aps.anl.gov/Accelerator_Systems_Division/Operations_Analysis/manuals/elegant_latest/elegant.html

[3] http://pbpl.physics.ucla.edu/~reiche/

[4] http://www.aps.anl.gov/Accelerator_Systems_Division/Operations_Analysis/software.shtml#SDDSbinaries

[5] http://www.ittvis.com/language/en-US/Downloads/ProductDownloads.aspx

[6] http://www.desy.de/xfel-beam/psviewer/files/ps_viewer_user_guide.pdf

[7] Elleaume P, 2000 *NIM-A* **455**, 503-523

AUTHOR INDEX

Aune, S ... 245, 254
Azizi, K ... 226, 293, 298
Barisonzi, M ... 273
Bashiry, V ... 226
Benekos, N C ... 51
Bingül, A ... 202
Carrettoni, M ... 265
Castel, J ... 254
Cebrián, S ... 254
Cerci, D S ... 283
Cervelli, A ... 135
Chevallay, E ... 307
Coco, V ... 175
Cornelissen, T ... 170
Csatari, M ... 307
Dabrowski, A ... 307
Dafni, T ... 245, 254
Davenport, M ... 245
Dell'Asta, L ... 181
Dettori, F ... 114
Di Crescenzo, A ... 80
Doebert, S ... 307
Egger, D ... 307
Espargilière, A ... 154
Fanourakis, G ... 245, 254
Fedosseev, V ... 307
Ferrer-Ribas, E ... 245, 254
Fortuño, D ... 254
Galán, J ... 245, 254
García, J A ... 245, 254
Gardikiotis, A ... 245, 254
Garza, J G ... 245, 254
Geralis, T ... 245, 254
Ghalenovi, Z ... 129
Giomataris, I ... 245, 254
Gómez, H ... 245, 254
Herrera, D C ... 254
Iguaz, F J ... 245, 254
Irastorza, I G ... 245, 254
Jabeen, S ... 19
Kaya, G ... 293
Khorramian, A ... 145, 150
Kurt, P ... 31
Lamont, M ... 1
Luzón, G ... 254
Mete, O ... 307
Mols, J P ... 254
Monfared, S T ... 150
Moosavinejad, M ... 145
Neri, N ... 70
Nishimura, Y ... 236
Olvegaard, M ... 307
Ortiz, A ... 254
Oyama, Y ... 41
Papaevangelou, T ... 245, 254
Petrarca, M ... 307
Pinci, D ... 164
Rajabi, A A ... 129
Rodríguez, A ... 245, 254
Ruz, J ... 245, 254
Schöfbeck, R ... 101
Segui, L ... 254
Simonetto, F ... 122
Soleymaninia, M ... 145
Spagnolo, P ... 9
Suleymanov, M K ... 195
Sultansoy, S ... 226
Sundu, H ... 293, 298
Süngü, J Y ... 298
Tavakolinezhad, A ... 129
Tehrani, S A ... 150
Tomás, A ... 245, 254
Türkan, A ... 293
Uysal, A K ... 109
Vafeiadis, T ... 245, 254
Veliev, E V ... 293
White, R M ... 209
Yamamura, T ... 87
Yavas, Ö ... 76
Yildiz, H D ... 318
Yildiz, S C ... 245, 254